Frontiers in Protein Structure Research

Frontiers in Protein Structure Research

Editors

Istvan Simon
Csaba Magyar

MDPI • Basel • Beijing • Wuhan • Barcelona • Belgrade • Manchester • Tokyo • Cluj • Tianjin

Editors

Istvan Simon
Institute of Enzymology
Research Centre for Natural Sciences
Budapest
Hungary

Csaba Magyar
Institute of Enzymology
Research Centre for Natural Sciences
Budapest
Hungary

Editorial Office
MDPI
St. Alban-Anlage 66
4052 Basel, Switzerland

This is a reprint of articles from the Special Issue published online in the open access journal *International Journal of Molecular Sciences* (ISSN 1422-0067) (available at: www.mdpi.com/journal/ijms/special_issues/f_protein_structure).

For citation purposes, cite each article independently as indicated on the article page online and as indicated below:

LastName, A.A.; LastName, B.B.; LastName, C.C. Article Title. *Journal Name* **Year**, *Volume Number*, Page Range.

ISBN 978-3-0365-6781-5 (Hbk)
ISBN 978-3-0365-6780-8 (PDF)

Contents

About the Editors

Istvan Simon

István Simon was born in Budapest, Hungary, in 1947. He graduated as a physicist and habilitated in biology and physics. He is a member of the Hungarian academy of Sciences, and currently a professor emeritus at the Institute of Enzymology of the Research Centre for Natural Sciences, where he has been since 1969. He turned his attention to computational analysis at Cornell University, where he spent several years in the group of Harold A. Scheraga. He continued his career in this field in Hungary, and pioneered computational protein structure research at the end of the 1970s. He has published 9 books chapters and 135 papers. The publication have been cited over 12,000 times, and he has twice been listed among the highly cited researchers according to the Web of Science. Together with his research group, he has provided 16 databases and prediction servers on the World Wide Web. These include the prediction of "stabilization centers", i.e., residue pairs that are responsible for keeping a proteins's structure intact, the prediction of disulfide-forming cysteines (CYSREDOX), and a number of top-cited transmembrane prediction algorithms (DAS, HMMTOP, and PDBTM). Recently, his group has uncovered the statistical thermodynamics forming the background of protein disorder, and provided the corresponding prediction server, IUPRED, followed by the prediction of functional regions of disordered proteins (ANCHOR). The strength of these methods is the groundbreaking discovery of principles underlying protein structure organization.

Csaba Magyar

Csaba Magyar was born in Budapest, Hungary, in 1972. He graduated as a physicist, his interest turned to computational chemistry already during his university years. He received the Ph.D. degree in 2001, and currently is a senior research fellow at the Institute of Enzymology of the Research Centre for Natural Sciences in Hungary. In the first year of his career he studied the structural background of thermal stability of proteins by utilizing homology modeling and molecular dynamics simulations. Protein stability remained in the focus of his research interest as a postdoc, he was working on the concept of stabilization centers in proteins, developed the concept of stabilizing residues, and set up the SRide web server. Gradually his interest turned to the investigation of protein–protein and protein–ligand complexes. In recent years, he has worked on a special subclass of disordered proteins, called Mutual Synergistic Folding proteins. He was awarded with the Youth Prize of the Hungarian Academy of Sciences, and the Bolyai János Research Fellowship.

International Journal of
Molecular Sciences

MDPI

Editorial

Assortment of Frontiers in Protein Science

István Simon * and Csaba Magyar *

Institute of Enzymology, Research Centre for Natural Sciences, Eötvös Loránd Research Network, 1117 Budapest, Hungary
* Correspondence: simon.istvan@ttk.hu (I.S.); magyar.csaba@ttk.hu (C.M.)

Citation: Simon, I.; Magyar, C. Assortment of Frontiers in Protein Science. *Int. J. Mol. Sci.* **2022**, *23*, 3685. https://doi.org/10.3390/ijms23073685

Received: 15 March 2022
Accepted: 25 March 2022
Published: 28 March 2022

Recent decades have brought significant changes to the protein structure research field. Thanks to the genome projects and advances in structure determination methods, the number of yearly released entries in the PDB database [1] has increased significantly. Protein structure research is experiencing a new renaissance, and in 2020 the number of deposited structures in the PDB database reached a new record of 14,022. Even in 2021, the number of new deposits was higher than ever before, with the exception of 2020. Most of these structures belong to globular proteins, but there are several transmembrane and even disordered proteins among them [2]. Moreover, there are also transmembrane proteins with disordered regions that have led to the emergence of new transmembrane specific disorder prediction methods [3]. Additionally, we cannot forget to mention the huge leap forward in the field of structure prediction methods achieved by the AlphaFold2 [4] method, which is able to predict protein structures with an error comparable with that of experimental methods. Our Special Issue features a research article connected to this research area, which evaluates a deep learning-based residue contact method [5]. With the development of Alphafold-Multimer [6], the accurate prediction of protein complexes is becoming a reality. One important application of this method would be the prediction of protein–protein interactions. On the way to this goal, this Special Issue presents a work dealing protein–protein docking [7]. Of course, the COVID-19 pandemic has left inevitable traces on our lives over the last two years and also on research. Not only did the development of vaccines arrive to the frontier, but structure research of the severe acute respiratory syndrome coronavirus 2 (SARS-CoV-2) proteins became an intensively researched field. One such piece of research has made its way into this Special Issue, as well [8]. There is an additional research article in our issue with medical relevance that is about model development to predict the phenotypic outcome of rare germline pathogenic TP53 missense variants [9]. An assortment of many new frontiers is presented in this collection. A single issue cannot give a comprehensive overview of a large field such as proteins science, but we aim to give a broad overview of current research. In this issue, there are 19 research articles, one review, and one commentary. The manuscripts could have been categorized according to the subject of the research into these major categories like structure and the folding of globular proteins, membrane proteins, and disordered proteins or we could have classified them into theoretical and experimental groups. However, since several publications would fit into more than one category, we discarded this classification system. As manuscripts were published within a very short time upon acceptance, we decided to introduce the published papers is chronological order, starting with the latest one and concluding with the very first accepted manuscript of this Special Issue. Interestingly, both the first and last manuscripts deal with protein folding. We may conclude that protein folding is still the alpha and the omega of protein science. It is still the ultimate question, and even though research articles dealing with protein folding were published as long as 75 years ago this fundamental problem has yet to be solved. Several other manuscripts in this Special Issue deal with the folding problem, for example some deal with glycoproteins and disordered proteins. In the following paragraphs, the manuscripts published in our

Special Issue *Frontiers in Protein Structure Research* are presented, which cover several fields of proteins science.

The fundamental problem of protein folding is reconsidered in the review of Sorokina et al. [10]. The generally accepted view of protein folding is the thermodynamic hypothesis, under which the native folded conformation of a protein corresponds to the global minimum of Gibbs free energy. The authors suggest that the evidence behind the thermodynamic hypothesis is not convincing. They argue that despite the continuous increase in computing power, only a few protein folds can be predicted by ab initio physics-based approaches. Furthermore, recent spectacular successes in protein structure prediction were achieved by deep learning-based evolution modeling methods. An alternative view of protein folding is proposed, implying that the native state of proteins lies in a local minimum of the fluctuating free energy surface. They also presume that the folding Gibbs free energy for numerous proteins is positive, and they are stabilized by the translation system and chaperones. Thus, folding should be modeled as a non-equilibrium energy-dependent in vivo process.

The conformational properties of covalently attached and bilayer contained carbohydrates influencing the structure of proteins were investigated by Guvench et al. [11] on a theoretical level. The ring puckering thermodynamics of the most common vertebrate monosaccharides were investigated by extended system-adaptive biasing force all-atom explicit-solvent molecular dynamics simulations. They found that the CHARMM [12] force field with proper parametrization is able to model the ring puckering of the investigated carbohydrates and could possibly be used widely for carbohydrate-containing vertebrate biomolecules. The accurate simulation of carbohydrates in glycoproteins, proteoglycans, and glycolipid-containing bilayer-embedded transmembrane proteins could help to narrow the gap in the number of suitable systems for theoretical and experimental methods and promote the in silico investigation of glycoproteins.

The possible structural background of the unusual behavior of mutual synergetic folding (MSF) proteins was analyzed by Magyar et al. [2]. These oligomeric proteins are disordered in their monomeric form but become almost completely ordered in their oligomeric form upon interacting with another disordered MSF protein chain. Solvent accessibility of the peptide bonds in the theoretical monomeric form seems to be a significant factor. Next to the local shielding effect of peptide bonds exerted by the side chains of the bond forming residues, nonlocal shielding also occurs upon oligomerization. To investigate these local and non-local shielding effects, Shannon information entropy calculations were performed on all available MSF and selected globular homodimeric proteins. According to the results, differences can be found in both local and non-local shielding. These findings open the possibility for a prediction method to distinguish MSF proteins from globular ones. The resulting larger dataset could be used to reveal the structural background of the MSF phenomenon.

The performance of the ProSPr distance prediction method, which is essentially an open-source alternative of the AlphaFold-1 contact prediction method, was evaluated by Stern et al. [5]. ProSPr is an accurate deep learning method to predict residue contacts based on amino acid sequence input. The authors tested the method on the CASP14 [13] test set and found that the ensemble predictions of short and mid contacts were reliable but that long contact prediction accuracy was only around 44%. They determined the useful multiple-sequence-alignment depth and found that amino acid sequence length did not correlate with contact prediction accuracy with the test set. The authors present a useful and accurate method with inference times two orders of magnitude faster than AlphaFold2. This tool could be helpful in many situations where the partial structural information of residue contacts is sufficient.

During the evolution of protein science, the discovery of transmembrane and later disordered proteins widened our view of the world of proteins. Dobson and Tusnady went one step further [3] and presented a novel method called MemDis, which is able to predict intrinsically disordered regions within transmembrane proteins. Although there are several protein disorder prediction methods, their accuracy is limited for membrane proteins, probably due to their special physicochemical properties. MemDis combines convolutional neural network and long short-term memory networks while adding transmembrane specific features to the prediction. The authors achieved an unprecedented level of disorder prediction accuracy on their transmembrane-specific test set. The method is publicly available at http://memdis.ttk.hu (accessed on 15 March 2022), providing an extremely useful tool for researchers to identify disordered regions within transmembrane proteins.

Phosphorylation-induced conformational change is a common way to regulate a protein's function and disordered proteins are no exception. Rieloff and Skepö [14] investigated the impact of phosphorylation on the conformation of disordered proteins using molecular dynamics simulations. Since this a relatively new and under-researched field, first they validated the method by comparing the results obtained with two different force fields. While these force fields were known to overestimate the compactness of the phosphorylated state of disordered proteins mainly because of overstabilized salt bridges, they concluded that this discrepancy can be resolved with the proper incorporation of the effect of salt into the simulations, corresponding to the ionic strength present in the experiments. They found that the effect of salt concentration on simulation results is small enough to be neglected; thus, simulations can be used to help understand the mechanisms behind the phosphorylation regulation of disordered proteins. After publishing the results of this validation, Rieloff and Skepö published a second paper in this Special Issue. In their subsequent paper [15], they applied the previously validated Amber ff99SB-ILDN force field with the TIP4P-D water model and performed all-atom molecular dynamics simulations to analyze the effect of phosphorylation on five disordered peptides originating from tau, statherin, and beta-casein proteins. Their results were in qualitative agreement with the experimental data. They found that some peptides contracted upon phosphorylation while others became more expanded and that the amount of charges does not account for the phosphorylation-induced changes. The sequential distribution of residues with positive charges is crucial to describe this behavior through the formation of salt bridges with phosphorylated residues. They are conducting an ongoing systematic investigation of several factors influencing the outcome of phosphorylation.

The transmembrane region of HokC was investigated by Ortiz et al. [16] using a systematic saturation mutagenesis study. HokC is a toxin produced by *Escherichia coli* to control its own population. They found that 92% of the single-site point mutations were tolerated and that all the non-tolerated mutations had compensatory mutations that reversed their effect. By utilizing the HokC family multiple sequence alignment, they found only a single invariant cysteine residue. Every site-directed mutagenesis of this residue performed was also tolerated. The authors concluded that maintaining function without conserving amino acids is possible by compensatory mutations. Because of the helical transmembrane structure, sequentially close residues are expected to be close spatially. Thus, they may be suitable to accommodate compensatory mutations. Their findings were in agreement with this hypothesis, and the authors found that transmembrane proteins favor the occurrence of multiple mutations between spatially neighboring residues more than globular proteins. A notable exception is the mutation of the only invariant cysteine residue to serine, which causes a change in the dimerization of HokC. A complementary mutation occurred at sequentially distant positions, suggesting a change in interactions between different monomers.

The imaginary "smoking gun" by Bocedi et al. [17] is a provocative commentary, which presents experimental data with an unusual interpretation of the role of the glutathione tripeptide. They line up data on the hyper-reactivity of structural cysteines, the dependence of the second-order kinetic constants on pK_a values, and the reactivity of protein cysteines towards natural disulfides. Their interpretation may change our assumptions regarding the role of glutathione in the early steps of oxidative folding that occur at the ribosomal exit tunnel at the interface of the endoplasmic reticulum.

The problem of the structural dynamics of proteins was investigated by Nehls et al. [18] using conformation-sensitive oxidative protein labelling, which may serve as a complementary technique to mass spectrometry for capturing conformational changes. They used a test set of proteins between 10 and 150 kDa and showed that conformational changes induced by ligand binding are reflected in the modification of the mass spectrometry pattern obtained by site-selective Fenton chemistry labelling. For smaller proteins, the extensive oxidation pattern correlates well with the protein structural dynamics while there are clear differences between the oxidation patterns of the ligand-bound and free forms. Despite its practical limitations, this method could become a valuable tool for conformational analysis alongside mass spectrometry.

The toxicity of tetrabromobisphenol-S (TBBPS) was investigated by Jarosiewicz et al. [19], in order to see whether it would be a proper replacement for the widely used flame retardant tetrabromobisphenol-A (TBBPA), which is potentially toxic. They used red blood cell membranes as a model system. They found that both TBBPA and TBBPS caused increases in the fluidity of the membranes, decreases in the ATP level, thiol group elevation, and conformational changes to the membrane proteins. Both substances also caused changes in the size and shape of red blood cells and with TBBPS an increase in lipid peroxidation also occurred. They determined that changes are observed at significantly lower concentrations in the case of TBBPA than with TBBPS. The published data indicate lower toxicity for TBBPS, which occurs only at very high concentrations in contrast to TBBPA.

The thermodynamical properties of the SARS-CoV-2 virus spike protein variants were analyzed by Kumar et al. [8] using molecular mechanics and dynamics calculations in complex with the human ACE2 receptor. They performed molecular dynamics simulations to estimate the stability of the complex and calculated ΔG_{bind} binding free energy values using molecular mechanics calculations to characterize the strength of the binding. They found that the mutations caused stronger binding in the alpha and kappa variants. In the case of the kappa and delta variants, the mutations mainly increased the stability and intra-chain interactions in the spike protein, possibly interfering with the neutralizing effect of the antibodies, which might be responsible for the higher transmissibility of these variants.

Holubowicz et al. [20] identified single-nucleotide variants of the trimeric structure of globular C1q-like otolin-1, a collagen-like scaffold protein responsible for the biomineralization of inner ear stones in vertebrates. The globular-like gC1q-like domain binds calcium and is responsible for trimerization. The stability of the variants was analyzed by thermal shift assay and the positions of the mutated residues were mapped on a small angle X-ray scattering-derived model structure of the hOtolC1q trimer. According to the experiments, most of the mutations caused decreased stability or aggregation, but in most cases the structure can be stabilized in the presence of Ca^{2+}. There is a Ca^{2+}-insensitive a mutation that disables trimerization. The mean allele frequency of these deleterious mutations is in the range of 10^{-4}. According to their results, these natural variants can cause pathological changes and affect one's sense of balance.

The quaternary structure of the iota carbonic anhydrase (CA) from the marine diatom *Thalassiosira pseudonana* was modeled by Jensen et al. [21]. The protein is built up from domains resembling a calcium–calmodulin protein kinase II association domain. The crystal structure of the single domain was recently uncovered, and comparing it with available CA structures reveals novel folding element; however, the quaternary structure of the four domain-containing homotetrameric protein is still unknown. The authors utilized biophysical techniques and modelling to build the homotetrameric structure, which is formed from a core structure from the first two domains of each monomer, while the arms are formed by the other domains. The authors discussed the role of a flexible linker between domain 3 and 4 and a possible relation of its atypical shape with its activity and metal coordination. They also proposed a possible structure for carbonic anhydrases with fewer domain repeats using experimental data.

Bifidobacterial α-L-Fucosidases (ALF) were investigated by Curiel et al. [22], which are important for the bifidobacterial colonization of the gut. Several ALFs have been identified by bioinformatical methods, which can be classified into three major families. Bifidobacterial ALFs show significant sequential differences, probably resulting from distinct phylogenetic evolution. The authors performed phylogenetic and comparative analyses of bifidobacterial ALFs utilizing existing physicochemical information. They revealed several ALF paralogue groups within two major ALF families. The authors suggest that because ALFs are phylogenetically related to other glycosyl hydrolase families they may exhibit additional glycosidase activities that utilize transfucosylate substrates other than lactose. This could have a substantial impact on the development of novel prebiotics.

The nuclear factor erythroid 2-related factor 2 (Nrf2) was studied by Karunatilleke et al. [23]. Nfr2 can interact with several proteins and mediates the transcription of cytoprotective genes in cellular responses to oxidative stress. Nrf2 is a promising target for anticancer drug design, but the limited information about its molecular details and interactions hinder rational drug design. The authors applied combined bioinformatics with experimental methods like CD and NMR spectroscopy approaches to characterize the structure of Nrf2, and hydrogen deuterium exchange mass spectrometry was used to analyze its interaction with the Kelch domain of an interaction partner. They found that Nrf2 is partially disordered with transiently ordered segments. Binding with the Kelch domain stabilizes the structure of other binding motifs while leaving other regions highly dynamic. According to their results, the conformational dynamics of full length Nrf2 have substantial consequences for its target recognition, enabling Nrf2 to bind to distinct targets with high specificity and low affinity.

Wesch et al. [24] examined the UFM1-activating enzyme 5 (UBA5) within the ufmylation cascade of the ubiquitin fold modifier 1 (UFM1) protein. This cascade reaction affects several cellular processes and plays a role in the pathogenicity of many human diseases, but the molecular mechanisms of the ufmylation cascade are still unclear. The authors focused on the biophysical and biochemical characterization of the interaction between UBA5 and the UFM1-conjugating enzyme 1 (UFC1). Their working hypothesis was that the unstructured C-terminal of UBA5 regulates the cellular localization of the elements in the ufmylation cascade. According to their results, the C-terminal 20 residues in UBA5 are crucial for UFC1 binding. They uncovered the NMR structure of UFC1 complexed with this C-terminal peptide and identified key residues in the UBA5–UFC1 interaction. The structural evidence augmented with isothermal titration calorimetry results revealed the mechanism of the interaction and confirmed the crucial role of the C-terminal unstructured region.

A novel protocol for protein–protein docking was developed by Kurcinski et al. [7], incorporating protein–protein orientation and backbone flexibility with a single simulation step instead of the traditional two-step procedure. Exhaustive sampling is required for this approach, which can be achieved using the CABS coarse-grained protein model and replica exchange Monte Carlo dynamics. In this proof of concept study, the new protocol was tested on 62 protein–protein complexes. They found that the modeling of large conformational changes was possible with acceptable computational costs within the range of 10 CPU hours. For low- and medium-flexibility cases, the acceptable accuracy can be achieved with an iRMSD of around 4 Å, but the selection of the most accurate model needs to be improved with a success rate of only around 50%. The current common approaches to taking flexibility into account have serious limitations. The proposed protocol is conceptually different and relies heavily on the exhaustive sampling capability of the simplified simulations, which is orders of magnitude faster than classical force field-based molecular dynamics simulations. While these kinds of simulations also have their limitations when reproducing the real free energy surface, the more exhaustive sampling could compensate for their weaknesses. Although this protocol opens new perspectives in flexible protein–protein docking applications, it also has limitations. Despite the high performance of the replica exchange annealing enhanced Monte Carlo dynamics, this method still scales poorly with the size of the "ligand" proteins setting a practical limit of about 150 residues. The simplified description of atomic interaction forces results in less sensitive docking energetics, which makes the above-described selection of the most accurate model difficult. The authors suggest that parallelizing the simulation may speed up the process. In this way, an automated public protein–protein docking server could be created, which could be a very useful tool for studying protein–protein interactions.

A logistic regression model was developed by Liu et al. [9] in order to predict the phenotypic outcome of rare germline pathogenic TP53 missense variants. They compiled non-overlapping datasets for the Li-Fraumeni syndrome and hereditary breast cancer outcomes. TP53 protein is a transcription factor that binds as a tetramer to DNA and activates a large number of genes that promote DNA repair mechanisms or apoptosis. Each monomer is built up from several domains, including a DNA binding domain and an oligomerization domain, among others. About two-thirds of reported germline TP53 variants are single-site missense changes. Predominantly located in the DNA binding domain, some of them result in the decreased thermal stability of this domain. By utilizing an X-ray structure of TP53, the conformational characteristics of the variants were included in the method. The models show a clear relationship between disease outcome for TP53 variants and their effects on aspects of protein conformation and function. The model could be helpful to avoid unnecessary examinations for a large proportion of TP53 variant carriers, which could relieve pressure on the medical system.

Pressure denaturation of the all-α GH2 domain of the GIPC1 protein adaptor was investigated using NMR spectroscopy by Dubois et al. [25]. To date, this method has been used mainly for small α/β and all-β single domain proteins. High-pressure perturbation was used with NMR spectroscopy to reveal the unfolding landscape at 10, 20, and 30 °C, and the results were compared with chemical denaturation experiments. While GIPC1-GH2 is most stable at 20 °C, it is more stable at 30 °C than at 10 °C. Their finding that the loss of tertiary and secondary structure was quasi-simultaneous was unexpected, meaning that helices are not stable outside the 3D scaffold. The unfolding was cooperative at high pressures and the highest temperatures but more progressive at the lowest temperatures. Although partial unfolding can occur at lower temperatures, at 30 °C the stability is decreased and thermal denaturation probably competes with high-pressure denaturation, sweeping away the partial unfolding that occurs at lower temperatures. The authors demonstrated the usefulness of pressure-induced unfolding experiments in exploring the unfolding landscape of proteins by monitoring the partial unfolding process, which could not have been followed by chemical denaturation.

Finally, we arrive at the first published paper in our Special Issue by Liu et al. [26], which emphasizes the importance of considering conformational entropy accurately for the simulation of disordered proteins. There are several pairwise additive force fields that were specifically modified to handle disordered proteins more accurately, yet they still often fail to reproduce experimental results. The authors propose the incorporation of configurational entropy for the development of universal force fields, which should be able to handle globular and disordered proteins and disorder to order transitions equally well. They compared pairwise additive force fields with the AMOEBA [27] many-body force field using experimental data on a set of disordered and medium-sized globular proteins. According to their results, fixed-charge force fields gave smaller yields, while the polarizable model yielded larger RMSD for ordered proteins. Force fields with the largest RMSD fluctuations are consistent with the results from the radius of gyration experiments. They argued that by exhibiting larger variations, they are better suited to describe the structural plasticity of disordered proteins. According their results, the polarizable AMOEBA many-body force field is beneficial for the simulation of disordered proteins and it can outperform specifically modified force fields without requiring problem-specific parametrization. By retaining its universality, it is well suited as a general force field for different types of disordered proteins and their complexes. They suggest, however, that in their evaluation the precision of the examined pairwise additive force fields was not adequate and that further efforts to reproduce the structural dynamics could be used as guidance for the development and validation of force fields. They concluded that force fields with the largest variations in the radius of gyration and universal Lindeman values for folded states describe disordered proteins and disorder to order transitions better and that a universal force field applicable to globular and disordered proteins should be able to describe the balance between energetics and configurational entropy.

In this Special Issue, we aim to represent the vibrant state of protein structure studies at the end of 2021 and the flowering of this field since the middle of the nineteenth century with this assortment of publications. The editors hope that the readers will welcome it!

Funding: This work was financially supported by the Ministry of Innovation and Technology of Hungary from the National Research, Development and Innovation Fund (OTKA grant No. K115698).

Institutional Review Board Statement: Not applicable.

Informed Consent Statement: Not applicable.

Conflicts of Interest: The authors declare no conflict of interest.

References

1. Berman, H.M.; Westbrook, J.; Feng, Z.; Gilliland, G.; Bhat, T.N.; Weissig, H.; Shindyalov, I.N.; Bourne, P.E. The Protein Data Bank. *Nucleic Acids Res.* **2000**, *28*, 235–242. [CrossRef]
2. Magyar, C.; Mentes, A.; Cserző, M.; Simon, I. Origin of Increased Solvent Accessibility of Peptide Bonds in Mutual Synergetic Folding Proteins. *Int. J. Mol. Sci.* **2021**, *22*, 13404. [CrossRef] [PubMed]
3. Dobson, L.; Tusnády, G.E. MemDis: Predicting Disordered Regions in Transmembrane Proteins. *Int. J. Mol. Sci.* **2021**, *22*, 12270. [CrossRef] [PubMed]
4. Jumper, J.; Evans, R.; Pritzel, A.; Green, T.; Figurnov, M.; Ronneberger, O.; Tunyasuvunakool, K.; Bates, R.; Žídek, A.; Potapenko, A.; et al. Highly accurate protein structure prediction with AlphaFold. *Nature* **2021**, *596*, 583. [CrossRef] [PubMed]
5. Stern, J.; Hedelius, B.; Fisher, O.; Billings, W.M.; Della Corte, D. Evaluation of Deep Neural Network ProSPr for Accurate Protein Distance Predictions on CASP14 Targets. *Int. J. Mol. Sci.* **2021**, *22*, 12835. [CrossRef] [PubMed]
6. Evans, R.; O'Neill, M.; Pritzel, A.; Antropova, N.; Senior, A.W.; Green, T.; Žídek, A.; Bates, R.; Blackwell, S.; Yim, J.; et al. Protein Complex Prediction with AlphaFold-Multimer. 2021. Available online: https://www.biorxiv.org/content/10.1101/2021.10.04.463034v2 (accessed on 15 March 2022).
7. Kurcinski, M.; Kmiecik, S.; Zalewski, M.; Kolinski, A. Protein–Protein Docking with Large-Scale Backbone Flexibility Using Coarse-Grained Monte-Carlo Simulations. *Int. J. Mol. Sci.* **2021**, *22*, 7341. [CrossRef] [PubMed]
8. Kumar, V.; Singh, J.; Hasnain, S.E.; Sundar, D. Possible Link between Higher Transmissibility of Alpha, Kappa and Delta Variants of SARS-CoV-2 and Increased Structural Stability of Its Spike Protein and hACE2 Affinity. *Int. J. Mol. Sci.* **2021**, *22*, 9131. [CrossRef] [PubMed]
9. Liu, Y.; Axell, O.; van Leeuwen, T.; Konrat, R.; Kharaziha, P.; Larsson, C.; Wright, A.P.H.; Bajalica-Lagercrantz, S. Association between Predicted Effects of *TP53* Missense Variants on Protein Conformation and Their Phenotypic Presentation as Li-Fraumeni Syndrome or Hereditary Breast Cancer. *Int. J. Mol. Sci.* **2021**, *22*, 6345. [CrossRef]
10. Sorokina, I.; Mushegian, A.R.; Koonin, E.V. Is Protein Folding a Thermodynamically Unfavorable, Active, Energy-Dependent Process? *Int. J. Mol. Sci.* **2022**, *23*, 521. [CrossRef] [PubMed]
11. Guvench, O.; Martin, D.; Greene, M. Pyranose Ring Puckering Thermodynamics for Glycan Monosaccharides Associated with Vertebrate Proteins. *Int. J. Mol. Sci.* **2022**, *23*, 473. [CrossRef]
12. Sairam, S.; Mallajosyula, S.; Guvench, O.; Hatcher, E.; MacKerell, A.D. CHARMM Additive All-Atom Force Field for Phosphate and Sulfate Linked to Carbohydrates. *J. Chem. Theory Comp.* **2012**, *8*, 759.
13. Kryshtafovych, A.; Schwede, T.; Topf, M.; Fidelis, K.; Moult, J. Critical assessment of methods of protein structure prediction (CASP)-Round XIV. *Proteins* **2021**, *89*, 1607. [CrossRef] [PubMed]
14. Rieloff, E.; Skepö, M. Molecular Dynamics Simulations of Phosphorylated Intrinsically Disordered Proteins: A Force Field Comparison. *Int. J. Mol. Sci.* **2021**, *22*, 10174. [CrossRef] [PubMed]
15. Rieloff, E.; Skepö, M. The Effect of Multisite Phosphorylation on the Conformational Properties of Intrinsically Disordered Proteins. *Int. J. Mol. Sci.* **2021**, *22*, 11058. [CrossRef] [PubMed]
16. Lara Ortiz, M.T.; Martinell García, V.; Del Rio, G. Saturation Mutagenesis of the Transmembrane Region of HokC in *Escherichia coli* Reveals Its High Tolerance to Mutations. *Int. J. Mol. Sci.* **2021**, *22*, 10359. [CrossRef] [PubMed]
17. Bocedi, A.; Cattani, G.; Gambardella, G.; Schulte, L.; Schwalbe, H.; Ricci, G. Oxidative Folding of Proteins: The "Smoking Gun" of Glutathione. *Int. J. Mol. Sci.* **2021**, *22*, 10148. [CrossRef] [PubMed]
18. Nehls, T.; Heymann, T.; Meyners, C.; Hausch, F.; Lermyte, F. Fenton-Chemistry-Based Oxidative Modification of Proteins Reflects Their Conformation. *Int. J. Mol. Sci.* **2021**, *22*, 9927. [CrossRef]
19. Jarosiewicz, M.; Duchnowicz, P.; Jarosiewicz, P.; Huras, B.; Bukowska, B. An In Vitro Comparative Study of the Effects of Tetrabromobisphenol A and Tetrabromobisphenol S on Human Erythrocyte Membranes—Changes in ATP Level, Perturbations in Membrane Fluidity, Alterations in Conformational State and Damage to Proteins. *Int. J. Mol. Sci.* **2021**, *22*, 9443. [CrossRef]

20. Hołubowicz, R.; Ożyhar, A.; Dobryszycki, P. Natural Mutations Affect Structure and Function of gC1q Domain of Otolin-1. *Int. J. Mol. Sci.* **2021**, *22*, 9085. [CrossRef]
21. Jensen, E.L.; Receveur-Brechot, V.; Hachemane, M.; Wils, L.; Barbier, P.; Parsiegla, G.; Gontero, B.; Launay, H. Structural Contour Map of the Iota Carbonic Anhydrase from the Diatom *Thalassiosira pseudonana* Using a Multiprong Approach. *Int. J. Mol. Sci.* **2021**, *22*, 8723. [CrossRef]
22. Curiel, J.A.; Peirotén, Á.; Landete, J.M.; Ruiz de la Bastida, A.; Langa, S.; Arqués, J.L. Architecture Insight of Bifidobacterial α-L-Fucosidases. *Int. J. Mol. Sci.* **2021**, *22*, 8462. [CrossRef] [PubMed]
23. Karunatilleke, N.C.; Fast, C.S.; Ngo, V.; Brickenden, A.; Duennwald, M.L.; Konermann, L.; Choy, W.-Y. Nrf2, the Major Regulator of the Cellular Oxidative Stress Response, is Partially Disordered. *Int. J. Mol. Sci.* **2021**, *22*, 7434. [CrossRef] [PubMed]
24. Wesch, N.; Löhr, F.; Rogova, N.; Dötsch, V.; Rogov, V.V. A Concerted Action of UBA5 C-Terminal Unstructured Regions Is Important for Transfer of Activated UFM1 to UFC1. *Int. J. Mol. Sci.* **2021**, *22*, 7390. [CrossRef] [PubMed]
25. Dubois, C.; Planelles-Herrero, V.J.; Tillatte-Tripodi, C.; Delbecq, S.; Mammri, L.; Sirkia, E.M.; Ropars, V.; Roumestand, C.; Barthe, P. Pressure and Chemical Unfolding of an α-Helical Bundle Protein: The GH2 Domain of the Protein Adaptor GIPC1. *Int. J. Mol. Sci.* **2021**, *22*, 3597. [CrossRef] [PubMed]
26. Liu, M.; Das, A.K.; Lincoff, J.; Sasmal, S.; Cheng, S.Y.; Vernon, R.M.; Forman-Kay, J.D.; Head-Gordon, T. Configurational Entropy of Folded Proteins and Its Importance for Intrinsically Disordered Proteins. *Int. J. Mol. Sci.* **2021**, *22*, 3420. [CrossRef]
27. Shi, Y.; Xia, Z.; Zhang, J.; Best, R.; Wu, C.; Ponder, J.W.; Ren, P. The Polarizable Atomic Multipole-based AMOEBA Force Field for Proteins. *J. Chem. Theory Comput.* **2013**, *9*, 4046. [CrossRef]

International Journal of
Molecular Sciences

MDPI

Article

Origin of Increased Solvent Accessibility of Peptide Bonds in Mutual Synergetic Folding Proteins

Csaba Magyar [1,†], Anikó Mentes [1,†], Miklós Cserző [1,2] and István Simon [1,*]

1 Institute of Enzymology, Research Centre for Natural Sciences, Eötvös Loránd Research Network, 1117 Budapest, Hungary; magyar.csaba@ttk.hu (C.M.); meaqaat@gmail.com (A.M.); cserzo.miklos@med.semmelweis-univ.hu (M.C.)
2 Department of Physiology, Faculty of Medicine, Semmelweis University, 1094 Budapest, Hungary
* Correspondence: simon.istvan@ttk.hu
† These authors contributed equally to the work.

Abstract: Mutual Synergetic Folding (MSF) proteins belong to a recently discovered class of proteins. These proteins are disordered in their monomeric but ordered in their oligomeric forms. Their amino acid composition is more similar to globular proteins than to disordered ones. Our preceding work shed light on important structural aspects of the structural organization of these proteins, but the background of this behavior is still unknown. We suggest that solvent accessibility is an important factor, especially solvent accessibility of the peptide bonds can be accounted for this phenomenon. The side chains of the amino acids which form a peptide bond have a high local contribution to the shielding of the peptide bond from the solvent. During the oligomerization step, other non-local residues contribute to the shielding. We investigated these local and non-local effects of shielding based on Shannon information entropy calculations. We found that MSF and globular homodimeric proteins have different local contributions resulting from different amino acid pair frequencies. Their non-local distribution is also different because of distinctive inter-subunit contacts.

Keywords: intrinsically disordered proteins; mutual synergetic folding; solvent accessibility of peptide bonds; inter-subunit interaction; solvent-accessible surface area; Shannon information entropy; amino acid composition

Citation: Magyar, C.; Mentes, A.; Cserző, M.; Simon, I. Origin of Increased Solvent Accessibility of Peptide Bonds in Mutual Synergetic Folding Proteins. *Int. J. Mol. Sci.* **2021**, 22, 13404. https://doi.org/10.3390/ijms222413404

Academic Editor: Alexande Baykov

Received: 14 November 2021
Accepted: 11 December 2021
Published: 14 December 2021

1. Introduction

The class of Mutual Synergetic Folding (MSF) proteins is a relatively newly discovered distinct class of oligomeric proteins, where a single polypeptide chain of an MSF protein is disordered, but all chains become ordered upon oligomerization [1,2]. In the case of traditional disordered proteins, there is a need for an already stable template structure. MSF proteins can fold into a stable structure without the presence of an already folded template, folding happens simultaneously with the association of the previously disordered subunits.

At the time of the discovery of the 3D structure of transmembrane proteins [3], and the "coupled folding and binding" mechanisms of disordered proteins [4,5], our knowledge of determinants of proteins structure has been expanded. The discovery of these new types of proteins was accompanied by a change of the perceivable average amino acid composition of proteins. However, the amino acid composition of MSF proteins is similar to that of globular proteins [6–8], which makes it difficult to identify them based on their primary structure. The different amino acid composition of disordered proteins leads to different energies, which causes the protein to have a disordered structure. A good example for the estimation of this energy is the use of statistical pairwise potentials as implemented in the IUPred prediction method, which reached recently its 3rd iteration [9]. Next to the energy, configurational entropy can also be important in disorder-to-order transitions, as shown by Liu et al. [10]. MSF proteins are special among disordered proteins because their oligomeric structure can be solved by traditional structure determination methods. In a recent publication [11] experimentally determined long intrinsically disordered protein regions were examined. The authors found that long disordered regions, which are present in MSF proteins, cannot be accurately predicted.

The tertiary structures of experimentally validated MSF proteins have been collected to the Mutual Folding Induced by Binding (MFIB) database. This database contains 205 MSF protein structures [2]. Our recent analyses [6,7] of this database aimed to find common properties among MSF proteins, which distinguish them from globular oligomeric proteins. We found that the most prominent change between MSF and globular proteins can be found in the change of solvent accessibility during the oligomerization step. The question arose as to what determines the hydration of these peptide bonds.

Sequence and structural studies of protein interactions have revealed that sequential and spatial neighboring residues often play important roles in the environmental hydrophobicity and long-term binding site interactions, thus determining the structural and functional behavior of proteins [12–14]. Based on this fact, the shielding effect (reducing hydration of peptide group) can be divided into local and non-local terms. The local contribution is provided by the side-chain atoms of the amino acid residues, which are connected by the peptide bond [15], while the non-local contribution is provided by the shielding effect of other sequentially distant residues [16].

In this work, we studied the solvent accessibility of peptide bonds in MSF and globular homodimeric proteins in light of relationship between dipeptide frequencies, and the ordered/disordered nature of the monomeric protein forms. Furthermore, we compared Shannon information entropy calculated from frequencies of local and spatial neighboring residues in MSF and globular proteins, which may indicate sequential differences between the two groups of proteins.

2. Results

We calculated the relative solvent accessible surface area (SASA) values for all peptide bonds in our MSF and globular homodimeric (MHOD and GHOD) protein datasets for both monomeric and dimeric forms using the FreeSASA program [17]. The monomeric form was modeled by taking only a single chain of the ordered dimeric structure into account. The distribution of peptide bonds with different relative SASA values can be seen in Figure 1 for the monomeric and dimeric forms. The results are presented as histograms, where the height of a bar denotes the ratio of entries with a property in a given range. For example, the height of the first red bars refers to the percentage of peptide bonds with a relative SASA value in the [0, 0.1] interval among MSF proteins in monomeric form. There is a clear tendency in homodimeric MSF proteins to have a lower percentage of highly buried peptide bonds with lower than 10% relative SASA values in monomeric form but a higher percentage in dimeric form, when compared to globular homodimers.

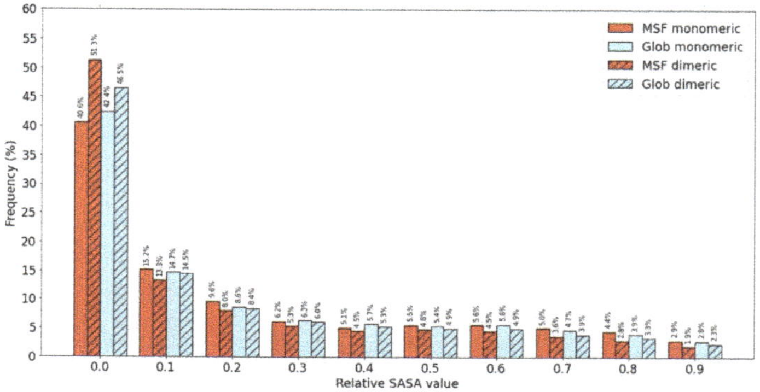

Figure 1. Occurrences of peptide bonds with different solvent accessibilities.

A higher number of peptides bonds become buried during dimerization in the case of MSF homodimers, than in the case of globular homodimers. Next, we calculated the solvent accessibility of peptide bonds averaged within individual proteins. We calculated the ratio of the SASA values summed over all peptide bonds divided with the sum of the reference values, thus representing the average solvent accessibility of peptide bonds within a protein. Results are presented in Figure 2 for both monomeric and dimeric calculations. The distribution of MSF average peptide bond accessibility is shifted towards higher values in the case of monomeric form, but towards lower values in dimeric form, when compared to globular proteins. These results underline our hypothesis about the importance of peptide bond solvent accessibility for MSF proteins.

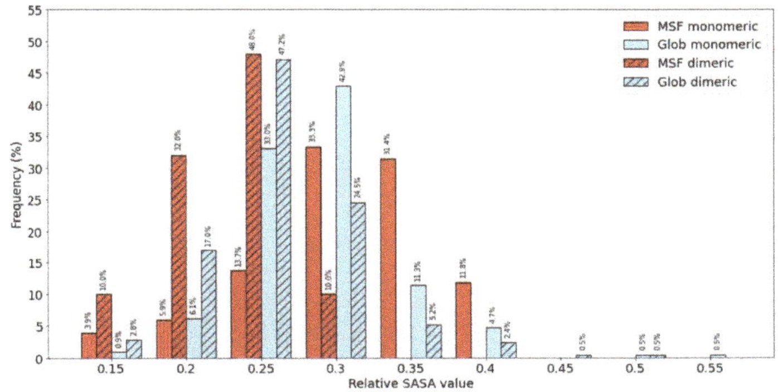

Figure 2. Distribution of individual proteins with different average peptide bond solvent accessibilities.

We created a measure of the increase of peptide bond shielding upon dimerization using the following protocol. A peptide bond was identified as accessible, that is not properly shielded if its relative solvent accessibility was larger or equal than a threshold value of 10%. It was identified as buried if its relative SASA value was below 10%. We found that there are 2.6 times more not properly shielded peptide bonds in monomeric MSF proteins, than in globular proteins, which become buried upon dimerization. The number of buried (B) and accessible (A) residues were counted and we calculated the B/A ratio for both monomeric and dimeric forms. Then we calculated the dimeric B/A over the monomeric B/A quotient. This value represents the increase of peptide bond burial upon dimerization. A value of 1 would indicate that the buried/accessible residue count ratio is the same in both dimeric and monomeric forms, a value of 2 means that in the dimeric form the buried/accessible ratio is twice the monomeric value. Results can be seen in Figure 3. The distribution of the MSF protein values is shifted toward higher values, meaning that in the case of MSF proteins the ratio of the buried/accessible residues is higher in the dimeric form.

These results implicate, that solvent accessibility of the peptide bonds is an important factor in the destabilization of the monomeric form of MSF proteins. We investigated if amino acid composition plays a role in the above-presented findings. Instead of the peptide bond relative SASA value, which involves two amino acids, the relative main-chain SASA values were compared for the 20 residue types. The data presented in Figure 4 shows increased average SASA values for the glycine, lysine, methionine, and tryptophan residues in MSF proteins, while slightly decreased values for proline, aspartic acid, serine and glutamine residues. However, most values are rather similar in the MHOD and the GHOD protein datasets.

Glycine and proline residues have the highest average main-chain accessibility in both MSF and globular proteins, while cysteine, leucine, valine, and isoleucine residues have the lowest average values in both datasets. Traditional disordered proteins can be distinguished from globular ones based on their amino acid composition. Because of the similar amino acid composition of MSF and globular proteins [6–8], statistical measures based on the residue composition did not help to separate the two types of proteins, thus we decided to compare their Shannon information entropy content [18]. The probability distribution of the individual amino acids was calculated as their observable frequency using Equation (1).

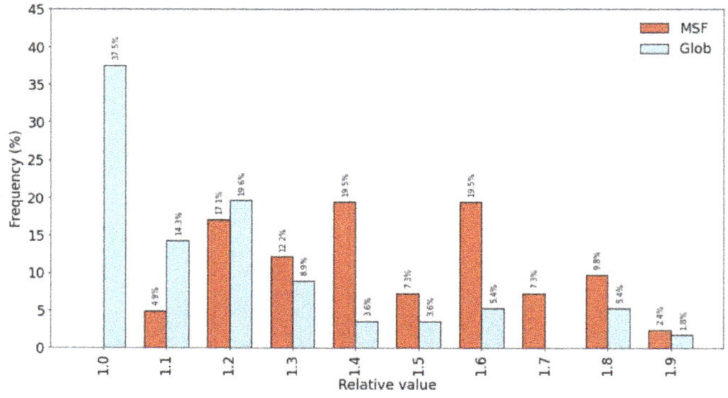

Figure 3. Distribution of individual proteins according to the increase of the buried/accessible peptide bond ratio.

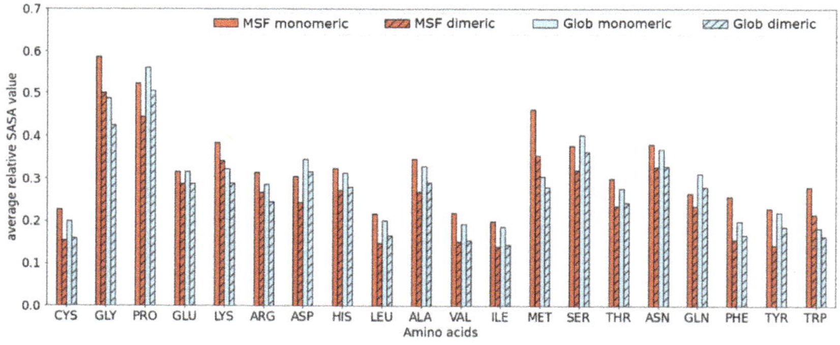

Figure 4. Average relative main-chain SASA values according to residue types.

Equation (1): Calculation of the frequencies of the ith amino acid type

$$P_i = \frac{N_i}{N_{tot}}, \tag{1}$$

where N_i is the number of the ith amino acid type and N_{tot} is the total number of amino acids in the actual dataset.

For calculating entropy values of individual proteins, we decided to use a normalized version of the relative Shannon information entropy, frequently referred to as the Kullback–Leibler divergence [19] using Equation (2). It measures how our P_i amino acid probability distribution differs from a reference Q_i probability distribution. We are using reference Q_i values obtained from a non-redundant subset of the PDB database [20] (see PDB codes in Table S1A), which is significantly larger than our homodimeric protein datasets. The $P_i log \frac{P_i}{Q_i}$ values obtained for the 20 amino acids are listed in Table S2A,B, calculated from the amino acid compositions of the GHOD and MHOD protein datasets, respectively. Entropy values for individual proteins were calculated using Equation (2), where entropy values were normalized using a division with log N to avoid size dependence of the values.

Equation (2): Calculation of the normalized Shannon information entropy for an individual protein based on the amino acid composition

$$S(j) = \frac{1}{\log N_j} \sum_{i=1}^{N_j} P_i log \frac{P_i}{Q_i}, \tag{2}$$

where N_j is the number of amino acids in protein j.

Entropy values were calculated for entries in our homodimeric protein datasets. Results are presented as a histogram in Figure 5.

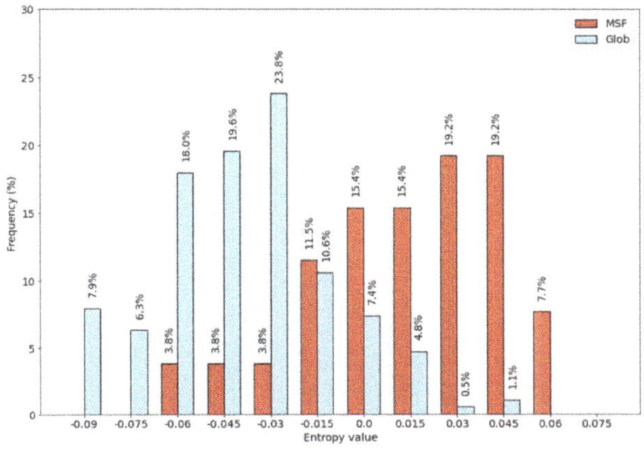

Figure 5. Entropy values calculated based on the amino acid composition.

There is a difference between the distributions obtained on MSF and globular homodimers, entropy values of the MSF proteins are shifted toward positive values. This result is somewhat unexpected in light of our previous results. Recently [6] we compared the amino acid composition of MSF and globular homodimers using principal component analysis and we found no significant difference. The Shannon entropy calculation seems to be more sensitive than our previous analysis.

As we have found a significant difference in the burial of peptide bonds between MSF and globular homodimers, we started to look for a possible reason. Since local shielding of the peptide bonds is mainly provided by the two amino acids creating the peptide bond, we calculated the dipeptide frequencies in our MHOD and GHOD protein datasets by dividing the count number of a specific dipeptide with the total number of peptide bonds using Equation (3).

Equation (3): Calculation of the frequencies of the ij dipeptides

$$P_{ij} = \frac{N_{ij}}{N_{pb}}, \tag{3}$$

where N_{pb} is the total number of peptide bonds.

To pinpoint the differences, we calculated relative entropy-like $P_{ij}log\frac{P_{ij}}{Q_{ij}}$ values for all dipeptides using P_{ij} values obtained on the MHOD protein dataset, and reference Q_{ij} values obtained on the GHOD protein dataset. The highest and lowest 10 values are plotted in Figure 6.

There are only a handful of dipeptides that have a strong preference for MSF or globular proteins, but most dipeptides have rather weak preference values. Though the amino acid composition of MSF homodimeric proteins is close to that of globular proteins [6,8], differences can be found in their sequence already at the dipeptide level.

Based on this observation dipeptide distribution might discriminate MSF proteins from globular ones, so we investigated the information content of the protein sequences. We calculated Shannon information entropy values based on dipeptide frequencies, similar to the previous case of amino acid compositions. The $P_{ij}log\frac{P_{ij}}{Q_{ij}}$ values calculated using our non-redundant sequence dataset derived Q_{ij} and P_{ij} values obtained on the GHOD and MHOD protein datasets can be found in Table S3A,B, respectively. Entropy values for individual proteins were calculated using Equation (4).

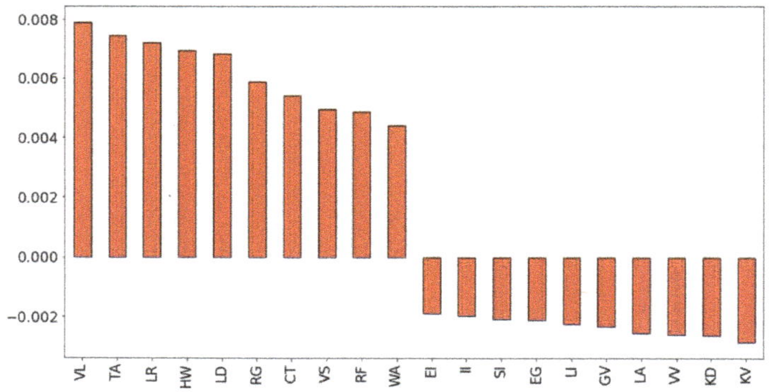

Figure 6. Dipeptides with the highest and lowest entropy-like contributions.

Equation (4): Calculation of the Shannon information entropy for an individual protein based on the dipeptide frequencies:

$$S(k) = \frac{1}{\log N_k} \sum_{1}^{N_k} P_{ij} log \frac{P_{ij}}{Q_{ij}}, \quad (4)$$

where N_k is the number of peptide bonds in protein k.

We plotted the entropy value distribution of individual proteins as histograms (see Figure 7). We can see a similar effect as in the case of entropy values calculated from the amino acid compositions. There is a shift in the distribution of MSF proteins towards positive values.

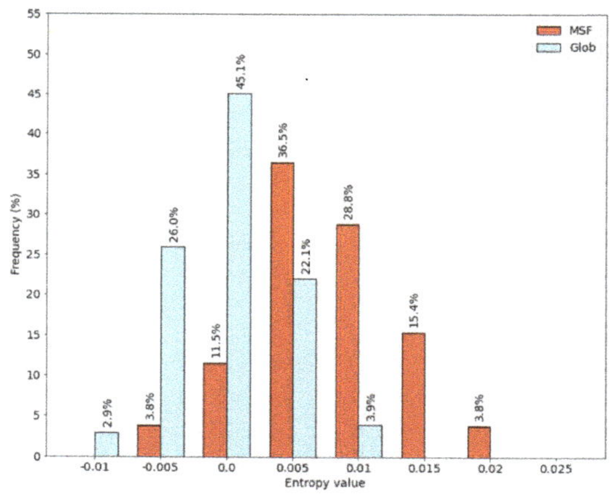

Figure 7. Entropy values calculated from the dipeptide frequencies.

To investigate the effect of the non-local long-range shielding of peptide bonds upon dimerization we created the following protocol. We identified inter-subunit residue contacts based on a simple distance criterion. Residues pairs were identified as important for long range shielding, if the participating residues are part of different protein chains and they have at least one heavy-atom pair with a distance shorter than 4 Ångströms. We identified all these residue pairs and calculated their frequencies using Equation (5).

Equation (5): Calculation of the frequencies of the *ij* residue pairs

$$P_{ij} = \frac{N_{ij}}{N_{tc}}, \quad (5)$$

where N_{tc} is the total number of contacting residue pairs.

Since reference Q_{ij} values can be derived only from dimeric structures, the GHOD protein dataset was used for this purpose. In order to produce entropy values also for globular proteins, both MHOD and GHOD protein datasets were scored using the MHOD derived P_{ij} values. The $P_{ij}log\frac{P_{ij}}{Q_{ij}}$ values for all residue pairs can be found in Table S4. Relative entropy values for individual proteins were calculated similarly to the previous cases, but the sum was created over all contacting residue pairs using Equation (6). The distribution of these values can be seen in Figure 8.

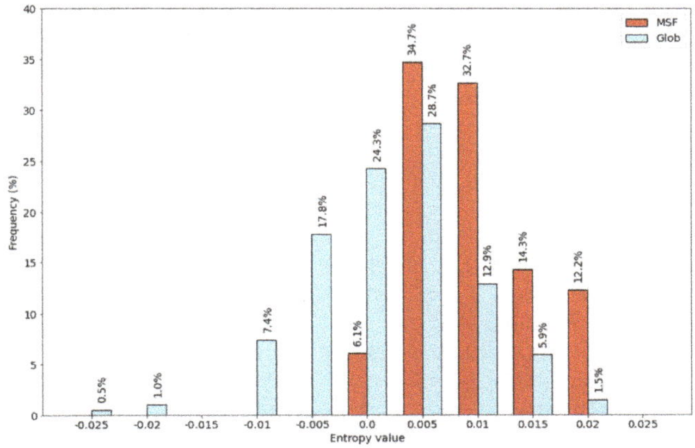

Figure 8. Entropy values calculated from the inter-subunit contacts.

Equation (6): Calculation of the Shannon information entropy for an individual protein based on inter-subunit contacts:

$$S(k) = \frac{1}{\log N_c} \sum_{1}^{N_c} P_{ij} log \frac{P_{ij}}{Q_{ij}}, \quad (6)$$

where N_c is the number of contacts within protein k.

Despite the same P_{ij} matrix was used for both datasets, there is a shift in the distribution of the MSF protein values towards positive values. There is an overlap of the two distributions, but more than a quarter of the globular proteins have negative values, while all MSF proteins have positive values. About one fifth of the globular proteins has a value larger than 0.01, while more than half of the MSF proteins are found in this range.

3. Discussion

MSF proteins are a relatively new class of proteins with little knowledge about their folding. Our current comparison of MSF and globular homodimeric proteins provided the following results. MSF proteins have a higher average relative solvent accessibility of the peptide bonds in their monomeric form. Upon dimerization, a higher proportion of accessible peptide bonds become buried in the case of MSF homodimers, when compared to globular ones. A significant increase in the number of both buried peptide bonds and buried residues upon oligomerization is characteristic for MSF proteins. Zhou et al. recently analyzed the normalized monomer surface area versus normalized interfacial surface area in a recent publication [21]. Their findings are in agreement with ours about the relevance of the increased inter-subunit surface area in MSF proteins.

The burial of the peptide bonds from solvent molecules is established by shielding through local and non-local residues, relative to the actual peptide bond. We found differences in both local and non-local dipeptide frequencies between MSF and globular homodimers using Shannon information entropy calculations. This behavior originates from the different dipeptide frequencies locally and different inter-subunit contacts non-locally.

Zhou et al. also emphasized the importance of intrinsic disorder in complex formation. Previously we found [6] that on our filtered homodimeric protein dataset, all seven tested methods predicted less than 30% of the residues as disordered, while the average value was around 14%. Our suggestion is that a simple physicho-chemical property may be responsible for the destabilization of MSF monomers. Our results indicate that despite the similar amino acid compositions [6], MSF and globular homodimeric proteins have different amino acid pair statistics, leading to different Shannon information entropy distributions. We suggest that this change in the dipeptide frequencies can be accounted for by the less efficient shielding of the peptide bonds of MSF proteins in their monomeric forms. This phenomenon can provide an important contribution to the destabilization of monomeric MSF protein chains, by disturbing the hydrogen bond network of the protein backbone, leading to the disruption of secondary structural elements. The different non-local entropy values may result from the increased necessity of proper peptide bond shielding during the dimerization step.

4. Materials and Methods

For the database analyses, we wrote our own Python programs using Biopython extensions [22] and created a sequence and two structural datasets of proteins with known three-dimensional structures. First, we created a larger sized non-redundant sequence dataset of PDB entries with less than 40% sequence identity using the Mufold-DB database [23] for reference purposes. PDB codes with the protein chain designation can be found in Table S1A. This low similarity cutoff value ensures that the entries are not too similar, which could have biased our residue pair statistics. We used a structural database as a starting point on purpose to include sequences of globular proteins. The resulting dataset contains around 23,000 entries, which is already large enough to provide reliable statistics. We created the MHOD protein dataset by collecting homodimeric MSF proteins found in the MFIB database [2]. We created as reference the GHOD protein dataset of globular homodimeric proteins based on the non-redundant PDB-Filter select 2017 database [24]. Since our aim is to understand which features differentiate MSF proteins from globular ones, we applied a volume/surface criterion as described in our previous publication [7]. This filtering step retains only compact globular like structures and eliminates rather one dimensional, rod-like proteins (like collagen), which would have biased our solvent accessibility calculations. Protein surfaces and volumes were calculated using the FreeSASA 2.03 [17] and the ProteinVolume 1.3 programs [25], respectively. SASA values were calculated for both dimeric and monomeric forms of all proteins. The monomeric form was modeled by deleting the second protein chain from the PDB files. During SASA calculation a couple of additional structures were excluded from the datasets because of structural problems influencing our calculations. A typical problem was that the different protein chains were not in contact, thus SASA values calculated from the dimeric and the monomeric forms were almost identical. The final list of the PDB codes of the remaining 52 entries, called the MHOD protein dataset, can be found is Table S1B. To match the size distribution of this dataset, only proteins with less than 300 residues were kept in the GHOD protein dataset. The list of the PDB codes of the remaining 203 GHOD proteins can be found in Table S1C.

In our previous publications, the solvent accessibility of the main chain was handled at the residue level [6,7]. In this work we focus on the solvent accessibility of peptide bonds, thus SASA values were calculated for atoms forming the peptide bonds. We utilized the absolute all-atom SASA values. The absolute SASA value of a peptide bond was the sum of the atomic absolute SASA values (N, CA, C and O) belonging to a peptide bond. To characterize the relative accessibility of peptide bonds, we created reference SASA values (see Table S5) for all 400 Ala-X-Y-Ala tetrapeptides, built in extended conformation using PyMOL [26]. The relative solvent accessibility of a peptide bond was calculated by dividing its absolute SASA value with the appropriate reference value. We calculated the relative

SASA values of peptide bonds of all MHOD and GHOD entries in both their monomeric and dimeric forms.

When available, modified PDB files were used from the MFIB database. In the case of NMR structures, the representative model structure was selected based on the OLDERADO NMR resource found in PDBe [27].

Supplementary Materials: The following are available online at https://www.mdpi.com/article/10.3390/ijms222413404/s1.

Author Contributions: Conceptualization, I.S., C.M.; methodology, A.M., C.M., M.C.; software, A.M., M.C.; validation, A.M., C.M.; formal analysis, I.S., C.M., M.C.; investigation, A.M., C.M.; resources, C.M., M.C.; data curation, C.M.; writing—original draft preparation, I.S., C.M.; writing—review and editing, I.S., C.M., A.M., M.C.; visualization, A.M., C.M.; supervision, I.S., C.M.; project administration, I.S.; funding acquisition, I.S. All authors have read and agreed to the published version of the manuscript.

Funding: This work was financially supported by the National Research, Development and Innovation Office (OTKA grant No. K115698). The work of A.M. was supported through the New National Excellence Program of the Ministry of Human Capacities (Hungary).

Institutional Review Board Statement: Not applicable.

Informed Consent Statement: Not applicable.

Data Availability Statement: Data supporting the reported results can be found in the Supplement.

Acknowledgments: The authors acknowledge the support of ELIXIR Hungary (www.elixir-hungary.org as accessed on 10 December 2021). On behalf of Project DPSA we thank for the usage of ELKH Cloud (https://science-cloud.hu/ as accessed on 10 December 2021) that significantly helped us achieving the results published in this paper.

Conflicts of Interest: The authors declare no conflict of interest. The funders had no role in the design of the study; in the collection, analysis, or interpretation of data; in the writing of the manuscript, or in the decision to publish the results.

Abbreviations

GHOD: Globular HomoDimeric; IDP: Intrinsically Disordered Protein; MFIB: Mutual Folding Induced by Binding; MHOD: MSF HomoDimeric; NMR: Nucleic Magnetic Resonance; MSF: Mutual Synergetic Folding; OLDERADO: On Line Database of Ensemble Representatives And Domains; PDB: Protein Data Bank; SASA: Solvent Accessible Surface Area.

References

1. Demarest, S.J.; Martinez-Yamout, M.; Chung, J.; Chen, H.; Xu, W.; Dyson, H.J.; Evans, R.M.; Wright, P.E. Mutual synergistic folding in recruitment of CBP/p300 by p160 nuclear receptor coactivators. *Nature* **2002**, *415*, 549–553. [CrossRef]
2. Fichó, E.; Reményi, I.; Simon, I.; Mészáros, B. MFIB: A repository of protein complexes with mutual folding induced by binding. *Bioinformatics* **2017**, *33*, 3682–3684. [CrossRef]
3. Hopf, T.A.; Colwell, L.J.; Sheridan, R.; Rost, B.; Sander, C.; Marks, D.S. Three-dimensional structures of membrane proteins from genomic sequencing. *Cell* **2012**, *149*, 1607–1621. [CrossRef]
4. Schad, E.; Fichó, E.; Pancsa, R.; Simon, I.; Dosztányi, Z.; Mészáros, B. DIBS: A repository of disordered binding sites mediating interactions with ordered proteins. *Bioinformatics* **2018**, *34*, 535–537. [CrossRef]
5. Fukuchi, S.; Sakamoto, S.; Nobe, Y.; Murakami, S.D.; Amemiya, T.; Hosoda, K.; Koike, R.; Hiroaki, H.; Ota, M. IDEAL: Intrinsically Disordered proteins with Extensive Annotations and Literature. *Nucleic Acids Res.* **2012**, *40*, D507–D511. [CrossRef]
6. Magyar, C.; Mentes, A.; Fichó, E.; Cserző, M.; Simon, I. Physical Background of the Disordered Nature of "Mutual Synergetic Folding" Proteins. *Int. J. Mol. Sci.* **2018**, *19*, 3340. [CrossRef] [PubMed]
7. Mentes, A.; Magyar, C.; Fichó, E.; Simon, I. Analysis of Heterodimeric "Mutual Synergistic Folding"-Complexes. *Int. J. Mol. Sci.* **2019**, *20*, 5136. [CrossRef] [PubMed]
8. Mészáros, B.; Dobson, L.; Fichó, E.; Simon, I. Sequence and Structure Properties Uncover the Natural Classification of Protein Complexes Formed by Intrinsically Disordered Proteins via Mutual Synergistic Folding. *Int. J. Mol. Sci.* **2019**, *20*, 5460. [CrossRef] [PubMed]

9. Erdős, G.; Pajkos, M.; Dosztányi, Z. IUPred3: Prediction of protein disorder enhanced with unambiguous experimental annotation and visualization of evolutionary conservation. *Nucleic Acids Res.* **2021**, *49*, W297–W303. [CrossRef] [PubMed]
10. Liu, M.; Das, A.K.; Lincoff, J.; Sasmal, S.; Cheng, S.Y.; Vernon, R.M.; Forman-Kay, J.D.; Head-Gordon, T. Configurational Entropy of Folded Proteins and Its Importance for Intrinsically Disordered Proteins. *Int. J. Mol. Sci.* **2021**, *22*, 3420. [CrossRef] [PubMed]
11. Monzon, A.M.; Necci, M.; Quaglia, F.; Walsh, I.; Zanotti, G.; Piovesan, D.; Tosatto, S.C.E. Experimentally Determined Long Intrinsically Disordered Protein Regions Are Now Abundant in the Protein Data Bank. *Int. J. Mol. Sci.* **2020**, *21*, 4496. [CrossRef] [PubMed]
12. Gromiha, M.M.; Yokota, K.; Fukui, K. Sequence and structural analysis of binding site residues in protein-protein complexes. *Int. J. Biol. Macromol.* **2010**, *46*, 187–192. [CrossRef] [PubMed]
13. Ofran, Y.; Rost, B. Predicted protein-protein interaction sites from local sequence information. *FEBS Lett.* **2003**, *544*, 236–239. [CrossRef]
14. Basu, S.; Bahadur, R.P. Do sequence neighbours of intrinsically disordered regions promote structural flexibility in intrinsically disordered proteins? *J. Struct. Biol.* **2020**, *209*, 107428. [CrossRef] [PubMed]
15. Bignucolo, O.; Leung, H.T.; Grzesiek, S.; Bernèche, S. Backbone hydration determines the folding signature of amino acid residues. *J. Am. Chem. Soc.* **2015**, *137*, 4300–4303. [CrossRef] [PubMed]
16. Vendruscolo, M.; Paci, E.; Dobson, C.M.; Karplus, M. Three key residues form a critical contact network in a protein folding transition state. *Nature* **2001**, *409*, 641–645. [CrossRef]
17. Mitternacht, S. FreeSASA: An open source C library for solvent accessible surface area calculations. *F1000Reserch* **2016**, *5*, 189. [CrossRef]
18. Shannon, C.E. Communication theory of secrecy systems 1945. *MD Comput.* **1998**, *15*, 57–64.
19. Kullback, B.J.; Leibler, R.A. On information and sufficiency. *Ann. Math. Statist.* **1951**, *22*, 79–86. [CrossRef]
20. Berman, H.M.; Westbrook, J.; Feng, Z.; Gilliland, G.; Bhat, T.N.; Weissig, H.; Shindyalov, I.N.; Bourne, P.E. The Protein Data Bank. *Nucleic Acids Res.* **2000**, *28*, 235–242. [CrossRef]
21. Zhou, J.; Oldfield, C.J.; Yan, W.; Shen, B.; Dunker, A.K. Identification of Intrinsic Disorder in Complexes from the Protein Data Bank. *ACS Omega* **2020**, *5*, 17883–17891. [CrossRef] [PubMed]
22. Cock, P.J.; Antao, T.; Chang, J.T.; Chapman, B.A.; Cox, C.J.; Dalke, A.; Friedberg, I.; Hamelryck, T.; Kauff, F.; Wilczynski, B.; et al. Biopython: Freely available Python tools for computational molecular biology and bioinformatics. *Bioinformatics* **2009**, *25*, 1422–1423. [CrossRef] [PubMed]
23. He, Z.; Zhang, C.; Xu, Y.; Zeng, S.; Zhang, J.; Xu, D. MUFOLD-DB: A processed protein structure database for protein structure prediction and analysis. *BMC Genom.* **2014**, *15* (Suppl. S11), S2. [CrossRef] [PubMed]
24. Griep, S.; Hobohm, U. PDBselect 1992-2009 and PDBfilter-select. *Nucleic Acids Res.* **2010**, *38*, D318–D319. [CrossRef] [PubMed]
25. Chen, C.R.; Makhatadze, G.I. ProteinVolume: Calculating molecular van der Waals and void volumes in proteins. *BMC Bioinform.* **2015**, *16*, 101. [CrossRef] [PubMed]
26. The PyMOL Molecular Graphics System, Version 2.0. Schrödinger, LLC. 2015. Available online: https://pymol.org/2/support.html (accessed on 14 November 2021).
27. Kelley, L.A.; Sutcliffe, M.J. OLDERADO: On-line database of ensemble representatives and domains. On Line Database of Ensemble Representatives and DOmains. *Protein Sci.* **1997**, *6*, 2628–2630. [CrossRef]

International Journal of *Molecular Sciences*

MDPI

Communication

MemDis: Predicting Disordered Regions in Transmembrane Proteins

Laszlo Dobson and Gábor E. Tusnády *

Institute of Enzymology, Research Centre for Natural Sciences, Magyar Tudósok Körútja 2, 1117 Budapest, Hungary; dobson.laszlo@ttk.hu

* Correspondence: tusnady.gabor@ttk.hu

Abstract: Transmembrane proteins (TMPs) play important roles in cells, ranging from transport processes and cell adhesion to communication. Many of these functions are mediated by intrinsically disordered regions (IDRs), flexible protein segments without a well-defined structure. Although a variety of prediction methods are available for predicting IDRs, their accuracy is very limited on TMPs due to their special physico-chemical properties. We prepared a dataset containing membrane proteins exclusively, using X-ray crystallography data. MemDis is a novel prediction method, utilizing convolutional neural network and long short-term memory networks for predicting disordered regions in TMPs. In addition to attributes commonly used in IDR predictors, we defined several TMP specific features to enhance the accuracy of our method further. MemDis achieved the highest prediction accuracy on TMP-specific dataset among other popular IDR prediction methods.

Keywords: transmembrane proteins; intrinsically disordered proteins; deep learning; convolutional neural network; bidirectional long-short term memory

Citation: Dobson, L.; Tusnády, G.E. MemDis: Predicting Disordered Regions in Transmembrane Proteins. *Int. J. Mol. Sci.* **2021**, *22*, 12270. https://doi.org/10.3390/ijms222212270

Academic Editor: Fabio Polticelli

Received: 8 October 2021
Accepted: 9 November 2021
Published: 12 November 2021

1. Introduction

Transmembrane proteins (TMPs) are located in different membranes and they provide gates between the inner and outer side of cells or organelles. Around 25% of the coded proteins in the human proteome contain one or more membrane regions [1]. These segments embedded in the lipid bilayer are structurally well defined; however, their tail and loop regions often contain unstructured segments. Such regions are aiding various functions from providing flexible linkers to binding motifs for other molecules [2]. Although intrinsically disordered regions (IDRs) are well studied in general, the currently available prediction methods have limited accuracy on membrane proteins for several reasons [3]. On the one hand, protein disorder is conditional [4] and heavily influenced by the environment; thus, membrane proteins, exposed on both outside and inside spaces, cannot be well described using a single function or machine learning algorithm. Moreover, lipid components of the membrane influence the charge and acidity near the transmembrane regions, further complicating the situation. On the other hand, these methods are generally trained on mixed protein sets predominantly containing non-TMPs, resulting in biased information from the perspective of TMPs. Here, we propose MemDis, a novel tool for predicting IDR regions in TMP proteins, which achieved the highest accuracy among tested methods. We utilized Convolutional Neural Networks (CNNs) to capture local features of the sequence represented by Position-Specific Scoring Matrix and Long Short-Term Memory (LSTM) Network to take advantage of the semantic properties of the protein sequence.

2. Results

To realistically capture the different flavors of disorder in membrane proteins, four different models were created according to different topological regions. CNNs were trained on extracellular-distant (distance from membrane > 15aa), proximal- (≤15aa) and intracellular-distant (distance > 15aa), proximal (≤15aa) residues separately. A bidirectional

LSTM network was also trained to "smooth" the prediction of CNNs on individual residues and achieve better sensitivity.

Based on the training and validation set, we found that the CNNs, with a slightly higher cutoff (0.65—notably this result is scaled so the web server will display 0.5 cut-off) and a ±4 residue smoothing achieved the best specificity, while also keeping other metric values considerably high. In contrast, the LSTM with a ±7 residue smoothing had the best sensitivity. Both versions (from now on referred to as specific and sensitive, respectively) achieved a remarkable 0.83–0.84 Area Under Curve (AUC) (Figure 1A, Supplementary Materials). We compared the results of our method to other popular algorithms [5–8] using metrics from the most recent CAID experiment [9] (Supplementary Table S1). We used the complete protein sequence for testing; however, we only considered fragments selected earlier for the evaluation. Some of the tested methods achieved slightly better specificity, at the cost of barely predicting disordered segments. The best sensitivity was achieved using the MemDis sensitive. Although dozens of IDR prediction methods are available, when selecting other methods, we aimed to select ones with slightly different methodology (machine learning, biophysical approaches) and training sets (X-ray, NMR, etc.). Both the sensitive and specific settings of MemDis achieved the highest balanced accuracy, Matthew's Correlation Coefficient (MCC) and AUC (Figure 1A, Supplementary Materials). Notably, MemDis uses different models to predict membrane-distant and proximal regions, and their separate performance also captures disorder better compared to other methods (Figure 1B,C; Supplementary Table S1, Supplementary Materials). When evaluating IUPred3 locally, experimental filtering was not used.

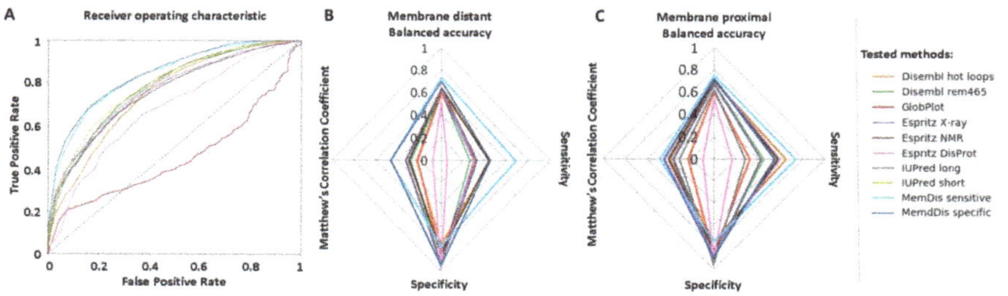

Figure 1. (**A**) Receiver operating characteristic of MemDis and other disorder prediction methods. (**B**) Averaged performance of membrane-distant predictors. (**C**) Average performance of membrane proximal predictors.

MemDis is available on GitHub at https://github.com/brgenzim/MemDis. Since the local installation is slightly complicated as users have to set up all dependencies as well, we also prepared a webserver (available at http://memdis.ttk.hu), where users can query their sequence(s). The webserver displays topology predicted by CCTOP and a graph for disordered prediction.

We also checked a handful of well-defined examples where the output of MemDis is supported by literature evidence. Phospholemman is a member of the FXYD family that regulates ion transport [10]. The cytosolic C-terminal tail was shown to associate with the micelle surface [11], forming a helical structure upon binding. MemDis predicts this region as disordered. The helical propensity prediction of FELLS [12] suggests that this region is likely helical (Figure 2A). Thus, combining the MemDis and other secondary structure prediction methods, lipid binding can be assumed for membrane proximal regions. Integrin alpha-IIIb is a receptor protein with a cytosolic disordered tail according to DisProt [13], exhibiting short linear motifs (SLiMs) proposed to play a role in SARS-CoV-2 infection [14]. Membrane proximal disordered regions are often missed by prediction methods, making it hard to find novel linear motif candidates; however, MemDis successfully detects these regions (Figure 2B). Mucolipin-1 is a cation channel, probably playing a role in membrane trafficking. The C-terminal cytosolic region has five cysteines, a residue that is often referred to as order-promoting (as they can form disulphide bridges in an extracellular environment), which deceives many predictors. MemDis has a built-in topology filter and predicts this region as disordered, in agreement with the electron-microscopy structure lacking coordinates for this region [15]. The C-terminal cytosolic tail of Mucolipin is also stacked with SLiMs: it has two di-leucine motifs [16], and phosphoserines [17] in the well-defined PKA phosphorylation site [18], further supporting that the C-terminal is disordered (Figure 2C).

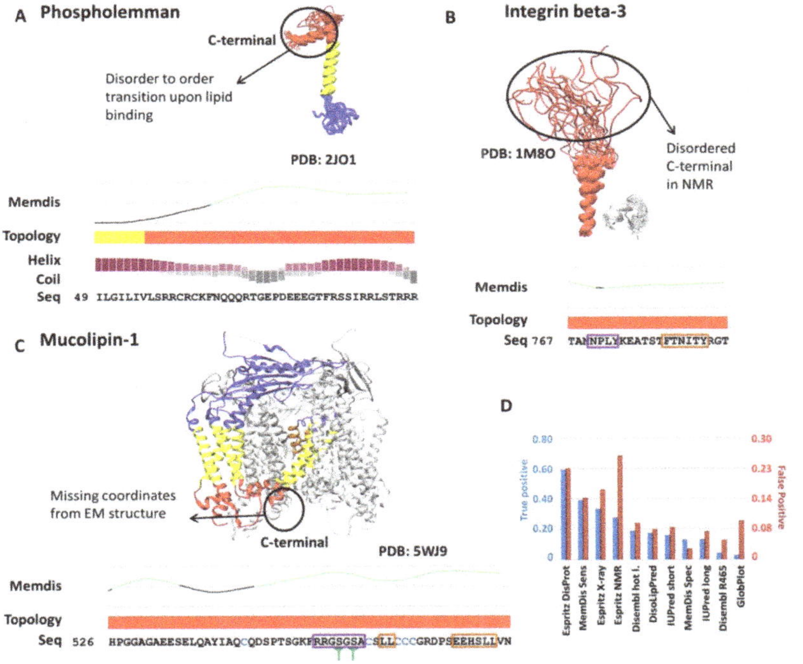

Figure 2. Interpretation of MemDis results. (**A**) Phospholemman: solution NMR structure, and representation of C-terminal by the prediction of MemDis, CCTOP and FELLS (helical propensity: purple, coil propensity: grey). (**B**) Integrin beta-3: solution NMR structure, MemDis and CCTOP predictions. The proposed NPxY endocytosis sorting signal is marked with purple, the LIR autophagy motif is marked with an orange box. (**C**) Mucopilin-1: Electron-microscopy structure, prediction from MemDis and CCTOP. Phosphoserines are marked with green cones below the sequence. The phosphorylation site is marked with a purple box, di-leucine motifs are marked with orange boxes. Cysteines have blue color. Topology is represented both in the structures and topology lines and structures are colored blue, red, yellow and orange (extracellular, cytosolic, transmembrane, and re-entrant loop regions, respectively). Disordered regions from MemDis are marked with green lines on the graphs. Note, only specific regions of the sequences are shown. (**D**) Detection rate of lipid-binding and non-lipid-binding disordered regions from the MemMoRF database.

We also assessed how predictors work to predict lipid-binding regions. MemMoRF is a novel database of disordered regions that undergo disorder-to-order transition upon membrane binding [19]. We measured the accuracy of different prediction methods on such regions. Unfortunately, all methods have poor performance (−0.19–0.03 MCC, Supplementary Table S1) on this dataset when measuring residue level accuracy. To overcome this, we counted the number of regions that have at least 60% of their residues predicted as disordered. In this comparison, Espritz DisProt had the highest hit rate, however, on the price of predicting many false positive regions too, while MemDis with sensitive settings was second, with somewhat fewer false positive regions (Figure 2D). We also evaluated DisoLipPred [20], which was developed specifically to find lipid-binding regions; however, it detected only 20% of lipid-binding disordered regions. In sum, none of the methods are capable of detecting such information reliably alone; however, introducing additional filters (topology, secondary structure) may increase their accuracy, as it was shown on MemDis in the case of Phospholemman.

3. Materials and Methods

We downloaded the MobiDB database [21] in 1April 2021, and selected the missing residues (th_90, used as disordered label) and observed (th_90, used as ordered label) subsets, defining regions from X-ray structures when there is 90% agreement between the observations. Next, we used CCTOP [22] to filter TMPs and used CD-HIT [23] to reduce redundancy to 40% sequence identity (Supplementary Table S2). In most cases, the full protein structure was not solved, so we used fragments of the protein sequences. First, we selected every IDR together with flanking ordered regions up to 15aa if they were included in the PDB. Next, we randomly selected ordered regions (Figure 3). The fragments were randomly selected into the train, validation and independent test set (Supplementary Table S3). We prepared Convolutional Neural Networks (CNNs) and a bidirectional Long Short-Term Memory (LSTM) network to predict IDRs.

For the CNNs, each non-membrane residue in this dataset belonged to one of the following four TMP topology categories: extracellular-distant (distance from membrane > 15aa), proximal ($\leq$15aa) and intracellular-distant (distance > 15aa), proximal ($\leq$15aa). Disordered and ordered residues were selected in a way that their distributions be roughly equal in each topological subset (max. 10% difference, Supplementary Table S4). We prepared four convolutional neural networks (CNNs) for the four topological regions (Figure 3). The features (Supplementary Table S5) include amino acid distribution, non-redundant AAIndex [24] categories (i.e., different amino acid scales), ProtParam [25] features (i.e., molecular weight, isoelectric point and instability index), topology information based on CCTOP and PSI-BLAST results. We also used Netsurfp [26] to predict accessibility of residues and SEG implemented in PlatoLoco [27] to detect low complexity regions. We used a $\pm$5 length window around each residue and calculated 39 features for them, this way producing a feature matrix of size 11×39 (Supplementary Table S5) that was fed into the appropriate CNN (this window may contain residues not included in PDB or transmembrane residues, as these residues are only used as features belonging to a properly labelled residue). The CNNs were trained until their validation loss stopped decreasing for a constitutive 10 epochs (this occurred roughly at 1000 epochs)—the training and the validation accuracy at this point did not show high differences (Supplementary Table S6).

The bidirectional Long Short-Term Memory (LSTM) was trained on the full length fragments (including membrane regions) and used the output of the CNNs with topology information to predict disordered regions. Since the CNNs can only predict residues in an aqueous environment, for membrane residues the LSTM received "0" value as input. The LSTM was set to consider the preceding 12 time steps (Figure 3). The parameters of the CNNs and LSTM are available in Supplementary Table S7.

For testing, we hold back each hit from PSI-BLAST that occurred during training to avoid data leakage. Since the redundancy filter was originally performed on full-length proteins, we ensured again that no fragment in the independent testing set shared 40% or higher sequence identity to any sequence in the training and validation sequence fragment sets.

To define lipid-binding regions, we used the MemMoRF [19] database. We used redundancy filtering to 40%, and excluded proteins from the training set of MemDis. The negative set was generated using fragments near to the membrane (15AA), that did not have lipid-binding annotation in MemMorRF.

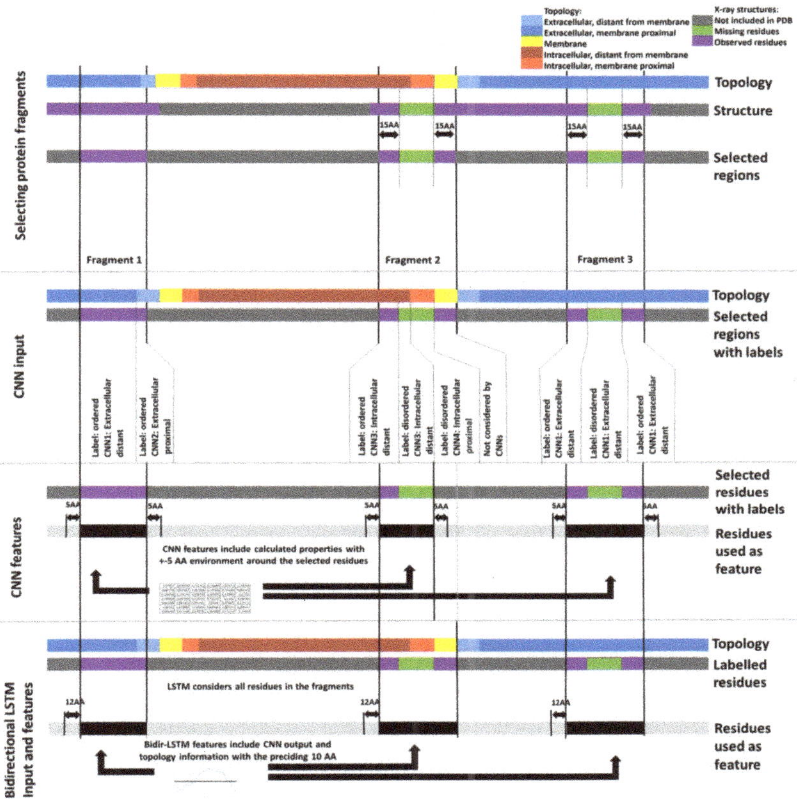

Figure 3. Data preparation for the training of MemDis. First, we selected protein fragments based on the available PDB information. Extracellular-distant (distance from membrane > 15AA), proximal (<15AA) and intracellular-distant, proximal residues from these fragments were fed into the appropriate CNN, also considering information from residues within 5AA from the residue of interest. The LSTM was trained on the full-length protein fragments considering the preceding 10AA.

Supplementary Materials: The following are available online at https://www.mdpi.com/article/10.3390/ijms222212270/s1.

Author Contributions: Conceptualization, L.D. and G.E.T.; methodology, L.D. and G.E.T.; software, L.D. and G.E.T.; validation, L.D. and G.E.T.; formal analysis, L.D. and G.E.T.; investigation, L.D. and G.E.T.; resources, L.D. and G.E.T.; data curation, L.D. and G.E.T.; writing—original draft preparation, L.D. and G.E.T.; writing—review and editing, L.D. and G.E.T.; visualization, L.D. and G.E.T.; supervision, L.D. and G.E.T.; project administration, L.D. and G.E.T.; funding acquisition, L.D. and G.E.T. All authors have read and agreed to the published version of the manuscript.

Funding: Project no. [K119287 and K132522] has been implemented with the support provided by the Ministry of Innovation and Technology of Hungary from the National Research, Development and Innovation Fund, financed under the OTKA funding scheme. G.E.T. was supported by the Momentum Grant of the Hungarian Academy of Sciences (LP2012-35).

Institutional Review Board Statement: Not applicable.

Informed Consent Statement: Not applicable.

Data Availability Statement: MemDis is available online at http://memdis.ttk.hu and for local use it can be downloaded from https://github.com/brgenzim/MemDis.

Acknowledgments: We thank András Zeke for helpful discussion of the manuscript.

Conflicts of Interest: The authors declare no conflict of interest.

References

1. Dobson, L.; Reményi, I.; Tusnády, G.E. The human transmembrane proteome. *Biol. Direct* **2015**, *10*, 31. [CrossRef] [PubMed]
2. Kjaergaard, M.; Kragelund, B.B. Functions of intrinsic disorder in transmembrane proteins. *Cell. Mol. Life Sci.* **2017**, *74*, 3205–3224. [CrossRef] [PubMed]
3. Tusnády, G.E.; Dobson, L.; Tompa, P. Disordered regions in transmembrane proteins. *Biochim. Biophys. Acta (BBA)-Biomembr.* **2015**, *1848*, 2839–2848. [CrossRef]
4. Reichmann, D.; Jakob, U. The roles of conditional disorder in redox proteins. *Curr. Opin. Struct. Biol.* **2013**, *23*, 436–442. [CrossRef] [PubMed]
5. Erdős, G.; Pajkos, M.; Dosztányi, Z. IUPred3: Prediction of protein disorder enhanced with unambiguous experimental annotation and visualization of evolutionary conservation. *Nucleic Acids Res.* **2021**, *49*, W297–W303. [CrossRef]
6. Linding, R.; Russell, R.B.; Neduva, V.; Gibson, T.J. GlobPlot: Exploring protein sequences for globularity and disorder. *Nucleic Acids Res.* **2003**, *31*, 3701–3708. [CrossRef] [PubMed]
7. Walsh, I.; Martin, A.J.M.; Di Domenico, T.; Tosatto, S.C.E. ESpritz: Accurate and fast prediction of protein disorder. *Bioinformatics* **2011**, *28*, 503–509. [CrossRef]
8. Linding, R.; Jensen, L.J.; Diella, F.; Bork, P.; Gibson, T.J.; Russell, R.B. Protein disorder prediction: Implications for structural proteomics. *Structure* **2003**, *11*, 1453–1459. [CrossRef]
9. Necci, M.; Piovesan, D.; Tosatto, S.C. Critical assessment of protein intrinsic disorder prediction. *Nat. Methods* **2021**, *18*, 472–481. [CrossRef]
10. Cheung, J.Y.; Zhang, X.-Q.; Song, J.; Gao, E.; Rabinowitz, J.E.; Chan, T.O.; Wang, J. Phospholemman: A Novel Cardiac Stress Protein. *Clin. Transl. Sci.* **2010**, *3*, 189–196. [CrossRef]
11. Teriete, P.; Franzin, C.M.; Choi, J.; Marassi, F.M. Structure of the Na, K-ATPase Regulatory Protein FXYD1 in Micelles. *Biochemistry* **2007**, *46*, 6774–6783. [CrossRef] [PubMed]
12. Piovesan, D.; Walsh, I.; Minervini, G.; Tosatto, S.C.E. FELLS: Fast estimator of latent local structure. *Bioinformatics* **2017**, *33*, 1889–1891. [CrossRef] [PubMed]
13. Quaglia, F.; Mészáros, B.; Salladini, E.; Hatos, A.; Pancsa, R.; Chemes, B.L.; Pajkos, M.; Lazar, T.; Pena-Diaz, S.; Santos, J.; et al. DisProt in 2022: Improved quality and accessibility of protein intrinsic disorder. *Nucleic Acids Res.* **2022**.
14. Mészáros, B.; Sámano-Sánchez, H.; Alvarado-Valverde, J.; Čalyševa, J.; Martínez-Pérez, E.; Alves, R.; Shields, D.C.; Kumar, M.; Rippmann, F.; Chemes, L.B.; et al. Short linear motif candidates in the cell entry system used by SARS-CoV-2 and their potential therapeutic implications. *Sci. Signal.* **2021**, *14*, eabd0334. [CrossRef]
15. Schmiege, P.; Fine, M.; Blobel, G.; Li, X. Human TRPML1 channel structures in open and closed conformations. *Nature* **2017**, *550*, 366–370. [CrossRef]
16. Vergarajauregui, S.; Puertollano, R. Two di-leucine motifs regulate trafficking of mucolipin-1 to lysosomes. *Traffic* **2006**, *7*, 337–353. [CrossRef]
17. Vergarajauregui, S.; Oberdick, R.; Kiselyov, K.; Puertollano, R. Mucolipin 1 channel activity is regulated by protein kinase A-mediated phosphorylation. *Biochem. J.* **2008**, *410*, 417–425. [CrossRef]
18. Kumar, M.; Michael, S.; Alvarado-Valverde, J.; Mészáros, B.; Sámano-Sánchez, H.; Zeke, A.; Dobson, L.; Lazar, T.; Örd, M.; Nagpal, A.; et al. The Eukaryotic Linear Motif resource: 2022 release. *Nucleic Acids Res.* **2021**, gkab975. [CrossRef]
19. Csizmadia, G.; Erdős, G.; Tordai, H.; Padányi, R.; Tosatto, S.; Dosztányi, Z.; Hegedűs, T. The MemMoRF database for recognizing disordered protein regions interacting with cellular membranes. *Nucleic Acids Res.* **2020**, *49*, D355–D360. [CrossRef]
20. Katuwawala, A.; Zhao, B.; Kurgan, L. DisoLipPred: Accurate prediction of disordered lipid binding residues in protein sequences with deep recurrent networks and transfer learning. *Bioinformatics* **2021**, *93*, btab640. [CrossRef]
21. Piovesan, D.; Tabaro, F.; Paladin, L.; Necci, M.; Mičetić, I.; Camilloni, C.; Davey, N.; Dosztányi, Z.; Mészáros, B.; Monzon, A.M.; et al. MobiDB 3.0: More annotations for intrinsic disorder, conformational diversity and interactions in proteins. *Nucleic Acids Res.* **2018**, *46*, D471–D476. [CrossRef]
22. Dobson, L.; Reményi, I.; Tusnády, G.E. CCTOP: A consensus constrained topology prediction web server. *Nucleic Acids Res.* **2015**, *43*, W408–W412. [CrossRef] [PubMed]
23. Huang, Y.; Niu, B.; Gao, Y.; Fu, L.; Li, W. CD-HIT Suite: A web server for clustering and comparing biological sequences. *Bioinformatics* **2010**, *26*, 680–682. [CrossRef] [PubMed]
24. Kawashima, S. AAindex: Amino acid index database. *Nucleic Acids Res.* **2000**, *28*, 374. [CrossRef]
25. Gasteiger, E.; Hoogland, C.; Gattiker, A.; Duvaud, S.; Wilkins, M.R.; Appel, R.D.; Bairoch, A. Protein identification and analysis tools on the ExPASy Server. In *The Proteomics Protocols Handbook*; Humana Press: Totowa, NJ, USA, 2005; pp. 571–607.

26. Petersen, B.; Petersen, T.N.; Andersen, P.; Nielsen, M.; Lundegaard, C. A generic method for assignment of reliability scores applied to solvent accessibility predictions. *BMC Struct. Biol.* **2009**, *9*, 51. [CrossRef]
27. Jarnot, P.; Ziemska-Legięcka, J.; Dobson, L.; Merski, M.; Mier, P.; Andrade-Navarro, M.A.; Hancock, J.M.; Dosztányi, Z.; Paladin, L.; Necci, M.; et al. PlaToLoCo: The first web meta-server for visualization and annotation of low complexity regions in proteins. *Nucleic Acids Res.* **2020**, *48*, W77–W84. [CrossRef] [PubMed]

International Journal of *Molecular Sciences*

MDPI

Article

Evaluation of Deep Neural Network ProSPr for Accurate Protein Distance Predictions on CASP14 Targets

Jacob Stern [1,2,†], Bryce Hedelius [1,†], Olivia Fisher [1], Wendy M. Billings [1] and Dennis Della Corte [1,*]

1 Department of Physics and Astronomy, Brigham Young University, Provo, UT 84602, USA; jastern33@gmail.com (J.S.); bhedelius@gmail.com (B.H.); oefish@gmail.com (O.F.); wendybillings7@gmail.com (W.M.B.)
2 Department of Computer Science, Brigham Young University, Provo, UT 84602, USA
* Correspondence: Dennis.DellaCorte@byu.edu
† Both authors contributed equally.

Abstract: The field of protein structure prediction has recently been revolutionized through the introduction of deep learning. The current state-of-the-art tool AlphaFold2 can predict highly accurate structures; however, it has a prohibitively long inference time for applications that require the folding of hundreds of sequences. The prediction of protein structure annotations, such as amino acid distances, can be achieved at a higher speed with existing tools, such as the ProSPr network. Here, we report on important updates to the ProSPr network, its performance in the recent Critical Assessment of Techniques for Protein Structure Prediction (CASP14) competition, and an evaluation of its accuracy dependency on sequence length and multiple sequence alignment depth. We also provide a detailed description of the architecture and the training process, accompanied by reusable code. This work is anticipated to provide a solid foundation for the further development of protein distance prediction tools.

Keywords: protein; prediction; contact; distance; deep learning; alphafold; ProSPr; CASP; dataset; retrainable

Citation: Stern, J.; Hedelius, B.; Fisher, O.; Billings, W.M.; Della Corte, D. Evaluation of Deep Neural Network ProSPr for Accurate Protein Distance Predictions on CASP14 Targets. *Int. J. Mol. Sci.* **2021**, *22*, 12835. https://doi.org/10.3390/ijms222312835

Academic Editors: Istvan Simon and Csaba Magyar

Received: 18 October 2021
Accepted: 25 November 2021
Published: 27 November 2021

1. Introduction

Proteins are among nature's smallest machines and fulfill a broad range of life-sustaining tasks. To fully understand the function of a protein, accurate knowledge of its folded structure is required. Protein structures can either be obtained from experiments, homology modeling, or computational structure prediction. Accurate structures can be used for the rational design of biosensors [1], the prediction of small-molecule docking [2], enzyme design [3], or simulation studies to explore protein dynamics [4].

Recent progress in the field of computational structure prediction includes the end-to-end deep learning models Alphafold2 [5] and RoseTTAfold [6] that are able to predict highly accurate protein structures from multiple sequence alignments. Alphafold2 has been used to predict the structures of many protein sequences found in nature, including the human proteome [7].

Despite these advancements, it is still not fully known if models such as Alphafold2 can extract dynamics or multiple conformations of proteins [8]. Furthermore, it is also not clear if Alphafold2 can be used effectively to support tasks in protein engineering, such as assessing if single point mutations in the amino acid sequence of a protein will alter stability or function.

A main bottleneck of Alphafold2 is the runtime for prediction. It can take multiple hours on a GPU cluster to predict the structure of a single protein. If thousands of sequences must be evaluated in a protein design study, this runtime can be prohibitive.

A valid alternative to full protein structure prediction is the prediction of structural features that provide sufficient information about conformational changes. The previous state-of-the-art tools Alphafold1 [9] and trRosetta [10] predict distances and contacts between amino acids. This task can be performed rapidly and allows for the comparison of differences between contact patterns of multiple sequences. We have developed ProSPr as an open-source alternative to enable the community to understand, train, and apply deep learning for the same tasks.

After Alphafold1 was initially presented during the Critical Assessment of Techniques for Protein Structure Prediction (CASP13) conference [11], many questions remained about its implementation. To demystify this process, our team developed and published ProSPr—a clone of Alphafold1 on GitHub and bioRxiv [12]. With the release of the Alphafold1 paper, we updated the ProSPr architecture and made new models available. After CASP14, it became apparent that ProSPr was used by multiple participating groups, as the Alphafold1 code was not easily usable by the community [13].

Deep learning methods are often complementary, and a variety of easy-to-use models can be very valuable to form ensembles that outperform single methods. In a previous study, we have shown that ProSPr contact predictions are of similar quality as Alphafold1 and trRosetta predictions but that an ensemble of all three methods is superior to any individual method [14]. We further showed that ProSPr can be used to rapidly predict large structural changes from small sequence variations, making it a useful tool for sequence assessment in protein engineering. [14]

Although the first ProSPr model has been used by multiple groups during CASP14 and shown its usefulness in driving improved contact predictions, this is the first detailed description of its updated architecture and the training process used. We did not use our original version of ProSPr in CASP14, but rather a completely distinct iteration with higher performance that drew from our growing expertise in the area. These updates were informed by the publication of Alphafold1 and trRosetta, which were not released until shortly before the CASP14 prediction season began, and so the models described here were still being trained during CASP14 and are distinct from those we used during the competition. Here, we present this improved ProSPr version and release the network code, training scripts, and related datasets.

Additionally, for those who are currently using the ProSPr network for protein distance prediction, it is important to know under which conditions the predictions are reliable. Two important factors upon which protein structure prediction accuracy depends are MSA depth and sequence length [5,15–17]. For example, AlphaFold 2 found that there was strong reliance on MSA depth up to about 30 alignments, after which the importance of additional aligned sequences was negligible. However, network dependence on MSA depth and sequence length can vary across networks architectures, so we investigate the dependence of the ProSPr network on these features.

2. Evaluation and Results

We evaluated the performance of three updated ProSPr models using the CASP14 target dataset. The CASP assessors provided access to full label information before it was publicly available (i.e., prior to PDB release) for many of the targets which enabled us to analyze our predictions across 61 protein targets. We evaluated these targets based on residue-residue contacts, which are defined by CASP as having a Cβ (or Cα for glycine) distance less than 8 Å [18]. Predicted contact probabilities were straightforward to derive from our binned distance predictions; we summed the probabilities of the first three bins since their distances correspond to those less than 8 Å.

Figure 1 shows results for two example targets from CASP14. For T1034, we were able to construct an MSA with a depth greater than 10,000 and the predicted accuracies (top of the diagonal) are in good agreement with the labels (bottom of the diagonal). The protein structure annotations on the right compare the prediction accuracy on top with the label on the bottom. This shows that even for an easy target, these predictions are not highly accurate, which is likely due to the small loss contribution assigned to auxiliary predictions (see Methods). For target T1042, no sequences could be found, and the corresponding predictions are without signal. The goal of training a contact prediction tool that can infer information from sequence alone is an open problem and will need to be addressed in future work.

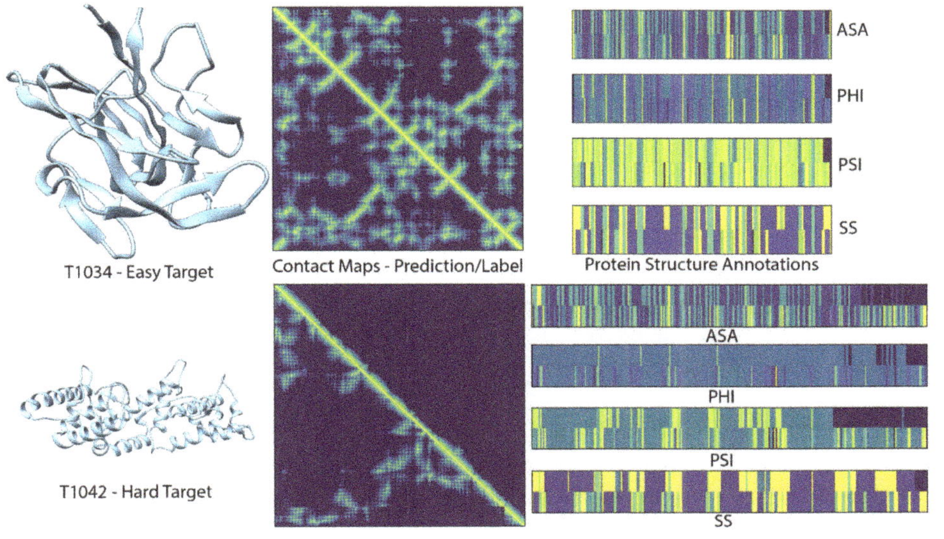

Figure 1. Two example targets from the CASP14 test set. Left: experimental structures from which labels were derived. Middle: contact maps predicted with ProSPr ensemble on top of the diagonal; label on bottom. Right: visualization of auxiliary loss predictions on top with labels at bottom. Accessible surface area (ASA), torsion angles (PHI, PSI), secondary structure (SS).

Table 1 shows the contact accuracies of the three ProSPr models evaluated at short, mid, and long contact ranges. These categories relate to the sequence separation of the two amino acids involved in each contact, where short-, mid-, and long-range pairs are separated by 6 to 11, 12 to 23, and 24+ residues, respectively [19]. All contact predictions in each of these ranges were ranked by probability and the top L (sequence length) pairs in each category were considered to be in contact. We then calculated contact accuracies using the following equation [20]:

$$Accuracy = \frac{TP+TN}{TP+FP+FN+TN} = Precision = \frac{TP}{TP+FP}$$

which reduces the precision since no negative predictions are made ($TN = FN = 0$). Furthermore, we normalized the accuracy scores for each target in each range so that the full range of 0–100% could be achieved (i.e., in some cases there may not be L true contacts, so the maximum score would otherwise be lower).

Table 1. CASP14 contact accuracies (see text for definition).

ProSPr Model	Contact Accuracy (%)			
	Short	Mid	Long	Average
A	81.09%	69.52%	41.63%	64.08%
B	81.15%	69.29%	42.41%	64.28%
C	81.94%	69.97%	43.59%	65.17%
Ensemble	**82.08%**	**70.55%**	**44.04%**	**65.56%**

The three ProSPr models shown in Table 1 have the same architecture and were trained on the same data (see Methods) but perform somewhat differently. By creating an ensemble of the three networks, the average results in all three areas are improved (for the ensemble performance on individual targets, see Table 2) which is in accordance with our previous work [14]. We have made all three models individually available, but in accordance with these results, the default inference setting of the code is to automatically ensemble all of them for the best performance.

Table 2. ProSPr ensemble contact accuracies (see text for definition).

Target	Contact Accuracy		
	Short	Mid	Long
T1045s2	0.833	0.924	0.694
T1046s1	1.000	1.000	0.536
T1046s2	0.892	0.574	0.303
T1047s1	0.907	0.985	0.639
T1047s2	1.000	0.983	0.852
T1060s2	0.857	0.575	0.282
T1060s3	0.976	0.955	0.793
T1065s1	1.000	0.973	0.518
T1065s2	1.000	1.000	0.870
T1024	1.000	1.000	0.809
T1026	0.750	0.425	0.494
T1027	0.485	0.278	0.054
T1029	0.891	0.818	0.220
T1030	0.804	0.792	0.333
T1031	0.686	0.457	0.105
T1032	0.889	0.851	0.580
T1033	0.750	0.316	0.216
T1034	0.988	0.874	0.885
T1035	0.412	0.080	0.000
T1037	0.690	0.455	0.030
T1038	0.720	0.538	0.407
T1039	0.269	0.000	0.007
T1040	0.318	0.222	0.027
T1041	0.644	0.357	0.021
T1042	0.487	0.441	0.058
T1043	0.431	0.216	0.014
T1049	1.000	0.939	0.440
T1050	0.964	0.821	0.705
T1052	0.728	0.600	0.417
T1053	0.796	0.521	0.093
T1054	1.000	1.000	0.710
T1055	0.932	0.860	0.200
T1056	0.823	0.829	0.661
T1057	1.000	0.987	0.815
T1058	0.821	0.678	0.678
T1061	0.807	0.687	0.511
T1064	0.615	0.500	0.094
T1067	0.865	0.824	0.466
T1068	0.926	0.813	0.204
T1070	0.941	0.707	0.579
T1073	1.000	1.000	1.000
T1074	0.845	0.700	0.328
T1076	0.970	0.947	0.911
T1078	0.984	0.892	0.587
T1079	0.956	0.964	0.739

Table 2. *Cont.*

Target	Contact Accuracy Short	Mid	Long
T1082	0.615	0.636	0.164
T1083	0.909	0.783	0.909
T1084	1.000	1.000	1.000
T1087	1.000	0.810	0.714
T1088	0.954	1.000	0.778
T1089	0.972	0.813	0.624
T1090	0.977	0.870	0.399
T1091	0.832	0.571	0.071
T1092	0.704	0.782	0.382
T1093	0.673	0.519	0.109
T1094	0.649	0.580	0.144
T1095	0.722	0.711	0.448
T1096	0.766	0.421	0.098
T1099	0.800	0.375	0.101
T1100	0.883	0.820	0.258
T1101	0.960	0.988	0.783

We also investigated the impact of alignment depth and sequence length on contact prediction using the CASP14 dataset. For this purpose, we segmented the targets into groups with either less than 400 sequences or between 400 and 15,000 sequences (threshold of maximum MSA depth). Figure 2 shows that a correlation between shallow MSAs and average prediction accuracy exists with a Pearson correlation coefficient of $r > 0.7$. However, for deeper MSAs this correlation is no longer observed. Furthermore, we compared the dependency of prediction accuracy on the sequence length of the target and found no correlation with $r = 0$. Based on this, we conclude that ProSPr is sequence-length-independent and that finding at least a few hundred sequences is helpful to increase the predictive performance of ProSPr, but deeper alignments hold no clear benefit.

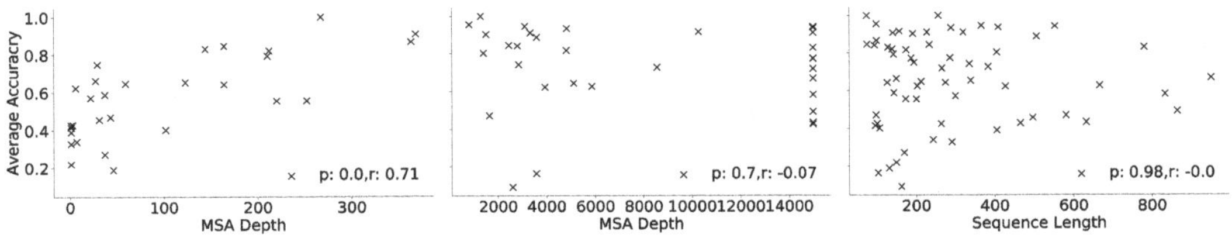

Figure 2. Left: correlation analysis of average accuracy (see text for definition) for CASP14 targets with MSA smaller than 400 sequences. Middle: correlation analysis for MSA deeper than 400 sequences. Right: correlation analysis of average accuracy and target amino acid sequence length.

Finally, we evaluated inference times for ProSPr and found that they scale linearly with the number of crops and quadratically with the sequence length. In comparison with AlphaFold 2 on a Tesla V100, for a sequence of length 256, one forward pass through our model takes 1.88 ± 0.18 s, compared to 4.8 min for an AlphaFold 2 prediction. The high-accuracy version of our model, which uses 10 overlapping offsets, takes 4.39 ± 0.44 s. For a sequence of length 384, one forward pass through our model takes 4.11 ± 0.35 s for low-accuracy and 40.32 ± 3.63 s for high-accuracy, compared to 9.2 min for AlphaFold 2. Note that these numbers are for a single model; the ensemble of three models takes three times as long.

3. Methods

3.1. ProSPr Overview

ProSPr predicts a series of features related to three-dimensional protein structures that can be referred to as protein structure annotations [21] (PSAs). The primary purpose of ProSPr is to predict the distances between pairs of residues for a given sequence. Specifically, this is defined as the distance between the Cβ atoms of two residues *i* and *j* (Cα is used in the case of glycine). ProSPr also predicts secondary structure (SS) classes, relative accessible surface area (ASA), and torsion angles for each residue in a sequence. However, these are included only as auxiliary features to improve the quality of the distance predictions (see Methods).

All ProSPr predictions are categorical in nature, and otherwise continuous values have been discretized into bins. For example, the inter-residue distances were divided into 10 bins: <4 Å, $4 \leq d < 6$ Å, $6 \leq d < 8$ Å, . . . , etc., up to the final bin, which included all distances greater than or equal to 20 Å. This specific format was developed in alignment with the distance prediction format announced for CASP14 [13].

ProSPr, as depicted in Figure 3, is a deep, two-dimensional convolutional residual neural network [22] of which the architecture was inspired by that of the 2018 version of AlphaFold1 [9]. After performing an initial BatchNorm [23] and 1×1 convolution on the input tensor, the result is fed through the 220 dilated residual blocks that make up the bulk of the network. Each block consists of a BatchNorm followed by an exponential linear unit (ELU) activation [24] and a 1×1 convolution, then another BatchNorm and ELU, a 3×3 dilated convolution [25], and finally another BatchNorm, ELU, a 1×1 projection, and an identity addition. The blocks cycle through 3×3 convolutions with dilation factors of 1, 2, 4, and 8. The first 28 of these blocks use 256 channels, but the last 192 only use 128. Once passed through all 220 blocks, a 1×1 convolution is applied to change the number of channels down to 10 for distance predictions, whereas 64×1 and 1×64 convolutions are applied to extract the *i* and *j* auxiliary predictions, respectively.

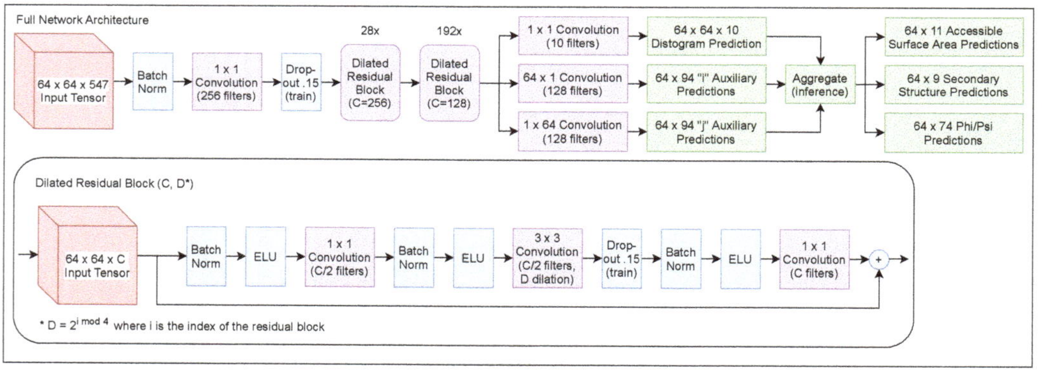

Figure 3. ProSPr network architecture and model architecture.

3.2. Input Features

The input tensor to ProSPr has dimensions $L \times L \times 547$ and contains both sequence- and MSA-derived features. The sequence information is provided as 21 one-hot encoded values; 20 for the natural amino acids; and another for unnatural residues, gaps, or padding. The residue index information is also included as integer values relative to the start of the sequence. A hidden Markov model is constructed from the MSA using HHBlits [26], for which numerical values are directly encoded as layers in the input tensor. Finally, 442 layers come from a custom direct-coupling analysis [10] (DCA), computed based on the raw MSA [27]. See Figure 4 for a detailed view of the data pipeline and find further details in the released code, which includes a function for constructing a full input from the sequence and MSA.

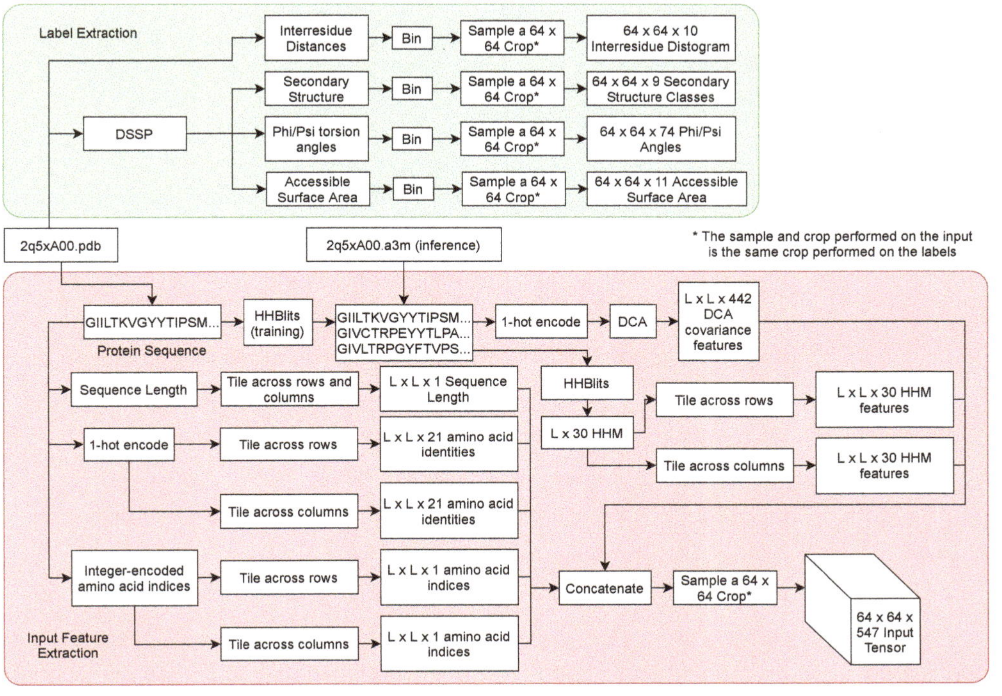

Figure 4. Detailed view of ProSPr data pipeline. For training a protein structure in the pdb file format is used to create inputs and labels. For inference, a multiple sequence alignment in the a3m file format is expected.

3.3. Training Data

We derived the data used to train these ProSPr models from the structures of protein domains in the CATH s35 dataset [28]. First, the sequences were extracted from the structure files. We then constructed multiple sequence alignments (MSAs) for each sequence using HHBlits [26] (E-value 0.001, 3 iterations, limit 15,000 sequences). Inter-residue distance labels were calculated from the CATH structure files and binned into 10 possible values, in accordance with CASP14 formatting, as described previously. We then used the DSSP algorithm [29] to extract labels for secondary structure (9 classes native to DSSP), torsion angles (phi and psi, each sorted into 36 10° bins from −180° to 180°, plus one for error/gap) and relative accessible surface area (ASA) (divided into 10 equal bins, plus another for N/A or a gap).

3.4. Training Strategy

After generating the input data and labels for the CATH s35 domains, we split them into training (27,330 domains) and validation sets (2067 domains). To augment the effective training set size, we used two strategies. First, we constructed ProSPr so that it predicted 64 × 64 residue crops of the final distance map. By doing this, we transformed ~27 k domains into over 3.4 million training crops. In each training epoch, we randomly applied a grid over every protein domain to divide it into a series of non-overlapping crops. Performing this step each epoch also increased the variety of the input since the crops were unlikely to be in the same positions each time. Second, we randomly subsampled 50% of the MSA for each domain in each epoch. Using this smaller MSA, we calculated the hidden Markov model and DCA features used in the input vector. This strategy also served to increase the variety of the training data used by the network to prevent overfitting.

All models were trained using a multicomponent cross-entropy loss function. The overall objective was to predict accurate inter-residue distances, the secondary structure (SS), torsion angles (phi/psi), and accessible surface area (ASA) tasks were included as auxiliary losses with the idea that adding components that require shared understanding with the main task could improve performance. Each of the cross-entropy losses was weighted by the following terms and summed to make up the overall loss: 0.5 SS, 0.25 phi, 0.25 psi, 0.5 ASA, and 15 for the distances.

All models used 15% dropout and an Adam optimizer with an initial learning rate (LR) of 0.001. The LR of model A decayed to 0.0005 at epoch 5 and further to 0.0001 at epoch 15. For model B the LR decreased to 0.0005 at epoch 10 and then to 0.0001 at epoch 25. Lastly, the LR of model C dropped to 0.0005 at epoch 8, and down to 0.0001 at epoch 20.

Each model trained on a single GPU (Nvidia Quadro RTX 5000 with 16 GB) with a batch size of 8 for between 100 and 140 epochs, which took about two months. The validation set was used as an early-stopping criterion (using static 64×64 crop grids to reduce noise) and the three checkpoints of each model with the lowest validation losses were selected for testing. The CASP13 test set was then used for final model selection, and the CASP14 predictions were made and analyzed as described earlier.

3.5. Inference

At inference time, we take crops that guarantee coverage of the entire sequence and take additional random crops to cover boundaries between the original crops. We then predict all features for each crop and average the aggregated predictions. The aggregation step consists of aggregating predictions across all crops for each pair i, j of indices (in the case of distance predictions), and each index i (in the case of auxiliary predictions), then taking the average prediction across all crops. Due to this cropping scheme, some crops will aggregate more predictions than others, which is corrected for through averaging.

The ensembling method first predicts a distance probability distribution with each of the three models. Next, the three distance probability distributions are averaged and normalized to yield the final prediction.

4. Conclusions

We developed an updated version of the ProSPr distance prediction network and trained three new models. We found that an ensemble of all three models yielded the best performance on the CASP14 test set, which agrees with our previous finding that deep learning models are frequently complimentary. We further investigated the dependency on multiple-sequence-alignment depth and found that very shallow alignments reduce the accuracy of the network but adding more sequences beyond a few hundred to an alignment does not result in further performance gains. We found that contact prediction accuracies for ProSPr on the CASP14 dataset are of high quality for short and mid contacts but lacking for long contacts. This is likely due to the strategy we used for creating multiple sequence alignments, which did not leverage genomic datasets and resulted frequently in very shallow alignments. We also found that amino acid sequence length did not correlate with contact prediction accuracy on the CASP14 test set. These findings suggest to ProSPr users that confidence in distance predictions is less dependent on sequence length and is maximized for MSAs with a depth of a few hundred sequences. Finally, we showed that the inference times of ProSPr are two orders of magnitude faster than those of AlphaFold2, allowing for feature predictions of protein libraries within a reasonable timeframe. This enables ProSPr to be used for tasks that require fast inference, such as protein design.

This work describes the comprehensive architecture of ProSPr and a training strategy, together with necessary scripts to enable rapid reproduction. To our knowledge, this is the first deep learning-based method for protein structure prediction for which the authors have publishes not only models but reproducible training scripts. As such, it might prove a very useful educational tool for students trying to understand the applications of deep learning in this rapidly evolving field [30]. The full training routine and necessary datasets are available to enable other groups to rapidly build on our networks. All necessary tools and datasets can be found at https://github.com/dellacortelab/prospr (last accessed 24 November 2021).

Author Contributions: D.D.C. and W.M.B. conceived this study. W.M.B., J.S., B.H. and D.D.C. trained the networks and performed the analysis. O.F. supported all other authors with the writing of the article. All authors have read and agreed to the published version of the manuscript.

Funding: We thank the department of Physics and Astronomy at BYU for providing start-up funds and the College of Physical and Mathematical Sciences at BYU for funding undergraduate wages for this project.

Institutional Review Board Statement: Not applicable.

Informed Consent Statement: Not applicable.

Data Availability Statement: All data is available on https://github.com/dellacortelab/prospr (last accessed 24 November 2021).

Conflicts of Interest: The authors declare no conflict of interest.

References

1. Della Corte, D.; van Beek, H.L.; Syberg, F.; Schallmey, M.; Tobola, F.; Cormann, K.U.; Schlicker, C.; Baumann, P.T.; Krumbach, K.; Sokolowsky, S. Engineering and application of a biosensor with focused ligand specificity. *Nat. Commun.* **2020**, *11*, 1–11. [CrossRef]
2. Morris, C.J.; Corte, D.D. Using molecular docking and molecular dynamics to investigate protein-ligand interactions. *Mod. Phys. Lett. B* **2021**, *35*, 2130002. [CrossRef]
3. Coates, T.L.; Young, N.; Jarrett, A.J.; Morris, C.J.; Moody, J.D.; Corte, D.D. Current computational methods for enzyme design. *Mod. Phys. Lett. B* **2021**, *35*, 2150155. [CrossRef]
4. Möckel, C.; Kubiak, J.; Schillinger, O.; Kühnemuth, R.; Della Corte, D.; Schröder, G.F.; Willbold, D.; Strodel, B.; Seidel, C.A.; Neudecker, P. Integrated NMR, fluorescence, and molecular dynamics benchmark study of protein mechanics and hydrodynamics. *J. Phys. Chem. B* **2018**, *123*, 1453–1480. [CrossRef]
5. Jumper, J.; Evans, R.; Pritzel, A.; Green, T.; Figurnov, M.; Ronneberger, O.; Tunyasuvunakool, K.; Bates, R.; Žídek, A.; Potapenko, A. Highly accurate protein structure prediction with AlphaFold. *Nature* **2021**, *596*, 583–589. [CrossRef]
6. Baek, M.; DiMaio, F.; Anishchenko, I.; Dauparas, J.; Ovchinnikovet, S.; Lee, G.R.; Wang, J.; Cong, Q.; Kinch, L.N.; Schaeffer, R.D.; et al. Accurate prediction of protein structures and interactions using a three-track neural network. *Science* **2021**, *373*, 871–876. [CrossRef] [PubMed]
7. Tunyasuvunakool, K.; Adler, J.; Wu, Z.; Green, T.; Zielinski, M.; Žídek, A.; Bridgland, A.; Cowie, A.; Meyer, C.; Laydon, A. Highly accurate protein structure prediction for the human proteome. *Nature* **2021**, *596*, 590–596. [CrossRef]
8. Fleishman, S.J.; Horovitz, A. Extending the new generation of structure predictors to account for dynamics and allostery. *J. Mol. Biol.* **2021**, *433*, 167007. [CrossRef]
9. Senior, A.W.; Evans, R.; Jumper, J.; Kirkpatrick, J.; Sifre, L.; Green, T.; Qin, C.; Žídek, A.; Nelson, A.W.; Bridgland, A. Improved protein structure prediction using potentials from deep learning. *Nature* **2020**, *577*, 706–710. [CrossRef] [PubMed]
10. Yang, J.; Anishchenko, I.; Park, H.; Peng, Z.; Ovchinnikov, S.; Baker, D. Improved protein structure prediction using predicted interresidue orientations. *Proc. Natl. Acad. Sci. USA* **2020**, *117*, 1496–1503. [CrossRef]
11. Senior, A.W.; Evans, R.; Jumper, J.; Kirkpatrick, J.; Sifre, L.; Green, T.; Qin, C.; Žídek, A.; Nelson, A.W.; Bridgland, A. Protein structure prediction using multiple deep neural networks in the 13th Critical Assessment of Protein Structure Prediction (CASP13). *Proteins Struct. Funct. Bioinform.* **2019**, *87*, 1141–1148. [CrossRef]
12. Billings, W.M.; Hedelius, B.; Millecam, T.; Wingate, D.; Della Corte, D. ProSPr: Democratized implementation of alphafold protein distance prediction network. *BioRxiv* **2019**, 830273. [CrossRef]
13. CASP. CASP14 Abstracts. Available online: https://predictioncenter.org/casp14/doc/CASP14_Abstracts.pdf (accessed on 24 November 2021).
14. Billings, W.M.; Morris, C.J.; Della Corte, D. The whole is greater than its parts: Ensembling improves protein contact prediction. *Sci. Rep.* **2021**, *11*, 1–7.
15. Xu, J.; Wang, S. Analysis of distance-based protein structure prediction by deep learning in CASP13. *Proteins Struct. Funct. Bioinform.* **2019**, *87*, 1069–1081. [CrossRef]
16. Jain, A.; Terashi, G.; Kagaya, Y.; Subramaniya, S.R.M.V.; Christoffer, C.; Kihara, D. Analyzing effect of quadruple multiple sequence alignments on deep learning based protein inter-residue distance prediction. *Sci. Rep.* **2021**, *11*, 1–13.

17. Li, Y.; Zhang, C.; Bell, E.W.; Yu, D.J.; Zhang, Y. Ensembling multiple raw coevolutionary features with deep residual neural networks for contact-map prediction in CASP13. *Proteins Struct. Funct. Bioinform.* **2019**, *87*, 1082–1091. [CrossRef]
18. Chen, X.; Liu, J.; Guo, Z.; Wu, T.; Hou, J.; Cheng, J. Protein model accuracy estimation empowered by deep learning and inter-residue distance prediction in CASP14. *Sci. Rep.* **2021**, *11*, 1–12.
19. Shrestha, R.; Fajardo, E.; Gil, N.; Fidelis, K.; Kryshtafovych, A.; Monastyrskyy, B.; Fiser, A. Assessing the accuracy of contact predictions in CASP13. *Proteins* **2019**, *87*, 1058–1068. [CrossRef] [PubMed]
20. Ji, S.; Oruc, T.; Mead, L.; Rehman, M.F.; Thomas, C.M.; Butterworth, S.; Winn, P.J. DeepCDpred: Inter-residue distance and contact prediction for improved prediction of protein structure. *PLoS ONE* **2019**, *14*, e0205214. [CrossRef]
21. Torrisi, M.; Pollastri, G. Protein structure annotations. In *Essentials of Bioinformatics*; Springer: Cham, Switzerland, 2019; Volume I, pp. 201–234.
22. He, K.; Zhang, X.; Ren, S.; Sun, J. Deep residual learning for image recognition. In Proceedings of the IEEE Conference on Computer Vision and Pattern Recognition, Las Vegas, NV, USA, 27–30 June 2016; pp. 770–778.
23. Santurkar, S.; Tsipras, D.; Ilyas, A.; Mądry, A. How does batch normalization help optimization? In Proceedings of the 32nd International Conference on Neural Information Processing Systems, Montreal, QC, Canada, 3–8 December 2018; pp. 2488–2498.
24. Clevert, D.-A.; Unterthiner, T.; Hochreiter, S. Fast and accurate deep network learning by exponential linear units (elus). *arXiv* **2015**, arXiv:1511.07289.
25. Yu, F.; Koltun, V. Multi-scale context aggregation by dilated convolutions. *arXiv* **2015**, arXiv:1511.07122.
26. Remmert, M.; Biegert, A.; Hauser, A.; Söding, J. HHblits: Lightning-fast iterative protein sequence searching by HMM-HMM alignment. *Nat. Methods* **2012**, *9*, 173–175. [CrossRef] [PubMed]
27. Morcos, F.; Pagnani, A.; Lunt, B.; Bertolino, A.; Marks, D.S.; Sander, C.; Zecchina, R.; Onuchic, J.N.; Hwa, T.; Weigt, M. Direct-coupling analysis of residue coevolution captures native contacts across many protein families. *Proc. Natl. Acad. Sci. USA* **2011**, *108*, E1293–E1301. [CrossRef] [PubMed]
28. Knudsen, M.; Wiuf, C. The CATH database. *Hum. Genom.* **2010**, *4*, 1–6. [CrossRef] [PubMed]
29. Kabsch, W.; Sander, C. Dictionary of protein secondary structure: Pattern recognition of hydrogen-bonded and geometrical features. *Biopolymers* **1983**, *22*, 2577–2637. [CrossRef]
30. Kryshtafovych, A.; Moult, J.; Billings, W.M.; Della Corte, D.; Fidelis, K.; Kwon, S.; Olechnovič, K.; Seok, C.; Venclovas, Č.; Won, J. Modeling SARS-CoV2 proteins in the CASP-commons experiment. *Proteins Struct. Funct. Bioinform.* **2021**, *89*, 1987–1996. [CrossRef]

International Journal of
Molecular Sciences

Article

Protein–Protein Docking with Large-Scale Backbone Flexibility Using Coarse-Grained Monte-Carlo Simulations

Mateusz Kurcinski *, Sebastian Kmiecik *, Mateusz Zalewski and Andrzej Kolinski

Biological and Chemical Research Centre, Faculty of Chemistry, University of Warsaw, 02-089 Warsaw, Poland; mateusz.zalewski.fuw@gmail.com (M.Z.); kolinski@chem.uw.edu.pl (A.K.)
* Correspondence: mkurc@chem.uw.edu.pl (M.K.); sekmi@chem.uw.edu.pl (S.K.)

Abstract: Most of the protein–protein docking methods treat proteins as almost rigid objects. Only the side-chains flexibility is usually taken into account. The few approaches enabling docking with a flexible backbone typically work in two steps, in which the search for protein–protein orientations and structure flexibility are simulated separately. In this work, we propose a new straightforward approach for docking sampling. It consists of a single simulation step during which a protein undergoes large-scale backbone rearrangements, rotations, and translations. Simultaneously, the other protein exhibits small backbone fluctuations. Such extensive sampling was possible using the CABS coarse-grained protein model and Replica Exchange Monte Carlo dynamics at a reasonable computational cost. In our proof-of-concept simulations of 62 protein–protein complexes, we obtained acceptable quality models for a significant number of cases.

Keywords: protein–protein interactions; protein–protein binding; protein–protein complex; coarse-grained modeling; multiscale modeling

Citation: Kurcinski, M.; Kmiecik, S.; Zalewski, M.; Kolinski, A. Protein–Protein Docking with Large-Scale Backbone Flexibility Using Coarse-Grained Monte-Carlo Simulations. *Int. J. Mol. Sci.* **2021**, *22*, 7341. https://doi.org/10.3390/ijms22147341

Academic Editors: Istvan Simon and Csaba Magyar

Received: 19 June 2021
Accepted: 4 July 2021
Published: 8 July 2021

1. Introduction

Protein–protein interactions are fundamental in many biological processes. Their structural characterization is one of the biggest challenges of computational biology. A variety of docking methods are currently available for structure prediction of protein–protein complexes [1,2]. They can be divided into free (global) and template-based docking. Free (global) docking methods are designed to generate many distinct binding configurations. Template-based methods restrict docking to a binding mode found in a structural template. As demonstrated in the blind docking challenge, Critical Assessment of Prediction of Interactions (CAPRI), template-based methods generate more accurate results but only if a good quality template exists [1–5]. In some cases lacking useful templates, free global docking can yield acceptable results. According to recent estimates, the best free docking methods find adequate models among the top 10 predictions for around 40% of the targets [1]. The CAPRI analysis also indicates that protein backbone flexibility is a big challenge; protein complexes that undergo substantial conformational changes upon docking get no successful predictions from any method [3–5].

Presently, most of the free docking methods treat the backbone of input protein structures as rigid. This approximation reduces the protein–protein docking problem to a 6D (three rotational and three translational degrees of freedom) search space. Rigid-body search for the binding site most often rely on the Fast Fourier Transform [6–8]. Other successful approaches include 3D Zernike descriptor-based docking [9,10] or geometric hashing [11]. These rigid-body methods are often used as a first docking step, followed by scoring [12–16], using experimental data [17] and/or structural refinement to capture backbone flexibility [5,18]. Molecular Dynamics is perhaps the most common refinement strategy, either in classic or enhanced sampling versions [17,19–22]. Other tools use rotamer libraries to address side-chain flexibility [23] and Elastic Network Models (ENMs) for modeling backbone rearrangements [24–28]. Accounting for backbone flexibility in the search for the binding site significantly increases the docking complexity and makes it practically intractable using conventional all-atom modeling approaches. This enormous computational complexity of flexible docking can be reduced using coarse-grained protein models [29–32]. The best-performing methods that can now include backbone flexibility during the docking calculations use coarse-grained models and/or ENM-driven simulations. These include RosettaDock combining coarse-grained generation of backbone ensembles and all-atom refinement [33–35]; ATTRACT combining coarse-grained docking with ENM and all-atom refinement [36,37]; and SwarmDock using all-atom ENM [25,38]. All these approaches show some advantages in modeling protein flexibility compared to rigid-body docking followed by structure refinements. However, effective modeling flexibility in protein–protein docking remains an unsolved problem, as demonstrated in the recent CAPRI round [25,35,37,39].

In this work, we use a well-established CABS coarse-grained protein model [29] for protein–protein docking. During the CABS docking simulation, one of the docking partners undergoes a long random process of rotations, translations, and extensive backbone conformational rearrangements that significantly modify its fold. Simultaneously, the backbone of the second protein undergoes small fluctuations.

2. Results

The most accurate models (out of the sets of 10,000 generated models and 10 top-scored) are characterized in Table 1. The table presents different metrics of similarity to the experimental structures for the set of 62 protein–peptide complexes (divided into three categories: low, medium and high flexibility cases). To assess the sampling performance, below we will use the iRMSD values for the best models out of all models. According to the iRMSD values the CABS-based docking algorithm produced a significant number of near-native protein–protein arrangements of acceptable quality (iRMSD < 4 Å, according to CAPRI criteria) for most protein–protein cases in the categories of low and medium flexibility cases. However, in the high-flexibility category, the best iRMSD values were noticeably higher (in the range of 4–12 Angstroms). This resulted from the adopted distance restraint scheme (see the Methods section), which was uniform for whole proteins and introduced a penalty for deviations of more than 1 Å from the input structures (unbound experimental structures). This penalty was very small for the protein ligands. Thus, the distance-restraints scheme allowed for the large-scale conformational changes, however, they might have prevented binding-induced conformational changes in the high-flexibility category. Therefore, there is the need to modify such a scheme for the most challenging targets.

Table 1. Summary of the docking simulations. The table characterizes X-ray data used in the docking, average ligand flexibility, and docking results. The table reports the best accuracy models out from all (10,000) and 10 top-scored models. The metrics definitions are provided in the Methods section. The table divides the presented cases on the three categories: low-flexibility, medium-flexibility and highly flexible cases.

X-ray Data (Number of Residues)				Ligand Flexibility	Results—Best from All Models			Results—Best from 10 Top-Scored Models		
Receptor *	Ligand *	Complex	RMSD **	Average LoRMSD	iRMSD	LRMSD	fNAT	iRMSD	LRMSD	fNAT
				Low-flexibility cases						
5CHA (238)	2OVO (53)	1CHO	0.62	4.84	2.65	6.95	0.48	2.96	10.93	0.18
2PKA (232)	6PTI (56)	2KAI	0.91	4.64	3.32	11.34	0.19	4.75	15.76	0.12
1CHG (245)	1HPT (56)	1CGI	1.53	5.24	2.76	4.13	0.37	6.18	14.15	0.09
2PTN (223)	6PTI (58)	2PTC	0.31	5.23	2.97	11.86	0.29	4.39	15.93	0.15

Table 1. *Cont.*

X-ray Data (Number of Residues)				Ligand Flexibility	Results—Best from All Models			Results—Best from 10 Top-Scored Models		
Receptor *	Ligand *	Complex	RMSD **	Average LoRMSD	iRMSD	LRMSD	fNAT	iRMSD	LRMSD	fNAT
1SUP (275)	2CI2 (64)	2SNI	0.37	3.89	1.09	3.86	0.69	2.81	9.09	0.46
2ACE (532)	1FSC (61)	1FSS	0.76	4.48	3.41	7.20	0.25	15.03	32.56	0.03
1MAA (533)	1FSC (61)	1MAH	0.60	4.58	2.49	3.89	0.45	11.25	24.43	0.06
1A2P (108)	1A19 (89)	1BRS	0.47	3.33	1.94	4.19	0.64	4.01	8.74	0.14
1CCP (294)	1YCC (103)	2PCC	0.39	4.18	3.13	10.19	0.25	11.89	26.68	0.08
1SUP (275)	3SSI (107)	2SIC	0.39	4.01	4.03	18.96	0.23	4.77	19.40	0.12
1VFA (223)	1LZA (129)	1VFB	0.59	3.72	4.61	15.07	0.11	17.45	37.15	0.00
1MLB (432)	1LZA (129)	1MLC	0.85	3.74	2.82	10.47	0.36	8.04	33.09	0.04
Medium-flexibility cases										
1CHG (226)	1HPT (56)	1CGI	2.02	5.80	2.46	3.21	0.44	5.86	10.72	0.12
5C2B (241)	4ZAI (80)	5CBA	1.49	4.51	2.48	7.64	0.42	9.34	16.01	0.10
5P2 (166)	1LXD (87)	1LFD	1.79	4.12	2.87	6.76	0.27	12.47	24.24	0.00
1R6C (142)	2W9R (97)	1R6Q	1.67	9.27	7.95	11.97	0.14	13.71	35.71	0.00
1JXQ (242)	2OPY (106)	1NW9	1.97	4.09	7.05	8.69	0.23	9.33	17.55	0.00
1IAS (330)	1D6O (107)	1B6C	1.96	4.65	4.72	10.74	0.14	12.24	23.99	0.00
5E56 (116)	5E03 (113)	5E5M	1.56	4.16	3.83	9.09	0.23	10.96	20.00	0.00
2HRA (180)	2HQT (115)	2HRK	2.03	7.27	3.55	10.17	0.26	10.81	32.54	0.00
4BLM (256)	4M3J (116)	4M3K	1.77	4.41	4.96	7.41	0.10	13.75	27.11	0.03
1E78 (578)	5VNV (120)	5VNW	1.49	3.81	5.93	22.23	0.10	23.89	70.83	0.00
3BX8 (167)	3OSK (121)	3BX7	1.63	4.63	4.94	17.46	0.28	6.22	20.32	0.12
6ETL (124)	4POY (121)	4POU	1.83	4.01	2.91	10.16	0.50	6.55	19.65	0.25
4FUD (246)	5HDO (126)	5HGG	0.84	4.22	3.59	12.52	0.19	13.00	29.3	0.00
3TGR (346)	3R0M (127)	3RJQ	0.79	4.00	5.32	16.98	0.13	12.77	33.94	0.00
6EY5 (585)	5FWO (129)	6EY6	1.90	3.86	3.83	6.03	0.14	12.89	27.61	0.00
1SZ7 (159)	2BJN (141)	2CFH	1.55	5.13	1.98	4.01	0.71	2.82	5.50	0.63
3V6F (437)	3KXS (142)	3V6Z	1.83	7.11	6.12	16.68	0.15	6.66	20.06	0.06

Table 1. *Cont.*

X-ray Data (Number of Residues)				Ligand Flexibility	Results—Best from All Models			Results—Best from 10 Top-Scored Models		
Receptor *	Ligand *	Complex	RMSD **	Average LoRMSD	iRMSD	LRMSD	fNAT	iRMSD	LRMSD	fNAT
3CPI (437)	1G16 (156)	3CPH	2.12	4.34	4.87	15.64	0.09	15.02	27.88	0.00
1QJB (460)	1KUY (166)	1IB1	2.09	4.22	6.56	14.83	0.13	16.10	46.26	0.00
1IAM (185)	1MQ9 (173)	1MQ8	1.76	4.22	4.93	14.99	0.21	26.17	70.50	0.00
3HI5 (430)	1MJN (179)	3HI6	1.65	3.77	5.79	23.30	0.21	19.38	49.77	0.00
2G75 (429)	2GHV (183)	2DD8	2.19	5.37	5.73	13.78	0.09	17.20	34.33	0.00
1A12 (401)	1QG4 (202)	1I2M	2.12	4.19	2.84	6.43	0.51	3.58	6.97	0.47
1N0V (825)	1XK9 (204)	1ZM4	2.11	3.54	8.82	28.17	0.04	11.05	48.14	0.00
4EBQ (429)	4E9O (230)	4ETQ	0.47	3.72	7.12	14.74	0.20	8.68	19.61	0.07
1S3X (380)	1XQR (259)	1XQS	1.77	5.44	5.63	26.14	0.11	15.88	30.51	0.00
3HEC (329)	3FYK (282)	2OZA	1.89	4.29	4.35	9.32	0.33	11.24	18.8	0.03
6A0X (437)	2FK0 (322)	6A0Z	1.28	5.75	5.75	25.59	0.16	11.43	31.39	0.00
Highly flexible cases										
1CL0 (316)	2TIR (108)	1F6M	4.9	3.83	7.02	11.34	0.10	11.92	18.06	0.00
1 × 9Y (346)	1NYC (110)	1PXV	2.63	4.86	5.74	14.10	0.07	7.46	16.31	0.02
1JZO (431)	1JPE (116)	1JZD	2.71	4.65	4.98	8.13	0.28	13.38	34.05	0.00
5D7S (423)	2GMF (121)	5C7X	2.26	4.17	4.12	13.61	0.34	4.69	16.74	0.20
1FCH (302)	1C44 (123)	2C0L	2.62	5.51	5.02	5.54	0.21	10.24	24.74	0.00
1YWH (268)	2I9A (123)	2I9B	3.79	7.14	5.79	17.59	0.14	6.92	33.23	0.05
3L88 (550)	1CKL (126)	3L89	2.51	9.86	4.83	10.90	0.17	17.84	31.87	0.00
1ZM8 (239)	1J57 (143)	2O3B	3.13	6.20	4.76	16.43	0.18	15.34	31.95	0.00
1G0Y (310)	1ILR (145)	1IRA	8.38	4.07	12.97	22.24	0.08	15.86	25.46	0.05
1QUP (219)	2JCW (153)	1JK9	2.51	9.40	8.07	13.85	0.10	17.41	30.74	0.00
1SYQ (259)	3MYI (163)	1RKE	4.25	4.15	5.26	6.43	0.38	16.11	34.67	0.00
2II0 (463)	1CTQ (166)	1BKD	2.86	4.51	4.80	7.33	0.14	19.96	39.32	0.00
1ERN (416)	1BUY (166)	1EER	2.44	5.22	12.97	13.18	0.02	17.12	30.73	0.00
3AVE (419)	1FNL (173)	1E4K	2.60	5.32	3.44	10.07	0.43	7.59	24.33	0.13

Table 1. *Cont.*

X-ray Data (Number of Residues)				Ligand Flexibility	Results—Best from All Models			Results—Best from 10 Top-Scored Models		
Receptor *	**Ligand ***	**Complex**	**RMSD ****	**Average LoRMSD**	**iRMSD**	**LRMSD**	**fNAT**	**iRMSD**	**LRMSD**	**fNAT**
1R8M (195)	1HUR (180)	1R8S	3.73	5.50	6.67	13.41	0.09	15.15	25.10	0.00
1QFK (348)	1TFH (182)	1FAK	6.18	5.64	8.97	15.57	0.16	15.59	34.46	0.00
1F59 (440)	1QG4 (202)	1IBR	2.54	5.01	6.65	14.36	0.14	16.41	33.07	0.00
4DVB (427)	4DVA (246)	4DW2	2.27	3.85	6.61	21.91	0.14	9.94	29.27	0.00
1NG1 (294)	2IYL (271)	2J7P	2.67	4.51	8.87	18.77	0.11	18.46	48.05	0.00
1UX5 (411)	2FXU (360)	1Y64	4.69	4.15	6.42	13.50	0.27	15.50	36.42	0.00
1D0N (729)	1IJJ (371)	1H1V	6.62	3.44	7.92	31.14	0.36	29.12	65.07	0.03

* 4-letter PDB code for the crystal structures used in this study. ** The RMSD (in Å) of the interface Cα atoms for input receptor and ligand after superposition onto the co-crystallized complex system.

The results analysis below focuses on the sampling performance for the selected low-flexibility barnase/barstar case. Figure 1 characterizes iRMSD versus CABS model energy values for the barnase/barstar (1BRS) and another low-flexibility case with clearly the lowest iRMSD value 1.09 Angstroms (2SNI).

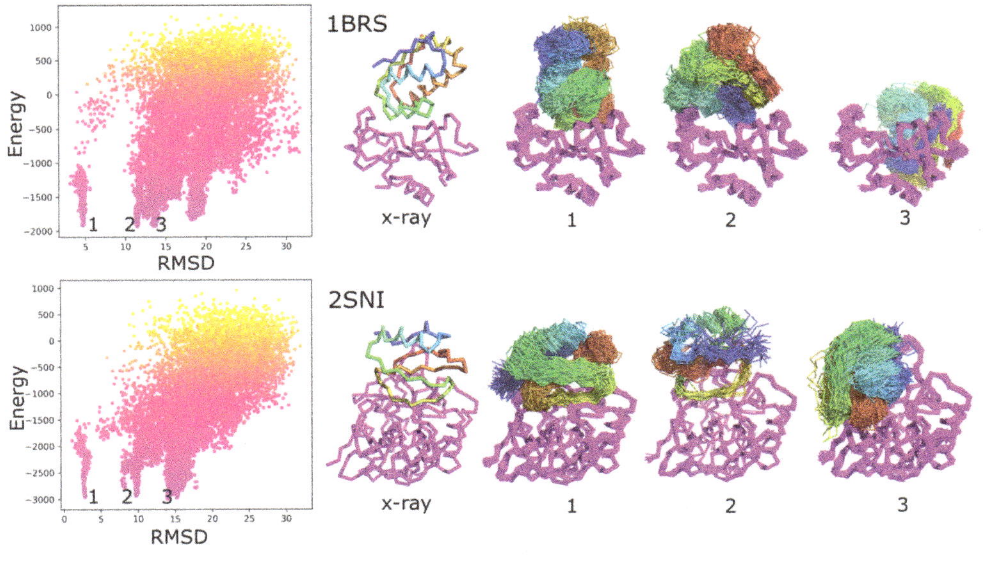

Figure 1. Characterization of docking results using RMSD to the X-ray structure and system energy. The left panels show the interface-RMSD versus CABS energy values. Point color represents the temperature—from yellow (**high**) to pink (**low**). The molecular visualizations show X-ray structures and ensembles of predicted models corresponding to selected energy minima (numbered in the picture from 1 to 3). As presented in the picture, the minima numbered as 1st corresponds to near-native protein–protein arrangements, others to non-native ensembles, as presented in the picture. The presented ensembles are the sets of similar models found in the structural clustering of contact maps (see Methods). The figure shows two modeling cases: 1BRS and 2SNI.

Figure 2 shows example ensembles of barnase/barstar models and the most accurate model (iRMSD 1.9 Å). A single system replica could explore an ample conformational space that involved significantly different binding configurations and protein-ligand conformations, as demonstrated in Figure 2c and Movie S1. Figure 3a further characterizes this single replica's using iRMSD and LoRMSD (RMSD for ligand only) values. As presented in the figure, the ligand structure fluctuated around 5 Å (the same fluctuations in the context of all replicas are shown in Figure 3b). The ligand became significantly more closer to the X-ray structure after binding to the native binding site as reflected by iRMSD values. Namely, after correct binding, LoRMSD values got noticeably lower to around 2 Å (see Figure 3a). In the following sections, we do not discuss this aspect of our method; however, it is worth mentioning that the proposed method enabled a detailed analysis of plausible docking trajectories. The described docking procedure uses REMC protocol enhanced by simulated annealing of all 20 replicas. Figure 3c shows their evolution through different temperatures. Figure 4 provide more detailed pictures of structural flexibility for protein "receptor" and "ligand". Protein–protein contacts defining the complex assembly are characterized in Figure 5. In the presented example, the most persistent protein–protein contacts occurred in about 15% of snapshots. Therefore, they were significantly less stable compared to intramolecular protein contacts (Figure 4).

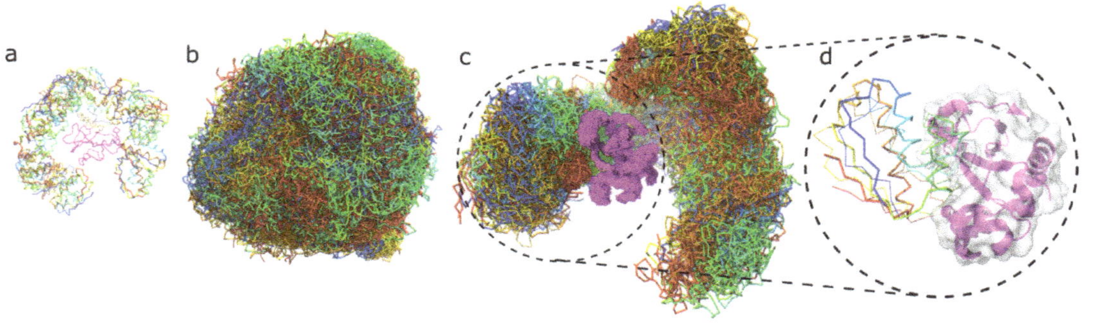

Figure 2. Protein–protein docking stages illustrated by barstar/barnase docking case. The figure shows the barnase receptor in magenta and the barstar ligand in rainbow colors. The respective panels show: (**a**) 20 starting structures for each replica of the system; (**b**) 10,000 models combined from 20 replicas (500 models per replica) in which the highly flexible ligand is covering the entire surface of the flexible receptor; (**c**) 500 models from one replica only, (**d**) the best model obtained for barnase/barstar system (the X-ray structure of the ligand is shown in thick ribbon, the modeled in thin ribbon).

An essential and unique feature of the presented docking simulations is the level of backbone flexibility during docking. In the example above, the ligand backbone fluctuations (LoRMSD) were in the range of 2–7 Å (Figure 3b), with the average LoRMSD value of 3.3 Å from the entire docking simulations. In other cases, the ligand fluctuations were at a similar level or higher (see LoRMSD values in Table 1).

Finally, using structural clustering of contact maps (see Methods), we attempted to select the set of 10 top-scored models for each case. Table 1 reports the most accurate models out of the 10 top-scored.

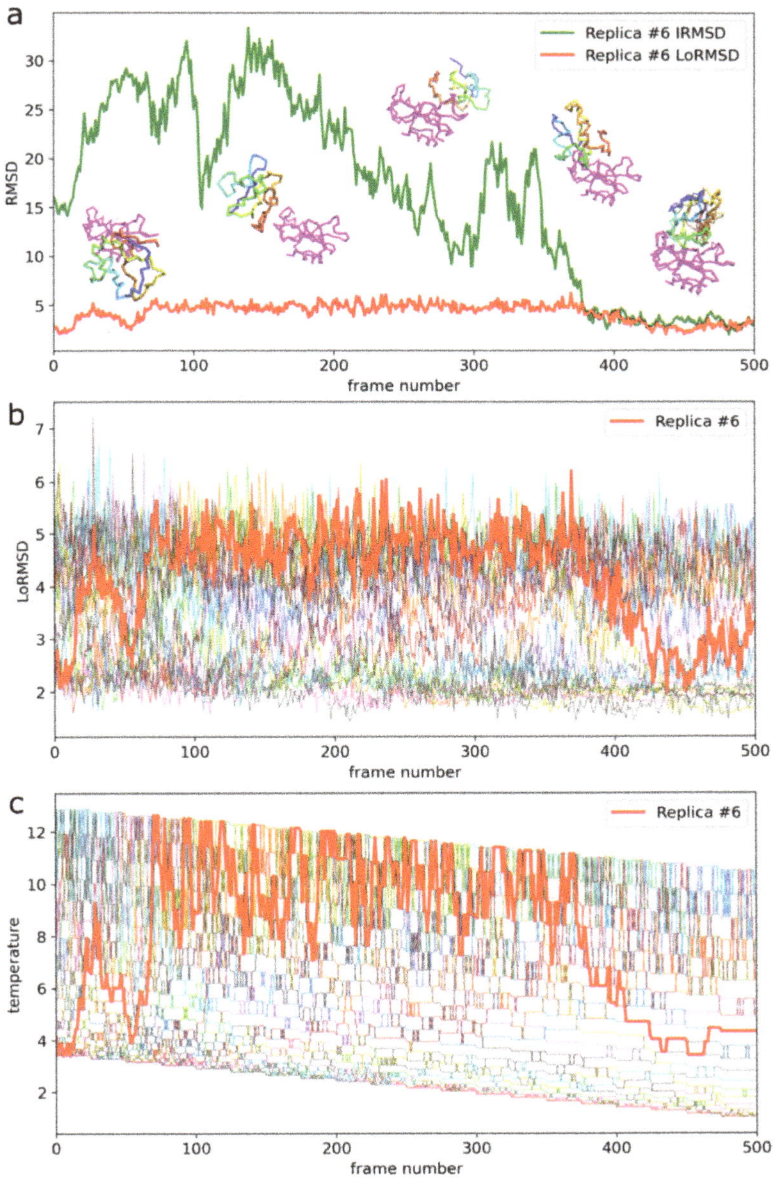

Figure 3. Docking trajectory for the selected replica of barnase/barstar system. The presented replica reached the most accurate barnase/barstar complex structure. (**a**) iRMSD (interface RMSD) and LoRMSD (ligand only RMSD) values. Example simulation snapshots illustrate the plot. The ligand is presented in rainbow colors, the receptor in magenta. The lowest iRMSD model (1.9 A from X-ray structure) is presented on the right lower corner superimposed on the X-ray structure (the X-ray structure is shown in thick lines, the predicted model in thin lines). (**b**) Ligand only RMSD (LoRMSD) values for all replicas. The thick red line presents selected replica. (**c**) Exchange of system replicas between different temperatures driven by Replica Exchange Monte Carlo (REMC) system. The thick red line presents selected replica. The replica trajectory is also presented in the Video S1.

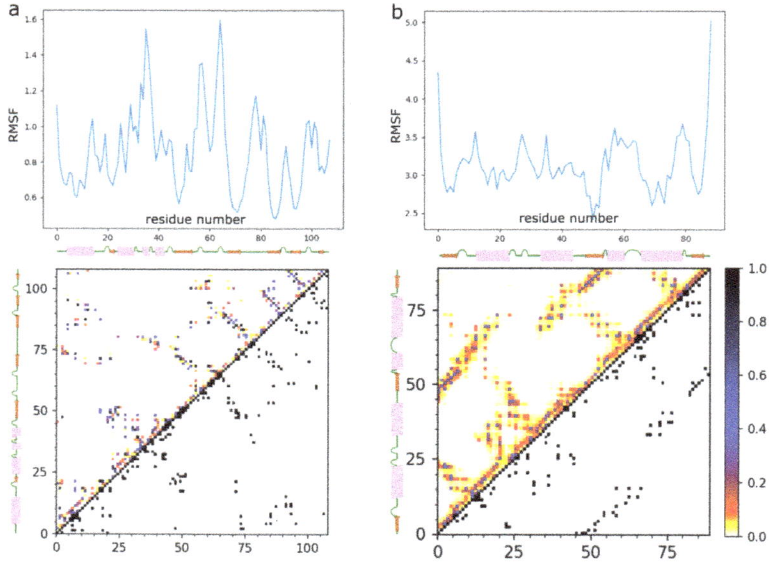

Figure 4. Characterization of barnase/barstar flexibility in the docking simulation. The figure shows RMSF plots (upper panels) and contact maps (lower panels) for (**a**) the barnase receptor and (**b**) the barstar ligand. The RMSF profile (see Methods) and contact maps showing the frequency of contacts are derived from the entire simulation (derived from 10,000 models).

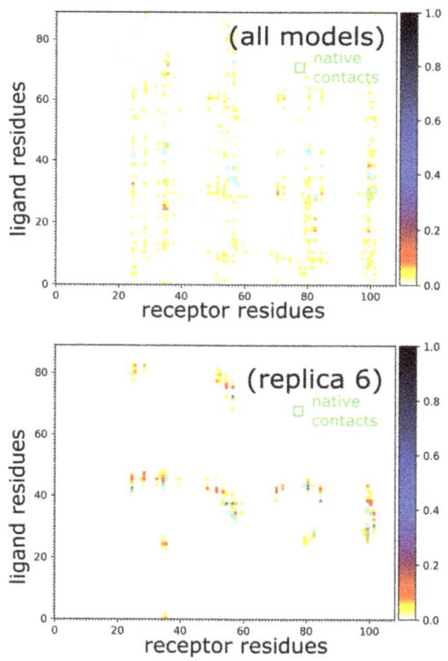

Figure 5. Characterization of barnase/barstar contacts. Panels show barnase/barstar models and contact maps for entire simulation (all models, 10,000 models) and single selected replica (replica 6, 500 models) that reached a near-native arrangement. In the maps, green circles mark the native contacts.

3. Discussion

This work demonstrates a significant improvement in the sampling of large-scale conformational transitions during global protein–protein docking compared to other state-of-the-art approaches. We show that modeling the large conformational changes is possible at a relatively low computational cost. The presented simulations took between 10 and 80 h (depending on the system size) using a single standard CPU. The proposed modeling protocol can be used as the docking engine in template-based and integrative docking protocols using experimental structural data and additional information from various sources [2,40]. We focused on the free docking of protein ligands with a highly flexible backbone in the present test simulations. Using unbound structures as the input, we produced acceptable accuracy models (iRMSD around 4 Å or lower) in low-flexibility and medium-flexibility cases. However, the selection procedure of the most accurate models needs further improvement. Namely, selecting the best-ranked models led to acceptable models in about half of the tested cases.

Presently, the most common approach to account for conformational changes in protein docking is using ENM [24–28,36–38]. The applicability of ENM to modeling protein flexibility is limited to specific systems and depends on how collective the protein motions are. Our method presents a conceptually different approach that seems to be more realistic (see review discussing coarse-grained CABS dynamics in the context of ENM approaches [24]). We demonstrated that it is possible to simulate effectively free docking of highly flexible protein ligands to quite elastic protein receptor structures. Such a significant degree of flexibility was achieved using a highly efficient simulation engine based on the coarse-grained representation of protein structures, Monte Carlo dynamics, and knowledge-based force field. CABS coarse-graining, enhanced by the discretized protein model and interaction patterns, significantly reduces the search space. Monte Carlo dynamics, enhanced by Replica Exchange annealing, leads to huge speed-up of the search procedures. Additionally, a significant (although acceptable for many problems) flattening of energy surfaces by statistical potentials of CABS model simplifies simulations. As a result the flexible docking using CABS-dock is orders of magnitude faster than equivalent simulations based on classical modeling methods. Obviously, the new method also has several limitations that have to be considered when designing new computational experiments. First, since the "ligand" protein is treated as a very elastic object (what is necessary to guarantee efficient search of the binding sites and poses) the cost of computations rapidly grows with the protein size. Thus, completely free global docking of protein ligands larger than 150 residues (see Table 1) may be impractical. Second, the coarse-graining of the sampling space and simplifying interaction patterns (so important for the huge acceleration of the simulations) makes the docking energetics less sensitive. For these reasons, the clustering procedures, refinement of the resulting structures, and final model selection become challenging and need further development. Additionally, speeding-up the entire protocol can be useful. We estimate that the simulations could be easily speeded-up at least 10 times or more through algorithm parallelization. The speed-up would enable making the protocol available as the publicly accessible and automated web service.

4. Methods

4.1. Docking Simulation Protocol

In this work, we present the protein–protein docking simulation protocol that relies on the CABS coarse-grained model. The CABS design and applications have been recently described in the reviews on protein coarse-grained [29] and protein flexibility [24,41] modeling. Here we outline only its main features. The CABS model uses a coarse-grained representation of protein chains (see Figure 6), Replica Exchange Monte Carlo (REMC) dynamics, and knowledge-based statistical potentials. Representation of protein chains is based on C-alpha traces, restricted to an underlying high-resolution lattice. The lattice spacing allows slight fluctuations of the C-alpha–C-alpha distances and many pseudo-bonds orientations. Virtual pseudo-atoms are placed in the centers of these C-alpha–C-alpha bonds and are used to locate the main-chain hydrogen bonds. Additionally, the positions of the two pseudo-atoms representing side chains are defined by the geometry of C-alpha traces and amino-acid identities. Such fixed positions of side chains (taken from the statistics of protein databases) reduce the model's resolution. However, this limitation is less serious than it may appear since even small movements of the main chain (allowed due to the soft nature of the assumed geometrical restrictions) leads to large moves of the side chains. This way, the packing of side chains can be quite accurate. The interaction scheme of CABS consists of statistical potentials mimicking effects of main chain rotational preferences, main-chain hydrogen bonds, and side-chain contacts. All statistical potentials, derived from structural regularities observed in PDB structures, have relatively broad minima compensating the low-resolution effects and allowing a fast search for global energy minima. The solvent is treated implicitly, and its averaged effects are encoded within the above-mentioned contact potentials. Energy computation for protein chain models is very fast since many interactions could be pre-computed (and coded in large tables) due to the discretized patterns of main chains geometry. The Monte Carlo sampling of CABS uses a set of local movers. The resulting model dynamics is quite realistic for large-scale distances, allowing coarse-grained modeling of protein structures, dynamics, and protein–protein interactions.

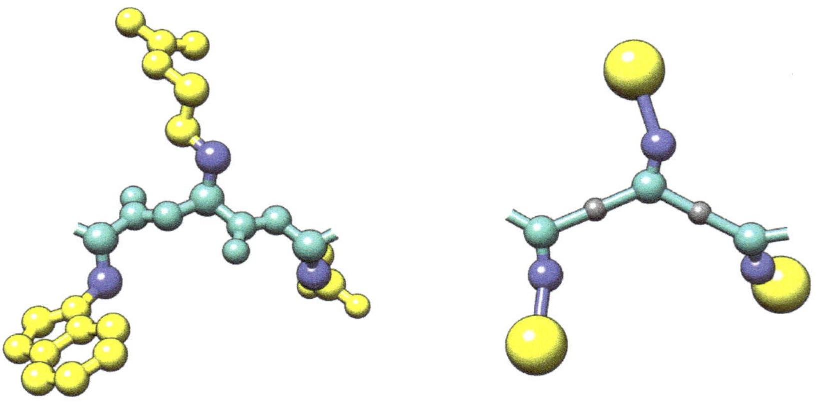

Figure 6. Comparison of the all-atom (**left**) and the CABS coarse-grained model representation (**right**) for an example tripeptide. In the CABS model, protein residues are represented using C-alpha, C-beta, united side-chain atom, and the peptide bond center [29].

The modeling protocol consists of the following steps:

1. **Preparing input structures of a protein-ligand and a protein-receptor**. The protocol requires the input of two protein structures (single- or multi-chain) in the PDB format. One of them has to be indicated as a ligand and the second as a receptor. The ligand undergoes large conformational fluctuations, translations, and rotations around the receptor within the proposed protocol. The "ligand" should be a smaller protein because the computational cost of searching its conformational space rapidly grows with the chain length. That is because the motion of the entire structure (including fold relaxation, rotation, and translation of the entire molecule) is simulated by a random sequence of local moves. The accuracy of such sampling is acceptable for not too-large proteins. On the other hand, treating the "ligand" as a fully flexible object allows approximate studies of entire docking trajectories. In some cases, it would be perhaps worth treating a larger protein (but not too large) as a flexible "ligand", although this was out of range of the present studies.
2. **Generating starting structures**. Starting conformations are built using C-alpha coordinates only (in the CABS model C-alpha traces define the position of other united

pseudo-atoms, see details [29]). The algorithm places the protein-ligand center at 20 random positions around the protein receptor at the approximate distance of 20 Å from the protein receptor's surface. Next, these protein-ligand systems are used as starting conformations for the 20 replicas in the REMC CABS sampling scheme (each replica starts from a different ligand-receptor arrangement).

3. **Docking simulations using CABS coarse-grained model and REMC dynamics**. During simulations, the protein receptor structure is kept close to the starting structure using distance restraints. Distance restraints are generated using the input coordinates of the C-alpha atoms. Two residues are automatically restrained if two conditions are met. First, their separation along the sequence has to be at least five residues. Second, the distance between their C-alpha atoms must be within the range of 5–15 Å. During simulations, the receptor restraints imply small-scale fluctuations of the protein receptor backbone in the range of 1 Å and, accordingly, more significant fluctuations of the side-chain atoms. A similar restraints scheme is applied to the protein-ligand but with tenfold weaker weights. During simulations, the ligand moves freely within the vicinity of the receptor and internal restraint allows for large-scale fluctuations of its structure. Usually, the ligand fluctuations are within the range of 2 and 12 Å to the input structure although folding-unfolding events are possible at highest temperatures. The docking simulation is conducted using CABS REMC pseudo-dynamics with simulated annealing. In this work, 20 replicas and 20 annealing steps have been used. All the REMC scheme parameters have been adjusted to allow for large-scale conformational transitions, rotations, and translations of the protein-ligand in a reasonable computational time. The modeling protocol collects trajectories from all 20 replicas. The protocol saves only a small fraction (2%) of the generated models for further analysis i.e., 500 models from each replica, thus 10,000 models in total.
4. **Reconstructing to CABS coarse-grained representation**. The set of 10,000 models in C-alpha traces are reconstructed to complete CABS model representation using CABS algorithm [29]. In CABS, positions of C-beta and Side-Chain united atoms are defined by the positions of the three consecutive C-alpha atoms and the amino acid identity (the most probable positions from the PDB database are used).
5. **Clustering of contact maps**. First, for all of the 10,000 models the contact maps between the receptor and the ligand proteins are calculated. Two residues are considered to form a contact if their Side Chain pseudoatoms are at most 6 Å apart (for Alanines the C-beta atoms are considered as the Side Chain; for Glycines—it's the C-alpha atoms). Next, the algorithm sorts the models according to the number of the receptor-ligand contacts, and the set of top 1000 is kept for further processing. This way the transient and weakly bound complexes are removed from the solutions pool. In the next step, the 1000 contact maps are clustered together to identify the most frequently occurring contact patterns. The complete link hierarchical clustering was used with the Jaccard index as the distance metric between contact maps. Finally, the identified clusters are ranked according to their density, defined as the number of the cluster members divided by the average metric between them.
6. **Reconstructing to all-atom representation**. Representative models from the ten most dense clusters are reconstructed to all-atom representation using Modeller-based rebuilding procedure [42] (or can be reconstructed using other rebuilding strategies, see review [43]).

In recent years, the CABS model has been used for modeling the flexibility of globular proteins [44–47] and various processes leading to large-scale conformational transitions. These included: ab initio simulations of protein folding mechanisms [48,49], folding and binding mechanisms [49,50], and free protein–peptide docking within the CABS-dock tool [51–57]. The CABS-dock is a well-established peptide docking tool that has been made available as a web server [51,52] and, most recently, as a standalone application [54]. Its distinctive feature among other tools is the possibility of fast simulation of the large backbone rearrangements of both peptide and protein receptors during binding (see the review on protein–peptide docking tools [58]). In addition, the CABS-dock has been used in multiple applications (recently reviewed [56]), including docking to receptors with disordered fragments [41,59], GPCRs [60], and modeling proteolysis mechanisms [61].

The presented protocol for protein–protein docking utilizes the CABS-dock standalone package [54] developed primarily for protein–peptide docking. In order to tackle the protein–protein docking problem, key changes have been made to the docking algorithm that aimed mainly at the improvement of the conformational sampling. First of all, the temperature distribution between replicas in the REMC scheme was adjusted. Instead of constant temperature increment between consecutive replicas, as in the original CABS-dock, here we've implemented progressive geometric raise of the temperature increment. Furthermore, the number of simulation replicas was increased to twenty versus ten in the original CABS-dock. Besides the sampling improvement, a new clustering protocol was introduced. The original CABS-dock used RMSD-based clustering, which worked well for peptides. For the protein–protein complexes, however, purely geometrical similarity condition such as the RMSD is too severe. Namely, for two binding poses, where the mobile protein was docked in the exact same pocket but is slightly tilted in one of them, the RMSD difference would be considerable. Despite representing similar binding poses, the two structures would end up in different clusters. To overcome this, the current protocol uses clustering based on the similarity between receptor-ligand contact maps.

4.2. Results Analysis and Quality Metrics

The docking simulation analysis was performed using Python and NumPy (Python library). Structural differences between experimentally determined structures and generated models were evaluated using Root Mean Square Deviations (RMSDs). Interface RMSD (iRMSD) is an RMSD calculated for interface residues of the receptor and the ligand separated by no more than 6 Angstroms. Ligand RMSD (LRMSD) is an RMSD computed for the ligands after the superimposition of the receptors. Ligand only RMSD (LoRMSD) is an RMSD computed for the ligand structure only. Root Mean Square Fluctuation (RMSF) is a measure of the amino acid's flexibility. It is calculated for every residue as the square root of this residue's variance around the reference residue position. The fraction of native contacts (fNAT) was calculated as a number of experimental structure contacts found in the generated structure divided by the total number of contacts found in the experimental structure. Rather restrictive contact criterion, distance up to 6 Å between side-chain centers, was used. All figures presented in this work were generated using PyMOL, UCSF Chimera, and Matplotlib (Python library).

4.3. Dataset

In this docking study, we used protein–protein cases from the ZDOCK benchmark set [62] (cases in which a smaller size protein—a protein-ligand—contained more than one protein chain, or chain gaps, were discarded from our set). The set comprises the three flexibility-based subsets: low-flexible (almost rigid), medium-flexible, and highly flexible with available unbound X-ray structures of both the protein-receptor and the protein-ligand. The unbound structures were used as the docking input. As the reference for calculating various similarity measures, we used the X-ray structures of the protein-ligand complexes. Table 1 lists all the PDB IDs of X-ray structures used in the study.

5. Conclusions

In summary, the described docking procedure accounts for large-scale protein structure fluctuations during unrestrained protein–protein docking search for the binding site. The exploration of such vast conformational space has not been demonstrated before to the best of our knowledge. The approach shows unprecedented sampling possibilities; however, the accuracy of the obtained complexes is still lower than observed for state-of-the-art docking tools. Definitely, the balancing of the structural restraints scheme needs further developments and tests. Therefore, this work is the first step towards a mature protein–protein docking tool. The next development steps would involve modifications of the distance restraints scheme, which allow for different degrees of flexibility for appropriate protein fragments (now the presented algorithm treats the entire protein-ligand as very flexible) and force-field improvements. The proposed approach is also very promising in the refinement applications when searching for the binding site is not needed, and only the protein–protein interface needs to be optimized.

Supplementary Materials: The following are available online at https://www.mdpi.com/article/10.3390/ijms22147341/s1, Video S1. The trajectory of a single replica from the protein-protein docking simulation of barnase/barstar system. The movie shows the barnase receptor in surface representation and the barstar ligand in ribbon. The presented replica reached the model with interface RMSD value 1.9 Angstrom from the complex X-ray structure, shown as transparent ribbon.

Author Contributions: Conceptualization, M.K., S.K. and A.K.; Methodology, M.K. and A.K.; Software, M.K. and M.Z.; Supervision, S.K.; Visualization, M.K. and S.K.; Writing—original draft, S.K.; Writing—review & editing, M.K., S.K., M.Z. and A.K. All authors have read and agreed to the published version of the manuscript.

Funding: SK: MZ and AK acknowledge funding by the National Science Centre, Poland [MAESTRO2014/14/A/ST6/00088]. MK acknowledges funding by the National Science Centre, Poland [501/D112/66 GR-6271]. SK also acknowledges funding in part by the National Science Centre, Poland [UMO-2020/39/B/NZ2/01301].

Institutional Review Board Statement: Not applicable.

Informed Consent Statement: Not applicable.

Data Availability Statement: Not applicable.

Conflicts of Interest: The authors declare no conflict of interest.

References

1. Porter, K.A.; Desta, I.; Kozakov, D.; Vajda, S. What method to use for protein–protein docking? *Curr. Opin. Struct. Biol.* **2019**, *55*, 1–7. [CrossRef]
2. Rosell, M.; Fernández-Recio, J. Docking approaches for modeling multi-molecular assemblies. *Curr. Opin. Struct. Biol.* **2020**, *64*, 59–65. [CrossRef]
3. Lensink, M.F.; Velankar, S.; Kryshtafovych, A.; Huang, S.; Schneidman-Duhovny, D.; Sali, A.; Segura, J.; Fernandez-Fuentes, N.; Viswanath, S.; Elber, R.; et al. Prediction of homoprotein and heteroprotein complexes by protein docking and template-based modeling: A CASP-CAPRI experiment. *Proteins Struct. Funct. Bioinf.* **2016**, *84*, 323–348. [CrossRef] [PubMed]
4. Lensink, M.F.; Velankar, S.; Wodak, S.J. Modeling protein–protein and protein–peptide complexes: CAPRI 6th edition. *Proteins Struct. Funct. Bioinf.* **2017**, *85*, 359–377. [CrossRef] [PubMed]
5. Lensink, M.F.; Nadzirin, N.; Velankar, S.; Wodak, S.J. Modeling protein-protein, protein-peptide, and protein-oligosaccharide complexes: CAPRI 7th edition. *Proteins Struct. Funct. Bioinf.* **2020**, *88*, 916–938. [CrossRef] [PubMed]
6. Pierce, B.G.; Wiehe, K.; Hwang, H.; Kim, B.-H.; Vreven, T.; Weng, Z. ZDOCK server: Interactive docking prediction of protein-protein complexes and symmetric multimers. *Bioinformatics* **2014**, *30*, 1771–1773. [CrossRef] [PubMed]
7. Kozakov, D.; Hall, D.R.; Xia, B.; Porter, K.A.; Padhorny, D.; Yueh, C.; Beglov, D.; Vajda, S. The ClusPro web server for protein–protein docking. *Nat. Protoc.* **2017**, *12*, 255–278. [CrossRef] [PubMed]
8. Yan, Y.; Tao, H.; He, J.; Huang, S.-Y. The HDOCK server for integrated protein–protein docking. *Nat. Protoc.* **2020**, *15*, 1829–1852. [CrossRef] [PubMed]
9. Venkatraman, V.; Yang, Y.D.; Sael, L.; Kihara, D. Protein-protein docking using region-based 3D Zernike descriptors. *BMC Bioinf.* **2009**, *10*, 407. [CrossRef] [PubMed]
10. Christoffer, C.; Terashi, G.; Shin, W.; Aderinwale, T.; Maddhuri Venkata Subramaniya, S.R.; Peterson, L.; Verburgt, J.; Kihara, D. Performance and enhancement of the LZerD protein assembly pipeline in CAPRI 38-46. *Proteins Struct. Funct. Bioinf.* **2020**, *88*, 948–961. [CrossRef]
11. Estrin, M.; Wolfson, H.J. SnapDock—template-based docking by Geometric Hashing. *Bioinformatics* **2017**, *33*, i30–i36. [CrossRef]
12. Gromiha, M.M.; Yugandhar, K.; Jemimah, S. Protein–protein interactions: Scoring schemes and binding affinity. *Curr. Opin. Struct. Biol.* **2017**, *44*, 31–38. [CrossRef]
13. Feng, T.; Chen, F.; Kang, Y.; Sun, H.; Liu, H.; Li, D.; Zhu, F.; Hou, T. HawkRank: A new scoring function for protein–protein docking based on weighted energy terms. *J. Cheminform.* **2017**, *9*, 66. [CrossRef] [PubMed]
14. Geng, C.; Jung, Y.; Renaud, N.; Honavar, V.; Bonvin, A.M.J.J.; Xue, L.C. iScore: A novel graph kernel-based function for scoring protein–protein docking models. *Bioinformatics* **2020**, *36*, 112–121. [CrossRef] [PubMed]

15. Yan, Y.; Huang, S.-Y. Pushing the accuracy limit of shape complementarity for protein-protein docking. *BMC Bioinf.* **2019**, *20*, 696. [CrossRef]
16. Siebenmorgen, T.; Zacharias, M. Evaluation of Predicted Protein–Protein Complexes by Binding Free Energy Simulations. *J. Chem. Theory Comput.* **2019**, *15*, 2071–2086. [CrossRef]
17. van Zundert, G.C.P.; Rodrigues, J.P.G.L.M.; Trellet, M.; Schmitz, C.; Kastritis, P.L.; Karaca, E.; Melquiond, A.S.J.; van Dijk, M.; de Vries, S.J.; Bonvin, A.M.J.J. The HADDOCK2.2 Web Server: User-Friendly Integrative Modeling of Biomolecular Complexes. *J. Mol. Biol.* **2016**, *428*, 720–725. [CrossRef] [PubMed]
18. Lensink, M.F.; Brysbaert, G.; Nadzirin, N.; Velankar, S.; Chaleil, R.A.G.; Gerguri, T.; Bates, P.A.; Laine, E.; Carbone, A.; Grudinin, S.; et al. Blind prediction of homo- and hetero-protein complexes: The CASP13-CAPRI experiment. *Proteins Struct. Funct. Bioinf.* **2019**, *87*, 1200–1221. [CrossRef] [PubMed]
19. Harmalkar, A.; Gray, J.J. Advances to tackle backbone flexibility in protein docking. *Curr. Opin. Struct. Biol.* **2021**, *67*, 178–186. [CrossRef] [PubMed]
20. Siebenmorgen, T.; Engelhard, M.; Zacharias, M. Prediction of protein–protein complexes using replica exchange with repulsive scaling. *J. Comput. Chem.* **2020**, *41*, 1436–1447. [CrossRef]
21. Park, T.; Woo, H.; Baek, M.; Yang, J.; Seok, C. Structure prediction of biological assemblies using GALAXY in CAPRI rounds 38-45. *Proteins Struct. Funct. Bioinf.* **2020**, *88*, 1009–1017. [CrossRef]
22. Zalewski, M.; Kmiecik, S.; Koliński, M. Molecular Dynamics Scoring of Protein–Peptide Models Derived from Coarse-Grained Docking. *Molecules* **2021**, *26*, 3293. [CrossRef]
23. Peterson, L.X.; Kang, X.; Kihara, D. Assessment of protein side-chain conformation prediction methods in different residue environments. *Proteins Struct. Funct. Bioinf.* **2014**, *82*, 1971–1984. [CrossRef] [PubMed]
24. Kmiecik, S.; Kouza, M.; Badaczewska-Dawid, A.; Kloczkowski, A.; Kolinski, A. Modeling of Protein Structural Flexibility and Large-Scale Dynamics: Coarse-Grained Simulations and Elastic Network Models. *Int. J. Mol. Sci.* **2018**, *19*, 3496. [CrossRef] [PubMed]
25. Torchala, M.; Gerguri, T.; Chaleil, R.A.G.; Gordon, P.; Russell, F.; Keshani, M.; Bates, P.A. Enhanced sampling of protein conformational states for dynamic cross-docking within the protein-protein docking server SwarmDock. *Proteins Struct. Funct. Bioinf.* **2020**, *88*, 962–972. [CrossRef]
26. Jiménez-García, B.; Roel-Touris, J.; Romero-Durana, M.; Vidal, M.; Jiménez-González, D.; Fernández-Recio, J. LightDock: A new multi-scale approach to protein–protein docking. *Bioinformatics* **2018**, *34*, 49–55. [CrossRef]
27. Kurkcuoglu, Z.; Bonvin, A.M.J.J. Pre- and post-docking sampling of conformational changes using ClustENM and HADDOCK for protein-protein and protein-DNA systems. *Proteins Struct. Funct. Bioinf.* **2020**, *88*, 292–306. [CrossRef] [PubMed]
28. Schindler, C.E.M.; de Vries, S.J.; Zacharias, M. iATTRACT: Simultaneous global and local interface optimization for protein-protein docking refinement. *Proteins Struct. Funct. Bioinf.* **2015**, *83*, 248–258. [CrossRef] [PubMed]
29. Kmiecik, S.; Gront, D.; Kolinski, M.; Wieteska, L.; Dawid, A.E.; Kolinski, A. Coarse-Grained Protein Models and Their Applications. *Chem. Rev.* **2016**, *116*, 7898–7936. [CrossRef] [PubMed]
30. Baaden, M.; Marrink, S.J. Coarse-grain modelling of protein–protein interactions. *Curr. Opin. Struct. Biol.* **2013**, *23*, 878–886. [CrossRef]
31. Roel-Touris, J.; Bonvin, A.M.J.J. Coarse-grained (hybrid) integrative modeling of biomolecular interactions. *Comput. Struct. Biotechnol. J.* **2020**, *18*, 1182–1190. [CrossRef]
32. Krupa, P.; Karczyńska, A.S.; Mozolewska, M.A.; Liwo, A.; Czaplewski, C. UNRES-Dock—protein–protein and peptide–protein docking by coarse-grained replica-exchange MD simulations. *Bioinformatics* **2020**. [CrossRef] [PubMed]
33. Kuroda, D.; Gray, J.J. Pushing the Backbone in Protein-Protein Docking. *Structure* **2016**, *24*, 1821–1829. [CrossRef] [PubMed]
34. Marze, N.A.; Roy Burman, S.S.; Sheffler, W.; Gray, J.J. Efficient flexible backbone protein–protein docking for challenging targets. *Bioinformatics* **2018**, *34*, 3461–3469. [CrossRef]
35. Roy Burman, S.S.; Nance, M.L.; Jeliazkov, J.R.; Labonte, J.W.; Lubin, J.H.; Biswas, N.; Gray, J.J. Novel sampling strategies and a coarse-grained score function for docking homomers, flexible heteromers, and oligosaccharides using Rosetta in CAPRI rounds 37–45. *Proteins Struct. Funct. Bioinf.* **2020**, *88*, 973–985. [CrossRef] [PubMed]
36. Zacharias, M. ATTRACT: Protein-protein docking in CAPRI using a reduced protein model. *Proteins Struct. Funct. Bioinf.* **2005**, *60*, 252–256. [CrossRef] [PubMed]
37. Glashagen, G.; Vries, S.; Uciechowska-Kaczmarzyk, U.; Samsonov, S.A.; Murail, S.; Tuffery, P.; Zacharias, M. Coarse-grained and atomic resolution biomolecular docking with the ATTRACT approach. *Proteins Struct. Funct. Bioinf.* **2020**, *88*, 1018–1028. [CrossRef] [PubMed]
38. Moal, I.H.; Bates, P.A. SwarmDock and the Use of Normal Modes in Protein-Protein Docking. *Int. J. Mol. Sci.* **2010**, *11*, 3623–3648. [CrossRef]
39. Yan, Y.; He, J.; Feng, Y.; Lin, P.; Tao, H.; Huang, S. Challenges and opportunities of automated protein-protein docking: HDOCK server vs. human predictions in CAPRI Rounds 38-46. *Proteins Struct. Funct. Bioinf.* **2020**, *88*, 1055–1069. [CrossRef] [PubMed]
40. Roel-Touris, J.; Don, C.G.V.; Honorato, R.; Rodrigues, J.P.G.L.M.; Bonvin, A.M.J.J. Less Is More: Coarse-Grained Integrative Modeling of Large Biomolecular Assemblies with HADDOCK. *J. Chem. Theory Comput.* **2019**, *15*, 6358–6367. [CrossRef]
41. Ciemny, M.; Badaczewska-Dawid, A.; Pikuzinska, M.; Kolinski, A.; Kmiecik, S. Modeling of Disordered Protein Structures Using Monte Carlo Simulations and Knowledge-Based Statistical Force Fields. *Int. J. Mol. Sci.* **2019**, *20*, 606. [CrossRef] [PubMed]

42. Badaczewska-Dawid, A.E.; Khramushin, A.; Kolinski, A.; Schueler-Furman, O.; Kmiecik, S. Protocols for All-Atom Reconstruction and High-Resolution Refinement of Protein–Peptide Complex Structures. *Methods Mol. Biol.* **2020**, *2165*, 273–287. [CrossRef] [PubMed]
43. Badaczewska-Dawid, A.E.; Kolinski, A.; Kmiecik, S. Computational reconstruction of atomistic protein structures from coarse-grained models. *Comput. Struct. Biotechnol. J.* **2020**, *18*, 162–176. [CrossRef] [PubMed]
44. Jamroz, M.; Orozco, M.; Kolinski, A.; Kmiecik, S. Consistent View of Protein Fluctuations from All-Atom Molecular Dynamics and Coarse-Grained Dynamics with Knowledge-Based Force-Field. *J. Chem. Theory Comput.* **2013**, *9*, 119–125. [CrossRef] [PubMed]
45. Kuriata, A.; Gierut, A.M.; Oleniecki, T.; Ciemny, M.P.; Kolinski, A.; Kurcinski, M.; Kmiecik, S. CABS-flex 2.0: A web server for fast simulations of flexibility of protein structures. *Nucleic Acids Res.* **2018**, *46*, W338–W343. [CrossRef] [PubMed]
46. Kurcinski, M.; Oleniecki, T.; Ciemny, M.P.; Kuriata, A.; Kolinski, A.; Kmiecik, S. CABS-flex standalone: A simulation environment for fast modeling of protein flexibility. *Bioinformatics* **2019**, *35*, 694–695. [CrossRef] [PubMed]
47. Kmiecik, S.; Gront, D.; Kouza, M.; Kolinski, A. From Coarse-Grained to Atomic-Level Characterization of Protein Dynamics: Transition State for the Folding of B Domain of Protein A. *J. Phys. Chem. B* **2012**, *116*, 7026–7032. [CrossRef]
48. Kmiecik, S.; Kolinski, A. Characterization of protein-folding pathways by reduced-space modeling. *Proc. Natl. Acad. Sci. USA* **2007**, *104*, 12330–12335. [CrossRef]
49. Kmiecik, S.; Kolinski, A. Simulation of Chaperonin Effect on Protein Folding: A Shift from Nucleation–Condensation to Framework Mechanism. *J. Am. Chem. Soc.* **2011**, *133*, 10283–10289. [CrossRef]
50. Kurcinski, M.; Kolinski, A.; Kmiecik, S. Mechanism of Folding and Binding of an Intrinsically Disordered Protein As Revealed by ab Initio Simulations. *J. Chem. Theory Comput.* **2014**, *10*, 2224–2231. [CrossRef]
51. Kurcinski, M.; Jamroz, M.; Blaszczyk, M.; Kolinski, A.; Kmiecik, S. CABS-dock web server for the flexible docking of peptides to proteins without prior knowledge of the binding site. *Nucleic Acids Res.* **2015**, *43*, W419–W424. [CrossRef]
52. Blaszczyk, M.; Kurcinski, M.; Kouza, M.; Wieteska, L.; Debinski, A.; Kolinski, A.; Kmiecik, S. Modeling of protein–peptide interactions using the CABS-dock web server for binding site search and flexible docking. *Methods* **2016**, *93*, 72–83. [CrossRef]
53. Ciemny, M.P.; Kurcinski, M.; Kozak, K.; Kolinski, A.; Kmiecik, S. Highly flexible protein-peptide docking using cabs-dock. *Methods Mol. Biol.* **2017**, *1561*, 69–94. [CrossRef]
54. Kurcinski, M.; Ciemny, M.P.; Oleniecki, T.; Kuriata, A.; Badaczewska-Dawid, A.E.; Kolinski, A.; Kmiecik, S. CABS-dock standalone: A toolbox for flexible protein–peptide docking. *Bioinformatics* **2019**, *35*, 4170–4172. [CrossRef]
55. Blaszczyk, M.; Ciemny, M.P.; Kolinski, A.; Kurcinski, M.; Kmiecik, S. Protein-peptide docking using CABS-dock and contact information. *Brief. Bioinf.* **2019**, *20*, 2299–2305. [CrossRef]
56. Kurcinski, M.; Badaczewska-Dawid, A.; Kolinski, M.; Kolinski, A.; Kmiecik, S. Flexible docking of peptides to proteins using CABS-dock. *Protein Sci.* **2020**, *29*, 211–222. [CrossRef] [PubMed]
57. Ciemny, M.P.; Kurcinski, M.; Blaszczyk, M.; Kolinski, A.; Kmiecik, S. Modeling EphB4-EphrinB2 protein–protein interaction using flexible docking of a short linear motif. *Biomed. Eng. Online* **2017**, *16*, 71. [CrossRef]
58. Ciemny, M.; Kurcinski, M.; Kamel, K.; Kolinski, A.; Alam, N.; Schueler-Furman, O.; Kmiecik, S. Protein–peptide docking: Opportunities and challenges. *Drug Discov. Today* **2018**, *23*, 1530–1537. [CrossRef] [PubMed]
59. Ciemny, M.P.; Debinski, A.; Paczkowska, M.; Kolinski, A.; Kurcinski, M.; Kmiecik, S. Protein-peptide molecular docking with large-scale conformational changes: The p53-MDM2 interaction. *Sci. Rep.* **2016**, *6*. [CrossRef] [PubMed]
60. Badaczewska-Dawid, A.E.; Kmiecik, S.; Koliński, M. Docking of peptides to GPCRs using a combination of CABS-dock with FlexPepDock refinement. *Brief. Bioinf.* **2020**. [CrossRef]
61. Koliński, M.; Kmiecik, S.; Dec, R.; Piejko, M.; Mak, P.; Dzwolak, W. Docking interactions determine early cleavage events in insulin proteolysis by pepsin: Experiment and simulation. *Int. J. Biol. Macromol.* **2020**, *149*, 1151–1160. [CrossRef] [PubMed]
62. Vreven, T.; Moal, I.H.; Vangone, A.; Pierce, B.G.; Kastritis, P.L.; Torchala, M.; Chaleil, R.; Jiménez-García, B.; Bates, P.A.; Fernandez-Recio, J. Updates to the Integrated Protein-Protein Interaction Benchmarks: Docking Benchmark Version 5 and Affinity Benchmark Version 2. *J. Mol. Biol.* **2015**. [CrossRef] [PubMed]

International Journal of *Molecular Sciences*

MDPI

Article

Possible Link between Higher Transmissibility of Alpha, Kappa and Delta Variants of SARS-CoV-2 and Increased Structural Stability of Its Spike Protein and hACE2 Affinity

Vipul Kumar [1], Jasdeep Singh [1], Seyed E. Hasnain [1,2,*] and Durai Sundar [1,*]

1 Department of Biochemical Engineering and Biotechnology, Indian Institute of Technology (IIT) Delhi, New Delhi 110016, India; vipul.kumar@dbeb.iitd.ac.in (V.K.); jasdeep002@gmail.com (J.S.)

2 Department of Life Science, School of Basic Sciences and Research, Sharda University, Greater Noida 201301, India

* Correspondence: seh@dbeb.iitd.ac.in (S.E.H.); sundar@dbeb.iitd.ac.in (D.S.)

Abstract: The Severe Acute Respiratory Syndrome Coronavirus 2 (SARS-CoV-2) outbreak in December 2019 has caused a global pandemic. The rapid mutation rate in the virus has created alarming situations worldwide and is being attributed to the false negativity in RT-PCR tests. It has also increased the chances of reinfection and immune escape. Recently various lineages namely, B.1.1.7 (Alpha), B.1.617.1 (Kappa), B.1.617.2 (Delta) and B.1.617.3 have caused rapid infection around the globe. To understand the biophysical perspective, we have performed molecular dynamic simulations of four different spikes (receptor binding domain)-hACE2 complexes, namely wildtype (WT), Alpha variant (N501Y spike mutant), Kappa (L452R, E484Q) and Delta (L452R, T478K), and compared their dynamics, binding energy and molecular interactions. Our results show that mutation has caused significant increase in the binding energy between the spike and hACE2 in Alpha and Kappa variants. In the case of Kappa and Delta variants, the mutations at L452R, T478K and E484Q increased the stability and intra-chain interactions in the spike protein, which may change the interaction ability of neutralizing antibodies to these spike variants. Further, we found that the Alpha variant had increased hydrogen interaction with Lys353 of hACE2 and more binding affinity in comparison to WT. The current study provides the biophysical basis for understanding the molecular mechanism and rationale behind the increase in the transmissivity and infectivity of the mutants compared to wild-type SARS-CoV-2.

Keywords: B.1.1.7; B.1.617.2; COVID-19; E484Q; T478K and L452R mutation; N501Y mutation; spike protein

Citation: Kumar, V.; Singh, J.; Hasnain, S.E.; Sundar, D. Possible Link between Higher Transmissibility of Alpha, Kappa and Delta Variants of SARS-CoV-2 and Increased Structural Stability of Its Spike Protein and hACE2 Affinity. *Int. J. Mol. Sci.* **2021**, *22*, 9131. https://doi.org/10.3390/ijms22179131

Academic Editors: Istvan Simon and Csaba Magyar

Received: 13 July 2021
Accepted: 22 August 2021
Published: 24 August 2021

Publisher's Note: MDPI stays neutral with regard to jurisdictional claims in published maps and institutional affiliations.

1. Introduction

The Severe Acute Respiratory Syndrome—Coronavirus-2 (SARS-CoV-2), first detected in December 2019 in the Wuhan province of China, has caused the COVID-19 pandemic. As of August 18, 2021, there are more than 208,470,375 confirmed cases, and 4,377,979 people have lost their lives (https://covid19.who.int/) (accessed on 18 August 2021). The SARS-CoV-2 belongs to the family of beta corona virus, the same class of viruses responsible for previous pandemics caused by SARS-CoV and MERS [1–3]. SARS-CoV-2 possesses a large single-stranded RNA as genetic material and has four main structural components, namely, Envelope protein, spike protein, membrane protein and nucleocapsid [4–6]. The main structural element that enables this virus to attach to the host receptor is the spike glycoprotein, and it also gives the crown-like appearance to the virus, hence it is named Coronavirus [7–9]. The spike glycoprotein of SARS-CoV-2 attaches to the human angiotensin converting enzyme (hACE2) receptor and is then activated by another human enzyme, transmembrane protease serine (TMPRSS2), to enter the host cells [9–11]. Since spike is the primary target receptor for the entry and the main virulence factor of the virus, various therapeutic drugs and vaccines are being made and tested against it [11,12]. Although multiple medications such as remdesivir or hydroxychloroquine, lopinavir and ritonavir have been recommended by the World Health Organization (WHO) against COVID-19, their efficacy is still the topic of debate [13–15]. Similarly, WHO has issued an emergency use listing for certain vaccines such as BNT162B2 from Pfizer, AstraZeneca/Oxford COVID-19 vaccine, manufactured by the Serum Institute of India and SKBio, and Ad26.COV2.S, developed by Janssen (Johnson & Johnson) (https://www.who.int/covid-19/vaccines) (accessed on 20 April 2021). However, the SARS-CoV-2 cases are still increasing at an alarming rate all over the globe, and the primary rationale behind it is the rapid accumulation of mutations in the SARS-CoV-2.

In the past few months, multiple variants of the SARS-CoV-2 have been reported. Some of them are the variant of concern (VOC), which have increased the infectivity or have the potential of immune escape. Almost all the VOCs reported till now have mutations in the spike glycoprotein of the virus, which has increased the binding affinity of the virus to hACE2 or has conferred immune escape potential [16,17]. The Lineage B.1.1.7 or 20I/501Y.V1 (Alpha variant) was detected in the United Kingdom in September 2020. This variant increased the transmissibility by 40–80% and has been partially correlated with N501Y mutation in the receptor binding domain (RBD) of spike protein [18] (Figure 1A). In October 2020, B.1.351 (Beta variant) was detected in the South African population, which could infect more younger people and had three primary mutations in the RBD of spike protein, namely, N501Y, K417N and E484K [19,20]. Similarly, the lineage P.1 (Gamma variant) detected in January 2021 in the Brazilian population had three mutations of concern in spike RBD, namely, N501Y, K417T and E484K [17,21]. In our previous study, we had reported that N501Y mutation could enhance the ACE2 affinity and possibly confer resistance towards the antibodies [17]. Our results also indicated the reinfection potential of P1 and N501Y.V2 variants. In another study, it has been reported that N501Y mutation increases (dissociation constant: 22 nM to 0.44 nM) the binding affinity with hACE2 [22]. In India, lineage B.1.617 and B.1.618 have been recently reported, which had caused a rapid increase in the COVID-19 cases in the country [23,24]. The B.1.617 lineage, has been further divided into three sub lineages namely, B.1.617.1 (Kappa), B.1.617.2 (Delta variant) and B.1.617.3 [25] (Figure 1B). Out of these three sub lineages of B.1.617, the Delta variant has been identified as a variant of concern (VOC) and reported to be the main variant behind the second wave in India by WHO (https://www.who.int/en/activities/tracking-SARS-CoV-2-variants/) (accessed on 6 June 2021). The Kappa is characterized by E154K, L452R, E484Q, D614G, P681R, Q1071H mutations in the spike protein and Delta by T19R, L452R, T478K, D614G, P681R, D950N mutations while the B.1.617.3 lineage has T19R, L452R, E484Q, D614G, P681R mutations in the spike protein. All these lineages have conserved L452R, D614G and P681R (https://www.who.int/en/activities/tracking-SARS-CoV-2-variants/) (accessed on 6 June 2021). While in this study we have focused on the mutations within the RBD of spike protein, the D614G mutation (present outside the RBD region) has already been reported to increase the binding affinity with hACE2 and is susceptible to neutralization by antibodies [26].

While this paper was ready for submission a new sub variant of B.1.617 was detected and named as Delta^{Plus}. It contains same mutations as Delta variant and two other mutations—K417N and W258L in the spike glycoprotein [27]. Further, the B.1.618 (triple mutant), recently detected in the four Indian states (Maharashtra, Delhi, West Bengal and Chhattisgarh), has been characterized by the deletion of Tyr145 and His146 as well as E484K and D614G mutation in the spike protein (https://cov-lineages.org/) (accessed on 10 July 2021) [24]. The sudden increase in COVID-19 cases in India is attributed to the Delta variant and its higher binding affinity towards hACE2 along with its immune escape ability [17,28]. In previous epidemiological and genomics study the sudden increase in the incidence of B.1.617.2 during February to April 2021 in India has been shown as the reason for the increased COVID-19 positivity rate [29]. A recent study focused on Delhi population sera survey, has reported that prevalence of B.1.617 lineage increased from 5 % in February to 60 % in April 2021 [29]. The loss of E484Q mutation and gain of T478K in the B.1.617.2 lineage directly correlated with increase in the positivity rate [29]. In another recent study, it has been reported that infection with B.1.617.2 variant could be controlled by antibodies induced due to prior infection or BNT162b2 vaccination, but with lower efficacy than the B.1.351 variant. This study further demonstrated that B.1.617.2 variant has greater lung cell entry and cell to cell fusion, indicating its higher lung infection capacity [30]. Although various studies have shown the phenotypic effect of the mutations, and increased transmissibility, limited data exist on comparative dynamics, molecular interactions, and changes in energetics due to these crucial mutations in the RBD domain of the spike protein of various mutants.

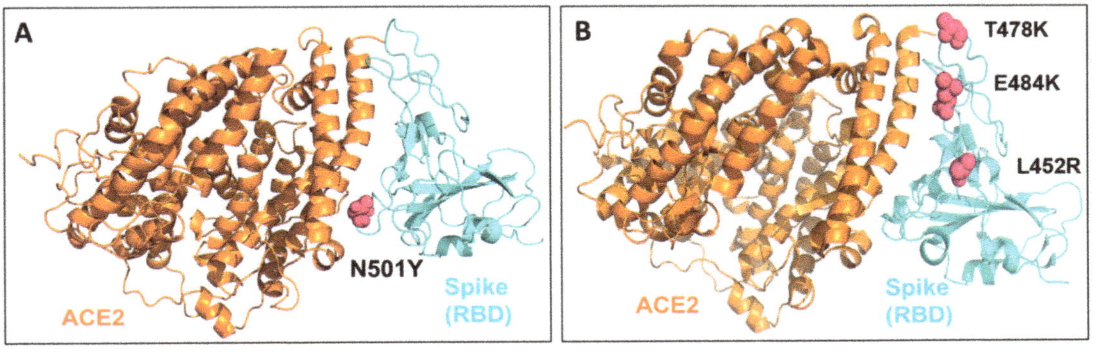

Figure 1. The structure of the receptor binding domain (RBD) of SARS-CoV-2 spike protein complexed with human angiotensin converting enzyme 2 (hACE2) receptor. (**A**) The sphere shape residues in hot pink colour show N501Y mutation in the spike protein of SARS-CoV-2. (**B**) At L452R, T478K and E484Q mutations in the spike protein (RBD) of B.1.617 lineage.

In the present study, we have aimed to investigate the thermodynamic effects of the mutations in the RBD region of the spike glycoprotein interacting with hACE2 and compare that with the wildtype. We accordingly studied two crucial variants Alpha, Kappa and Delta, which caused an increase in COVID-19 cases in various countries, including India. As in lineage B.1.617, Kappa and B.1.617.3 have same L452R and E484Q mutation in RBD of spike, while Delta has L452R and T478K, we have only considered Kappa and Delta in this study. These three variants (Alpha, Kappa and Delta) possess significant mutations in the RBD domain of the spike glycoprotein and have a higher infectivity rate. Therefore, to study and compare the dynamics, interactions and binding free energy of wildtype and spike protein variants with hACE-2 at the molecular level, we have performed the classical molecular dynamic (MD) simulations.

2. Results

The Mutant Spike Proteins Have a Better Binding Affinity with hACE2 in Comparison to Wildtype

The wildtype (WT) spike-hACE2 complex, along with the prepared and equilibrated Alpha, Kappa and Delta spike variants, were simulated for 200 ns. All the four structure complexes were first analysed for investigating the dynamics. In RMSD analysis, we found that the three complexes had a similar deviation around 2.5 Å from the initial structure over the 200 ns of simulations; Kappa_Spike-hACE2 (2.36 ± 0.27 Å), Alpha_Spike-hACE2 (2.62 ± 0.67 Å) and WT_Spike-ACE2 (2.82 ± 0.84 Å), while more around 3 Å deviation was found in the Delta_ Spike-hACE2 (3.12 ± 0.69 Å), as shown in Figure 2A. When RMSF of the simulated complexes was analysed, it was found that the residues number Arg355 to Phe400 of the spike protein was more flexible, especially in Delta. In addition, the fluctuation in the mutant residues was not high in the case of Kappa, although, it was found that residue Val445 had more fluctuations than the WT and N501Y mutants (Figure 2B). The average RMSF for WT was 2.95 ± 0.86 Å, for B.1.617 it was 2.66 ± 0.94 Å, for Delta it was 4.44 ± 1.62 Å, and for the Alpha spike protein it was 3.02 ± 1.09 Å. Though the RMSD and RMSF analysis suggested lesser stability of Delta in comparison to other studied complexes, no significant higher fluctuation was seen in the mutated residues in comparison to its overall structure. After analysing the fluctuation and deviations in the structures, the number of hydrogen bond count was calculated between the spike protein and hACE2 for all three structures. It was found that WT (12.23 ± 2.58) and Kappa (11.81 ± 2.07) had similar number of hydrogen bonds, followed by Delta (9.78 ± 2.40) and Alpha (9.19 ± 1.81) (Figure 2C). We further analysed the significant residues, to find out which of them has greater than 30% of the occupancy of hydrogen bond throughout the simulation. It was found that Alpha and Kappa spike mutants had more residues interaction with hACE2 than WT and Delta. In the case of WT and Delta, three residues (Tyr453, Thr500 and Gly502) and (Lys417, Gln493 and Gly502) of the spike protein were making a hydrogen bond with hACE2 for more than 30% of the simulation time, respectively. In comparison, in the Alpha spike mutant, five residues (Lys417, Ala475, Asn487, Thr500 and Gly502) were involved, and in Kappa, there were six residues (Lys417, Tyr449, Asn487, Tyr489, Tht500 and Gly502) of the spike protein that had significant hydrogen bond interactions with hACE2 (Figure S1). When the hydrogen bond interaction of mutated residues was checked, it was found that in Kappa, Q484 had only 0.1 fraction time interaction with E75 of hACE2 and in Delta, T478 had 1.5 fraction of time of interaction with Q353 of hACE2 throughout the simulations. Similarly, in case of Alpha, N501 had only 1.5 fraction of time of interaction with K353 of hACE2. Hence, the hydrogen bond analysis suggested that this mutation did not have any direct significant in terms of interaction with hACE2. It was observed that Gly502 was the critical residue interacting significantly with hACE2 in all four complexes. None of the mutated residues in Alpha, Kappa and Delta were found to be making significant hydrogen bonding with hACE2. Hence, it was essential to investigate if these mutated residues of the spike protein had interaction with any other spike residues or other interactions with hACE2 for any fraction of time. To analyse the changes in the interaction due to mutation, we extracted the three structures at the 50 ns interval from all the three simulated complexes. It was found that in the case of Kappa variant, in the 50th ns frame, neither Arg452 nor Gln484 were involved in any polar contact with other residues, while in 100th ns and 150th ns frame, it was found that Gln484 was making hydrogen bond contact with Ser349 and Asn450 of the spike protein itself, while in WT spike protein, Glu484 was making a hydrogen bond only with Ser349. It was observed that there was an increase in intra-chain interaction Spike protein due to mutation of E484Q. In Delta, no major interactions of mutated residues were found in comparison to WT spike, however in 100th ns frame, Lys478 was making intra-chain interaction with Ser476 of spike that was not found in case of Thr478 of WT spike. Similarly, when the Alpha variant was compared with WT, it was found that due to Asn to Tyr mutation at 501st residue, there was an increase in the hydrogen bonding with Lys353 of hACE2 (Figure 3).

The increase in the intra-chain interaction in case of B.1.617 indicated that it may interfere in the human antibodies' interaction with the spike protein. In the case of Alpha, the increase in hydrogen bond contact with hACE2 indicated higher binding affinity of this mutant with hACE2 in comparison to WT. The MM/GBSA binding free energy has been earlier reported to correlate with the binding affinity between the complexes [31,32]. However, it is mainly used for comparing the binding energies of the studied complexes, not for absolute free energy calculations. Therefore, to assess and compare the binding affinity of the spike protein towards hACE2, we calculated the MM/GBSA binding free energy by extracting twenty structures in equal spans from 50th to 200 ns of the simulated trajectories. It was found that Alpha (−103.35 ± 16.31 kcal/mol) and Kappa (−101.90 ± 18.40 kcal/mol) spike proteins had a similar and higher binding affinity with hACE2 in comparison to WT (−96.87 ± 14.57 kcal/mol) (Figure 2D). Surprisingly, the MM/GBSA binding free energy of Delta with hACE2 was far less (−37.03 ± 22.79 kcal/mol) in comparison to all the studied complexes. Further, to calculate the energy contribution of individual mutated residues, prime energy was calculated for the twenty extracted structures, which showed that Kappa and Delta had a more stabilizing effect on the spike protein compared to WT, a recent study also shows similar results [33]. The average energy contribution of Arg (−50.90 ± 3.99 kcal/mol) in comparison to Leu (−22.15 ± 2.60 kcal/mol) at 452nd position of the spike protein was found to be high. The energy contribution of Gln (−56.03 ± 2.31 kcal/mol) in comparison to Glu (−47.12 ± 2.25 kcal/mol) at 484th position of spike protein was relatively higher. Similarly, Lys (−9.40 ± 2.66 kcal/mol) was favourable than Thr (−3.20 ± 2.93 kcal/mol) at 478th position. However, in the case of the Alpha mutant, it was noticed that Asn (−64.31 ± 3.59 kcal/mol) at 501st position was more energetically favourable than Tyr (−31.55 ± 3.26 kcal/mol) (Table 1).

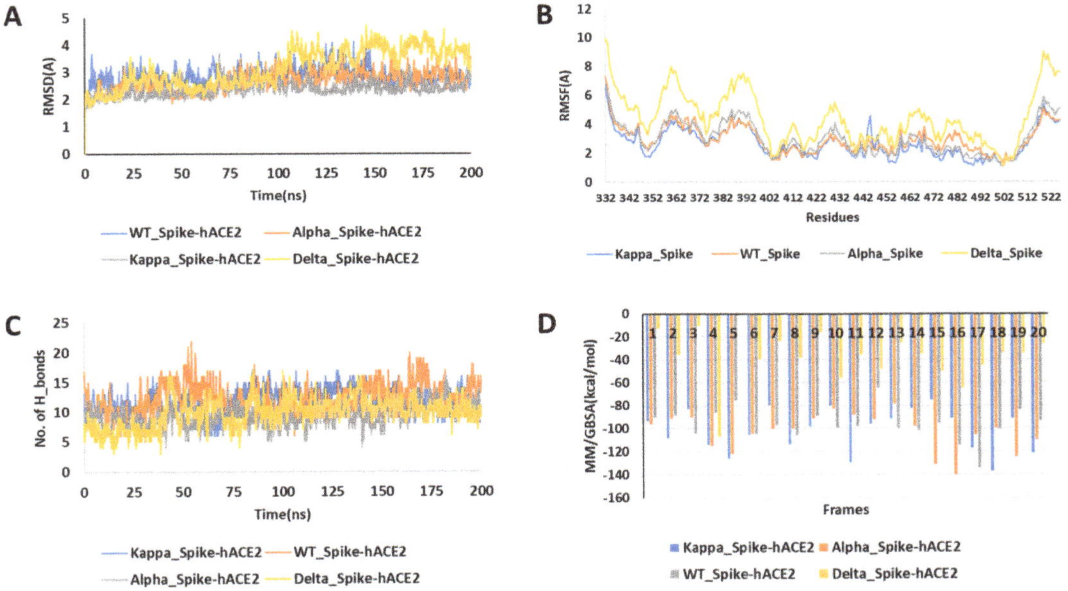

Figure 2. MD simulation analysis of the three simulated complexes. (**A**) RMSD plot showing similar deviation of all the simulated structures. (**B**) RMSF plot reveals that Residues 350-400 of the spike receptor binding domain (RBD) are more flexible, while the mutated residues have lesser fluctuation and are also comparable in all three structures. (**C**) The number of hydrogen bond count indicates that WT and Kappa variant have similar and higher number hydrogen bonds compared to Delta and Alpha variants. (**D**) MM/GBSA binding free energy of the 20 structure complexes extracted from each trajectory at equal span, suggesting that Kappa and Alpha spike variants have higher affinity for hACE2 in comparison to Delta and WT.

Although these binding energy calculations are theoretical and cannot be taken as absolute values, however, they are typically used for the comparison of binding affinity of the complexes with respect to each other. The interactions and binding energy calculations showed that in B.1.617 variant, there is a decrease of energy due to mutations as well as change in intra-chain interactions, which may lead to stabilization and interference with neutralizing antibodies interactions.

Overall, a significant increase in the binding affinity was observed in case of Kappa and Alpha variant in comparison to WT. However, the MM/GBSA binding energy of Delta with hACE2 was less in comparison to WT, suggesting that there must be some other ways these spike RBD mutations of Delta variant are helping in increased transmission but not by increasing the affinity with hACE2. While the Delta and Kappa mutations were found to be stabilizing the spike protein, but not N501Y of Alpha, increase/change in the intrachain interaction in the spike protein was observed in all the studied variants. Therefore, it can be interpreted that stabilization of the spike protein, increase of binding energy and increase in intra-chain interactions are crucial and are somehow aiding the Kappa variant, whereas in the Delta variant, it is the stabilizing of spike and increases in the intra-chain interactions. In the Alpha spike mutant, increases in the hydrogen-bond interaction and binding affinity with hACE2 could be the reason for more transmissivity of this mutant.

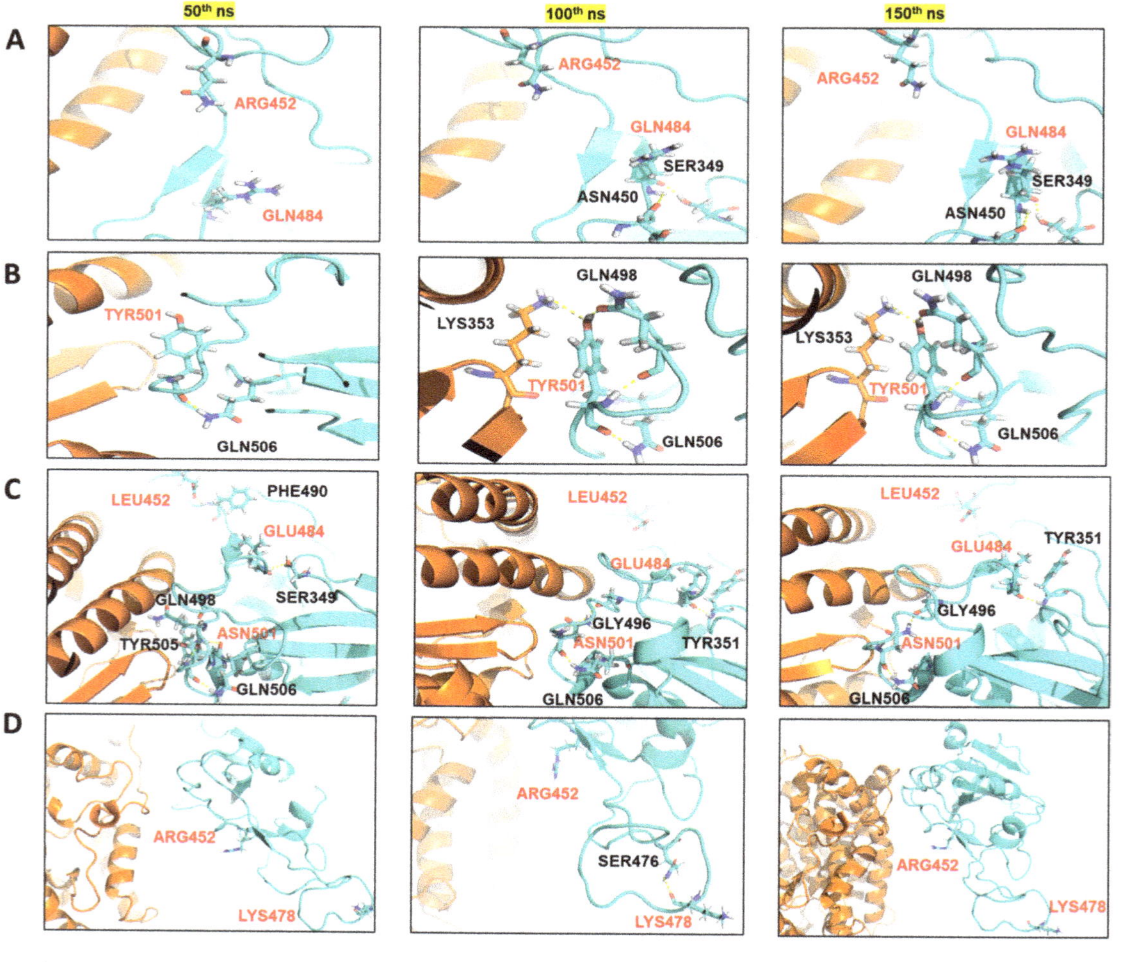

Figure 3. Comparing the interaction of the mutated residues and wild-type residues in the three structures extracted at 50 ns span from the simulated trajectories: spike protein (turquois color), and hACE2 (orange color). (**A**) Kappa spike variant and its interactions; Gln484 of the spike protein making intra-chain hydrogen bonding with Ser349 and Asn450 in the 100th and 150th ns frame. (**B**) Alpha spike variant interactions. The hydrogen bond interaction of Tyr501 of mutant spike protein with Lys353 of hACE2 in the 100th and 150th ns frame. (**C**) The interaction of wild type residues at 50th, 100th and 150th ns of the simulation shows that Glu484 of spike protein had only one hydrogen bond interaction with Ser349 or Tyr351 of spike itself. Similarly, Asn501 of spike was making hydrogen bond interactions with its residues only. (**D**) In the Delta spike variant, at 100th ns frame, an addition of a hydrogen bond of Lys478 with Ser476 was observed.

Table 1. Residue wise energy contribution of the mutated residues compared with the wildtype for the twenty structures extracted from 50 to 200 ns of the simulation for all the three complexes.

Kappa (kcal/mol)		Delta (kcal/mol)		WT (kcal/mol)		Alpha (kcal/mol)		
R452	**Q484**	**R452**	**K478**	**L452**	**E484**	**N501**	**T478**	**Y501**
−47.16	−51.43	−48.67	−7.51	−23.45	−47.06	3.44	3.44	−31.72
−50.69	−57.57	−49.32	−7.05	−23.04	−48.53	9.92	9.92	−27.6
−42.13	−56.15	−47.15	−10.92	−23.89	−45.52	3.81	3.81	−32.73
−52.43	−56.51	−45.49	−11.73	−20.05	−48.94	2.89	2.89	−31.96
−54.38	−58.47	−47.68	−9.31	−23.68	−50.2	2.03	2.03	−26.71
−54.05	−56.86	−46.76	−12.17	−25.88	−47.91	2.60	2.60	−29.47
−56.11	−53.98	−50.68	−7.42	−21.65	−47.86	4.45	4.45	−37.94
−47.14	−56.4	−45.18	−10.38	−22.44	−44.1	7.27	7.27	−33.03
−56.47	−57.21	−50.03	−13.46	−22.87	−43.55	−3.05	−3.05	−31.87
−54.79	−53.25	−48.51	−10.35	−22.05	−48.18	3.55	3.55	−33.2
−50.07	−51.51	−52.83	−6.85	−20.9	−46.46	7.14	7.14	−33.69
−45.38	−53.04	−55.77	−9.25	−17.67	−45.08	1.14	1.14	−25.84
−49.22	−59.17	−50.34	−11.01	−25.26	−49.82	4.00	4.00	−34.98
−51.75	−58.8	−47.65	−9.99	−16.19	−47.49	−0.80	−0.80	−29.93
−45.95	−55.64	−45.07	−14.89	−18.58	−47.46	2.51	2.51	−34.11
−48.85	−56.21	−45.52	−11.16	−20.55	−49.06	4.97	4.97	−34.51
−48.93	−56.69	−49.63	−6.93	−24.27	−46.24	5.18	5.18	−31.49
−54.01	−56.35	−48.91	−4.89	−23.78	−50.55	1.66	1.66	−32.64
−55.01	−59.25	−54.44	−7.28	−21.35	−41.95	0.17	0.17	−32.55
−53.5	−56.27	−53.66	−5.53	−25.55	−46.55	1.23	1.23	−25.06
−50.90 ± 3.99	**−56.03 ± 2.31**	**−49.16 ± 3.11**	**−9.40 ± 2.66**	**−22.15 ± 2.60**	**−47.12 ± 2.25**	**−64.31 ± 3.59**	**−3.20 ± 2.93**	**−31.55 ± 3.26**

3. Discussion

The recent variants of SARS-CoV-2 are cause of the second wave of the COVID-19 around the world and setback to healthcare infrastructure specially in India [28,34]. The transmission of the three variant of concerns (VOC), namely, Alpha, Beta and Delta identified in UK, South Africa and India, respectively were drivers of subsequent infection waves in these nations (https://www.who.int/en/activities/tracking-SARS-CoV-2-variants/) (accessed on 6 June 2021). Alpha, the first VOC initially discovered in September 2020 in the UK population has four main mutations (H69-, V70-, N501Y and D614G) in the spike protein [35]. These mutations are reported to be the mutations of concern, in other words, these mutations are positively selected by the virus for its higher transmission. These mutations were found in various other SARS-CoV-2 variants as well, which emerged after it. After Alpha, the second main VOC detected was Beta in October 2020 in the South African population, and it had five main mutations, which were reported to be beneficial for transmission—L18F, K417N, E484K, N501Y and D614G [35]. The N501Y and D614G was conserved in both Alpha and Beta VOC and are believed to be crucial mutations for their higher transmission and infectivity. In a recent study, where 12 monoclonal antibodies were tested for their neutralizing activity against Alpha and Beta variants, it was found that N501Y of Alpha variant modulated interaction of neutralizing antibodies only, while in case of Beta, complete loss of activity was observed in most of the antibodies, mediated by K417N and E484K, in comparison to wildtype [36]. The same study further reported that when convalescent plasma from the 20 patients infected before the emergence of Alpha was investigated, it lost >2.5-fold neutralizing activity against Beta, while maintaining activity against Alpha. Additionally, when the efficacy of Moderna and Pfizer vaccines were tested, it was found that there was no loss of neutralizing activity against Alpha, whereas every sample lost activity against Beta [36]. Another similar study, where convalescent sera from infected people and vaccine recipients were tested against Alpha, suggested that it is not a neutralization escape VOC in terms of vaccine efficacy. Several studies including the current one has indicated that N501Y mutation is the main reason behind the increase of Alpha transmission [22,37]. Overall, recent studies suggest that though the Alpha variant has higher transmission, it is not an escape variant and could be neutralized by the vaccines available and will be available in the near future [36]. The VOC next to Alpha, i.e., Beta has been found to be greater concern than Alpha in terms of its neutralization by convalescent plasma of the infected individuals and Moderna and Novavax vaccines [38].

Earlier this year, in March 2021, the B.1.617 lineage found in India transmitted rapidly and is being investigated for its role in severity and mortality [34]. Recently, sub lineages of the B.1.617 - Kappa, Delta and B.1.617.3 were reported and characterized. The B.1.617.1 is characterized by E154K, L452R, E484Q, D614G, P681R, Q1071H mutations in the spike protein. In a recent study on B.1.617 lineages, it has been shown that the P681R has highest impact in increasing the fusion activity, followed by E484K and L452R [39]. Further, when the Kappa spike mutant was tried to be neutralized with Pfizer vaccine sera, it was found that E484K conferred a ten-fold reduction in neutralisation, E484Q had a slightly milder yet significant impact, however, with E484Q and L452R combined, there was a statistically significant loss of sensitivity [39]. In another study, two-fold reduction in the neutralization efficacy of Covaxin vaccine (BBV152) was observed against B.1.617.2 variant [40]. Combining previously published literature with our current observations, the alpha and delta SARS-CoV-2 variants with their mutations have optimally struck balance between higher transmission and immune evasive capabilities. Overall, the previous studies reported against the B.1.617 variant have indicated a slight decrease of neutralizing activity of vaccines in comparison to wildtype, however, they still provided significant protection. Similarly, previous studies have reported that L452R, E484Q/K, P681R and T478K might have role in the increased transmissibility, while the molecular level rationale is not clear [39–41]. This was investigated in this study.

In the current study, we described the interactions of mutant spike RBD with hACE2 of wildtype, Alpha, Kappa and Delta variants. The binding affinity was found to be least in case of Delta, while Kappa and Alpha spike RBD had higher binding affinity with hACE2 in comparison to wildtype. The results of binding free energy calculations suggested that E484Q and N501Y mutations are crucial for increasing the binding affinity. The comparative MM/GBSA binding energy calculations of N501Y reported here positively correlate with the available experimental absolute binding free energy reported elsewhere [22]. Further, it was found that the L45R, E484Q and T478K mutations are highly energetically favourable for the spike protein based on the prime energy calculations of the mutated residues. Though the mutations do not change, the molecular interactions between the hACE2 and spike significantly, the snapshots from the MD simulations clearly indicated the change and increase of the intra-chain interactions in the mutated spike proteins, possibly interfering with the neutralising antibodies. Further analysis of these mutants with neutralizing antibodies is expected to provide more mechanistic insights.

4. Materials and Methods

4.1. MD Simulations

The X-ray crystal structure of SARS-CoV-2, spike RBD bound with hACE2 was retrieved from Protein Data Bank (PDB) having PDB ID 6M0J. Along with wildtype, three mutants of the spike protein were created, namely, Alpha (N501Y), Kappa (L452R and E484Q) and Delta (L45R and T478K) using the Maestro Suite of Schrodinger software (2020-3, NY, USA) [31]. All the four structures were then pre-processed for missing side chains, deleting waters, the addition of hydrogens, hydrogen bond optimization and restrained minimization using the protein preparation wizard of Schrodinger software (2020-3, NY, USA) [31]. The prepared mutated structures were then subjected to classical molecular dynamics for 50 ns for the stabilization of the mutated structures and the last frame structure was taken for further studies. The following protocol was adopted for the MD simulations of all four prepared structures—each system was solvated with the TIP3P water model in an orthorhombic periodic boundary box. To prevent interaction of the protein complex with its own periodic image, the distance between the complex and the box wall was kept at 10 Å. The system was then neutralized by the addition of appropriate number of Na^+/Cl^- ions depending on the complex using OPLS3e forcefield. Then the energy of the prepared systems was minimized by running 100 ps low-temperature (10 K) Brownian motion MD simulation (NVT ensemble) to remove steric clashes and move the system away from an unfavourable high-energy conformation. Further, the minimized systems were equilibrated in NVT and NPT ensembles using the "relax model system before simulation" option in the Desmond Schrodinger suite [31]. The equilibrated systems were then subjected to 200 ns unrestrained MD simulations in NPT ensemble with 300 K temperature maintained by Nose–Hoover chain thermostat constant pressure of 1 atm maintained by Martyna–Tobias–Kelin barostatand an integration time step of 2 fs with a recording interval of 200 ps.

4.2. Analysis of the MD Simulation

The root mean square deviation (RMSD), root mean square fluctuation (RMSF), number of hydrogen bonding was calculated using the simulation event analysis tool of the Desmond Suite integrated into Schrodinger software. Further, the occupancy of the hydrogen bonding between the spike protein and hACE2 was calculated using visual molecular dynamics (VMD) (1.9.4, UIUC, Champaign, IL, USA) [42]. The molecular mechanics generalized born surface area (MM/GBSA) free binding energy between spike proteins and hACE2 was calculated using the prime module of Schrodinger software [31]. Twenty structures extracted from 50 ns to 200 ns from each of the trajectories were used for this computation using the following equation:

$$\frac{\mathrm{MM}}{\mathrm{GBSA}}\Delta G_{\mathrm{bind}} = \Delta G_{\mathrm{complex}} - \left(\Delta G_{\mathrm{receptor}} + \Delta G_{\mathrm{ligand}}\right)$$

$$\Delta G = \Delta E_{\mathrm{gas}} + \Delta G_{\mathrm{sol}} - T\Delta S_{\mathrm{gas}}$$

$$\Delta E_{\mathrm{gas}} = \Delta E_{\mathrm{int}} + \Delta E_{\mathrm{ele}} + \Delta E_{\mathrm{vdw}}$$

$$\Delta G_{\mathrm{sol}} = \Delta G_{\mathrm{gb}} + \Delta G_{\mathrm{surf}}$$

The binding free energy (ΔG_{bind}) is dissociated into binding free energy of the complex, spike and hACE2. The gas–phase interaction energy (ΔE_{gas}) was calculated as the sum of electrostatic (ΔE_{elec}) and Van der Waal (ΔE_{vdw}) interaction energies, while internal energy was neglected. The solvation free energy (ΔG_{sol}) contains non-polar (ΔG_{surf}) and polar solvation energy (ΔG_{gb}), which was calculated by using the VSGB solvation model and OPL3e force field, while the entropy term was neglected by default [31,43].

The energy contribution of the mutated residues was then compared with wildtype residues. The Prime module of Schrodinger software (2020-3, NY, USA) was used for calculation of the energy contribution of the individual residues. The solvent model used here was surface generalized born (SGB), with variable dielectric enabled, the internal dielectric was 1.00 and solvent dielectric was 80.00 [31].

The following equation was used for the calculation of prime energy of individual residues:

$$\text{Total energy} = \text{covalent total} + \text{non} - \text{bonded total} + \text{other} - \text{SGB14}|\text{SGB} - \text{torsinol}$$

$$\text{Here, other energy} = \text{SGB self, nonpolar, hydrogen bond, packing, self} - \text{contact.}$$

5. Conclusions

In this study, MD simulations were performed to compare the binding energy, interactions and change in dynamics of spike (RBD)–hACE2 complexes, namely WT, Alpha, Kappa and Delta. It has been shown that mutants have a higher number of significant hydrogen bond interactions with hACE2, and the binding free energy of the mutants is also higher in comparison to WT, except in case of Delta. In the B.1.617 lineage, the mutations were favourable in terms of making the spike energetically stable as well as in terms of intra–chain residue interactions. In alpha spike, the mutation led to an extra interaction and higher binding affinity with hACE2 compared to WT. The increased molecular level interaction dynamics of spike–hACE2 and the predicted increased structural stability of its spike protein and hACE2 affinity can be possibly linked to higher transmissibility of B.1.617 and Alpha variants of SARS-CoV-2.

Supplementary Materials: The following are available online at https://www.mdpi.com/article/10.3390/ijms22179131/s1, Figure S1. (A) The hydrogen bond occupancy in WT spike–hACE2 complex, (B) Alpha variant (C) Kappa and (D) Delta variants throughout the 200 ns of MD simulations.

Author Contributions: Conceptualization, V.K. and D.S.; methodology, V.K.; software, V.K.; validation V.K.; formal analysis, V.K. and J.S.; investigation, V.K. and J.S.; resources, D.S. and S.E.H.; data curation, V.K. and J.S.; writing—original draft preparation, V.K., J.S., S.E.H. and D.S.; writing—review and editing, V.K., J.S., S.E.H. and D.S.; visualization, V.K and D.S.; supervision, S.E.H. and D.S.; project administration, S.E.H. and D.S.; funding acquisition, S.E.H. and D.S. All authors have read and agreed to the submitted version of the manuscript.

Funding: Department of Science and Technology, Government of India for J.C. Bose National Fellowship to S.E.H.

Data Availability Statement: The authors confirm that the data supporting the findings of this study are available within the article and/or its Supplementary Materials.

Conflicts of Interest: The authors declare no conflict of interest.

References

1. Zheng, J. SARS-CoV-2: An Emerging Coronavirus that Causes a Global Threat. *Int. J. Biol. Sci.* **2020**, *16*, 1678–1685. [CrossRef] [PubMed]
2. Zhu, Z.; Lian, X.; Su, X.; Wu, W.; Marraro, G.A.; Zeng, Y. From SARS and MERS to COVID-19: A brief summary and comparison of severe acute respiratory infections caused by three highly pathogenic human coronaviruses. *Respir. Res.* **2020**, *21*, 224. [CrossRef]
3. Rahman, S.; Singh, H.; Singh, J.; Khubaib, M.; Jamal, S.; Sheikh, J.; Kohli, S.; Hasnain, S. Mapping the genomic landscape & diversity of COVID-19 based on >3950 clinical isolates of SARS-CoV-2: Likely origin & transmission dynamics of isolates sequenced in India. *Indian J. Med Res.* **2020**, *151*, 474–478. [CrossRef]
4. Naqvi, A.A.T.; Fatima, K.; Mohammad, T.; Fatima, U.; Singh, I.K.; Singh, A.; Atif, S.M.; Hariprasad, G.; Hasan, G.M.; Hassan, I. Insights into SARS-CoV-2 genome, structure, evolution, pathogenesis and therapies: Structural genomics approach. *Biochim. Biophys. Acta Mol. Basis Dis.* **2020**, *1866*, 165878. [CrossRef]
5. Sironi, M.; Hasnain, S.E.; Rosenthal, B.; Phan, T.; Luciani, F.; Shaw, M.-A.; Sallum, M.A.; Mirhashemi, M.E.; Morand, S.; González-Candelas, F. SARS-CoV-2 and COVID-19: A genetic, epidemiological, and evolutionary perspective. *Infect. Genet. Evol.* **2020**, *84*, 104384. [CrossRef]
6. Sheikh, J.A.; Singh, J.; Singh, H.; Jamal, S.; Khubaib, M.; Kohli, S.; Dobrindt, U.; Rahman, S.A.; Ehtesham, N.Z.; Hasnain, S.E. Emerging genetic diversity among clinical isolates of SARS-CoV-2: Lessons for today. *Infect. Genet. Evol.* **2020**, *84*, 104330. [CrossRef]
7. Li, F. Structure, Function, and Evolution of Coronavirus Spike Proteins. *Annu. Rev. Virol.* **2016**, *3*, 237–261. [CrossRef]
8. Mittal, A.; Manjunath, K.; Ranjan, R.K.; Kaushik, S.; Kumar, S.; Verma, V. COVID-19 pandemic: Insights into structure, function, and hACE2 receptor recognition by SARS-CoV-2. *PLoS Pathog.* **2020**, *16*, e1008762. [CrossRef]
9. Shang, J.; Wan, Y.; Luo, C.; Ye, G.; Geng, Q.; Auerbach, A.; Li, F. Cell entry mechanisms of SARS-CoV-2. *Proc. Natl. Acad. Sci. USA* **2020**, *117*, 11727–11734. [CrossRef]
10. Hoffmann, M.; Kleine-Weber, H.; Schroeder, S.; Krüger, N.; Herrler, T.; Erichsen, S.; Schiergens, T.S.; Herrler, G.; Wu, N.-H.; Nitsche, A.; et al. SARS-CoV-2 Cell Entry Depends on ACE2 and TMPRSS2 and Is Blocked by a Clinically Proven Protease Inhibitor. *Cell* **2020**, *181*, 271–280.e278. [CrossRef] [PubMed]
11. Huang, Y.; Yang, C.; Xu, X.-F.; Xu, W.; Liu, S.-W. Structural and functional properties of SARS-CoV-2 spike protein: Potential antivirus drug development for COVID-19. *Acta Pharmacol. Sin.* **2020**, *41*, 1141–1149. [CrossRef]

12. Du, L.; He, Y.; Zhou, Y.; Liu, S.; Zheng, B.-J.; Jiang, S. The spike protein of SARS-CoV—A target for vaccine and therapeutic development. *Nat. Rev. Microbiol.* **2009**, *7*, 226–236. [CrossRef]
13. Beigel, J.H.; Tomashek, K.M.; Dodd, L.E.; Mehta, A.K.; Zingman, B.S.; Kalil, A.C.; Hohmann, E.; Chu, H.Y.; Luetkemeyer, A.; Kline, S.; et al. Remdesivir for the Treatment of COVID-19—Final Report. *N. Engl. J. Med.* **2020**, *383*, 1813–1826. [CrossRef]
14. Meini, S.; Pagotto, A.; Longo, B.; Vendramin, I.; Pecori, D.; Tascini, C. Role of Lopinavir/Ritonavir in the Treatment of COVID-19: A Review of Current Evidence, Guideline Recommendations, and Perspectives. *J. Clin. Med.* **2020**, *9*, 2050. [CrossRef]
15. Kumar, R.; Sharma, A.; Srivastava, J.K.; Siddiqui, M.H.; Uddin, S.; Aleya, L. Hydroxychloroquine in COVID-19: Therapeutic promises, current status, and environmental implications. *Environ. Sci. Pollut. Res.* **2021**, *28*, 40431–40444. [CrossRef]
16. Korber, B.; Fischer, W.M.; Gnanakaran, S.; Yoon, H.; Theiler, J.; Abfalterer, W.; Hengartner, N.; Giorgi, E.E.; Bhattacharya, T.; Foley, B.; et al. Tracking Changes in SARS-CoV-2 Spike: Evidence that D614G Increases Infectivity of the COVID-19 Virus. *Cell* **2020**, *182*, 812–827.e19. [CrossRef]
17. Singh, J.; Samal, J.; Kumar, V.; Sharma, J.; Agrawal, U.; Ehtesham, N.; Sundar, D.; Rahman, S.; Hira, S.; Hasnain, S. Structure-Function Analyses of New SARS-CoV-2 Variants B.1.1.7, B.1.351 and B.1.1.28.1: Clinical, Diagnostic, Therapeutic and Public Health Implications. *Viruses* **2021**, *13*, 439. [CrossRef]
18. La Rosa, G.; Mancini, P.; Ferraro, G.B.; Veneri, C.; Iaconelli, M.; Lucentini, L.; Bonadonna, L.; Brusaferro, S.; Brandtner, D.; Fasanella, A.; et al. Rapid screening for SARS-CoV-2 variants of concern in clinical and environmental samples using nested RT-PCR assays targeting key mutations of the spike protein. *Water Res.* **2021**, *197*, 117104. [CrossRef]
19. Tang, J.W.; Toovey, O.T.R.; Harvey, K.N.; Hui, D.D.S. Introduction of the South African SARS-CoV-2 variant 501Y.V2 into the UK. *J. Infect.* **2021**, *82*, e8–e10. [CrossRef] [PubMed]
20. Liu, Y.; Liu, J.; Plante, K.S.; Plante, J.A.; Xie, X.; Zhang, X.; Ku, Z.; An, Z.; Scharton, D.; Schindewolf, C. The N501Y spike substitution enhances SARS-CoV-2 transmission. *BioRxiv* **2021**. [CrossRef]
21. Hirotsu, Y.; Omata, M. Discovery of a SARS-CoV-2 variant 1 from the P.1 lineage harboring K417T/E484K/N501Y mutations in Kofu, Japan. *J. Infect.* **2021**, *82*, 276–316. [CrossRef]
22. Williams, A.H.; Zhan, C.-G. Fast Prediction of Binding Affinities of the SARS-CoV-2 Spike Protein Mutant N501Y (UK Variant) with ACE2 and Miniprotein Drug Candidates. *J. Phys. Chem. B* **2021**, *125*, 4330–4336. [CrossRef]
23. Bernal, J.L.; Andrews, N.; Gower, C.; Gallagher, E.; Simmons, R.; Thelwall, S.; Stowe, J.; Tessier, E.; Groves, N.; Dabrera, G.; et al. Effectiveness of COVID-19 Vaccines against the B.1.617.2 (Delta) Variant. *N. Engl. J. Med.* **2021**, *385*, 585–594. [CrossRef]
24. Sahoo, J.P.; Mishra, A.P.; Samal, K.C. Triple Mutant Bengal Strain (B. 1.618) of Coronavirus and the Worst COVID Outbreak in India. *Biot. Res. Today* **2021**, *3*, 261–265.
25. Mallapaty, S. India's massive COVID surge puzzles scientists. *Nat. Cell Biol.* **2021**, *592*, 667–668. [CrossRef]
26. Ozono, S.; Zhang, Y.; Ode, H.; Sano, K.; Tan, T.S.; Imai, K.; Miyoshi, K.; Kishigami, S.; Ueno, T.; Iwatani, Y.; et al. SARS-CoV-2 D614G spike mutation increases entry efficiency with enhanced ACE2-binding affinity. *Nat. Commun.* **2021**, *12*, 1–9. [CrossRef]
27. Kannan, S.R.; Spratt, A.N.; Cohen, A.R.; Naqvi, S.H.; Chand, H.S.; Quinn, T.P.; Lorson, C.L.; Byrareddy, S.N.; Singh, K. Evolutionary analysis of the Delta and Delta Plus variants of the SARS-CoV-2 viruses. *J. Autoimmun.* **2021**, *124*, 102715. [CrossRef]
28. Singh, J.; Rahman, S.A.; Ehtesham, N.Z.; Hira, S.; Hasnain, S.E. SARS-CoV-2 variants of concern are emerging in India. *Nat. Med.* **2021**, *27*, 1131–1133. [CrossRef]
29. Dhar, M.S.; Marwal, R.; Radhakrishnan, V.; Ponnusamy, K.; Jolly, B.; Bhoyar, R.C.; Fatihi, S.; Datta, M.; Singh, P.; Sharma, U.; et al. Genomic characterization and Epidemiology of an emerging SARS-CoV-2 variant in Delhi, India. *medRxiv* **2021**. [CrossRef]
30. Arora, P.; Kempf, A.; Nehlmeier, I.; Sidarovich, A.; Krüger, N.; Graichen, L.; Moldenhauer, A.-S.; Winkler, M.S.; Schulz, S.; Jäck, H.-M.; et al. Increased lung cell entry of B.1.617.2 and evasion of antibodies induced by infection and BNT162b2 vaccination. *BioRxiv* **2021**. [CrossRef]
31. *Schrödinger Protein Preparation Wizard, Epik, Impact, Prime, LigPrep, Glide; Desmond Molecular Dynamics System (Developed by D.E. Shaw Research); Maestro-Desmond Interoperability Tools*; Schrödinger, LLC.: New York, NY, USA, 2020.
32. Ylilauri, M.; Pentikäinen, O.T. MMGBSA As a Tool to Understand the Binding Affinities of Filamin–Peptide Interactions. *J. Chem. Inf. Model.* **2013**, *53*, 2626–2633. [CrossRef]
33. Pascarella, S.; Ciccozzi, M.; Zella, D.; Bianchi, M.; Benedetti, F.; Benvenuto, D.; Broccolo, F.; Cauda, R.; Caruso, A.; Angeletti, S.; et al. SARS-CoV-2 B.1.617 Indian variants: Are electrostatic potential changes responsible for a higher transmission rate? *J. Med. Virol.* **2021**, 1–6. [CrossRef]
34. Kar, S.K.; Ransing, R.; Arafat, S.; Menon, V. Second wave of COVID-19 pandemic in India: Barriers to effective governmental response. *EClinicalMedicine* **2021**, *36*, 100915. [CrossRef]
35. Aleem, A.; Akbar Samad, A.B.; Slenker, A.K. *Emerging Variants of SARS-CoV-2 and Novel Therapeutics Against Coronavirus (COVID-19)*; StatPearls: Treasure Island, FL, USA, 2021.
36. Wang, P.; Liu, L.; Iketani, S.; Luo, Y.; Guo, Y.; Wang, M.; Yu, J.; Zhang, B.; Kwong, P.D.; Graham, B.S.; et al. Increased Resistance of SARS-CoV-2 Variants B.1.351 and B.1.1.7 to Antibody Neutralization. *BioRxiv* **2021**. [CrossRef]
37. Sabino, E.C.; Buss, L.F.; Carvalho, M.P.S.; Prete, C.A.; Crispim, M.A.E.; Fraiji, N.A.; Pereira, R.H.M.; Parag, K.V.; Peixoto, P.D.S.; Kraemer, M.U.G.; et al. Resurgence of COVID-19 in Manaus, Brazil, despite high seroprevalence. *Lancet* **2021**, *397*, 452–455. [CrossRef]
38. Shen, X.; Tang, H.; Pajon, R.; Smith, G.; Glenn, G.M.; Shi, W.; Korber, B.; Montefiori, D.C. Neutralization of SARS-CoV-2 Variants B.1.429 and B.1.351. *N. Engl. J. Med.* **2021**, *384*, 2352–2354. [CrossRef] [PubMed]

39. Ferreira, I.; Datir, R.; Kemp, S.; Papa, G.; Rakshit, P.; Singh, S.; Meng, B.; Pandey, R.; Ponnusamy, K.; Radhakrishnan, V.S.; et al. SARS-CoV-2 B.1.617 emergence and sensitivity to vaccine-elicited antibodies. *BioRxiv* **2021**. [CrossRef]
40. Yadav, P.D.; Sapkal, G.N.; Abraham, P.; Ella, R.; Deshpande, G.; Patil, D.Y.; A Nyayanit, D.; Gupta, N.; Sahay, R.R.; Shete, A.M.; et al. Neutralization of Variant Under Investigation B.1.617.1 With Sera of BBV152 Vaccinees. *Clin. Infect. Dis.* **2021**, ciab411. [CrossRef]
41. Winger, A.; Caspari, T. The Spike of Concern—The Novel Variants of SARS-CoV-2. *Viruses* **2021**, *13*, 1002. [CrossRef]
42. Humphrey, W.; Dalke, A.; Schulten, K. VMD: Visual molecular dynamics. *J. Mol. Graph.* **1996**, *14*, 33–38. [CrossRef]
43. Greenidge, P.A.; Kramer, C.; Mozziconacci, J.-C.; Wolf, R.M. MM/GBSA Binding Energy Prediction on the PDBbind Data Set: Successes, Failures, and Directions for Further Improvement. *J. Chem. Inf. Model.* **2013**, *53*, 201–209. [CrossRef] [PubMed]

International Journal of *Molecular Sciences*

MDPI

Article

Association between Predicted Effects of *TP53* Missense Variants on Protein Conformation and Their Phenotypic Presentation as Li-Fraumeni Syndrome or Hereditary Breast Cancer

Yaxuan Liu [1,*], Olga Axell [1], Tom van Leeuwen [2], Robert Konrat [3], Pedram Kharaziha [1], Catharina Larsson [1], Anthony P. H. Wright [2,†] and Svetlana Bajalica-Lagercrantz [1,†]

1 Department of Oncology-Pathology, Karolinska Institutet, Bioclinicum, Karolinska University Hospital, 17164 Stockholm, Sweden; olga.axell@outlook.com (O.A.); kharaziha@gmail.com (P.K.); Catharina.Larsson@ki.se (C.L.); svetlana.lagercrantz@ki.se (S.B.-L.)

2 Department of Laboratory Medicine, Division of Biomolecular and Cellular Medicine, Karolinska Institutet, 17177 Stockholm, Sweden; tom.van.leeuwen@stud.ki.se (T.v.L.); anthony.wright@ki.se (A.P.H.W.)

3 Christian Doppler Laboratory for High-Content Structural Biology and Biotechnology, Department of Structural and Computational Biology, Max F. Perutz Laboratories, University of Vienna, 1030 Vienna, Austria; robert.konrat@univie.ac.at

* Correspondence: yaxuan.liu@ki.se; Tel.: +46-076-070-6720

† Authors contributed equally.

Citation: Liu, Y.; Axell, O.; van Leeuwen, T.; Konrat, R.; Kharaziha, P.; Larsson, C.; Wright, A.P.H.; Bajalica-Lagercrantz, S. Association between Predicted Effects of *TP53* Missense Variants on Protein Conformation and Their Phenotypic Presentation as Li-Fraumeni Syndrome or Hereditary Breast Cancer. *Int. J. Mol. Sci.* **2021**, *22*, 6345. https://doi.org/10.3390/ijms22126345

Academic Editors: Istvan Simon and Csaba Magyar

Received: 20 April 2021
Accepted: 11 June 2021
Published: 14 June 2021

Abstract: Rare germline pathogenic *TP53* missense variants often predispose to a wide spectrum of tumors characterized by Li-Fraumeni syndrome (LFS) but a subset of variants is also seen in families with exclusively hereditary breast cancer (HBC) outcomes. We have developed a logistic regression model with the aim of predicting LFS and HBC outcomes, based on the predicted effects of individual *TP53* variants on aspects of protein conformation. A total of 48 missense variants either unique for LFS (n = 24) or exclusively reported in HBC (n = 24) were included. LFS-variants were over-represented in residues tending to be buried in the core of the tertiary structure of TP53 (p = 0.0014). The favored logistic regression model describes disease outcome in terms of explanatory variables related to the surface or buried status of residues as well as their propensity to contribute to protein compactness or protein-protein interactions. Reduced, internally validated models discriminated well between LFS and HBC (C-statistic = 0.78−0.84; equivalent to the area under the ROC (receiver operating characteristic) curve), had a low risk for over-fitting and were well calibrated in relation to the known outcome risk. In conclusion, this study presents a phenotypic prediction model of LFS and HBC risk for germline *TP53* missense variants, in an attempt to provide a complementary tool for future decision making and clinical handling.

Keywords: Li-Fraumeni syndrome; hereditary breast cancer; germline *TP53* missense variants; quantitative prediction model; protein conformation

1. Introduction

Li-Fraumeni syndrome (LFS) is a rare heritable extreme tumor risk syndrome characterized mainly by premenopausal breast cancer, soft tissue sarcoma, brain tumors, osteosarcoma and adrenocortical carcinoma, and was first described in 1969 [1]. LFS was subsequently shown to be associated with a germline *TP53* variant [2]. As more families with a variety of tumors were reported, less restricted criteria became used to define Li-Fraumeni-like (LFL) families [3] that did not meet the classical LFS criteria but were suggestive of LFS, with a detection rate for germline *TP53* alterations of 20–40% in LFL as compared to 70% in LFS [4]. At present, the most commonly used screening criteria are the Chompret criteria, with a detection rate of 29%, since they include a large group of patients for screening [5]. For example, according to these criteria a patient with breast cancer below 31 years, should be screened irrespective of family history. With the increased use of cancer gene panels in genetic testing, the detection of pathogenic *TP53* variants has increased, and up to 1% of families with exclusively hereditary breast cancer (HBC) have been shown to carry a germline *TP53* variant [6].

The wide range of phenotypic presentation associated with germline *TP53* variants makes tumor risk assessment difficult and genetic counseling challenging in these patients and families. Moreover, 7–20% of constitutional *TP53* alterations are *de novo* [7], and thus presented in individuals without a family history of the disease. Due to the lack of knowledge about predicting genotype-phenotype association, all germline carriers are recommended a thorough surveillance program including yearly whole-body magnetic resonance imaging (MRI) examinations with the result that a large proportion of *TP53* variant carriers are exposed to unnecessary examinations [8,9].

The TP53 protein is a transcription factor that binds as a tetramer to DNA, and activates a large number of genes that promote DNA repair mechanisms or apoptosis including cell cycle regulatory proteins and members of the Bcl-2 family [10,11]. Each monomer is divided into different structural and functional domains, including a transactivation domain, a proline-rich region, a DNA binding domain (DBD), a oligomerization domain, a nuclear localization signal and a C-terminal regulatory domain [12]. TP53 plays a critical role in genomic homeostasis, and its activities are tightly regulated by a network of protein-protein interactions, microRNAs, and a range of post-translational modifications, including phosphorylation, acetylation, methylation and ubiquitination [13].

About two thirds of reported germline *TP53* variants are single site missense changes, predominantly located in the DBD [14]. Carriers are heterozygous for the *TP53* variant thus possessing both wild-type and variant monomers allowing formation of hetero-tetramers that result in a dominant-negative functional effect of some variants [15]. It has been suggested that patients with missense variants have earlier age of tumor onset (23.8 years), compared to those with loss of function variants (28.5 years) [16]. Moreover, unequal penetrance of missense variants is known in LFS where, for example, 58% of carriers with R248W (amino acid change at residue 248 from arginine to tryptophan) compared to only 21% of carriers with R231Q develop tumors before 30 years of age [17].

The TP53 DBD consists of a beta-sandwich tertiary structure with two antiparallel beta-sheets, that serve to orientate and stabilize the loop-sheet-helix DNA-binding motif [18]. Contacts with DNA are mainly to the sugar-phosphate backbone of the DNA helix (K120, S241, R248, R273, A276, R283) as well as a smaller number of contacts to specific bases within the consensus pentamer binding sequence (C277, R280 and K120). Other residues that are also mutated in sporadic tumors are important for anchoring the DNA binding motif to the beta-sandwich structure (e.g., R175, G245, R249, R282) or for stabilizing the beta-sandwich structure (e.g., V143, V157, Y220, F270). Although the TP53 DBD folds into a compact tertiary structure at body temperature, it is thermally unstable and unfolds at only slightly higher temperatures (>40 °C) or in response to tumor associated *TP53* variants [19,20]. Interestingly, some novel pharmaceutical agents (e.g., CP-31398 and APR-246) have been shown to restore wild-type functionality to mutant TP53 proteins by increasing their thermal stability [21]. Different missense variants of the DBD have different effects on protein conformation and its mechanistic characteristics and, interestingly, for some sporadic tumors a relationship between the effect of variants on mechanistic aspects of TP53 function and the type of tumor has been observed (e.g., glioblastoma vs. adrenocortical carcinoma) [22]. It is therefore possible that differential effects of germline *TP53* variants on conformational aspects of TP53 and its functionality could contribute to differences in phenotypes (e.g., LFS vs. HBC).

In this study we investigate whether the phenotypic outcome observed for different *TP53* variants can be accounted for by the differential effects of the variants on TP53 protein conformation as well as whether variant associated protein conformation changes can be used to predict disease outcome.

2. Materials and Methods

2.1. Selection of TP53 Variants

Included variants and their clinical characteristics were selected from publicly available databases and publications as described below. All *TP53* variants that were defined as LFS in our cohort were identified through the IARC database, and were not found to be reported in a HBC-family there or elsewhere. According to the IARC database the families thus fulfilled the classic LFS criteria [23] upon screening. For the HBC cohort, 17 variants were identified through the IARC database, 17 from the meta-analysis by Fortuno et al. [24] and 2 from Kharaziha et al. [25]. However, 12 of the HBC variants were reported both in the IARC database and in Fortuno et al., thus resulting in 24 unique HBC variants.

The selection process for the *TP53* missense variants used in this study is summarized in Figure S1. A total of 24 germline *TP53* variants unique for LFS were selected by evaluation of 408 *TP53* variants in the IARC database (R20, January 2020) [26]. Out of 296 missense variants, 62 variants were LFS-class, while 58 were LFL-class and 78 variants were TP53-Chompret-class, according to the terminology used in the IARC database. 117 variants were present in the FH-class (family history of cancer which does not fulfill LFS or any of the LFL definitions), noFH-class (no family history of cancer) or the other-class (variants that were not included in other classes). Many variants were present in more than one class. We selected the variants uniquely classified as LFS ($n = 24$) to represent the LFS-variants used in the study and the amino acids involved are referred to as LFS-residues (Table S1).

The non-redundant unified group of 24 HBC-specific germline *TP53* variants was selected from the IARC database, from the Fortuno et al. [24] meta-analysis of *TP53*-related HBC without a history of LFS and from the Kharaziha et al. [25] Swedish germline *TP53* cohort. Since there was no specified HBC-class in the IARC database, we selected the HBC-variants from FH-class, noFH-class and other-class and excluded those that overlapped with the LFS-class, LFL-class and TP53-Chompret-class. Further, the selected HBC *TP53* missense variants were exclusively reported in breast cancer. Out of 73 identified germline variants in Fortuno et al. [24], 41 were missense variants, of which 17 were also not present in the LFS-class, LFL-class or TP53-Chompret-class in the IARC database. In the Swedish cohort, reported by Kharaziha et al. [25], 24 germline *TP53* variants were identified, of which 6 missense variants were specifically found in HBC including two that were not present in the LFS-class, LFL-class or TP53-Chompret-class in the IARC database. The 24 resulting *TP53* variants were included in the study and the amino acids involved are referred to as HBC-residues (Table S1).

2.2. Analysis of Protein Structure

A published X-ray crystal structure of a tetrameric TP53 derivative containing the DBD fused to the oligomerization domain bound to the natural p21 TP53-response element (PDB accession number 3TS8) was used [27]. Details of the p53CR2 derivative used for crystallography have been fully described [28]. P53CR2 contains protein regions equivalent to residues 94–292 and 324–355 of TP53 and includes some stabilizing amino acid substitutions that distinguish the protein from the equivalent wild-type TP53 residues. PyMOL software (Schrödinger, https://pymol.org, 23 April 2019) was used to display HBC- and LFS-residues in the context of the tertiary structure of the protein and to identify different residue classes, using customized scripts adapted from those available in the PyMOL script library [29]. The findSurfaceResidues script [30] was adapted to allow identification of Buried (non-Surface) residues, defined as residues with a Solvent-Accessible Surface Area (SASA) below user-defined cutoff values. A cutoff value of 11Å^2 defined approximately 30% of residues as Buried and was used to classify residues for statistical analysis. The interfaceResidues script [31] was adapted to allow identification of residues at the interface between TP53 monomers in the tetrameric structure or between TP53 monomers and DNA. Interactions are defined as regions where the overlapping Surface area between atoms from different molecules exceeds a cutoff area. The default cutoff value of 1.0Å^2 was used. The ss script [32] was adapted to list residues in different types of secondary structure, defined as alpha-helix, beta-sheet or loop.

2.3. Explanatory Variables

Explanatory variables were a priori restricted to variables reflecting the effects of missense variants on different aspects of TP53 protein conformation (Table 1). Bur is a categorical variable defining Buried and Surface residues (as described above). A third Bur category (unknown) describes residues that are not included in the TP53 tertiary structure used. The remaining explanatory variables are continuous and are calculated using prediction algorithms for different aspects of protein conformation, which generate residue-by-residue scores for wild-type TP53 and each of the included TP53 variants. The value of these variables is defined as the difference between the variant and wild-type scores at the position of the substituted residue (variant score minus wild-type score). The prediction algorithms predict values that reflect propensity for intrinsic protein disorder, peptide-backbone flexibility, secondary structure, protein tertiary structure/compactness and protein interaction (Table 1). Protein interaction site prediction in the TP53 sequence was performed using a meta-structure-based homology method [33] as already reported [34] and summarized in the Supplementary Materials and methods. The analysis results in a residue specific score that is proportional to the propensity of a given residue to be part of a protein interaction site. The PPI6_dif variable used in regression models was calculated using more sensitive settings for predicting protein-interaction regions (minimum query segment = 6 amino acids), while less sensitive settings (minimum query segment = 10 amino acids) were used to identify and plot the most prominent TP53 protein-interaction regions (see Supplementary Materials and methods for details).

Table 1. Explanatory variables related to protein conformation.

Protein Characteristics	Variables	Predictor Algorithm [Ref]
Tertiary structure propensity		
buried/surface	Bur	Pymol-findSurfaceResidues script [28]
Intrinsic protein disorder propensity		
disorder (trained on Disprot DB)	disprot_dif	Espritz [35]
disorder (trained on NMR structures)	nmr_dif	Espritz [35]
disorder (trained on X-ray structures)	xray_dif	Espritz [35]
disorder (longer regions)	iupl_dif	IUPred2A [36]
disorder (short regions)	iups_dif	IUPred2A [36]
Predicted protein backbone flexibility		
protein backbone dynamics	dyn_dif	Dynamine [37]
Secondary structure propensity		
alpha-helix/beta-sheet	Sec_dif	Meta-structure [33,34]
Compactness propensity		
protein compactness	comp_dif	Meta-structure [33,34]
protein globularity	iupstr_dif	IUPred2A [36]
Protein interaction propensity		
protein protein interaction	PPI6_dif	Meta-structure-PPI [33,34]
protein protein interaction	anc_dif	IUPred2A [36]

The findSurfaceResidues script from Pymol software, version 2.3.1 [28]. was adapted to allow identification of Buried residues (<cutoff at 11Å^2). Bur is a categorical variable including Buried, Surface (>cutoff at 11Å^2) and unknown (not included in the TP53 tertiary structure used). The Espritz predictor [35] was trained using proteins in the Disprot database (disprot) as well as tertiary structures determined by nuclear magnetic resonance (nmr) or X-ray diffraction (xray). The IUPred2A [36] algorithm was run with the long disorder (iupl), short disorder (iups), structured domain (iupstr) and anchor (anc) arguments. Dynamine [37] predicts protein backbone flexibility (dyn). Meta-structure analysis [33,34] predicts values for two parameters, compactness (comp) and secondary structure (Sec).

PPI6 uses Meta-structure values to predict residues in regions with propensity for protein-protein interactions.

The _dif suffix indicates that the variable is the difference between the value for variant TP53 and wild-type TP53 at the position of the substituted residue (variant score minus wild type score).

Abbreviations: DB, database; NMR, nuclear magnetic resonance; X-ray, Xray diffraction.

2.4. Statistical Analysis

Data were collected and processed using R software, version 3.6. Fisher's exact test was used to evaluate whether LFS- or HBC-residues were over- or under-represented in residue sets reflecting different structural aspects of the TP53 protein structure. *p*-values were for two-sided tests and were adjusted for multiple testing, where appropriate, using the false discovery rate method. Associations between disease outcome (LFS or HBC) and the explanatory variables for different *TP53* variants as well as their predictive potential were evaluated by logistic regression using internal 1000-fold bootstrapped validation and a backwards step-down approach to variable number reduction as implemented in the rms-package (validate function: method = "boot", B = 1000, bw = TRUE, rule = "p", type = "individual", sls = 0.13). The validate function delivers values for a number of parameters relevant for assessing the discrimination performance of models and the risk for overfitting, including the concordance statistic (C-statistic). Similarly, the calibrate function (method = "boot", B = 1000) was used to test the quality of model calibration. Further validation was performed using leave-one-out cross validation by using the validate function (arguments as above except sls = 0.16) to produce reduced models for each combination of n-1 variants. Predictions were expressed as probability of LFS (predict function, type = "fitted"). ROC curves were produced using the ROCit package. The favored model was described visually using a nomogram and its potential utility was evaluated by decision curve analysis (rmda package) [38]. $p < 0.05$ was considered as the threshold for statistical significance unless stated otherwise.

3. Results

3.1. Characteristics of LFS and HBC Germline TP53 Variants

A total of 48 germline *TP53* missense variants were selected from the IARC database, Fortuno et al. [24] and Kharaziha et al. [25] including 24 uniquely observed in LFS and 24 exclusively reported in HBC (Figure S1). The source, number of patients and families as well as the type of LFS-core tumor types obsrved for each of the 48 variants are detailed in Table S1.

The vast majority of variants were mapped to the DBD of the TP53 protein (Figure 1a). Specifically, 23 variants in the LFS-group were located in the DBD and one in the oligomerization domain. In the HBC-group 22 variants were in the DBD, and two in the C-terminal regulatory domain. Figure 1b shows that most of the LFS- and HBC-residues are located in regions predicted to have an ordered protein conformation, however, the prediction values are generally close to the threshold of 0.5 for transition to predicted conformational disorder. This is consistent with previous reports showing low conformational stability of the DNA-binding domain tertiary structure [12,20]. Consistently, the meta-structure prediction method [33] predicts a higher degree of compactness in the DBD (Figure 1c) and correctly predicts a predominance of beta-sheet conformation in the DBD as well as the alpha-helical nature of the oligomerization domain (Figure 1d). Finally, the three most predominant predicted protein interaction domains are in the DBD and the two most C-terminal of these coincide with regions containing clusters of variant LFS- and HBC-residues (Figure 1e).

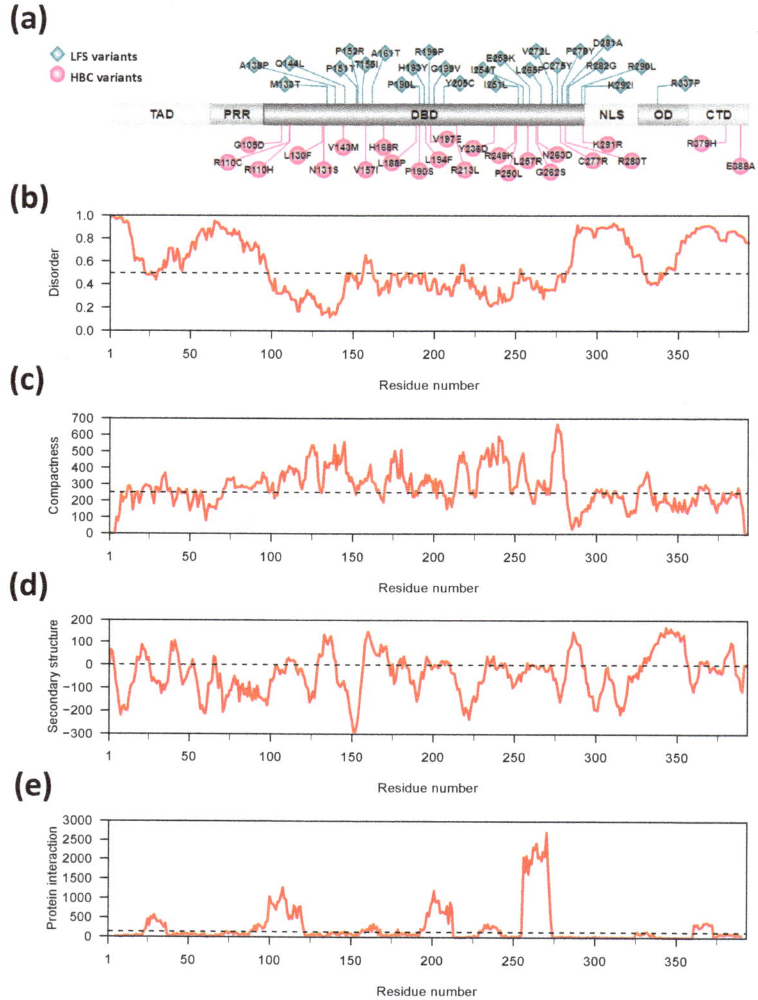

Figure 1. Location of *TP53* missense variants in the TP53 protein sequence and in relation to its predicted disorder. (**a**). Schematic illustration of the TP53 amino acid sequence and protein domains with the location of the 24 LFS variants shown above (cyan green-blue rhombus) and the 24 HBC-variants indicated below (magenta purple-red circles) [39]. The TP53 domains are illustrated for the transactivation domain (TAD), the proline-rich region (PRR), the DNA binding domain (DBD), the nuclear localization signal (NLS), the oligomerization domain (OD) and the C-terminal regulatory domain (CTD). (**b**) Predicted disorder profile of wild type TP53. The IUPred2A predictor was used with the "long" argument. Scores > 0.5 (above dotted line) indicate disordered regions. The approximate location of the DBD and OD are shown (grey shading). (**c**) Predicted compactness of wild type TP53. The dotted line at a value of 250 (*y*-axis) emphasizes the higher compactness values predicted for the DBD. The approximate location of the DBD and OD are shown (grey shading). (**d**) Predicted secondary structure of wild type TP53. Values > 0 (dotted line) are predicted to be alpha-helical and values < 0 are predicted to have beta-strand conformation. The approximate location of the DBD and OD are shown (grey shading). (**e**) Predicted regions with protein interaction propensity in wild type TP53. The dotted line shows a level equivalent to 5% of the maximum value. Apparently artefactual values for the first 4 residues and last 3 residues of TP53 were omitted. The approximate location of the DBD and OD are shown (grey shading).

3.2. Location of LFS- and HBC-Residues in Relation to the TP53 Protein Structure

Variant LFS- and HBC-residues are distributed and juxtaposed throughout the DBD located in the central part of TP53 with no apparent pattern associated with either disease outcome. We considered whether there might be associations with secondary-structure elements (alpha-helix, beta-sheet or disordered regions), tertiary structure aspects (such as Surface or Buried locations) or quaternary structure aspects (DNA interacting residues or inter-monomer protein interacting residues in the context of the TP53 tetramer). For each of the TP53 monomers (Chain A, B, C and D), there was a significant enrichment of variant LFS-residues in the approximately 30% of residues that are Buried (i.e., least Surface exposed) in the crystal structure of the wild-type TP53 tetramer bound to DNA (Table S2). A similar result was also obtained when the Buried (non-Surface) classification of residues was combined for all four monomers (Figure 2a). No other significant associations were found (Table S2).

Figure 2b shows the location of LFS- and HBC-residues in relation to the location of Buried residues defined by different cutoff values used to define the threshold Surface-exposed area per residue. As summarized in Figure 2a, there is a clear tendency for LFS-residues to be Buried, while no such association was found for HBC-residues. The same pattern is seen in Figure 2c, which shows LFS- and HBC-residues in the context of the tertiary structure of the TP53 tetramer bound to DNA.

3.3. Association between TP53-Variant-Induced Changes in Protein Conformation Characteristics and Disease Outcome

The enrichment of LFS-residues in the set of most Buried TP53 residues suggests that the disease associated variants might tend to alter the folded conformation of TP53 in LFS patients. The more even distribution of HBC-residues on the Surface and core of the protein structure would give a greater possibility for disease associated variants to disrupt or modify interactions between TP53 and its DNA or protein ligands in HBC patients. To investigate these aspects further we made multivariate models to predict disease outcome as a function of Buried vs. Surface status for LFS- and HBC-residues, together with a range of variables predicting the effect of the variant residues on protein conformation aspects, such as intrinsic disorder, protein backbone flexibility, propensity for tertiary structure formation, propensity for secondary structure formation and protein interaction propensity (Table 1). Several of these protein conformation aspects appear to be of potential relevance (Figure 1b–d) and values for all variables are listed in Table S3 and Table S3 Appendix.

Since the disease outcome is defined by a binary variable (LFS or HBC) we used a logistic regression approach. The relatively small number of variants for each outcome (n = 24 in each group) imposed limitations on the number of explanatory variables ($n = 2$ to 4) that could reasonably be included in a final model. First a full model (mod_full), likely associated with overfitting problems, was made using 12 explanatory variables describing the effects of the LFS- and HBC-variants on different aspects of protein conformation. An internal bootstrap cross-validation procedure was then used to produce a reduced model by removing less useful variables in a stepwise manner. The reduced model produced during the cross-validation procedure contained 4 variables (mod_4v). Further variable number reduction was done manually by successively removing the variable with the least significant beta coefficient to produce models with three and two explanatory variables, mod_3v and mod_2v, respectively (Figure 3a, Table S4). The performance of the models, as shown by the C-statistic, is lower in the reduced models than the full model as expected, but they are still in the vicinity of the level required for a useful predictive model. The corresponding ROC curves are shown in Figure S2.

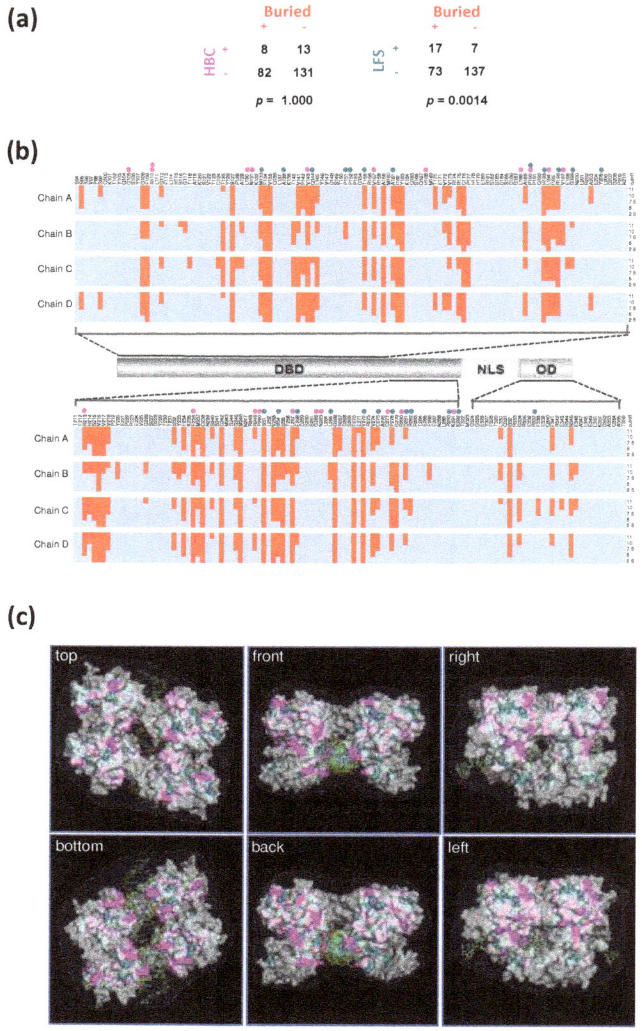

Figure 2. Comparison of surface exposure of *TP53* missense variants in LFS and HBC. (**a**) LFS-residues are significantly enriched in Buried residues with lower surface exposure (<cutoff at 11Å^2) than expected by chance while HBC-residues are not. Contingency tables and respective *p*-values are shown (Fisher's Exact Test, two-sided). Residues were defined as Buried if their surface area was below the cutoff value in any of the 4 TP53 monomers in the TP53 structure (PDB name = 3TS8). The HBC-residues R379 and E388 are not included in this TP53 structure, and the R110 was calculated once, therefore only 21 HBC-residues were included. (**b**) Location of Buried residues (red shading) in a TP53 derivative containing the DBD and OD that was used for tertiary structure determination (3TS8). Cutoff values (Å^2) to distinguish between Surface (>cutoff) and Buried residues (<cutoff) were 11 (used in the statistical test in (**a**), 10, 7.5, 5 and 2.5 (see right side of panel). Results are shown for each of the monomers (Chains A–D) within the TP53 tetramer bound to DNA. LFS-residues and HBC-residues are indicated by cyan and magenta filled circles respectively. (**c**) Localization of LFS-residues (cyan) and HBC-residues (magenta) in relation to the surface of the TP53 tetramer bound to DNA (3TS8, transparent grey). Stronger color indicates Surface exposure while weaker color indicates parts of residues that are below the protein surface (Buried). The 6 panels show all views of the protein caused by stepwise 90° rotations of the "top" structure.

(a)

Model name	Model formula	C-statistic
mod_full	Cancer type ~ Bur + xray_dif + comp_dif + iups_dif + anc_dif + disprot_dif + dyn_dif + iupl_dif + iupstr_dif + nmr_dif + PPI6_dif + Sec_dif	0.92
mod_4v	Cancer type ~ Bur + comp_dif + nmr_dif + PPI6_dif	0.84
mod_3v	Cancer type ~ Bur + comp_dif + PPI6_dif	0.81
mod_2v	Cancer type ~ Bur + comp_dif	0.78

(b)

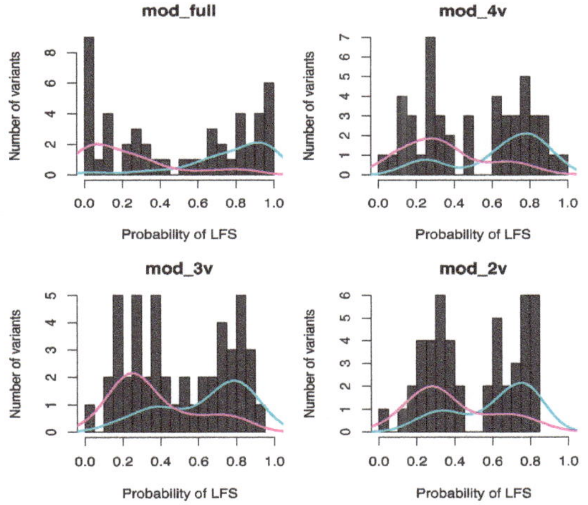

(c)

LFS

	mod_full	mod_4v	mod_3v	mod_2v	loo_cv	
M133T	0.891	0.812	0.742	0.708	0.745	
A138P	0.701	0.338	0.215	0.245	0.179	
Q144L	0.915	0.814	0.851	0.848	0.798	*
P151T	0.695	0.746	0.789	0.762	0.730	*
P152R	0.034	0.159	0.201	0.230	0.002	
T155I	0.663	0.755	0.826	0.804	0.529	
A161T	0.936	0.869	0.780	0.736	0.861	*
P190L	0.844	0.908	0.886	0.371	0.036	
H193Y	0.793	0.753	0.797	0.781	0.462	
R196P	0.826	0.793	0.738	0.762	0.861	
G199V	0.999	0.986	0.949	0.831	0.985	*
Y205C	0.817	0.703	0.559	0.647	0.639	*
I251L	0.637	0.724	0.694	0.648	0.672	
I254T	0.706	0.595	0.628	0.603	0.743	
E258K	0.271	0.453	0.590	0.759	0.103	
L265P	0.597	0.681	0.376	0.616	0.508	*
V272L	0.457	0.667	0.690	0.603	0.581	
C275Y	0.975	0.246	0.357	0.343	0.117	*
P278T	0.970	0.827	0.805	0.762	0.871	
D281A	0.934	0.875	0.862	0.829	0.868	*
R282G	0.942	0.922	0.853	0.808	0.917	*
R290L	0.967	0.180	0.387	0.427	0.087	*
K292I	0.998	0.456	0.517	0.561	0.386	*
R337P	0.954	0.744	0.415	0.327	0.658	*

HBC

	mod_full	mod_4v	mod_3v	mod_2v	loo_cv	
G105D	0.085	0.119	0.148	0.143	0.129	*
R110C	0.376	0.524	0.584	0.812	0.941	
R110H	0.265	0.334	0.315	0.387	0.310	
L130F	0.104	0.323	0.265	0.294	0.348	*
N131S	0.197	0.338	0.293	0.330	0.363	*
V143M	0.320	0.640	0.598	0.566	0.625	
V157I	0.345	0.649	0.708	0.673	0.567	
H168R	0.012	0.082	0.215	0.187	0.090	*
L188P	0.057	0.255	0.140	0.198	0.220	
P190S	0.033	0.177	0.312	0.283	0.195	*
L194F	0.838	0.854	0.806	0.736	0.902	*
V197E	0.007	0.083	0.205	0.325	0.096	*
R213L	0.523	0.692	0.852	0.831	0.905	
Y236D	0.657	0.480	0.358	0.442	0.575	*
R249K	0.816	0.721	0.703	0.667	0.760	*
P250L	0.021	0.337	0.484	0.371	0.395	*
L257R	0.286	0.363	0.442	0.559	0.876	
G262S	0.036	0.128	0.198	0.228	0.136	*
N263D	0.223	0.209	0.192	0.221	0.225	*
C277R	0.006	0.017	0.024	0.032	0.018	*
R280T	0.139	0.334	0.340	0.381	0.233	
K291R	0.129	0.334	0.308	0.324	0.344	
R379H	0.000	0.001	0.001	0.001	0.002	*
E388A	0.002	0.001	0.001	0.001	0.002	*

Figure 3. Protein conformation parameters are associated with disease phenotype and may have predictive value. (**a**) Multivariate logistic regression models for prediction of phenotype class (LFS or HBC) using a range of available protein conformation related explanatory variables describing different protein conformation aspects (full model, mod_full). A reduced model (mod_4v) was produced by stepwise variable exclusion from the full model (rms package). Further reduction was done by progressive manual removal of the least well performing variable to produce models with

3 and 2 explanatory variables, respectively (mod_3v and mod_2v). Indicators of model performance (C-statistic) are shown. (**b**) Bimodal probability distributions for models, showing the overall separation of output variables (LFS and HBC). Histograms show the distribution of the predicted probabilities of residues causing LFS, for the different models. The overall separation of variants as LFS (cyan line) or HBC (magenta line) are shown for each model. (**c**) Probability values for individual LFS (cyan) and HBC (magenta) variants produced by the full model and reduced models (columns 1–4 in each panel) as well as leave-one-out cross validation results (loo_cv) in which each respective variant is left out from a reduced model that is then used to predict the outcome associated with the left-out variable (column 5 in each panel). An asterisk (*) indicates that the loo_cv model contains the same variables as the mod_4v model (model details and results for each of the loo_cv model are tabulated in Table S5). Based on the binomial distribution minima in part b, probability values >0.5 for LFS and <0.5 for HBC are colored darker to give an indication of the relative performance (correct predictions) of the different models as well as how performance is affected in the cross-validation procedure in which predictions are made for each individual variant by models excluding data for the predicted variant. Variant/model combinations with lighter color indicate incorrect predictions.

Figure 3b shows the binomial distribution of prediction probabilities for the models with all showing a minima close to a probability of 0.5. The figure also shows the relative distribution of the known disease outcomes in relation to the probability distributions. Even though the full model shows a higher degree of discrimination between the LFS and HBC outcomes than the reduced models, the full model is likely associated with overfitting issues. The reduced models none-the-less show correct prediction of most variants with a much lower risk of potential for overfitting (Figure S3).

Figure 3c shows correct (dark color) and incorrect (light color) predictions for the different models if a threshold value for prediction of LFS or HBC status is arbitrarily placed at a probability value of 0.5. Predictions from leave-one-out cross validation are also shown, where reduced models were produced for all combinations of n-1 variants and then used to predict the disease outcome for the left-out variant (see also Table S5). All 4 models predicted the correct outcome for 16 HBC-variants and 14 LFS-variants. Only 4 variants were incorrectly predicted by all four models, and for these variants the same result was obtained by leave-one-out cross validation. Consistent with the progressive reduction in the C-statistic (Figure 3a), the models make progressively fewer correct predictions as the number of variables in the models was reduced (41, 36, 36 and 35 correct predictions for mod_full, mod_4v, mod_3v and mod_2v, respectively). The reduction in correct predictions is accompanied by a reduced risk of over-fitting (Figure S3) and therefore the reduced models would be expected to perform better than the full model on an independent data set.

Reduced tendency for overfitting was also associated with improved calibration of the reduced models, such that calibration of the mod_2v and mod_3v models was much better than for mod_full and mod_4v (Figure S4). The beta coefficients, *p*-values and odds rations for the reduced models are shown in Table 2. Only two variables in each of the three reduced variable models reach statistical significance, namely the Buried status of variant residues and the predicted difference in their compactness characteristics. In choosing between the models, mod_2v and mod_3v show better calibration compared to mod_4v and mod_full. Since mod_3v gave a slightly higher C-statistic (0.81) compared to mod_2v (0.78) it was selected as the most favored model.

Table 2. Reduced models for multivariate logistic regression analysis of disease outcome.

Intercept and Variable	β	OR (95% CI)	*p*-Value
Four variables model (mod_4v)			
Intercept	1.282		**0.017** *
Bur			
Surface vs. Buried	−2.465	0.09 (0.02–0.44)	**0.003** *
unknown vs. Buried	−9.897	5.03×10^{-5} (4.31×10^{-27}–5.87×10^{17})	0.703
comp_dif	0.023	3.88 (1.30–11.59)	**0.015** *
PPI6_dif	0.001	1.29 (0.99–1.67)	0.059
nmr_dif	10.461	2.20 (0.79–6.11)	0.132
Three variables model (mod_3v)			
Intercept	1.094		**0.030** *
Bur			
Surface vs. Buried	−2.229	0.11 (0.02–0.50)	**0.005** *
unknown vs. Buried	−9.03	0.00012 (1.18×10^{-26}–1.22×10^{18})	0.727
comp_dif	0.015	2.40 (1.08–5.30)	**0.031** *
PPI6_dif	0.0006	1.15 (0.95–1.39)	0.165
Two variables models (mod_2v)			
Intercept	0.914		**0.048** *
Bur			
Surface vs. Buried	−1.886	0.15 (0.04–0.62)	**0.009** *
unknown vs. Buried	−8.804	0.0002 (1.48×10^{-26}–1.53×10^{18})	0.733
comp_dif	0.014	2.30 (1.04–5.09)	**0.039** *

β = beta coefficient; OR = odds ratio; CI = confidence interval; Bur = categorical variable describing whether residues are burried, surface or of unknown location in the TP53 tertiary structure; comp_dif = a continuous variable showing the effect of each variant on the predicted compactness of TP53 at the location of each variant residue (variant value minus wild type value); PPI6_dif and nmr_dif are continuous variables calculated as for comp_dif but reflecting the effect of variant residues on the predicted protein interaction protensity and the predicted intrinsic disorder of TP53, respectively; * = $p < 0.05$.

3.4. Potential for the Most Favored Model

To estimate the potential value of the most favored model (mod_3v) for use in developing improved tools for clinical decision making we used decision curve analysis. The net benefit of using the mod_3v to predict LFS disease outcome at different risk thresholds is shown in Figure 4a. The decision curve for mod_3v provides a higher net benefit than assuming that all potential patients will develop LFS (grey line) at a risk threshold of about 0.2 and out-performs the assumption that no patients will develop LFS (black line) up to a risk threshold of about 0.8. Thus, under conditions where relative LFS prevalence is not extremely low or high, the model would be expected to provide a net benefit if used in the clinical decision-making process.

Figure 4b shows the mod_3v model in the form of a nomogram that visualizes the prediction model with the respect to the relative importance of the included explanatory variables as well as the way they contribute to a prediction of risk for the alternative disease outcomes. Most important is the classification of Buried or Surface status for the variant residue in the tertiary structure of the TP53 tetramer bound to DNA but the predicted effect of variants on the compactness of the TP53 conformation is also important, with increased compactness of variants increasing the risk for LFS. Changed predicted propensity for protein interaction plays a lesser role with increased interaction propensity of variants increasing the risk for LFS. The values of the Bur variable are already known for all residues in the tertiary TP53 structure used here and it would of course be possible to determine compactness effects (comp_dif) and protein interaction propensity effects (PPI6_dif) values for all possible substitutions of all TP53 residues. Thus, it would be possible to calculate disease outcome risk probabilities for all possible substitutions of residues in the DNA-binding and oligomerization domains if, after due external validation and model development, models similar to those described here were judged to be useful in a clinical setting.

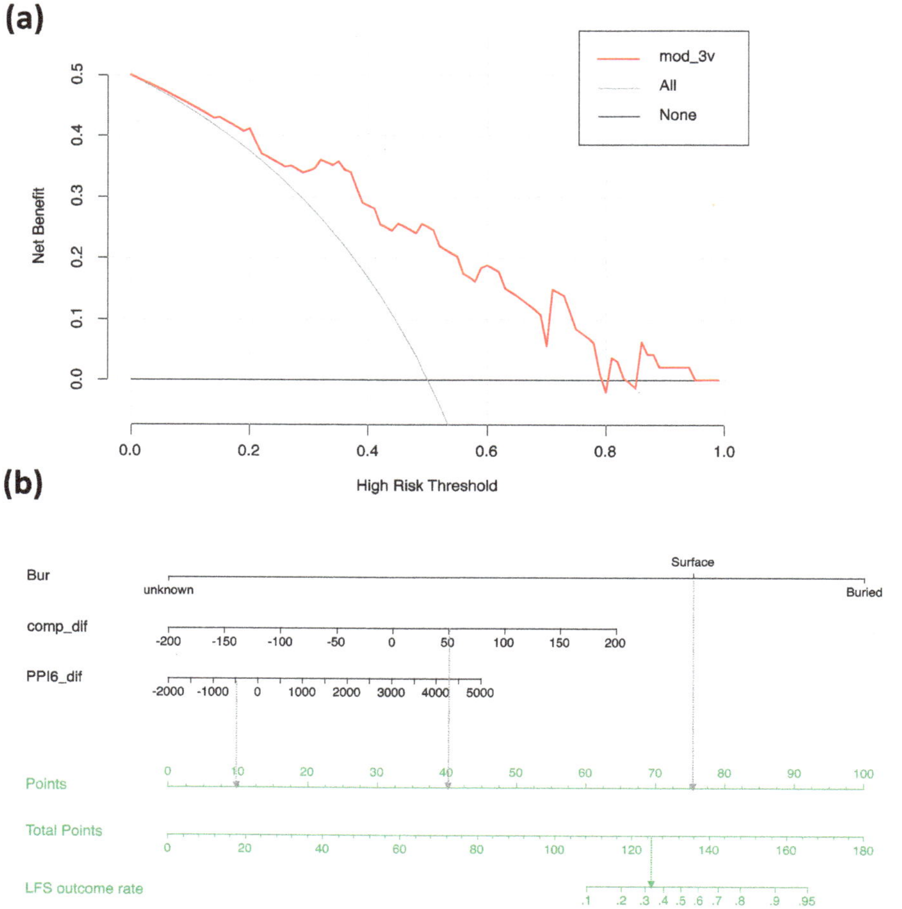

Figure 4. Potential for the most favored model. (**a**) Decision curve analysis of models for prediction of phenotypic outcome (LFS or HBC). The y-axis indicates the net benefit of using mod_3v model (red line). The thin gray line (All) shows net benefit values expected assuming all assessed variants are LFS. The darker gray line (None) shows net benefit values expected assuming no assessed variants are LFS. The net benefit for prediction of LFS variants is regarded as positive for probability values exceeding those for the "All" and "None" values. (**b**) The nomogram that facilitates manual estimation of the risk of LFS disease outcome using the mod_3v model. For the value of each explanatory variable the equivalent value on the "Points" scale is assessed. The sum of all Points values is then located on the "Total Points" scale (middle green row) so that the corresponding probability value can be read from the "LFS outcome rate" scale (upper green row). The dotted arrows show a hypothetical example for a surface residue with a comp_dif value of 50 and a PPI6_dif value of −500, for which the equivalent "Point" values (10, 40 and 75) summate to 125 ("Total Points"), giving a LFS outcome risk of slightly over 0.3.

4. Discussion

Pathogenic germline variants in *TP53* have classically been associated with Li-Fraumeni syndrome (LFS), a tumor predisposition syndrome with high risk of various childhood as well as adult onset tumors. Increased genetic testing has however revealed that germline

TP53 variants are associated with a broader range of phenotypes, from classical LFS to hereditary breast cancer (HBC), and the outcome may be dependent on both variant characteristics and modifier gene variants elsewhere in the genome [40]. Differential expression of TP53 isoforms has also been discussed to have an impact on cancer risk profile [41,42], but this has mainly been studied in sporadic cancers [43]. The wide variation in the phenotypic outcome in families carrying *TP53* variants creates challenges for the genetic counseling and clinical handling of these individuals.

In an attempt to better understand the molecular basis for the differential disease outcomes associated with different variants and to develop a prediction tool, we studied the impact of germline *TP53* missense variants on protein conformation and their association to disease phenotype. We present a quantitative model that predicts disease outcome (LFS or HBC) as a function of localization of variant residues in the tertiary structure of the TP53 DBD and oligomerization domain together with predicted variant-associated effects on conformation of the full-length protein.

Our results demonstrate that LFS-variants were enriched in Buried regions ($p = 0.0014$) of the tertiary structure of one or more TP53 monomers in the DNA-bound tetramer, indicating that the set of variant LFS-residues may hypothetically have a larger impact on the folding and overall conformation of the TP53 protein than the set of variant HBC-residues. While the Buried/Surface variable relates to the predisposition of affected residues to lead to LFS or HBC, the compactness (comp_dif) variable is related to how the substitution of the variant residues is predicted to affect compactness, with enhanced compactness favoring the LFS outcome. The protein interaction propensity variable (PPI6_dif) is also positively correlated with the probability of LFS outcome suggesting the importance of protein interactions for the LFS phenotype. These protein interactions could in principle be interactions between monomers within the TP53 tetramer or interactions between the TP53 tetramer and other proteins. Comparison with the positions of residues forming intra-tetramer interactions shows that the major DBD regions predicted to have protein interaction propensity (Figure 1d) contain interface residues between monomers within the TP53 tetramer, suggesting that the PPI6_dif variable may be a measure of effects of variants on the tetrameric integrity of TP53. We cannot of course exclude a role of these regions in other protein interactions. Thus, our favored quantitative model (mod_3v) incorporated three variables encompassing overall topological effects, protein chain related effects and residue-level effects, and it performed acceptably well with a C-statistic of 0.81 as well as having acceptable calibration characteristics and strongly reduced risk of overfitting compared to more complex models.

Extrapolation of the modelling results suggests that variants that tend to strengthen the tertiary and quaternary structure of the TP53 tetramer would tend to favor the LFS disease outcome. This may be related to the dominant-negative phenotype associated with *TP53* variants that are particularly strongly associated with the LFS phenotype (see Introduction). It could be speculated that a variant which stabilized structural aspects of TP53 monomers and their propensity for tetrameric interactions in relation to wild-type would facilitate the formation of hetero-tetrameric TP53 tetramers in heterozygous individuals, thereby resulting in the dominant-negative phenotype that is observed for many missense variants that are associated with LFS [15].

The HBC-residues are not significantly associated with Buried or Surface status in the structure of wild-type TP53. For the compactness and protein interaction propensity variables, the risk for HBC shows the opposite trend to LFS, since the HBC risk is increased by a decrease in predicted compactness and protein interaction propensity in the mutant proteins. A reasonable speculation would be that the TP53 proteins encoded by HBC-variants are still functional but that the variants cause subtle qualitative or quantitative functional changes that alter the transcriptional output in a way that predisposes carriers to breast cancer but not to other LFS phenotypes. Other explanations are also possible. For example, we cannot exclude that the HBC variants are linked to modifier loci that cause the HBC phenotype and that the HBC outcome is not linked to effects of the *TP53* variants at all. Since the number of patients and families displaying some HBC variants is limited, it is also possible that some variants may subsequently be coupled to LFS. The V157I variant, for example, that is reported in 7 individuals in 2 families was classified as HBC by both IARC and Fortuno et al. (Table S1) although one case with sarcoma was reported in addition to all breast cancer cases in these families. We have not been able to obtain more pedigree information to verify if the family fulfills the LFS-criteria and thereby misclassified. Notably, this variant was by our prediction model predicted as 0.71 likely hood to belong to the LFS-group. While evaluating this model one must be aware of that there is a greater risk that variants within the HBC cohort are misclassified than within the LFS cohort.

None of the included variants affect the main residues involved in interactions with the DNA backbone but several coincide with base-interacting residues as well as with residues important for stabilizing the TP53 tertiary structure. For example, the C277R and R280T variants that are clearly HBC-associated affect residues that make specific interactions with bases in the TP53 binding site and would be likely to affect qualitative or quantitative aspects of DNA binding. Similarly, R249K, V143M and V157I are variants characterized by conservative amino acid substitutions, which are also associated with HBC and affect residues important for the stability of the TP53 tertiary structure. Conceivably, these variants could cause qualitative or quantitative changes to the function of TP53 without having a major negative impact on function. The R282G variant is associated with LFS and affects a residue important for stabilizing the DNA-binding surface in relation to the rest of the TP53 tertiary structure. It could be speculated that this variant disrupts DNA binding activity, which would be likely to cause a dominant-negative phenotype if TP53 hetero-tetramers formed in heterozygous patients.

A limitation of our prediction model is the small cohort of only 24 unique missense variants in each group. The challenge has been to identify cohorts of families with exclusively HBC, especially in the case of *de novo* alterations in breast cancer patients that thus lack information of family history. However, we made an effort to select as clean groups of LFS and HBC variants as possible by following a strict selection procedure (Figure S1) but with the consequence of a limited cohort size. Therefore, there is a need to further evaluate the model in an independent cohort, and if possible with more reliable pedigree information concerning tumor panorama and age of onset, before it is used as a tool for clinical counseling and clinical management, perhaps in combination with other modelling approaches [44].

Amadou et al. [17] tried to stratify clinical management according to dominant negative variants and loss of function variants. As families with loss of function variants tend to develop tumors later, they suggested it may be considered to test and screen adults instead of children in those families for the consideration of psychological and financial burdens. Nichols et al. [45] discussed that there were however many cases having the same tumor onset age in families with dominant negative variants as in those with loss of function variants. Therefore, this distinction of *TP53* variants can apparently not be used as a sole guidance for further clinical handling. Instead, we made an attempt to provide a tool for improving genetic counselling and clinical management of these patients and families by creating a prediction nomogram based on the protein conformational impact of the germline missense *TP53* variants. The prediction nomogram (Figure 4b) may support psychological issues in genetic counselling especially in families were the model predicts HBC rather than LFS. However, this model cannot yet be used to stratify for example surveillance programs, as it requires validation in an independent cohort.

5. Conclusions

This study explored the relationship between germline *TP53* missense variants and their phenotypic impact, with regard to LFS and HBC, based on a quantitative model combining conformational characteristics of the TP53 protein. Logistic regression models show a clear relationship between disease outcome (LFS or HBC) for *TP53* variants with their effects on aspects of protein conformation and function. The models also appear to have a predictive capacity that may be of practical future use in genetic counselling and management of missense variant carriers. However, there is a need to evaluate the prediction model in an independent cohort prior to any implementation in clinical practice.

Supplementary Materials: The Supplementary Materials are available online at https://www.mdpi.com/article/10.3390/ijms22126345/s1.

Author Contributions: Conceptualization, S.B.-L., A.P.H.W. and Y.L.; methodology, A.P.H.W. and Y.L.; software, A.P.H.W., Y.L., and T.v.L.; validation, Y.L. and A.P.H.W.; formal analysis, O.A. and T.v.L.; investigation, Y.L.; resources, R.K., P.K. and S.B.-L.; data curation, A.P.H.W. and Y.L.; writing—original draft preparation, Y.L. and A.P.H.W.; writing—review and editing, S.B.-L., C.L. and R.K.; visualization, Y.L. and A.P.H.W.; supervision, A.P.H.W. and S.B.-L.; project administration, Y.L. All authors have read and agreed to the published version of the manuscript.

Funding: The Swedish Cancer Society (Cancerfonden), the Cancer Research Funds of Radiumhemmet, Stiftelsen Längmanska Kulturfonden, Karolinska Institutet, Cancer Research KI, and Stockholm County Council.

Institutional Review Board Statement: All *TP53* variants that were defined as LFS in our cohort were identified through the IARC database.

Informed Consent Statement: Not applicable.

Data Availability Statement: Data available in article Supplementary Files.

Acknowledgments: We are grateful to Mattias Rantalainen from the Department of Medical Epidemiology and Biostatistics, Karolinska Institutet for helpful advice.

Conflicts of Interest: The authors declare no conflict of interest.

References

1. Li, F.P. Soft-Tissue Sarcomas, Breast Cancer, and Other Neoplasms. A Familial Syndrome? *Ann. Intern. Med.* **1969**, *71*, 747–752. [CrossRef]
2. Malkin, D.; Jolly, K.W.; Barbier, N.; Look, A.T.; Friend, S.H.; Gebhardt, M.C.; Andersen, T.I.; Børresen, A.-L.; Li, F.P.; Garber, J.; et al. Germline Mutations of the p53 Tumor-Suppressor Gene in Children and Young Adults with Second Malignant Neoplasms. *N. Engl. J. Med.* **1992**, *326*, 1309–1315. [CrossRef]
3. Birch, J.M.; Hartley, A.L.; Tricker, K.J.; Prosser, J.; Condie, A.; Kelsey, A.M.; Harris, M.; Jones, P.H.; Binchy, A.; Crowther, D. Prevalence and diversity of constitutional mutations in the p53 gene among 21 Li-Fraumeni families. *Cancer Res.* **1994**, *54*, 1298–1304.
4. Kleihues, P.; Schäuble, B.; Hausen, A.Z.; Estève, J.; Ohgaki, H. Tumors associated with p53 germline mutations: A synopsis of 91 families. *Am. J. Pathol.* **1997**, *150*, 1.
5. Chompret, A.; Abel, A.; Stoppa-Lyonnet, D.; Brugières, L.; Pagès, S.; Feunteun, J.; Bonaïti-Pellié, C. Sensitivity and predictive value of criteria for p53germline mutation screening. *J. Med. Genet.* **2001**, *38*, 43–47. [CrossRef]
6. Martin, A.-M.; Kanetsky, A.P.; Amirimani, B.; Colligon, A.T.; Athanasiadis, G.; Shih, A.H.; Gerrero, M.R.; Calzone, K.; Rebbeck, T.R.; Weber, B.L. Germline TP53 mutations in breast cancer families with multiple primary cancers: Is TP53 a modifier of BRCA1? *J. Med. Genet.* **2003**, *40*, e34. [CrossRef]
7. Gonzalez, K.D.; Noltner, K.A.; Buzin, C.H.; Gu, D.; Wen-Fong, C.Y.; Nguyen, V.Q.; Han, J.H.; Lowstuter, K.; Longmate, J.; Sommer, S.S.; et al. Beyond Li Fraumeni Syndrome: Clinical Characteristics of Families With p53 Germline Mutations. *J. Clin. Oncol.* **2009**, *27*, 1250–1256. [CrossRef] [PubMed]
8. Frebourg, T.; Genturis, T.E.R.N.; Lagercrantz, S.B.; Oliveira, C.; Magenheim, R.; Evans, D.G. Guidelines for the Li–Fraumeni and heritable TP53-related cancer syndromes. *Eur. J. Hum. Genet.* **2020**, *28*, 1379–1386. [CrossRef] [PubMed]
9. Villani, A.; Shore, A.; Wasserman, J.; Stephens, D.; Kim, R.H.; Druker, H.; Gallinger, B.; Naumer, A.; Kohlmann, W.; Novokmet, A.; et al. Biochemical and imaging surveillance in germline TP53 mutation carriers with Li-Fraumeni syndrome: 11 year follow-up of a prospective observational study. *Lancet Oncol.* **2016**, *17*, 1295–1305. [CrossRef]
10. Levine, A.J. p53, the Cellular Gatekeeper for Growth and Division. *Cell* **1997**, *88*, 323–331. [CrossRef]
11. Petros, A.M.; Gunasekera, A.; Xu, N.; Olejniczak, E.T.; Fesik, S.W. Defining the p53 DNA-binding domain/Bcl-xL-binding interface using NMR. *FEBS Lett.* **2004**, *559*, 171–174. [CrossRef]
12. Joerger, A.C.; Fersht, A.R. Structural Biology of the Tumor Suppressor p53. *Annu. Rev. Biochem.* **2008**, *77*, 557–582. [CrossRef]
13. Uversky, V.N. p53 Proteoforms and Intrinsic Disorder: An Illustration of the Protein Structure–Function Continuum Concept. *Int. J. Mol. Sci.* **2016**, *17*, 1874. [CrossRef]

14. Wasserman, J.D.; Novokmet, A.; Eichler-Jonsson, C.; Ribeiro, R.C.; Rodriguez-Galindo, C.; Zambetti, G.P.; Malkin, D. Prevalence and Functional Consequence of TP53 Mutations in Pediatric Adrenocortical Carcinoma: A Children's Oncology Group Study. *J. Clin. Oncol.* **2015**, *33*, 602–609. [CrossRef]
15. Freed-Pastor, W.; Prives, C. Mutant p53: One name, many proteins. *Genes Dev.* **2012**, *26*, 1268–1286. [CrossRef]
16. Bougeard, G.; Renaux-Petel, M.; Flaman, J.-M.; Charbonnier, C.; Fermey, P.; Belotti, M.; Gauthier-Villars, M.; Stoppa-Lyonnet, D.; Consolino, E.; Brugières, L.; et al. Revisiting Li-Fraumeni Syndrome from TP53 Mutation Carriers. *J. Clin. Oncol.* **2015**, *33*, 2345–2352. [CrossRef] [PubMed]
17. Amadou, A.; Achatz, M.I.; Hainaut, P. Revisiting tumor patterns and penetrance in germline TP53 mutation carriers: Temporal phases of Li–Fraumeni syndrome. *Curr. Opin. Oncol.* **2018**, *30*, 23–29. [CrossRef]
18. Cho, Y.; Gorina, S.; Jeffrey, P.; Pavletich, N. Crystal structure of a p53 tumor suppressor-DNA complex: Understanding tumorigenic mutations. *Science* **1994**, *265*, 346–355. [CrossRef] [PubMed]
19. Bullock, A.N.; Henckel, J.; Fersht, A.R. Quantitative analysis of residual folding and DNA binding in mutant p53 core domain: Definition of mutant states for rescue in cancer therapy. *Oncogene* **2000**, *19*, 1245–1256. [CrossRef]
20. Jurneczko, E.; Cruickshank, F.L.; Porrini, M.; Clarke, D.J.; Campuzano, I.D.G.; Morris, M.; Nikolova, P.V.; Barran, P.E. Probing the Conformational Diversity of Cancer-Associated Mutations in p53 with Ion-Mobility Mass Spectrometry. *Angew. Chem. Int. Ed.* **2013**, *52*, 4370–4374. [CrossRef] [PubMed]
21. Bykov, V.J.N.; Eriksson, S.E.; Bianchi, J.; Wiman, K. Targeting mutant p53 for efficient cancer therapy. *Nat. Rev. Cancer* **2018**, *18*, 89–102. [CrossRef]
22. Olivier, M.; Goldgar, E.D.; Sodha, N.; Ohgaki, H.; Kleihues, P.; Hainaut, P.; Eeles, A.R. Li-Fraumeni and related syndromes: Correlation between tumor type, family structure, and TP53 genotype. *Cancer Res.* **2003**, *63*, 6643–6650.
23. Li, F.P.; Fraumeni, J.F.; Mulvihill, J.J.; Blattner, A.W.; Dreyfus, M.G.; Tucker, A.M.; Miller, R.W. A cancer family syndrome in twenty-four kindreds. *Cancer Res.* **1988**, *48*, 5358–5362.
24. Fortuno, C.; James, P.A.; Spurdle, A.B. Current review ofTP53pathogenic germline variants in breast cancer patients outside Li-Fraumeni syndrome. *Hum. Mutat.* **2018**, *39*, 1764–1773. [CrossRef] [PubMed]
25. Kharaziha, P.; Ceder, S.; Axell, O.; Krall, M.; Fotouhi, O.; Böhm, S.; Lain, S.; Borg, Å.; Larsson, C.; Wiman, K.G.; et al. Functional characterization of novel germline TP53 variants in Swedish families. *Clin. Genet.* **2019**, *96*, 216–225. [CrossRef] [PubMed]
26. Bouaoun, L.; Sonkin, D.; Ardin, M.; Hollstein, M.; Byrnes, G.; Zavadil, J.; Olivier, M. TP53Variations in Human Cancers: New Lessons from the IARC TP53 Database and Genomics Data. *Hum. Mutat.* **2016**, *37*, 865–876. [CrossRef] [PubMed]
27. Emamzadah, S.; Tropia, L.; Halazonetis, T.D. Crystal Structure of a Multidomain Human p53 Tetramer Bound to the Natural CDKN1A (p21) p53-Response Element. *Mol. Cancer Res.* **2011**, *9*, 1493–1499. [CrossRef] [PubMed]
28. Petty, T.J.; Emamzadah, S.; Costantino, L.; Petkova, I.D.; Stavridi, E.S.; Saven, J.G.; Vauthey, E.; Halazonetis, T.D. An induced fit mechanism regulates p53 DNA binding kinetics to confer sequence specificity. *EMBO J.* **2011**, *30*, 2167–2176. [CrossRef]
29. Script Library. PyMOLWiki. Available online: https://pymolwiki.org/index.php/Category:Script_Library (accessed on 29 April 2019).
30. Vertrees, J. FindSurfaceResidues—PyMOLWiki. Available online: https://pymolwiki.org/index.php/FindSurfaceResidues (accessed on 26 April 2019).
31. Vertrees, J. InterfaceResidues—PyMOLWiki. Available online: https://pymolwiki.org/index.php/InterfaceResidues (accessed on 30 April 2019).
32. Torrance, G. Ss–PyMOLWiki. Available online: https://pymolwiki.org/index.php/Ss (accessed on 30 April 2019).
33. Konrat, R. The protein meta-structure: A novel concept for chemical and molecular biology. *Cell. Mol. Life Sci.* **2009**, *66*, 3625–3639. [CrossRef]
34. Mayer, C.; Slater, L.; Erat, M.C.; Konrat, R.; Vakonakis, I. Structural Analysis of the Plasmodium falciparum Erythrocyte Membrane Protein 1 (PfEMP1) Intracellular Domain Reveals a Conserved Interaction Epitope. *J. Biol. Chem.* **2012**, *287*, 7182–7189. [CrossRef]
35. Walsh, I.; Martin, A.; Di Domenico, T.; Tosatto, S.C.E. ESpritz: Accurate and fast prediction of protein disorder. *Bioinformatics* **2011**, *28*, 503–509. [CrossRef]
36. Meszaros, B.; Erdos, G.; Dosztanyi, Z. IUPred2A: Context-Dependent Prediction of Protein Disorder as a Function of Redox State and Protein Binding. *Nucleic Acids Res.* **2018**, *46*, W329–W337. Available online: https://academic.oup.com/nar/article/46/W1/W329/5026265 (accessed on 24 September 2020). [CrossRef] [PubMed]
37. Cilia, E.; Pancsa, R.; Tompa, P.; Lenaerts, T.; Vranken, W.F. The DynaMine webserver: Predicting protein dynamics from sequence. *Nucleic Acids Res.* **2014**, *42*, W264–W270. [CrossRef]
38. Vickers, A.J.; Cronin, A.M.; Elkin, E.B.; Gonen, M. Extensions to decision curve analysis, a novel method for evaluating diagnostic tests, prediction models and molecular markers. *BMC Med. Inform. Decis. Mak.* **2008**, *8*, 53. [CrossRef] [PubMed]
39. Liu, W.; Xie, Y.; Ma, J.; Luo, X.; Nie, P.; Zuo, Z.; Lahrmann, U.; Zhao, Q.; Zheng, Y.; Zhao, Y.; et al. IBS: An illustrator for the presentation and visualization of biological sequences: Fig. 1. *Bioinformatics* **2015**, *31*, 3359–3361. [CrossRef] [PubMed]
40. Gargallo, P.; Yáñez, Y.; Segura, V.; Juan, A.; Torres, B.; Balaguer, J.; Oltra, S.; Castel, V.; Cañete, A. Li–Fraumeni syndrome heterogeneity. *Clin. Transl. Oncol.* **2020**, *22*, 978–988. [CrossRef]
41. Khoury, M.P.; Bourdon, J.-C. p53 Isoforms: An Intracellular Microprocessor? *Genes Cancer* **2011**, *2*, 453–465. [CrossRef] [PubMed]
42. Schubert, S.; De Miranda, N.; Ruano, D.; Barge-Schaapveld, D.; Hes, F.; Tops, C.; Joruiz, S.; Diot, A.; Bourdon, J.; Van Wezel, T. PO-059 Cancer-predisposing variants in alternatively spliced TP53 exons. *ESMO Open* **2018**, *3*, A249–A250. [CrossRef]

43. Eiholzer, R.A.; Mehta, S.; Kazantseva, M.; Drummond, C.J.; McKinney, C.; Young, K.; Slater, D.; Morten, B.C.; Avery-Kiejda, K.A.; Lasham, A.; et al. Intronic *TP53* Polymorphisms Are Associated with Increased *Δ133TP53* Transcript, Immune Infiltration and Cancer Risk. *Cancers* **2020**, *12*, 2472. [CrossRef]
44. Fortuno, C.; Cipponi, A.; Ballinger, M.L.; Tavtigian, S.V.; Olivier, M.; Ruparel, V.; Haupt, Y.; Haupt, S.; International Sarcoma Kindred Study; Tucker, K.; et al. A quantitative model to predict pathogenicity of missense variants in the TP53 gene. *Hum. Mutat.* **2019**, *40*, 788–800. [CrossRef] [PubMed]
45. Nichols, K.E.; Malkin, D. Genotype Versus Phenotype: The Yin and Yang of Germline TP53 Mutations in Li-Fraumeni Syndrome. *J. Clin. Oncol.* **2015**, *33*, 2331–2333. [CrossRef] [PubMed]

International Journal of *Molecular Sciences*

MDPI

Review

Is Protein Folding a Thermodynamically Unfavorable, Active, Energy-Dependent Process?

Irina Sorokina [1,*], Arcady R. Mushegian [2,3] and Eugene V. Koonin [4,*]

1 Strenic LLC, McLean, VA 22102, USA
2 Division of Molecular and Cellular Biosciences, National Science Foundation, Alexandria, VA 22314, USA; mushegian2@gmail.com
3 Clare Hall College, University of Cambridge, Cambridge CB3 9AL, UK
4 National Center for Biotechnology Information, National Library of Medicine, National Institutes of Health, Bethesda, MD 20894, USA
* Correspondence: irinsoro@gmail.com (I.S.); koonin@ncbi.nlm.nih.gov (E.V.K.)

Abstract: The prevailing current view of protein folding is the thermodynamic hypothesis, under which the native folded conformation of a protein corresponds to the global minimum of Gibbs free energy *G*. We question this concept and show that the empirical evidence behind the thermodynamic hypothesis of folding is far from strong. Furthermore, physical theory-based approaches to the prediction of protein folds and their folding pathways so far have invariably failed except for some very small proteins, despite decades of intensive theory development and the enormous increase of computer power. The recent spectacular successes in protein structure prediction owe to evolutionary modeling of amino acid sequence substitutions enhanced by deep learning methods, but even these breakthroughs provide no information on the protein folding mechanisms and pathways. We discuss an alternative view of protein folding, under which the native state of most proteins does not occupy the global free energy minimum, but rather, a local minimum on a fluctuating free energy landscape. We further argue that ΔG of folding is likely to be positive for the majority of proteins, which therefore fold into their native conformations only through interactions with the energy-dependent molecular machinery of living cells, in particular, the translation system and chaperones. Accordingly, protein folding should be modeled as it occurs in vivo, that is, as a non-equilibrium, active, energy-dependent process.

Keywords: protein folding; entropy; free energy; free energy landscape; energy-dependent protein folding; co-translational protein folding; molecular chaperones; physical model of protein folding

Citation: Sorokina, I.; Mushegian, A.R.; Koonin, E.V. Is Protein Folding a Thermodynamically Unfavorable, Active, Energy-Dependent Process? *Int. J. Mol. Sci.* **2022**, *23*, 521. https://doi.org/10.3390/ijms23010521

Academic Editors: Istvan Simon and Csaba Magyar

Received: 6 December 2021
Accepted: 31 December 2021
Published: 4 January 2022

Publisher's Note: MDPI stays neutral with regard to jurisdictional claims in published maps and institutional affiliations.

1. Introduction

For the last six decades, the general understanding in the protein folding field has been that proteins fold into their native conformations driven by decrease in Gibbs free energy (negative ΔG). This thermodynamic hypothesis of protein folding stems from the iconic experiments of Anfinsen on in vitro folding of RNase A. Based on the successful refolding of this enzyme into the active, native conformation, Haber and Anfinsen concluded in a seminal 1962 paper that "*the unique secondary and tertiary structure of RNase is, thermodynamically, the most stable configuration*" [1]. Codified in Anfinsen's 1973 Nobel lecture-based review [2], the thermodynamic hypothesis has become the default physical description of protein folding.

The thermodynamic hypothesis of folding, and in particular, the idea that the native state is the most stable one, that is, the global *G* minimum, is indeed highly attractive and appears natural. Furthermore, this view drastically simplifies theory development and modeling by effectively avoiding the need to explain how and why a protein reaches the unique native conformation: indeed, the global minimum is unique by definition. Assuming that the native conformation occupies a local rather than the global minimum of

G immediately complicates the problem because this demands an explanation of how this particular minimum is selected among the many other local minima.

In the last two decades, the protein folding problem has been addressed primarily in terms of a free energy landscape that is usually represented as containing a funnel, the bottom of which corresponds to the global G minimum, that is, the native conformation; many different shapes of this hypothetical funnel have been considered [3–11].

Volcano-shaped landscapes have been also proposed, where during folding, the protein initially has to overcome a barrier of positive ΔG due to entropy decrease, but the native structure still occupies the global G minimum [12,13] (Figure 1a). However, there is effectively no information on the actual structure of the landscape, and the possibility that the native conformation represents a local minimum in a rugged landscape that is generated and continuously affected by dynamic interactions within the cell environment, rather than the global minimum (Figure 1b), has not been systematically addressed. The distinction between the two classes of models can be formulated, in general terms, as thermodynamic vs. kinetic control of protein folding. Indeed, the early work of Wetlaufer and others (reviewed in [14]) emphasized that the native conformation would be the one with the minimum G among the kinetically accessible structures. However, this approach to the study of protein folding has not received much attention or further development, arguably, because it dramatically complicates modeling compared to the straightforward thermodynamic approach.

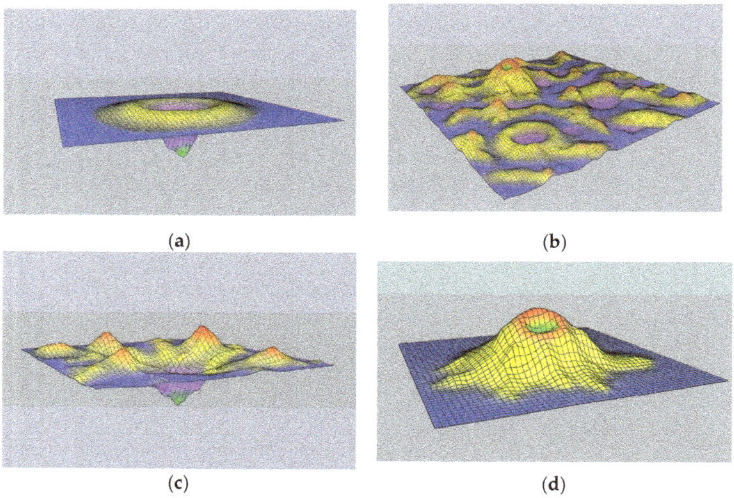

Figure 1. Protein folding energy landscapes in vitro and in vivo. Blue areas are occupied by "perfectly unfolded" conformations with no stable interactions between non-contiguous residues. Yellow and purple areas are populated by more compact protein conformations. Red zones are thought to arise mostly as the result of interactions between the protein and cellular components in a crowded environment. Green zones correspond to proteins in native conformation. (**a**) Canonical funnel-shaped energy landscape that most likely applies only to folding of small, thermodynamically stable proteins as it occurs spontaneously, in vitro, in isolation from all cellular compounds. (**b**) Folding energy landscape for a protein that folds in vivo is poorly understood, but most likely, is complex, rugged, dynamic, and shaped by interactions of the folding polypeptide with multiple cellular components. (**c**) Folding energy landscape of the same small protein as in (**a**) is most likely substantially different and far more complex when folding occurs in a crowded cellular environment. (**d**) Native conformations of most proteins are likely to occupy local thermodynamic minima with higher Gibbs free energy than their unfolded conformations (positive ΔG of folding). Such native conformation can only arise as a result of active, energy dependent folding process.

Although the funnel landscape concept dominates the protein folding field, it is not without its critics. As argued in detail by Ben-Naim, the funnel folding landscape is effectively a metaphor that lacks substantial support [15]. Furthermore, as pointed out by Shakhnovich [16], in simulations, the shape of the landscape in low-dimensional spaces is sensitive to the procedure used for dimensionality reduction, and the procedures that yield the funnel landscapes tend to be physically unrealistic.

Despite the attractive simplicity of the notion that the native conformation of a protein occupies the global minimum of *G*, it remains a hypothesis. Several lines of evidence can be and often are construed as supporting this thermodynamic hypothesis, including direct measurements of the ΔG of folding for multiple proteins, refolding of numerous proteins after denaturation, and spontaneous folding of proteins that were produced by complete chemical synthesis. Crucially, however, all this data pertains to a small number of small, highly stable proteins that have been studied in vitro, in isolation. Even apart from problems with the quantity and quality of this data, the question remains how generalizable these results are and how relevant are they for protein folding under native conditions, that is, in the crowded cell environment (compare panels (a) and (c) in Figure 1).

In this article, we critically assess the empirical data behind the thermodynamic hypothesis of protein folding and discuss an alternative, non-equilibrium folding hypothesis.

2. Review of Protein Folding

2.1. Experimental Data on Free Energy of Protein Folding

The ΔG of protein folding can be determined from denaturation-renaturation experiments under the basic assumptions that proteins are completely denatured in the well-controlled experimental conditions and that such denaturation is fully reversible. A careful examination of the methodology of these experiments, however, reveals a complicated picture, with each class of methods employed for assessing the degree of denaturation rife with its own assumptions and biases (see, for example, [17] on chemical denaturation methods assayed by spectrophotometry, [18,19] for updates on urea- and guanidinium chloride-mediated denaturation methods, and [20,21] for thermal denaturation methods and microscanning calorimetry assays). A recent discussion of the biases, sensitivity issues, and other concerns in the analysis of denaturation-renaturation data can be found in [22]. All told, the results of such experiments. that have been reported for only a handful of proteins, have led to the general consensus that ΔG of folding is a small negative value, that is, proteins (at least, single domain ones) fold spontaneously, but are only marginally stable reviewed in [23–25].

There are several reasons why, in our view, the experimental evidence in support of the thermodynamic hypothesis of folding is far less compelling than it is usually perceived to be. In particular, only in very few folding experiments, the completeness of protein unfolding at the start of the experiment has been convincingly demonstrated. Although it is often claimed that proteins in such experiments were completely denatured, a closer examination shows that typically this is an assumption rather than an experimentally validated observation. In early work (1950s–1970s), the extent of denaturation was typically assessed using indirect methods, such as circular dichroism (CD), which yields a general measure of the proportion of secondary structure in a protein, or fluorescence, which assesses the exposure of individual aromatic residues to the solute, or other, similarly indirect, approaches. However, reanalysis of a subset of cases with more advanced, direct methods has shown that proteins that have been initially characterized as completely denatured often turn out to be only partially unfolded [26]. For example, an NMR analysis of staphylococcal nuclease, the second enzyme extensively studied by Anfinsen after the seminal experiments with RNase A, has demonstrated persistence of native-like structure in the protein that was denatured in 8 M urea [27]. Subsequent NMR analysis of multiple, diverse proteins has similarly revealed preservation of extensive structure in 10 M urea [28]. Strikingly, for the paradigmatic case for the thermodynamic hypothesis, RNase A itself, advanced methods, such as small-angle X-ray scattering (SAXS) and time-resolved fluorescence energy transfer, have demonstrated that compact regions survive many thermal and chemical denaturation regimes [29]. Furthermore, it has been shown that both the degree and the character of unfolding of the same protein can substantially differ depending on the denaturation protocol (e.g., [30,31]). A computational study of protein conformers, in which backbone torsion angles were randomly varied for only 8% of the residues, while the remaining 92% of the residues remained fixed in their native conformations, has shown that the vast majority of these ensembles had end-to-end distances and mean radii of gyration that were within the range of the random-coil expectations. Therefore, it has been concluded that observation of random-coil statistics for denatured proteins cannot be taken as evidence of the absence of residual structure [32].

Measurements of ΔG of folding have been collected in protein thermodynamics databases, of which the most comprehensive one is ProTherm/ProThermDB. The latest release of this database [33] contains more than 30,000 entries. It would be important to determine how many records in the database are informative for estimating ranges of folding ΔG. We analyzed the slightly smaller, 2017 release of ProTherm (available at [34]) that contained 26,045 records representing ~700 unique wild-type proteins, of which less than 500 were annotated as reversibly denatured. Strikingly, only for 18 of these proteins, denaturation was monitored using a rigorous method, such as NMR, and for three of those the actual values of ΔG for the entire molecule were not reported (Supplementary Files S1 and S2). Thus, in actuality, the "vast body of curated literature" does not amount to much. Nearly all experimental data that are cited in support of the hypothesis of spontaneous refolding was obtained on a limited set of compact, stable, globular proteins. Most of these are small, single-domain monomers that are marginally stable (ΔG of folding between -3.5 and 7 kcal/mol) and have been shown to fold rapidly- typically, on the millisecond time scale [35]. Furthermore, the most thoroughly studied set of spontaneous refolders is enriched in extracellular proteins, often containing disulfide bonds, which are likely to dominate the fold stabilization mechanisms (see more on this below where we discuss total protein synthesis).

With all these caveats in mind, the reported ΔG values are within the range of -1 to -20 kcal/mol, with a Poisson-like distribution peaking around -5 kcal/mol [36]. The common range that is pervasively quoted in the literature is -5–15 kcal/mol, which is typically interpreted qualitatively as "proteins are marginally stable", or in other words, the folding funnel is thought to be extremely shallow (e.g., [37–40]).

New opportunities to study the thermodynamics of protein folding/unfolding could be provided by single-molecule methods; for an overview of these methods as applied to protein folding, see [41,42], and for ΔG measurement using these methods, see [43–45]. However, these studies face the same major problem as bulk denaturation experiments discussed above: most proteins do not unfold completely in single-molecule manipulations, such as atomic force microscopy or optical tweezers. Almost always, there is an uncertainty about the amount of residual structure, as indicated by the fact that the stretched form of a protein is often measured to be shorter, or occasionally, paradoxically longer than the theoretically predicted length (e.g., [46–48]). There also indications that the theoretical length of a polypeptide chain can be sequence- and structure-dependent [48]. Taken together, these data suggest that single-molecule methods are not yet sufficiently reliable for a confident determination of the state of protein unfolding.

Overall, the survey of the experimental study of protein folding/unfolding shows that ΔG has been measured only for a highly biased set of a few small, compact, single domain proteins, and even for most of these, the obtained values cannot be considered reliable due to the lack of evidence of complete unfolding or, worse, presence of evidence of persisting, residual secondary structure. Even for those few proteins, for which reliable experimental data have been obtained, the negative ΔG values were low, in many cases, not far above the level of thermal fluctuations.

2.2. Chemically Synthesized Proteins Folding into Native Conformations

Another major argument in support of the thermodynamic hypothesis of folding is thought to come from experiments on proteins obtained by complete chemical synthesis. By and large, ΔG of folding for these proteins has not been measured directly, but is strongly believed to be negative because these proteins were produced by ligation of individual amino acids or peptides in a chemical reaction, in the complete absence of ribosomes, chaperones or other cell components, and then folded into native conformations in solution. Such spontaneous folding from the denatured form is commonly seen as direct, highly convincing evidence in support of the thermodynamic hypothesis (notably, the Nobel Prize has been awarded to Anfinsen only after the appearance of papers from Hirschmann and Merrifield groups on the complete synthesis of RnaseA [49,50]).

To assess the evidence obtained from this type of experiments, we performed a literature search on the proteins that, in the last 50 years or so, were produced by complete chemical synthesis and refolded to the biologically active form. In the set of about 60 unique proteins (not counting mutants and variants) studied in these experiments, only two were longer than 200 aa; the mean protein length in this group was 94 amino acids, which is at least 3–5-fold less than the proteome-wide mean lengths in Archaea, Bacteria, and Eukarya (Table 1). The proportions of secreted proteins and proteins containing disulfide bonds (DSB) in this dataset was several-fold higher than in the complete proteomes of various organisms (cysteine preference in these sequences is built in because modern methods of complete chemical synthesis assemble proteins from peptides, which usually requires internal cysteine residues for conjugation chemistry). Thus, recapitulating the properties of spontaneously refolded proteins discussed in the preceding section, the set of chemically synthesized proteins is heavily biased and not at all representative of real proteomes. Furthermore, there is no reliable data on those targets that were synthesized but could not be refolded. Even apart from these limitations, successful folding of chemically synthesized proteins requires non-physiological renaturation times in the hours' range. The yields of the native conformations are often omitted from the reports, but vary widely when reported (5–48%; Supplementary File S3); the folding protocols are complex and are developed on a case-by-case basis. Thus, even for this collection of privileged proteins, folding to the active forms in vitro is not straightforward and likely proceeds differently than in vivo.

Table 1. Properties of 59 proteins produced by total chemical synthesis and refolded to their active forms, as compared to the properties of whole proteomes.

	Total Chemical Synthesis [1]	Archaea	Bacteria	Eukarya	Data Sources for Archaea, Bacteria and Eukarya
mean protein length, amino acids	94	283	320	472	[51]
% secreted	62	6–19	18–30	13 (humans)	[52–54]
% with DSB in the known 3-D structures	57	15	11	30	[55]

[1] For the full data compilation from the literature, see Supplementary File S3.

Then, there is an even deeper problem with the folding of chemically synthesized proteins as the ultimate argument for the thermodynamics hypothesis. Through the course of the chemical synthesis, these proteins remain attached to solid phase, with limited degrees of freedom for the main chain rearrangement. The structure of the Gibbs energy landscapes (or other landscapes) for such proteins and their precursor peptides, before or after they are released from the solid phase into the solution, is completely unknown. These landscapes would be important to study, not only to resolve this open question as such, but also because this might yield clues both to the mechanisms of protein folding inside cells and to the folding of primordial peptides during the early evolution of life (see discussion towards the end of this paper).

2.3. Refolding of Insoluble Overexpressed Proteins from Denatured Bacterial Inclusion Bodies into Soluble Active Proteins in Native Conformations

A widely used approach to protein production is based on the fact that, when a recombinant protein is overexpressed in bacterial cells, it often forms insoluble inclusion bodies that are easy to isolate from other cellular components. Such aggregates of overexpressed proteins can be collected, further purified, denatured in vitro and often can be successfully refolded into soluble, active proteins. Because of the numerous industrial applications, protein purification and refolding from bacterial inclusion bodies have been extensively studied (for overview, see [56–59]).

If indeed the proteins that are purified from inclusion bodies were shown to unfold completely and then routinely refold to the native, active conformation, this would comprise strong evidence that spontaneous protein folding is common, in accord with the thermodynamic hypothesis. However, experiments with overexpressed proteins that form inclusion bodies resulted in a key observation that suggests quite different conclusions. Typically, proteins within the aggregates that form the inclusion bodies are neither disordered nor unfolded, but have specific secondary and tertiary structures, which are substantially ordered and are often enriched in in-register beta-sheets [60–63]. Detailed analysis of the refolding process shows that some of the ordered structure is preserved throughout the purification stages ([64–66], reviewed in [67]). Moreover, harsh denaturing conditions tend to be detrimental for protein refolding to the native conformation, so that new protocols for unfolding–refolding under mild conditions are constantly proposed in attempts to improve the yield of functional proteins (e.g., [68–70]).

Thus, experiments on refolding of overexpressed proteins demonstrate the key role of the residual secondary and tertiary structures, which are generated in the first place by the cell during protein expression in vivo and apparently have to be retained by the protein for efficient refolding. Furthermore, even when refolding occurs, it barely resembles the folding processes that occur in living cells. Indeed, purification and refolding of nearly every protein requires extensive protocol development, which often includes solutions and treatments that are far from physiological conditions and refolding times that are typically much longer than the biologically relevant time scale. All these efforts notwithstanding, the yields of the refolded native proteins vary widely [57,59]. Therefore, in general, refolding of proteins from inclusion bodies cannot be counted as a showcase for spontaneous refolding of completely denatured proteins and hence does not provide clear support for the thermodynamic hypothesis.

2.4. Scarcity of Data on ΔG of Protein Folding Reflects Pervasive Non-Refoldability and Instability of Proteomes

A general conclusion from all of the above is that the evidence for the negative ΔG of folding is quite thin, at best, and that the data on the ability of proteins to refold from a completely denatured state is fragmentary and comes from small, heavily biased datasets. What causes this scarcity of data, especially for larger proteins? The principal cause appears to be quite simple: most proteins actually cannot refold once completely unfolded, but the negative results of this type, that is, failed attempts to refold proteins, are almost never published. To our knowledge, no representative samples of proteins have been studied under this angle until very recently. However, in a recent proteome-wide study, protease-resistance assay was used in combination with quantitative mass-spectrometry to show that about 50% of the proteins in *E. coli* cell lysates could not refold into their native states following chemical denaturation, even when the conditions were optimized for refolding [71]. These findings indicate that non-refoldability in vitro is a general characteristic of at least this bacterial proteome, especially, taking into consideration that the completeness of unfolding was not monitored in these experiments.

Another widely observed but often misinterpreted phenomenon is the general proteome instability, that is, pervasive spontaneous loss of native structure and activity in proteins that have been originally properly folded in the living cells. This loss of native conformation and functional activity is commonly observed both in vitro, in preparations of isolated proteins, and in vivo. Indeed, it is well known that all cells maintain elaborate proteostasis machineries that functions to repair or destroy any proteins that have irreversibly lost their active conformations [72,73].

Spontaneous protein unfolding (denaturation) in vitro is an extremely common observation which suggests that at equilibrium many if not most proteins are in unfolded conformations. Unfortunately, to our knowledge, there is no published comprehensive statistics on protein (in)stability in vitro. Studies on protein stability and approaches to stabilization are a major expense in the pharma/biotech/industrial enzymology industry, and apparently, much of the results comprise intellectual property of these companies. The physical and chemical processes that are associated with instability have been thoroughly studied for a relatively small number of proteins [74–76]. The principal take-home message from these experiments is that even correctly folded proteins are often not intrinsically stable, as it would have been expected if they were in a deep free energy minimum, either global or local; instead, many lose native conformation easily.

Protein engineering experiments indicate that many proteins are easily destabilized with small sequence changes. However, despite years of research, predicting the effect of mutations on protein stability remains challenging. Nevertheless, the general conclusion is that most proteins are only barely stable, such that there are many destabilizing mutations (reviewed in [77,78]). Poorly understood tradeoffs seem to exist between protein stability and solubility such that a mutation on the exterior of a protein that increases its solubility is often destabilizing [79–82]. Similarly, there are tradeoffs between protein activity and stability such that mutations that enhance enzyme activity often destabilize the protein, and vice versa, stabilizing mutations often decrease activity [83].

Over the decades, many ad hoc explanations have been given for the spontaneous unfolding, denaturation, and destabilization that proteins typically undergo. Mostly, some irreversible events are postulated to occur during unfolding, such as protein oxidation, other chemical modifications, and/or aggregation, and such secondary effects are claimed to shift the equilibrium towards the unfolded state, preventing thermodynamically driven folding [74,76,84,85]. However, few targeted studies of protein denaturation mechanisms have been published. Usually, the loss of the native conformation and consequently activity by an isolated protein is perceived as a (often major) nuisance and is rarely seen as an opportunity to study the mechanisms of irreversible denaturation, and apparently, for this reason, not much systematic research has been done in this field. A notable exception are experiments of Klibanov and colleagues on the mechanisms of amylase denaturation. In these studies, the processes involved in thermal inactivation of this enzyme were dissected, showing that denaturation (unfolding), chemical modification, and aggregation are all distinct processes separated in time, and irreversible denaturation of this enzyme precedes chemical modifications and aggregation [86,87]. Several studies on other enzymes have also demonstrated that denaturation by irreversible chain unfolding is a process distinct from protein aggregation [88–90].

There is a call in the literature to apply modern approaches, such as new methods of spectroscopy and mass-spectrometry, for the proteome scale analysis of protein stability [91–94], but the actual experiments of this type remain to be performed.

A rough estimate of the failure rate of attempts on isolation of proteins in the native conformations can be extracted from large-scale structural genomics projects, which publish some statistics of protein production and purification. For example, Page et al. [95] reported that of more than 1800 proteins encoded in the genome of the hyperthermophilic bacterium *Thermotoga maritima* and cloned into expression plasmids, only 539 (~29%) could be purified in the form suitable for crystallization. In the Northeast Structural Genomics Consortium project [96], 6493 proteins could be purified out of the total 16,992 expressed (34%). The

New York Structural Genomics consortium has not reported consolidated statistics, but ~30% purified-to-expressed ratio seems to be a general trend across many participating projects [97]. It should be noted that in all these efforts, except for the Thermotoga case, the set of targets is strongly biased by pre-selection for predicted solubility, globularity, and evolutionary conservation. Even in these privileged sets of proteins, two-thirds could not be purified in the native conformation.

Generally, for the vast protein space, the thermodynamic parameters of protein folding and unfolding remain effectively unknown. There is no strong evidence that negative ΔG of folding is a general property of many proteins. On the contrary, a wealth of data seem to present evidence against this possibility, showing instead that unfolding of most proteins occurs spontaneously, whereas folding does not. Thus, protein folding appears to be a non-equilibrium process that is accompanied by free energy increase.

2.5. Special Features of Protein Folding In Vivo

Complex proteostasis systems operate in every cell, and malfunction of these systems leads to (often lethal) accumulation of unfolded and misfolded proteins in the cell [72,73]. Proteomics shows that more than two thirds of all proteins in yeast specifically interact with one, or often more than one, of the proteostasis systems, and more specifically, with molecular chaperones [98]. Hundreds of proteins in *E. coli* interact with the chaperone GroEL-GroES alone [99]. The most abundant proteins in eukaryotes (actin, tubulin) do not fold in vitro at all, and to fold in vivo, they require, in addition to general chaperone systems, also the specialized co-chaperone prefoldin [100,101]. Apparently, proteostasis mechanisms consume a substantial fraction of the cellular energy supply; although the estimates do not seem to achieve high precision, the fraction of the energy budget dedicated to these processes is thought to be greater than 10% [102].

Chaperone clients are classified based on how closely they interact with the chaperones (for example, obligately-dependent vs. partially-dependent clients, based on the occupancy of the client-chaperone complexes [99]) and what, specifically, do they need chaperones for some proteins aggregate in the absence of chaperones, others stay soluble but are inactive, yet others need chaperones only under stress [103,104].

How do chaperones facilitate protein folding? The dominant view is that they help client proteins to quickly reach the minimum of free energy, that is, the chaperones create conditions for the thermodynamically driven folding of a substrate protein molecule into the native conformation. Some specific mechanisms include: (1) holding the client in isolation so that it does not aggregate with other proteins and folds correctly by itself, a mechanism known as "Anfinsen's cage" in the case of GroES/GroEL [105–107], (2) preventing client proteins from getting stuck in kinetic traps during folding, conceivably, via partial unfolding [108–110], (3) unfolding misfolded or aggregated substrates before proceeding with mechanisms (1) or (2) [111–114], (4) reshaping the folding landscape in ways different from mechanism 2, known as "kinetic assistance", but typically not specified further [115–118].

Some chaperones, known as foldases, are ATPases, whereas others, dubbed holdases, are not [119], but the distinction does not appear to map well onto the mechanisms listed above. Indeed, some chaperones from each functional class seem to exercise mechanisms 1–4 (see, e.g., [120]), whereas some appear to combine properties of foldases and holdases, as argued for the ATP-independent chaperone trigger factor [121] as well as the ATP-dependent HSP70 [122,123].

Crucially, all proposed chaperone mechanisms are predicated on the thermodynamic hypothesis, and to our knowledge, the relevance of these mechanisms has not been tested against the alternatives. A different view of the chaperone mechanisms will be discussed below.

2.6. Is Protein Folding In Vivo an Active, Energy-Dependent Process?

In our view, the above discussion shows that there is very little experimental evidence that ΔG of folding is negative for most proteins. Conversely, a massive amount of experimental observations indicates that native conformations of proteins are only metastable. Taken together, these lines of evidence compel us, in the least, to seriously consider the possibility that, for the majority of proteins, ΔG of folding is positive (Figure 1d). The key implication of this hypothesis is that protein folding in vivo does not occur spontaneously, but rather, is an active, energy-dependent process.

This conceptual shift in our understanding of protein folding further implies that the ribosome itself is likely to act as a giant chaperone and the most important part of the protein folding machinery [124–126]. Clearly, this possibility is fully compatible with the numerous observations indicating that folding of most if not all proteins occurs co-translationally [127–151].

The most obvious way the ribosome could cause the increase in the Gibbs free energy that seems to accompany protein folding is by lowering the entropy of the protein by reducing the number of possible conformations of the peptide backbone. Some strained conformations with elevated enthalpy appear possible, too.

A common assumption in modeling of protein folding and in theoretical discussions is that the protein backbone can be well approximated by a freely jointed chain (FJC), so that all energy that could be applied to it would rapidly dissipate because of unrestricted rotation around each psi and phi bond. However, theoretical argument against this assumption, based on the available data on excluded volume effects and steric hindrances, has been brought up (e.g., [152]). Recently, our all-atom molecular dynamics simulations have revealed the situations when the backbone indeed is not FJC. When rotational force is applied to the protein backbone during the simulation, diverse helical peptides, despite their purported freedom to rotate about the psi and phi bonds, rapidly fold into the native structure, which remains stable [153].

It is important to recall that translation is coupled to the hydrolysis of massive amounts of GTP, but there is no clarity as to what this energy is actually expended on [154]. If ΔG of protein folding is positive, it appears likely that at least some fraction of the energy of GTP hydrolysis contributes to active, non-equilibrium, co-translational folding. Apart from the ribosome, other molecular players are likely to be involved in active co-translational (and "co-translocational") folding as well, in particular, the signal recognition particle (SRP) that contains two GTPases of its own, while the role of GTP hydrolysis is no better understood than it is in the case of the ribosome [155].

The energy-coupling machine framework has been suggested also for chaperone mechanisms as an alternative to the Anfinsen's cage. Once again, it is unclear what the energy of ATP hydrolysis by ATP-dependent chaperones is actually spent on. Most studies link the ATPase activity with rearrangements of the chaperone itself [156,157]. However, the energy balance of these reactions remains unknown, and the possibility of coupling between ATP hydrolysis and the client protein rearrangements is typically not even considered because folding is assumed to be spontaneous. In contrast, a series of studies pioneered by Lorimer, De Los Rios and Goloubinoff argue that ATP-dependent chaperones, such as HSP60, HSP70 or HSP90, might expend at least part of the energy of ATP hydrolysis to manipulate the substrate directly ("non-equilibrium activation") although the mechanistic details remain unclear [158–163].

Apart from the empirical evidence and thermodynamic considerations, the notion of active, non-equilibrium protein folding also appears to be better compatible with the evolutionary history of the relevant cellular components than the thermodynamic hypothesis of spontaneous folding. Indeed, the ribosome, translation factors with GTPase activity, and the SRP are universal to all cellular life, and several key chaperones also are among the most highly conserved proteins. All these molecular machines are likely to antedate the Last Universal Cellular Ancestor [164–166]. During all the 4 billion years or so of their existence, natural selection (including purifying selection for most of this time) would have acted primarily on the foldability of proteins on these machines, rather than their ability to fold/re-fold spontaneously. Perhaps, spontaneous folding could be a factor only occasionally, in particular, for secreted proteins that have little access to chaperones once outside the cell.

2.7. Towards a Realistic Physical Model of Active Protein Folding

If the thermodynamic concept of protein folding generally fails, a new physical model of protein folding as an active, energy-dependent process is needed. Where to start? To begin with, a better definition of a "perfectly unfolded" protein conformation is essential. In such a fully unfolded conformation, there are no stable contacts between any two amino acids that are not adjacent in the polypeptide chain. It is currently unclear whether a perfectly unfolded conformation actually exists in vitro or in vivo for any particular protein, but this definition will be an appropriate starting point for building a physical model and recreating the folding process in silico.

Depending on the length of the polypeptide chain, there are theoretically on the order of 10^{100} perfectly unfolded conformations for each protein [167,168]; more recently, suggestions have been made for a more conservative upper bound which, however, remains astronomically high [169]. From this vast set of perfectly unfolded conformations, one can build up and arrive at all kinds of structures: active intrinsically disordered ("natively unfolded") conformations, stable misfolded conformations, conformations that only emerge through interaction with other proteins, classic globular native conformations with hydrophobic cores, and more. What do we know about "perfectly unfolded" conformations? If we measure (calculate) Gibbs free energy for these conformations, we should observe approximately the same value for each of them because these are random conformations with no interactions other than with the solvent. Even a single contact that forms within the polypeptide chain, whether it is a short or long distance one, makes the polypeptide more compact and increases the Gibbs free energy both due to the entropy reduction resulting from limiting the degrees of freedom and to changing enthalpy if, for example, the contact is hydrophobic or ionic.

Considering that most of protein folding in vivo takes place co-translationally, while the polypeptide chain is built up one amino acid at a time, simultaneously exploring the shifting folding landscape while interacting with multiple other molecules in a crowded environment, the task of incorporating all known cellular biochemistry and structural biology into the physical model of non-equilibrium protein folding as it occurs in vivo seems daunting. Nevertheless, this goal no longer appears to be out of reach. Advanced methods for quantitative measurement of various energy inputs, molecular motions, heat transfers and other relevant quantities should provide the values, or at least the bounds, of many parameters that determine protein folding as in vivo. Such work has already started although it is notable that many crucial parameters of the relevant processes, even some basic ones, such as the translation rate, rely on estimates obtained decades ago [170,171] and refined only very recently [172].

A complementary class of approaches involves building, both in silico and in vitro, simplified artificial protein folding machines that apply various forces to the folding polypeptide in an attempt to directly manipulate the peptide backbone into the desired conformations, imposing various kinds of physical constraints on the folding process, and thus, causing shifts and introducing kinetic barriers into the folding landscape. Work in this direction has already started as well. In the next section, we provide a brief overview of several advanced techniques and some recent observations, which suggest a more sophisticated understanding of the mechanisms of protein folding than what was provided by the canonical models of spontaneous protein folding in vitro.

2.8. Non-Equilibrium Protein Folding: New Approaches and Recent Results

In recent years, a variety of novel experimental techniques have been applied to study co-translational and chaperone-assisted protein folding. Particularly informative are methods that can manipulate a defined single molecule using a specific force probe, such as atomic force microscope, optical tweezers, or magnetic tweezers; these methods are sometimes collectively referred to as single-molecule force spectroscopy methods (SMFS; reviewed in [173]). The SMFS methods have been recently applied to the study of co-translational folding of nascent protein chains, using "life-like" in vitro translation systems (reviewed in [149]). Although SMFS approaches have not yet reached the precision required to infer the thermodynamic parameters of protein folding (see Section 2.1 above), these approaches are well-suited to address questions about the effects of specific treatments and interactions on the folding process. Recent observations made using such methods include, for example, detection of co-translational folding intermediates, suggesting a defined folding pathway for a small domain that had been thought to fold in a two-step fashion in vitro [174], and observation of a direct accelerating and stabilizing effect of the ribosomal tunnel on the co-translational folding of another small domain [151]. Although often discussed within the conventional framework of thermodynamically-driven folding, these and similar results can be productively exploited to develop the non-equilibrium protein folding model. Furthermore, with regard to chaperone-assisted protein refolding, SMFS methods have revealed that the chaperones of the Hsp90 family use the energy of ATP hydrolysis to perform mechanical work, which is applied to compact unfolded chains against the counteracting denaturing forces [175], in an apparent contradiction to the traditionally envisaged, Anfinsen chamber-like mechanisms of chaperone action.

Another group of powerful methods are modern structural biology approaches, including cryo-electron microscopy, solid state nuclear magnetic resonance, SAXS and others, which reveal the structure of nascent polypeptide chains during protein synthesis. Many of these methods are focused on the kinetics and regulation of protein synthesis [176,177] and on the functions of nascent chain, such as sensing the state of regulatory metabolites in the environment and communicating the results to the peptidyltransferase center of the translating ribosome [178,179]. These studies also highly informative for the study of co-translational protein folding, and have already illuminated defined secondary and tertiary structures adopted by nascent peptides in the ribosome tunnel and exit vestibule [180–182]. In the forthcoming years, we expect to see more explicit investigation of the interactions of the nascent peptide with the peptidyl transferase center, ribosome exit tunnel, and other components of the protein folding machinery.

Computational modeling of protein folding also is taking a new direction towards a closer mimicking of the folding environment encountered by proteins in vivo. We recently reported the results of all-atom molecular dynamics simulations, in which the standard force field was augmented by the application of a mechanical force that rotated a single N-terminal amino acid of peptides, while simultaneously restricting the movements of a distal amino acid. Such directional rotation changed the peptide backbone behavior, facilitating rapid formation of native structures in several diverse alpha-helical peptides [153]. Apparently, steric clashes arising due to the forced directional rotation resulted in the behavior of the peptide backbone that no longer resembled an FJC. Further studies are needed to determine whether such an effect can be observed in single-molecule experiments in vitro as well. Other attempts to build simplified folding machines to model aspects of co-translational peptide folding in vivo include the molecular-dynamics studies of folding in a tubular chamber representing the ribosome exit tunnel, either with uncharged elastic walls or with charged walls [183–186]. Finally, sophisticated methods of visualization and analysis of the massive dynamic data on protein folding, unfolding, and refolding are also undergoing active development (see [187] for a recent review). Such methods should greatly aid our understanding of the complex mechanisms of protein folding in vivo.

3. Conclusions

The cornerstone assumption in the field of protein folding is that proteins spontaneously fold into their native conformations driven by negative ΔG. Furthermore, it is generally assumed that the native conformation of a protein is the global minimum of Gibbs free energy. However, a survey of the available data on spontaneous protein folding and refolding, in particular, for chemically synthesized and over-expressed proteins, presents little evidence in support of this thermodynamic hypothesis of folding. On the contrary, the majority of proteins appear not to be spontaneously foldable and are only marginally stable, at best. The totality of these observations along with thermodynamic considerations suggest that across the protein world, there is a wide variety of rugged, dynamic landscapes of folding free energy, resulting in a broad range of thermodynamic and kinetic stability, and refoldability of proteins. For different proteins, ΔG of folding can be either negative or positive, conceivably, for the majority of the proteins. Even regardless of the specific value of ΔG, folding of most proteins is likely to be an active, non-equilibrium, energy-dependent process. This conceptual shift in our understanding of protein folding appears to be best compatible with the extensive molecular data on the universal translation and proteostasis machineries that operate in all cells, and with the evolutionary history of these molecular machines that is traced to the earliest stages of life evolution. We believe that this change in perspective on protein folding can and should stimulate a dedicated program of theoretical, modeling, and experimental studies.

Supplementary Materials: The following are available online at https://www.mdpi.com/article/10.3390/ijms23010521/s1.

Author Contributions: Conceptualization, I.S., A.R.M. and E.V.K.; methodology, I.S. and A.R.M.; investigation, I.S., A.R.M. and E.V.K.; resources, I.S.; data curation, I.S., A.R.M.; writing—original draft, I.S. and A.R.M.; writing—review and editing, I.S, A.R.M. and E.V.K.; visualization, I.S.; funding acquisition, I.S., A.R.M. and E.V.K. All authors have read and agreed to the published version of the manuscript.

Funding: A.R.M.'s stay at Clare Hall College was partially supported by the U.S. National Science Foundation Long-Term Professional Development program; E.V.K. is supported by the Intramural Research Program of the National Institutes of Health of the USA (National Library of Medicine).

Institutional Review Board Statement: Not applicable.

Informed Consent Statement: Not applicable.

Data Availability Statement: All data supporting reported results have been submitted as Supplementary Files to this manuscript, except the ProTherm data that have been converted by Protabit LLC with permission from ProTherm and deposited at https://github.com/protabit/protherm-conversion (accessed on 6 December 2021).

Acknowledgments: We are grateful to Connie Wang (Protabit, LLC) for the advice on the converted ProTherm dataset, to Natalia Mushegian for help with data analysis, and to Sophia Belkin for the artwork.

Conflicts of Interest: A.R.M. is a program director at the National Science Foundation (NSF), an agency of the U.S. Government, but the statements and opinions expressed herein are made in the personal capacity and do not constitute the endorsement by NSF or the government of the United States. The funders had no role in the design of the study; in the collection, analyses, or interpretation of data; in the writing of the manuscript, or in the decision to publish the results.

References

1. Haber, E.; Anfinsen, C.B. Side-Chain Interactions Governing the Pairing of Half-Cystine Residues in Ribonuclease. *J. Biol. Chem.* **1962**, *237*, 1839–1844. [CrossRef]
2. Anfinsen, C.B. Principles That Govern the Folding of Protein Chains. *Science* **1973**, *181*, 223–230. [CrossRef] [PubMed]
3. Bryngelson, J.D.; Wolynes, P.G. Intermediates and Barrier Crossing in a Random Energy Model (with Applications to Protein Folding). *J. Phys. Chem.* **1989**, *93*, 6902–6915. [CrossRef]
4. Zwanzig, R.; Szabo, A.; Bagchi, B. Levinthal's Paradox. *Proc. Natl. Acad. Sci. USA* **1992**, *89*, 20–22. [CrossRef]
5. Leopold, P.E.; Montal, M.; Onuchic, J.N. Protein Folding Funnels: A Kinetic Approach to the Sequence-Structure Relationship. *Proc. Natl. Acad. Sci. USA* **1992**, *89*, 8721–8725. [CrossRef]
6. Bryngelson, J.D.; Onuchic, J.N.; Socci, N.D.; Wolynes, P.G. Funnels, Pathways, and the Energy Landscape of Protein Folding: A Synthesis. *Proteins* **1995**, *21*, 167–195. [CrossRef] [PubMed]
7. Wolynes, P.G. Energy Landscapes and Solved Protein-Folding Problems. *Philos. Trans. A Math. Phys. Eng. Sci.* **2005**, *363*, 453–464; discussion 464–467. [CrossRef] [PubMed]

8. Dill, K.A.; Ozkan, S.B.; Shell, M.S.; Weikl, T.R. The Protein Folding Problem. *Annu. Rev. Biophys.* **2008**, *37*, 289–316. [CrossRef] [PubMed]
9. Dill, K.A.; MacCallum, J.L. The Protein-Folding Problem, 50 Years On. *Science* **2012**, *338*, 1042–1046. [CrossRef] [PubMed]
10. Schafer, N.P.; Kim, B.L.; Zheng, W.; Wolynes, P.G. Learning To Fold Proteins Using Energy Landscape Theory. *Isr. J. Chem.* **2014**, *54*, 1311–1337. [CrossRef] [PubMed]
11. Nassar, R.; Dignon, G.L.; Razban, R.M.; Dill, K.A. The Protein Folding Problem: The Role of Theory. *J. Mol. Biol.* **2021**, *433*, 167126. [CrossRef]
12. Rollins, G.C.; Dill, K.A. General Mechanism of Two-State Protein Folding Kinetics. *J. Am. Chem. Soc.* **2014**, *136*, 11420–11427. [CrossRef]
13. Finkelstein, A.V.; Badretdin, A.J.; Galzitskaya, O.V.; Ivankov, D.N.; Bogatyreva, N.S.; Garbuzynskiy, S.O. There and Back Again: Two Views on the Protein Folding Puzzle. *Phys. Life Rev.* **2017**, *21*, 56–71. [CrossRef]
14. Wetlaufer, D.B.; Ristow, S. Acquisition of Three-Dimensional Structure of Proteins. *Annu. Rev. Biochem.* **1973**, *42*, 135–158. [CrossRef]
15. Ben-Naim, A. *Myths and Verities in Protein Folding Theories;* World Scientific: Singapore, 2015; ISBN 978-981-4725-98-9.
16. Shakhnovich, E. Protein Folding Thermodynamics and Dynamics: Where Physics, Chemistry, and Biology Meet. *Chem. Rev.* **2006**, *106*, 1559–1588. [CrossRef] [PubMed]
17. Santoro, M.M.; Bolen, D.W. Unfolding Free Energy Changes Determined by the Linear Extrapolation Method. 1. Unfolding of Phenylmethanesulfonyl Alpha-Chymotrypsin Using Different Denaturants. *Biochemistry* **1988**, *27*, 8063–8068. [CrossRef] [PubMed]
18. Grimsley, G.R.; Huyghues-Despointes, B.M.P.; Pace, C.N.; Scholtz, J.M. Determining a Urea or Guanidinium Chloride Unfolding Curve. *CSH Protoc.* **2006**, *2006*, pdb-prot4242. [CrossRef]
19. Shaw, K.L.; Scholtz, J.M.; Pace, C.N.; Grimsley, G.R. Determining the Conformational Stability of a Protein Using Urea Denaturation Curves. *Methods Mol. Biol.* **2009**, *490*, 41–55. [CrossRef] [PubMed]
20. Privalov, P.L. Microcalorimetry of Proteins and Their Complexes. *Methods Mol. Biol.* **2009**, *490*, 1–39. [CrossRef]
21. Ibarra-Molero, B.; Naganathan, A.N.; Sanchez-Ruiz, J.M.; Muñoz, V. Modern Analysis of Protein Folding by Differential Scanning Calorimetry. *Methods Enzymol.* **2016**, *567*, 281–318. [CrossRef]
22. Naganathan, A.N.; Perez-Jimenez, R.; Muñoz, V.; Sanchez-Ruiz, J.M. Estimation of Protein Folding Free Energy Barriers from Calorimetric Data by Multi-Model Bayesian Analysis. *Phys. Chem. Chem. Phys.* **2011**, *13*, 17064–17076. [CrossRef]
23. Makhatadze, G.I.; Privalov, P.L. Energetics of Protein Structure. *Adv. Protein Chem.* **1995**, *47*, 307–425. [PubMed]
24. Baldwin, R.L. Energetics of Protein Folding. *J. Mol. Biol.* **2007**, *371*, 283–301. [CrossRef] [PubMed]
25. Bedouelle, H. Principles and Equations for Measuring and Interpreting Protein Stability: From Monomer to Tetramer. *Biochimie* **2016**, *121*, 29–37. [CrossRef] [PubMed]
26. Basharov, M.A. Residual Ordered Structure in Denatured Proteins and the Problem of Protein Folding. *Indian J. Biochem. Biophys.* **2012**, *49*, 7–17.
27. Shortle, D.; Ackerman, M.S. Persistence of Native-like Topology in a Denatured Protein in 8 M Urea. *Science* **2001**, *293*, 487–489. [CrossRef]
28. Vendruscolo, M.; Paci, E.; Karplus, M.; Dobson, C.M. Structures and Relative Free Energies of Partially Folded States of Proteins. *Proc. Natl. Acad. Sci. USA* **2003**, *100*, 14817–14821. [CrossRef]
29. Sosnick, T.R.; Trewhella, J. Denatured States of Ribonuclease A Have Compact Dimensions and Residual Secondary Structure. *Biochemistry* **1992**, *31*, 8329–8335. [CrossRef]
30. Lim, W.K.; Rösgen, J.; Englander, S.W. Urea, but Not Guanidinium, Destabilizes Proteins by Forming Hydrogen Bonds to the Peptide Group. *Proc. Natl. Acad. Sci. USA* **2009**, *106*, 2595–2600. [CrossRef]
31. Lapidus, L.J. Protein Unfolding Mechanisms and Their Effects on Folding Experiments. *F1000Research* **2017**, *6*, 1723. [CrossRef]
32. Fitzkee, N.C.; Rose, G.D. Reassessing Random-Coil Statistics in Unfolded Proteins. *Proc. Natl. Acad. Sci. USA* **2004**, *101*, 12497–12502. [CrossRef] [PubMed]
33. Nikam, R.; Kulandaisamy, A.; Harini, K.; Sharma, D.; Gromiha, M.M. ProThermDB: Thermodynamic Database for Proteins and Mutants Revisited after 15 Years. *Nucleic Acids. Res.* **2021**, *49*, D420–D424. [CrossRef] [PubMed]
34. ProTherm Conversion. 2017. Available online: https://github.com/protabit/protherm-conversion (accessed on 6 December 2021).
35. Braselmann, E.; Chaney, J.L.; Clark, P.L. Folding the Proteome. *Trends Biochem. Sci.* **2013**, *38*, 337–344. [CrossRef] [PubMed]
36. Zeldovich, K.B.; Chen, P.; Shakhnovich, E.I. Protein Stability Imposes Limits on Organism Complexity and Speed of Molecular Evolution. *Proc. Natl. Acad. Sci. USA* **2007**, *104*, 16152–16157. [CrossRef]
37. Taverna, D.M.; Goldstein, R.A. Why Are Proteins Marginally Stable? *Proteins* **2002**, *46*, 105–109. [CrossRef] [PubMed]
38. Godoy-Ruiz, R.; Perez-Jimenez, R.; Ibarra-Molero, B.; Sanchez-Ruiz, J.M. Relation between Protein Stability, Evolution and Structure, as Probed by Carboxylic Acid Mutations. *J. Mol. Biol.* **2004**, *336*, 313–318. [CrossRef]
39. Williams, P.D.; Pollock, D.D.; Goldstein, R.A. Functionality and the Evolution of Marginal Stability in Proteins: Inferences from Lattice Simulations. *Evol. Bioinform. Online* **2007**, *2*, 91–101. [CrossRef]
40. Wilson, A.E.; Kosater, W.M.; Liberles, D.A. Evolutionary Processes and Biophysical Mechanisms: Revisiting Why Evolved Proteins Are Marginally Stable. *J. Mol. Evol.* **2020**, *88*, 415–417. [CrossRef]

41. Borgia, A.; Williams, P.M.; Clarke, J. Single-Molecule Studies of Protein Folding. *Annu. Rev. Biochem.* **2008**, *77*, 101–125. [CrossRef]
42. Bustamante, C.; Alexander, L.; Maciuba, K.; Kaiser, C.M. Single-Molecule Studies of Protein Folding with Optical Tweezers. *Annu. Rev. Biochem.* **2020**, *89*, 443–470. [CrossRef] [PubMed]
43. Tinoco, I., Jr.; Bustamante, C. The Effect of Force on Thermodynamics and Kinetics of Single Molecule Reactions. *Biophys. Chem.* **2002**, *101–102*, 513–533. [CrossRef]
44. Junier, I.; Mossa, A.; Manosas, M.; Ritort, F. Recovery of Free Energy Branches in Single Molecule Experiments. *Phys. Rev. Lett.* **2009**, *102*, 070602. [CrossRef]
45. Wang, J.; Ferguson, A.L. Nonlinear Reconstruction of Single-Molecule Free-Energy Surfaces from Univariate Time Series. *Phys. Rev. E* **2016**, *93*, 032412. [CrossRef] [PubMed]
46. Yang, G.; Cecconi, C.; Baase, W.A.; Vetter, I.R.; Breyer, W.A.; Haack, J.A.; Matthews, B.W.; Dahlquist, F.W.; Bustamante, C. Solid-State Synthesis and Mechanical Unfolding of Polymers of T4 Lysozyme. *Proc. Natl. Acad. Sci. USA* **2000**, *97*, 139–144. [CrossRef]
47. Dietz, H.; Rief, M. Exploring the Energy Landscape of GFP by Single-Molecule Mechanical Experiments. *Proc. Natl. Acad. Sci. USA* **2004**, *101*, 16192–16197. [CrossRef] [PubMed]
48. Ainavarapu, S.R.K.; Brujic, J.; Huang, H.H.; Wiita, A.P.; Lu, H.; Li, L.; Walther, K.A.; Carrion-Vazquez, M.; Li, H.; Fernandez, J.M. Contour Length and Refolding Rate of a Small Protein Controlled by Engineered Disulfide Bonds. *Biophys. J.* **2007**, *92*, 225–233. [CrossRef]
49. Hirschmann, R.; Nutt, R.F.; Veber, D.F.; Vitali, R.A.; Varga, S.L.; Jacob, T.A.; Holly, F.W.; Denkewalter, R.G. Studies on the Total Synthesis of an Enzyme. V. The Preparation of Enzymatically Active Material. *J. Am. Chem. Soc.* **1969**, *91*, 507–508. [CrossRef] [PubMed]
50. Gutte, B.; Merrifield, R.B. The Synthesis of Ribonuclease A. *J. Biol. Chem.* **1971**, *246*, 1922–1941. [CrossRef]
51. Tiessen, A.; Pérez-Rodríguez, P.; Delaye-Arredondo, L.J. Mathematical Modeling and Comparison of Protein Size Distribution in Different Plant, Animal, Fungal and Microbial Species Reveals a Negative Correlation between Protein Size and Protein Number, Thus Providing Insight into the Evolution of Proteomes. *BMC Res. Notes* **2012**, *5*, 85. [CrossRef]
52. Saleh, M.T.; Fillon, M.; Brennan, P.J.; Belisle, J.T. Identification of Putative Exported/Secreted Proteins in Prokaryotic Proteomes. *Gene* **2001**, *269*, 195–204. [CrossRef]
53. Saleh, M.; Song, C.; Nasserulla, S.; Leduc, L.G. Indicators from Archaeal Secretomes. *Microbiol. Res.* **2010**, *165*, 1–10. [CrossRef] [PubMed]
54. Uhlén, M.; Karlsson, M.J.; Hober, A.; Svensson, A.-S.; Scheffel, J.; Kotol, D.; Zhong, W.; Tebani, A.; Strandberg, L.; Edfors, F.; et al. The Human Secretome. *Sci. Signal.* **2019**, *12*, eaaz0274. [CrossRef] [PubMed]
55. Bošnjak, I.; Bojović, V.; Šegvić-Bubić, T.; Bielen, A. Occurrence of Protein Disulfide Bonds in Different Domains of Life: A Comparison of Proteins from the Protein Data Bank. *Protein Eng. Des. Sel.* **2014**, *27*, 65–72. [CrossRef]
56. Guise, A.D.; West, S.M.; Chaudhuri, J.B. Protein Folding in Vivo and Renaturation of Recombinant Proteins from Inclusion Bodies. *Mol. Biotechnol.* **1996**, *6*, 53–64. [CrossRef] [PubMed]
57. Panda, A.K. Bioprocessing of Therapeutic Proteins from the Inclusion Bodies of Escherichia Coli. *Adv. Biochem. Eng. Biotechnol.* **2003**, *85*, 43–93. [PubMed]
58. Cabrita, L.D.; Bottomley, S.P. Protein Expression and Refolding–A Practical Guide to Getting the Most out of Inclusion Bodies. *Biotechnol. Annu. Rev.* **2004**, *10*, 31–50.
59. Vera, A.; González-Montalbán, N.; Arís, A.; Villaverde, A. The Conformational Quality of Insoluble Recombinant Proteins Is Enhanced at Low Growth Temperatures. *Biotechnol. Bioeng.* **2007**, *96*, 1101–1106. [CrossRef]
60. Georgiou, G.; Valax, P.; Ostermeier, M.; Horowitz, P.M. Folding and Aggregation of TEM Beta-Lactamase: Analogies with the Formation of Inclusion Bodies in Escherichia Coli. *Protein Sci.* **1994**, *3*, 1953–1960. [CrossRef]
61. Przybycien, T.M.; Dunn, J.P.; Valax, P.; Georgiou, G. Secondary Structure Characterization of Beta-Lactamase Inclusion Bodies. *Protein Eng.* **1994**, *7*, 131–136. [CrossRef]
62. de Groot, N.S.; Sabate, R.; Ventura, S. Amyloids in Bacterial Inclusion Bodies. *Trends Biochem. Sci.* **2009**, *34*, 408–416. [CrossRef]
63. Ramón, A.; Señorale-Pose, M.; Marín, M. Inclusion Bodies: Not That Bad. *Front. Microbiol.* **2014**, *5*, 56. [CrossRef]
64. Bowden, G.A.; Paredes, A.M.; Georgiou, G. Structure and Morphology of Protein Inclusion Bodies in Escherichia Coli. *Biotechnology* **1991**, *9*, 725–730. [CrossRef]
65. Chaffotte, A.F.; Guillou, Y.; Goldberg, M.E. Inclusion Bodies of the Thermophilic Endoglucanase D from Clostridium Thermocellum Are Made of Native Enzyme That Resists 8 M Urea. *Eur. J. Biochem.* **1992**, *205*, 369–373. [CrossRef]
66. Vandenbroeck, K.; Martens, E.; D'Andrea, S.; Billiau, A. Refolding and Single-Step Purification of Porcine Interferon-Gamma from Escherichia Coli Inclusion Bodies. Conditions for Reconstitution of Dimeric IFN-Gamma. *Eur. J. Biochem.* **1993**, *215*, 481–486. [CrossRef] [PubMed]
67. Doglia, S.M.; Ami, D.; Natalello, A.; Gatti-Lafranconi, P.; Lotti, M. Fourier Transform Infrared Spectroscopy Analysis of the Conformational Quality of Recombinant Proteins within Inclusion Bodies. *Biotechnol. J.* **2008**, *3*, 193–201. [CrossRef]
68. Kudou, M.; Yumioka, R.; Ejima, D.; Arakawa, T.; Tsumoto, K. A Novel Protein Refolding System Using Lauroyl-l-Glutamate as a Solubilizing Detergent and Arginine as a Folding Assisting Agent. *Protein Expr. Purif.* **2011**, *75*, 46–54. [CrossRef] [PubMed]
69. Singh, S.M.; Sharma, A.; Upadhyay, A.K.; Singh, A.; Garg, L.C.; Panda, A.K. Solubilization of Inclusion Body Proteins Using n -Propanol and Its Refolding into Bioactive Form. *Protein Expr. Purif.* **2012**, *81*, 75–82. [CrossRef]

70. Singh, A.; Upadhyay, V.; Upadhyay, A.K.; Singh, S.M.; Panda, A.K. Protein Recovery from Inclusion Bodies of Escherichia Coli Using Mild Solubilization Process. *Microb. Cell Fact.* **2015**, *14*, 41. [CrossRef] [PubMed]
71. To, P.; Whitehead, B.; Tarbox, H.E.; Fried, S.D. Nonrefoldability Is Pervasive Across the E. Coli Proteome. *J. Am. Chem. Soc.* **2021**, *143*, 11435–11448. [CrossRef]
72. Hartl, F.U.; Bracher, A.; Hayer-Hartl, M. Molecular Chaperones in Protein Folding and Proteostasis. *Nature* **2011**, *475*, 324–332. [CrossRef]
73. Saibil, H. Chaperone Machines for Protein Folding, Unfolding and Disaggregation. *Nat. Rev. Mol. Cell Biol.* **2013**, *14*, 630–642. [CrossRef] [PubMed]
74. Manning, M.C.; Patel, K.; Borchardt, R.T. Stability of Protein Pharmaceuticals. *Pharm. Res.* **1989**, *6*, 903–918. [CrossRef] [PubMed]
75. Manning, M.C.; Chou, D.K.; Murphy, B.M.; Payne, R.W.; Katayama, D.S. Stability of Protein Pharmaceuticals: An Update. *Pharm. Res.* **2010**, *27*, 544–575. [CrossRef] [PubMed]
76. Wang, W. Advanced Protein Formulations. *Protein Sci.* **2015**, *24*, 1031–1039. [CrossRef]
77. Magliery, T.J.; Lavinder, J.J.; Sullivan, B.J. Protein Stability by Number: High-Throughput and Statistical Approaches to One of Protein Science's Most Difficult Problems. *Curr. Opin. Chem. Biol.* **2011**, *15*, 443–451. [CrossRef]
78. Magliery, T.J. Protein Stability: Computation, Sequence Statistics, and New Experimental Methods. *Curr. Opin. Struct. Biol.* **2015**, *33*, 161–168. [CrossRef]
79. Klesmith, J.R.; Bacik, J.-P.; Wrenbeck, E.E.; Michalczyk, R.; Whitehead, T.A. Trade-Offs between Enzyme Fitness and Solubility Illuminated by Deep Mutational Scanning. *Proc. Natl. Acad. Sci. USA* **2017**, *114*, 2265–2270. [CrossRef]
80. García-Fruitós, E.; Arís, A.; Villaverde, A. Localization of Functional Polypeptides in Bacterial Inclusion Bodies. *Appl. Environ. Microbiol.* **2007**, *73*, 289–294. [CrossRef] [PubMed]
81. Broom, A.; Jacobi, Z.; Trainor, K.; Meiering, E.M. Computational Tools Help Improve Protein Stability but with a Solubility Tradeoff. *J. Biol. Chem.* **2017**, *292*, 14349–14361. [CrossRef] [PubMed]
82. Broom, A.; Trainor, K.; Jacobi, Z.; Meiering, E.M. Computational Modeling of Protein Stability: Quantitative Analysis Reveals Solutions to Pervasive Problems. *Structure* **2020**, *28*, 717–726.e3. [CrossRef] [PubMed]
83. Siddiqui, K.S. Defying the Activity-Stability Trade-off in Enzymes: Taking Advantage of Entropy to Enhance Activity and Thermostability. *Crit. Rev. Biotechnol.* **2017**, *37*, 309–322. [CrossRef]
84. Manning, M.; Colón, W. Structural Basis of Protein Kinetic Stability: Resistance to Sodium Dodecyl Sulfate Suggests a Central Role for Rigidity and a Bias Toward β-Sheet Structure. *Biochemistry* **2004**, *43*, 11248–11254. [CrossRef]
85. Kazlauskas, R. Engineering More Stable Proteins. *Chem. Soc. Rev.* **2018**, *47*, 9026–9045. [CrossRef]
86. Ahern, T.J.; Klibanov, A.M. Analysis of Processes Causing Thermal Inactivation of Enzymes. *Methods Biochem. Anal.* **1988**, *33*, 91–127.
87. Tomazic, S.J.; Klibanov, A.M. Mechanisms of Irreversible Thermal Inactivation of Bacillus Alpha-Amylases. *J. Biol. Chem.* **1988**, *263*, 3086–3091. [CrossRef]
88. Nury, S.; Meunier, J.C. Molecular Mechanisms of the Irreversible Thermal Denaturation of Guinea-Pig Liver Transglutaminase. *Biochem. J.* **1990**, *266*, 487–490. [CrossRef]
89. Blaber, S.I.; Culajay, J.F.; Khurana, A.; Blaber, M. Reversible Thermal Denaturation of Human FGF-1 Induced by Low Concentrations of Guanidine Hydrochloride. *Biophys. J.* **1999**, *77*, 470–477. [CrossRef]
90. Jahromi, R.R.F.; Morris, P.; Martinez-Torres, R.J.; Dalby, P.A. Structural Stability of E. Coli Transketolase to Temperature and PH Denaturation. *J. Biotechnol.* **2011**, *155*, 209–216. [CrossRef]
91. Leurs, U.; Mistarz, U.H.; Rand, K.D. Getting to the Core of Protein Pharmaceuticals–Comprehensive Structure Analysis by Mass Spectrometry. *Eur. J. Pharm. Biopharm.* **2015**, *93*, 95–109. [CrossRef]
92. Gan, J.; Ben-Nissan, G.; Arkind, G.; Tarnavsky, M.; Trudeau, D.; Noda Garcia, L.; Tawfik, D.S.; Sharon, M. Native Mass Spectrometry of Recombinant Proteins from Crude Cell Lysates. *Anal. Chem.* **2017**, *89*, 4398–4404. [CrossRef]
93. Kaur, U.; Meng, H.; Lui, F.; Ma, R.; Ogburn, R.N.; Johnson, J.H.R.; Fitzgerald, M.C.; Jones, L.M. Proteome-Wide Structural Biology: An Emerging Field for the Structural Analysis of Proteins on the Proteomic Scale. *J. Proteome Res.* **2018**, *17*, 3614–3627. [CrossRef]
94. Atsavapranee, B.; Stark, C.D.; Sunden, F.; Thompson, S.; Fordyce, P.M. Fundamentals to Function: Quantitative and Scalable Approaches for Measuring Protein Stability. *Cell Syst.* **2021**, *12*, 547–560. [CrossRef] [PubMed]
95. Page, R.; Grzechnik, S.K.; Canaves, J.M.; Spraggon, G.; Kreusch, A.; Kuhn, P.; Stevens, R.C.; Lesley, S.A. Shotgun Crystallization Strategy for Structural Genomics: An Optimized Two-Tiered Crystallization Screen against the Thermotoga Maritima Proteome. *Acta Crystallogr. Biol. Crystallogr.* **2003**, *59*, 1028–1037. [CrossRef] [PubMed]
96. Northeast Structural Genomics Consortium Statistics. 2021. Available online: https://www.nesg.org/statistics_00.html (accessed on 6 December 2021).
97. New York Structural Genomics Research Consortium. 2021. Available online: http://www.nysgxrc.org/psi3/progress_statistics.html (accessed on 6 December 2021).
98. Gong, Y.; Kakihara, Y.; Krogan, N.; Greenblatt, J.; Emili, A.; Zhang, Z.; Houry, W.A. An Atlas of Chaperone-Protein Interactions in Saccharomyces Cerevisiae: Implications to Protein Folding Pathways in the Cell. *Mol. Syst. Biol.* **2009**, *5*, 275. [CrossRef] [PubMed]

99. Kerner, M.J.; Naylor, D.J.; Ishihama, Y.; Maier, T.; Chang, H.-C.; Stines, A.P.; Georgopoulos, C.; Frishman, D.; Hayer-Hartl, M.; Mann, M.; et al. Proteome-Wide Analysis of Chaperonin-Dependent Protein Folding in Escherichia Coli. *Cell* **2005**, *122*, 209–220. [CrossRef]
100. Lopez-Fanarraga, M.; Avila, J.; Guasch, A.; Coll, M.; Zabala, J.C. Review: Postchaperonin Tubulin Folding Cofactors and Their Role in Microtubule Dynamics. *J. Struct. Biol.* **2001**, *135*, 219–229. [CrossRef] [PubMed]
101. Povarova, O.I.; Uversky, V.N.; Kuznetsova, I.M.; Turoverov, K.K. Actinous Enigma or Enigmatic Actin: Folding, Structure, and Functions of the Most Abundant Eukaryotic Protein. *Intrinsically Disord. Proteins* **2014**, *2*, e34500. [CrossRef]
102. Finka, A.; Goloubinoff, P. Proteomic Data from Human Cell Cultures Refine Mechanisms of Chaperone-Mediated Protein Homeostasis. *Cell Stress Chaperones* **2013**, *18*, 591–605. [CrossRef]
103. Fujiwara, K.; Ishihama, Y.; Nakahigashi, K.; Soga, T.; Taguchi, H. A Systematic Survey of in Vivo Obligate Chaperonin-Dependent Substrates. *EMBO J.* **2010**, *29*, 1552–1564. [CrossRef]
104. Azia, A.; Unger, R.; Horovitz, A. What Distinguishes GroEL Substrates from Other Escherichia Coli Proteins? *FEBS J.* **2012**, *279*, 543–550. [CrossRef]
105. Hartl, F.U.; Hayer-Hartl, M. Molecular Chaperones in the Cytosol: From Nascent Chain to Folded Protein. *Science* **2002**, *295*, 1852–1858. [CrossRef]
106. Gupta, A.J.; Haldar, S.; Miličić, G.; Hartl, F.U.; Hayer-Hartl, M. Active Cage Mechanism of Chaperonin-Assisted Protein Folding Demonstrated at Single-Molecule Level. *J. Mol. Biol.* **2014**, *426*, 2739–2754. [CrossRef]
107. Singhal, K.; Vreede, J.; Mashaghi, A.; Tans, S.J.; Bolhuis, P.G. The Trigger Factor Chaperone Encapsulates and Stabilizes Partial Folds of Substrate Proteins. *PLoS Comput. Biol.* **2015**, *11*, e1004444. [CrossRef]
108. Grantcharova, V.; Alm, E.J.; Baker, D.; Horwich, A.L. Mechanisms of Protein Folding. *Curr. Opin. Struct. Biol.* **2001**, *11*, 70–82. [CrossRef]
109. Balchin, D.; Hayer-Hartl, M.; Hartl, F.U. In Vivo Aspects of Protein Folding and Quality Control. *Science* **2016**, *353*, aac4354. [CrossRef] [PubMed]
110. Balchin, D.; Miličić, G.; Strauss, M.; Hayer-Hartl, M.; Hartl, F.U. Pathway of Actin Folding Directed by the Eukaryotic Chaperonin TRiC. *Cell* **2018**, *174*, 1507–1521. [CrossRef]
111. Shtilerman, M.; Lorimer, G.H.; Englander, S.W. Chaperonin Function: Folding by Forced Unfolding. *Science* **1999**, *284*, 822–825. [CrossRef]
112. Sousa, R. Structural Mechanisms of Chaperone Mediated Protein Disaggregation. *Front. Mol. Biosci.* **2014**, *1*, 12. [CrossRef] [PubMed]
113. Nillegoda, N.B.; Bukau, B. Metazoan Hsp70-Based Protein Disaggregases: Emergence and Mechanisms. *Front. Mol. Biosci.* **2015**, *2*, 57. [CrossRef] [PubMed]
114. Deville, C.; Carroni, M.; Franke, K.B.; Topf, M.; Bukau, B.; Mogk, A.; Saibil, H.R. Structural Pathway of Regulated Substrate Transfer and Threading through an Hsp100 Disaggregase. *Sci. Adv.* **2017**, *3*, e1701726. [CrossRef] [PubMed]
115. Bukau, B.; Horwich, A.L. The Hsp70 and Hsp60 Chaperone Machines. *Cell* **1998**, *92*, 351–366. [CrossRef]
116. Zhang, X.; Kelly, J.W. Chaperonins Resculpt Folding Free Energy Landscapes to Avoid Kinetic Traps and Accelerate Protein Folding. *J. Mol. Biol.* **2014**, *426*, 2736–2738. [CrossRef]
117. Sekhar, A.; Rosenzweig, R.; Bouvignies, G.; Kay, L.E. Hsp70 Biases the Folding Pathways of Client Proteins. *Proc. Natl. Acad. Sci. USA* **2016**, *113*, E2794–E2801. [CrossRef]
118. Bagdany, M.; Veit, G.; Fukuda, R.; Avramescu, R.G.; Okiyoneda, T.; Baaklini, I.; Singh, J.; Sovak, G.; Xu, H.; Apaja, P.M.; et al. Chaperones Rescue the Energetic Landscape of Mutant CFTR at Single Molecule and in Cell. *Nat. Commun.* **2017**, *8*, 444. [CrossRef] [PubMed]
119. Çetinbaş, M.; Shakhnovich, E.I. Is Catalytic Activity of Chaperones a Selectable Trait for the Emergence of Heat Shock Response? *Biophys. J.* **2015**, *108*, 438–448. [CrossRef] [PubMed]
120. Suss, O.; Reichmann, D. Protein Plasticity Underlines Activation and Function of ATP-Independent Chaperones. *Front. Mol. Biosci.* **2015**, *2*, 43. [CrossRef]
121. Hebert, D.N.; Chandrasekhar, K.D.; Gierasch, L.M. You Got to Know When to Hold (or Unfold) 'Em. *Mol. Cell* **2012**, *48*, 3–4. [CrossRef]
122. Guisbert, E.; Yura, T.; Rhodius, V.A.; Gross, C.A. Convergence of Molecular, Modeling, and Systems Approaches for an Understanding of the *Escherichia Coli* Heat Shock Response. *Microbiol. Mol. Biol. Rev.* **2008**, *72*, 545–554. [CrossRef]
123. Fernandez-Funez, P.; Sanchez-Garcia, J.; de Mena, L.; Zhang, Y.; Levites, Y.; Khare, S.; Golde, T.E.; Rincon-Limas, D.E. Holdase Activity of Secreted Hsp70 Masks Amyloid-B42 Neurotoxicity in *Drosophila*. *Proc. Natl. Acad. Sci. USA* **2016**, *113*, E5212–E5221. [CrossRef]
124. Sorokina, I.; Mushegian, A. The Role of the Backbone Torsion in Protein Folding. *Biol. Direct* **2016**, *11*, 64. [CrossRef]
125. Sorokina, I.; Mushegian, A. Rotational Restriction of Nascent Peptides as an Essential Element of Co-Translational Protein Folding: Possible Molecular Players and Structural Consequences. *Biol. Direct* **2017**, *12*, 14. [CrossRef]
126. Sorokina, I.; Mushegian, A. Modeling Protein Folding in Vivo. *Biol. Direct* **2018**, *13*, 13. [CrossRef]
127. Netzer, W.J.; Hartl, F.U. Recombination of Protein Domains Facilitated by Co-Translational Folding in Eukaryotes. *Nature* **1997**, *388*, 343–349. [CrossRef]
128. Basharov, M.A. Cotranslational Folding of Proteins. *Biochemistry* **2000**, *65*, 1380–1384.

129. Lorimer, G.H. A Personal Account of Chaperonin History. *Plant Physiol.* **2001**, *125*, 38–41. [CrossRef]
130. Bashan, A.; Yonath, A. Ribosome Crystallography: Catalysis and Evolution of Peptide-Bond Formation, Nascent Chain Elongation and Its Co-Translational Folding. *Biochem. Soc. Trans.* **2005**, *33*, 488–492. [CrossRef]
131. Lim, V.I.; Curran, J.F.; Garber, M.B. Ribosomal Elongation Cycle: Energetic, Kinetic and Stereochemical Aspects. *J. Mol. Biol.* **2005**, *351*, 470–480. [CrossRef] [PubMed]
132. Ziv, G.; Haran, G.; Thirumalai, D. Ribosome Exit Tunnel Can Entropically Stabilize Alpha-Helices. *Proc. Natl. Acad. Sci. USA* **2005**, *102*, 18956–18961. [CrossRef] [PubMed]
133. Krishna, M.M.G.; Englander, S.W. A Unified Mechanism for Protein Folding: Predetermined Pathways with Optional Errors. *Protein Sci.* **2007**, *16*, 449–464. [CrossRef] [PubMed]
134. Steitz, T.A. A Structural Understanding of the Dynamic Ribosome Machine. *Nat. Rev. Mol. Cell Biol.* **2008**, *9*, 242–253. [CrossRef]
135. Kramer, G.; Boehringer, D.; Ban, N.; Bukau, B. The Ribosome as a Platform for Co-Translational Processing, Folding and Targeting of Newly Synthesized Proteins. *Nat. Struct. Mol. Biol.* **2009**, *16*, 589–597. [CrossRef]
136. Kaiser, C.M.; Goldman, D.H.; Chodera, J.D.; Tinoco, I.; Bustamante, C. The Ribosome Modulates Nascent Protein Folding. *Science* **2011**, *334*, 1723–1727. [CrossRef]
137. Wilson, D.N.; Beckmann, R. The Ribosomal Tunnel as a Functional Environment for Nascent Polypeptide Folding and Translational Stalling. *Curr. Opin. Struct. Biol.* **2011**, *21*, 274–282. [CrossRef]
138. Choi, S.I.; Kwon, S.; Son, A.; Jeong, H.; Kim, K.-H.; Seong, B.L. Protein Folding in Vivo Revisited. *Curr. Protein Pept. Sci.* **2013**, *14*, 721–733. [CrossRef]
139. Krobath, H.; Shakhnovich, E.I.; Faísca, P.F.N. Structural and Energetic Determinants of Co-Translational Folding. *J. Chem. Phys.* **2013**, *138*, 215101. [CrossRef]
140. Gloge, F.; Becker, A.H.; Kramer, G.; Bukau, B. Co-Translational Mechanisms of Protein Maturation. *Curr. Opin. Struct. Biol.* **2014**, *24*, 24–33. [CrossRef] [PubMed]
141. Javed, A.; Christodoulou, J.; Cabrita, L.D.; Orlova, E.V. The Ribosome and Its Role in Protein Folding: Looking through a Magnifying Glass. *Acta Crystallogr. Sect. D Struct. Biol.* **2017**, *73*, 509–521. [CrossRef]
142. Thommen, M.; Holtkamp, W.; Rodnina, M.V. Co-Translational Protein Folding: Progress and Methods. *Curr. Opin. Struct. Biol.* **2017**, *42*, 83–89. [CrossRef] [PubMed]
143. Sharma, A.K.; O'Brien, E.P. Non-Equilibrium Coupling of Protein Structure and Function to Translation-Elongation Kinetics. *Curr. Opin. Struct. Biol.* **2018**, *49*, 94–103. [CrossRef] [PubMed]
144. Wruck, F.; Avellaneda, M.J.; Koers, E.J.; Minde, D.P.; Mayer, M.P.; Kramer, G.; Mashaghi, A.; Tans, S.J. Protein Folding Mediated by Trigger Factor and Hsp70: New Insights from Single-Molecule Approaches. *J. Mol. Biol.* **2018**, *430*, 438–449. [CrossRef]
145. Alexander, L.M.; Goldman, D.H.; Wee, L.M.; Bustamante, C. Non-Equilibrium Dynamics of a Nascent Polypeptide during Translation Suppress Its Misfolding. *Nat. Commun.* **2019**, *10*, 2709. [CrossRef] [PubMed]
146. Waudby, C.A.; Dobson, C.M.; Christodoulou, J. Nature and Regulation of Protein Folding on the Ribosome. *Trends Biochem. Sci.* **2019**, *6*, 8. [CrossRef] [PubMed]
147. Cassaignau, A.M.E.; Cabrita, L.D.; Christodoulou, J. How Does the Ribosome Fold the Proteome? *Annu. Rev. Biochem.* **2020**, *89*, 389–415. [CrossRef]
148. Cassaignau, A.M.E.; Włodarski, T.; Chan, S.H.S.; Woodburn, L.F.; Bukvin, I.V.; Streit, J.O.; Cabrita, L.D.; Waudby, C.A.; Christodoulou, J. Interactions between Nascent Proteins and the Ribosome Surface Inhibit Co-Translational Folding. *Nat. Chem.* **2021**, *21*, 796. [CrossRef] [PubMed]
149. Maciuba, K.; Rajasekaran, N.; Chen, X.; Kaiser, C.M. Co-Translational Folding of Nascent Polypeptides: Multi-Layered Mechanisms for the Efficient Biogenesis of Functional Proteins. *Bioessays* **2021**, *43*, e2100042. [CrossRef]
150. Plessa, E.; Chu, L.P.; Chan, S.H.S.; Thomas, O.L.; Cassaignau, A.M.E.; Waudby, C.A.; Christodoulou, J.; Cabrita, L.D. Nascent Chains Can Form Co-Translational Folding Intermediates That Promote Post-Translational Folding Outcomes in a Disease-Causing Protein. *Nat. Commun.* **2021**, *12*, 6447. [CrossRef] [PubMed]
151. Wruck, F.; Tian, P.; Kudva, R.; Best, R.B.; von Heijne, G.; Tans, S.J.; Katranidis, A. The Ribosome Modulates Folding inside the Ribosomal Exit Tunnel. *Commun. Biol.* **2021**, *4*, 523. [CrossRef] [PubMed]
152. Fleming, P.J.; Rose, G.D. Conformational Properties of Unfolded Proteins. In *Protein Science Encyclopedia*; Fersht, A.R., Ed.; Wiley-VCH Verlag GmbH & Co. KGaA: Weinheim, Germany, 2008; p. sf06. ISBN 978-3-527-61075-4.
153. Sahakyan, H.; Nazaryan, K.; Mushegian, A.; Sorokina, I. Energy-Dependent Protein Folding: Modeling How a Protein Folding Machine May Work. *F1000Res* **2021**, *10*, 3. [CrossRef]
154. Spirin, A.S. *Ribosomes*; Kluwer Academic/Plenum Publishers: New York, NY, USA, 1999.
155. Doudna, J.A.; Batey, R.T. Structural Insights into the Signal Recognition Particle. *Annu. Rev. Biochem.* **2004**, *73*, 539–557. [CrossRef]
156. Lavery, L.A.; Partridge, J.R.; Ramelot, T.A.; Elnatan, D.; Kennedy, M.A.; Agard, D.A. Structural Asymmetry in the Closed State of Mitochondrial Hsp90 (TRAP1) Supports a Two-Step ATP Hydrolysis Mechanism. *Mol. Cell* **2014**, *53*, 330–343. [CrossRef]
157. Clerico, E.M.; Meng, W.; Pozhidaeva, A.; Bhasne, K.; Petridis, C.; Gierasch, L.M. Hsp70 Molecular Chaperones: Multifunctional Allosteric Holding and Unfolding Machines. *Biochem. J.* **2019**, *476*, 1653–1677. [CrossRef] [PubMed]
158. De Los Rios, P.; Barducci, A. Hsp70 Chaperones Are Non-Equilibrium Machines That Achieve Ultra-Affinity by Energy Consumption. *Elife* **2014**, *3*, e02218. [CrossRef] [PubMed]

159. Barducci, A.; De Los Rios, P. Non-Equilibrium Conformational Dynamics in the Function of Molecular Chaperones. *Curr. Opin. Struct. Biol.* **2015**, *30*, 161–169. [CrossRef] [PubMed]
160. Goloubinoff, P.; Sassi, A.; Fauvet, B.; Barducci, A.; De Los Rios, P. Molecular Chaperones Inject Energy from ATP Hydrolysis into the Nonequilibrium Stabilisation of Native Proteins. *Nat. Chem. Biol.* **2018**, *14*, 388–395. [CrossRef]
161. Assenza, S.; Sassi, A.S.; Kellner, R.; Schuler, B.; De Los Rios, P.; Barducci, A. Efficient Conversion of Chemical Energy into Mechanical Work by Hsp70 Chaperones. *Elife* **2019**, *8*, 8491. [CrossRef]
162. Chakrabarti, S.; Hyeon, C.; Ye, X.; Lorimer, G.H.; Thirumalai, D. Molecular Chaperones Maximize the Native State Yield on Biological Times by Driving Substrates out of Equilibrium. *Proc. Natl. Acad. Sci. USA* **2017**, *114*, E10919–E10927. [CrossRef]
163. Xu, H. ATP-Driven Nonequilibrium Activation of Kinase Clients by the Molecular Chaperone Hsp90. *Biophys. J.* **2020**, *119*, 1538–1549. [CrossRef]
164. Mirkin, B.G.; Fenner, T.I.; Galperin, M.Y.; Koonin, E.V. Algorithms for Computing Parsimonious Evolutionary Scenarios for Genome Evolution, the Last Universal Common Ancestor and Dominance of Horizontal Gene Transfer in the Evolution of Prokaryotes. *BMC Evol. Biol.* **2003**, *3*, 2. [CrossRef]
165. Mushegian, A. Gene Content of LUCA, the Last Universal Common Ancestor. *Front. Biosci.* **2008**, *13*, 4657–4666. [CrossRef]
166. Kannan, L.; Li, H.; Rubinstein, B.; Mushegian, A. Models of Gene Gain and Gene Loss for Probabilistic Reconstruction of Gene Content in the Last Universal Common Ancestor of Life. *Biol. Direct* **2013**, *8*, 32. [CrossRef]
167. Levinthal, C. Are There Pathways for Protein Folding? *J. Chim. Phys. Phys. Chim. Biol.* **1968**, *65*, 44–45. [CrossRef]
168. Levinthal, C. How to Fold Graciously. In *Mössbauer Spectroscopy in Biological Systems Proceedings. University of Illinois Bulletin*; Forgotten Books: London, UK, 1969; Volume 67, pp. 22–24.
169. Grosberg, A. A Few Disconnected Notes Related to Levinthal Paradox. *J. Biomol. Struct. Dyn.* **2002**, *20*, 317–321. [CrossRef]
170. Talkad, V.; Schneider, E.; Kennell, D. Evidence for Variable Rates of Ribosome Movement in Escherichia Coli. *J. Mol. Biol.* **1976**, *104*, 299–303. [CrossRef]
171. Olofsson, S.-O.; Boström, K.; Carlsson, P.; Borén, J.; Wettesten, M.; Bjursell, G.; Wiklund, O.; Bondjers, G. Structure and Biosynthesis of Apolipoprotein B. *Am. Heart J.* **1987**, *113*, 446–452. [CrossRef]
172. Li, G.-W.; Burkhardt, D.; Gross, C.; Weissman, J.S. Quantifying Absolute Protein Synthesis Rates Reveals Principles Underlying Allocation of Cellular Resources. *Cell* **2014**, *157*, 624–635. [CrossRef]
173. Petrosyan, R.; Narayan, A.; Woodside, M.T. Single-Molecule Force Spectroscopy of Protein Folding. *J. Mol. Biol.* **2021**, *433*, 167207. [CrossRef]
174. Liutkute, M.; Maiti, M.; Samatova, E.; Enderlein, J.; Rodnina, M.V. Gradual Compaction of the Nascent Peptide during Cotranslational Folding on the Ribosome. *Elife* **2020**, *9*, e60895. [CrossRef] [PubMed]
175. Mashaghi, A.; Moayed, F.; Koers, E.J.; Kramer, G.; Mayer, M.P.; Tans, S.J. Direct Observation of Hsp90-Induced Compaction in a Protein Chain. Available online: https://www.biorxiv.org/content/10.1101/2021.08.08.455546v1#page (accessed on 6 December 2021).
176. Julián, P.; Milon, P.; Agirrezabala, X.; Lasso, G.; Gil, D.; Rodnina, M.V.; Valle, M. The Cryo-EM Structure of a Complete 30S Translation Initiation Complex from *Escherichia coli*. *PLoS Biol.* **2011**, *9*, 1001095. [CrossRef] [PubMed]
177. Bögeholz, L.A.K.; Mercier, E.; Wintermeyer, W.; Rodnina, M.V. Kinetic Control of Nascent Protein Biogenesis by Peptide Deformylase. *Sci. Rep.* **2021**, *11*, 24457. [CrossRef] [PubMed]
178. Herrero Del Valle, A.; Seip, B.; Cervera-Marzal, I.; Sacheau, G.; Seefeldt, A.C.; Innis, C.A. Ornithine Capture by a Translating Ribosome Controls Bacterial Polyamine Synthesis. *Nat. Microbiol.* **2020**, *5*, 554–561. [CrossRef]
179. van der Stel, A.X.; Gordon, E.R.; Sengupta, A.; Martínez, A.K.; Klepacki, D.; Perry, T.N.; Herrero Del Valle, A.; Vázquez-Laslop, N.; Sachs, M.S.; Cruz-Vera, L.R.; et al. Structural Basis for the Tryptophan Sensitivity of TnaC-mediated Ribosome Stalling. *Nat. Commun.* **2021**, *12*, 5340. [CrossRef]
180. Nilsson, O.B.; Hedman, R.; Marino, J.; Wickles, S.; Bischoff, L.; Johansson, M.; Müller-Lucks, A.; Trovato, F.; Puglisi, J.D.; O'Brien, E.P.; et al. Cotranslational Protein Folding inside the Ribosome Exit Tunnel. *Cell Rep.* **2015**, *12*, 1533–1540. [CrossRef]
181. Bañó-Polo, M.; Baeza-Delgado, C.; Tamborero, S.; Hazel, A.; Grau, B.; Nilsson, I.; Whitley, P.; Gumbart, J.C.; von Heijne, G.; Mingarro, I. Transmembrane but not Soluble Helices Fold inside the Ribosome Tunnel. *Nat. Commun.* **2018**, *9*, 5246. [CrossRef] [PubMed]
182. Schulte, L.; Mao, J.; Reitz, J.; Sreeramulu, S.; Kudlinzki, D.; Hodirnau, V.V.; Meier-Credo, J.; Saxena, K.; Buhr, F.; Langer, J.D.; et al. Cysteine Oxidation and Disulfide Formation in the Ribosomal Exit Tunnel. *Nat. Commun.* **2020**, *11*, 5569. [CrossRef] [PubMed]
183. Bui, P.T.; Hoang, T.X. Protein Escape at the Ribosomal Exit Tunnel: Effects of Native Interactions, Tunnel Length, and Macromolecular Crowding. *J. Chem. Phys.* **2018**, *149*, 045102. [CrossRef] [PubMed]
184. Bui, P.T.; Hoang, T.X. Protein Escape at the Ribosomal Exit Tunnel: Effect of the Tunnel Shape. *J. Chem. Phys.* **2020**, *153*, 045105. [CrossRef]
185. Bui, P.T.; Hoang, T.X. Hydrophobic and Electrostatic Interactions Modulate Protein Escape at the Ribosomal Exit Tunnel. *Biophys. J.* **2021**, *9*, 27. [CrossRef]
186. Joiret, M.; Rapino, F.; Close, P.; Geris, L. Ribosome Exit Tunnel Electrostatics. Available online: https://www.biorxiv.org/content/10.1101/2020.10.20.346684v1.full (accessed on 6 December 2021).
187. Ferina, J.; Daggett, V. Visualizing Protein Folding and Unfolding. *J. Mol. Biol.* **2019**, *431*, 1540–1564. [CrossRef]

International Journal of *Molecular Sciences*

MDPI

Article

Pyranose Ring Puckering Thermodynamics for Glycan Monosaccharides Associated with Vertebrate Proteins

Olgun Guvench [1,2,*], Devon Martin [1,2] and Megan Greene [1]

1 Department of Pharmaceutical Sciences and Administration, School of Pharmacy, University of New England, 716 Stevens Avenue, Portland, ME 04103, USA; dmartin11@une.edu (D.M.); mgreene3@une.edu (M.G.)
2 Graduate School of Biomedical Science and Engineering, University of Maine, 5775 Stodder Hall, Orono, ME 04469, USA
* Correspondence: oguvench@une.edu; Tel.: +1-207-221-4171

Abstract: The conformational properties of carbohydrates can contribute to protein structure directly through covalent conjugation in the cases of glycoproteins and proteoglycans and indirectly in the case of transmembrane proteins embedded in glycolipid-containing bilayers. However, there continue to be significant challenges associated with experimental structural biology of such carbohydrate-containing systems. All-atom explicit-solvent molecular dynamics simulations provide a direct atomic resolution view of biomolecular dynamics and thermodynamics, but the accuracy of the results depends on the quality of the force field parametrization used in the simulations. A key determinant of the conformational properties of carbohydrates is ring puckering. Here, we applied extended system adaptive biasing force (eABF) all-atom explicit-solvent molecular dynamics simulations to characterize the ring puckering thermodynamics of the ten common pyranose monosaccharides found in vertebrate biology (as represented by the CHARMM carbohydrate force field). The results, along with those for idose, demonstrate that the CHARMM force field reliably models ring puckering across this diverse set of molecules, including accurately capturing the subtle balance between $^4C_1$ and $^1C_4$ chair conformations in the cases of iduronate and of idose. This suggests the broad applicability of the force field for accurate modeling of carbohydrate-containing vertebrate biomolecules such as glycoproteins, proteoglycans, and glycolipids.

Keywords: glucose; GlcNAc; galactose; GalNAc; mannose; xylose; fucose; Neu5Ac; glucuronate; iduronate; tetrahydropyran

Citation: Guvench, O.; Martin, D.; Greene, M. Pyranose Ring Puckering Thermodynamics for Glycan Monosaccharides Associated with Vertebrate Proteins. *Int. J. Mol. Sci.* **2022**, *23*, 473. https://doi.org/10.3390/ijms23010473

Academic Editors: Istvan Simon and Csaba Magyar

Received: 6 December 2021
Accepted: 28 December 2021
Published: 31 December 2021

Publisher's Note: MDPI stays neutral with regard to jurisdictional claims in published maps and institutional affiliations.

1. Introduction

Glycosylation is a common and important post-translational modification to proteins in eukaryotic biology. Additionally, carbohydrates are key components of eukaryotic lipids that make up the bilayers in which transmembrane proteins are embedded [1]. The carbohydrate portions of glycosylated proteins and glycolipids are called glycans. Naturally occurring glycans in vertebrates, including in humans, are composed of the monosaccharides D-glucose (Glc), *N*-acetyl-D-glucosamine (GlcNAc), D-galactose (Gal), *N*-acetyl-D-galactosamine (GalNAc), D-mannose (Man), D-xylose (Xyl), L-fucose (Fuc), *N*-acetyl-D-neuraminic acid (Neu5Ac), D-glucuronic acid (GlcA), and L-iduronic acid (IdoA), all in their pyranose forms [2] (Figure 1). As Neu5Ac, GlcA, and IdoA are expected to be deprotonated under typical physiological conditions, Figure 1 shows their conjugate base forms, *N*-acetyl-D-neuraminate, D-glucuronate, and L-iduronate, and it is these forms that are exclusively considered in what follows. Examples of glycans as components of glycosylated proteins are the *N*-glycans [3] and O-glycans [4] attached to glycoproteins and the glycosaminoglycans attached to proteoglycans [5]. Experimental atomic-resolution structural biology on glycosylated proteins is complicated by the non-template based synthesis of the attached glycans [6], which precludes a convenient source of homogeneous sample from biological sources, the intrinsic flexibility of glycans, which hinders conformational analysis by X-ray crystallography and NMR spectroscopy [7], and the covalent linkage of proteins with glycans, which can affect the structural properties of both the glycan and protein components [8–10]. In the context of membrane proteins, experimental atomic-resolution structural biology using X-ray crystallography entails extracting the membrane protein from its native lipid environment in order to create protein crystals [11], which means the effects of natural glycolipids in the native bilayer are not included in the structure determination. Therefore, the impact of glycans on protein structure continues to be at the frontiers of protein structure research.

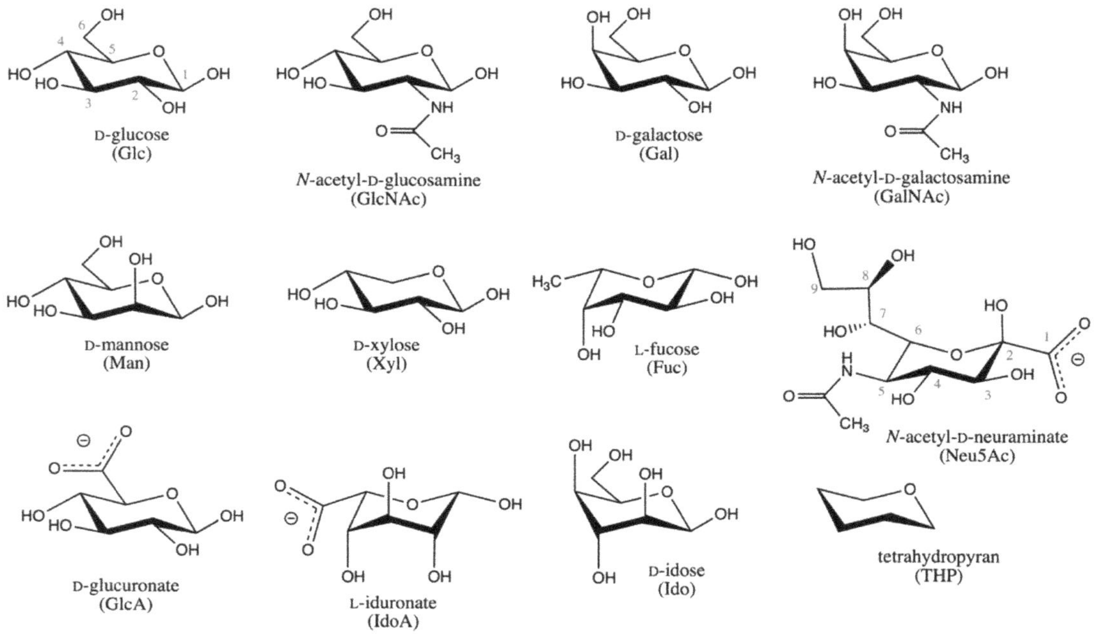

Figure 1. Compounds considered in the current study. Glc carbon atoms are numbered in blue. All other monosaccharides follow the same numbering scheme, except for Neu5Ac, which is numbered as pictured. All monosaccharides are drawn as the β anomer. The α anomer is created by inversion of the configuration at carbon 2 for Neu5Ac and at carbon 1 for all other monosaccharides. Both anomers for each monosaccharide as well as the corresponding O-methyl glycosides, formed by methylation at the anomeric carbon hydroxyl, were studied, for a total of 45 compounds (44 monosaccharides + THP).

Computational approaches for three-dimensional modeling of the atomic-resolution conformational properties of glycans have been developed and applied to help bridge the gaps in experimental methods [12–26]. Widely used among these computational approaches are explicit solvent molecular dynamics simulations employing atomistic force fields such as GLYCAM06 [27,28], GROMOS 53A6GLYC [29,30], GROMOS 56a6CARBO/CARBO_R [31–33], OPLS-AA [34,35], and CHARMM [36–39]. The quality of the results from such molecular dynamics simulations depends upon the quality of the force field parametrization. The conformational properties of glycans are determined principally by flexibility in the rings of the constituent monosaccharides and in the glycosidic linkages connecting them (Figure 2) [12,40], and thus it is important for force field parametrizations to accurately capture the physics of these sources of flexibility in order to ensure reliable modeling results.

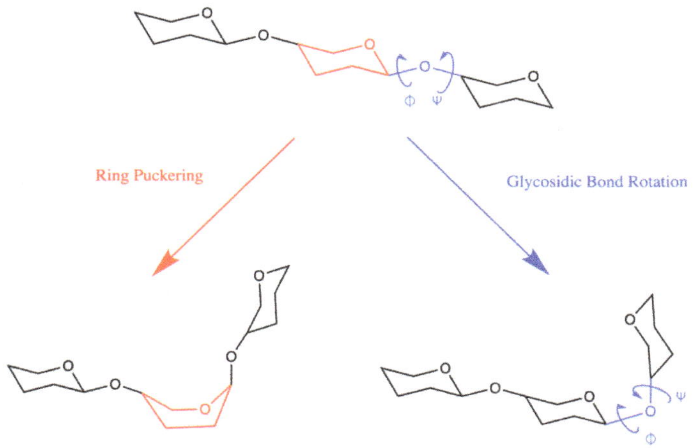

Figure 2. Pyranose ring puckering (red) and glycosidic bond rotation (blue) are the major sources of polymer flexibility in vertebrate glycans.

Since pyranose ring puckering occurs at the microsecond and beyond timescale [40–42], which is near the upper limit of typical present-day all-atom explicit-solvent molecular dynamics simulations, limitations in force field accuracy may not be readily apparent simply based on analysis of such simulation results. Here, we systematically determine the ring puckering thermodynamics of all compounds in Figure 2, including both the α and the β anomers and their corresponding O-methyl glycosides for the ten monosaccharides (i.e., 45 systems total), with the widely-used CHARMM force field. Extended System Adaptive Biasing Force (eABF) [43,44] is applied to achieve well-converged equilibrium statistics for ring puckering probabilities, with error estimates from triplicate 200-ns simulations for each system. The ring puckering thermodynamics from these simulations are in line with expected behavior, including for the highly flexible IdoA, and imply that the CHARMM force field can be used with confidence to correctly capture pyranose ring puckering contributions to glycan conformational heterogeneity in the context of the vertebrate glycans such as N-glycans, O-glycans, glycosaminoglycans, and glycolipids.

We additionally consider idose in its pyranose form, since, like IdoA, idose has a close balance between $^4C_1$ and $^1C_4$ chair probabilities, which makes it a useful test of force field accuracy. In agreement with prior computational results [45] and recent NMR data [46], the CHARMM carbohydrate force field performs very well in capturing the close balance for idose as well as for IdoA. Finally, for completeness, we include tetrahydropyran, which is the basic six-membered ring scaffold common to all of the monosaccharides considered here (Figure 2). As expected, there is an exact 50:50 balance for chair-chair interconversion for THP.

It is possible to tune ring puckering thermodynamics by selectively refining specific force field parameters and by using ring puckering thermodynamics as target data in the parametrization process. In the case of the GROMOS force field, such an approach was taken as a force field revision [31,32,47], and has yielded excellent results for ring puckering across a wide variety of pyranoses [32,33,45,48]. In the case of CHARMM, ring puckering thermodynamics in solution were not used as target data for CHARMM parametrization, and both bonded and nonbonded force field parameters, which built upon quantum mechanical gas phase puckering energetics for tetrahydropyran [36,49], are conserved across all of the different monosaccharides considered here. This demonstrates it is possible to correctly account for pyranose monosaccharide ring puckering thermodynamics in solution with a general transferable bonded and nonbonded force field parameter set. In the case of CHARMM, combining this parameter set with CHARMM force field parameters for proteins [50–52] can enable accurate modeling of glycoproteins and proteoglycans, and combining these parameters set with CHARMM force field parameters for lipids can do the same for glycolipids [53,54], which in turn can enable accurate modeling of transmembrane proteins embedded in complex bilayers composed of natural lipids.

2. Results and Discussion

2.1. Reaction Coordinate and Sampling Approach

Ring puckering for pyranose monosaccharides is commonly described using the Cremer-Pople (C-P) parameters (Q, θ, ϕ) [55], which provide a convenient quantitative means to identify both the extent and the nature of the puckering using spherical coordinates. The puckering amplitude Q describes the extent or magnitude of the puckering, while the angular values $0° \leq \theta \leq 180°$ and $0° \leq \phi < 360°$ describe the nature of the puckering. "Polar" values of θ near 0° and 180° correspond to $^4C_1$ and $^1C_4$ chair conformations, respectively, while "equatorial" values of θ near 90° correspond to boat and skew boat conformations, with the ϕ value indicating the specific boat or skew boat (e.g., ^2S_O). Intermediate or "tropical" values of θ, which are between the poles and the equator, correspond to envelope and half-envelope conformations, with the ϕ value indicating the specific envelope or half envelope [56].

Due to the long timescale for interconversion between $^4C_1$ and $^1C_4$, it is impractical to precisely determine the balance of probabilities and, hence, free energy difference, ΔG, between these conformations for pyranose monosaccharides on a routine basis using standard all-atom explicit-solvent molecular dynamics simulations. This is true for pyranoses where $\Delta G \approx 0$ due to the energy barrier separating the conformations [41,42], and the situation is even more difficult in cases where ΔG is substantially different than zero due to the difficulty in achieving equilibrium sampling of the unfavored conformation.

A logical means to address this issue is to apply a bias to θ during a simulation and either to reweight the sampling distribution to get unbiased conformational probabilities or to directly compute ΔG from the bias. Such an approach employing metadynamics [57,58] has enabled a number of studies to this end [29,32,33,45,48,59]. As demonstrated in these studies, this approach allows one to obtain a good estimate for ΔG with much less computation time than through standard (non-biased) molecular dynamics. There are two potential downsides to using a bias on θ. The first is the need to develop specialized computer code for the bias since the C-P θ is not a standard cartesian or internal coordinate. The second is that, while the single parameter θ can differentiate the two chair conformations from each other and also non-chair conformations from chair conformations, it cannot differentiate one non-chair conformation from another non-chair conformation. This second potential downside can be addressed by introducing a second simultaneous bias on ϕ but at the expense of further complicating the first downside.

For these reasons, direct use of dihedral angles is an attractive alternative. For example, Pickett and Strauss (P-S) defined three out-of-plane dihedrals constructed as various combinations of atoms in the pyranose ring [60], and it has been shown that simultaneous biases on all three of these angles can be effectively used to sample pyranose ring puckering [61]. In fact, the P-S and C-P approaches are mathematically equivalent [62]. However, there is an important practical difference with regard to applying biases on C-P parameters versus P-S out-of-plane dihedrals: only the two angular C-P parameters are required to uniquely identify the pucker nature (as opposed to magnitude) of a particular conformation whereas all three P-S out-of-plane dihedrals are required to do the same [59,63].

Babin and Sagui (B-S) have also proposed using dihedral angles for biased sampling of pyranose ring pucker [64]. In contrast to the P-S approach, only two dihedral angles are used in their scheme, $\alpha_1 \equiv$ O5–C1–C2–C3 and $\alpha_2 \equiv$ C3–C4–C5–O5, and the dihedral angles are real dihedrals determined by sequentially bonded atoms. Babin and Sagui have shown that biased sampling of (α_1, α_2) is an effective approach for sampling IdoA and GlcA puckering, and Alibay and Bryce have extended on these two monosaccharides to sulfated variants, as well as to non-sulfated and sulfated variants of GlcNAc, Gal, and

GalNAc [65]. In what follows, we demonstrate that major minima in $\Delta G(\alpha_1, \alpha_2)$ are populated by unique conformations. As such, it is possible to do direct integration of regions of $\Delta G(\alpha_1, \alpha_2)$ to determine ΔG not only between the $^4C_1$ and $^1C_4$ chairs, but also between specific boat/skew-boat conformations.

2.2. Extended System Adaptive Biasing Force (eABF) Sampling of the B-S (α_1, α_2) Reaction Coordinate

Methyl α-L-idopyranosiduronic acid (MeαIdoA) (Figure 2 "IdoA" with a methylated axial C1 hydroxyl) serves as a good test system to demonstrate the efficacy of eABF sampling of (α_1, α_2) owing to a small (<1 kcal/mol [46]) ΔG for conversion between the $^4C_1$ and $^1C_4$ chair conformations and a large energy barrier (~10 kcal/mol from the present work based on transition path saddle points in Figure 3), and therefore, slow kinetics, for this transition. Triplicate 200-ns eABF simulations with simultaneous biases on α_1 and α_2 and seeded with different randomized initial velocities yield essentially identical results across the entire $\Delta G(\alpha_1, \alpha_2)$ surface (Figure 3). Not only are the thermodynamic minima equal in both value and location, but so are the saddle regions and even the maxima, which demonstrates the excellent convergence properties of eABF for this system. $\Delta G(\alpha_1, \alpha_2)$ data are similarly well-converged for all 45 systems in this study (four different anomerization/methylation states for each of the 11 monosaccharides in Figure 2 plus tetrahydropyran; Supplementary Material Figures S1–S12).

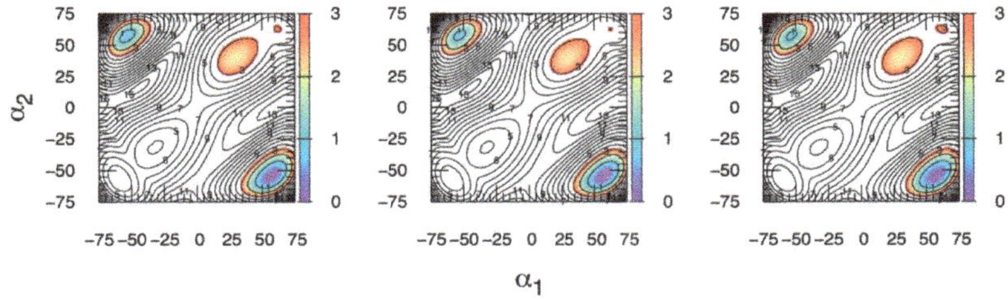

Figure 3. MeαIdoA $\Delta G(\alpha_1, \alpha_2)$ from eABF simulation. Each panel is from a separate 200-ns simulation seeded with different initial random velocities. α_1 and α_2 are in degrees. $\Delta G(\alpha_1, \alpha_2)$ is in kcal/mol, with contours drawn every 1 kcal/mol, colored from 0–3 kcal/mol, and labeled every 2 kcal/mol.

Additionally, each major thermodynamic minimum, that is, where $\Delta G(\alpha_1, \alpha_2) < 3$ kcal/mol, is populated by a single type of ring puckering conformation (Figure 4). We have chosen 3 kcal/mol as a cutoff value for the definition of major thermodynamic minimum since, at the simulation temperature of 298 K, values greater than 3 kcal/mol correspond to small probabilities, specifically, less than 0.64%. This association between a single ring puckering conformation and each major thermodynamic minimum in $\Delta G(\alpha_1, \alpha_2)$ holds for all 44 monosaccharides in this study, which illustrates the practical utility of the B-S reaction coordinate for characterizing pyranose ring puckering not only for chair conformations but also for specific non-chair conformations.

Kinetic data from the simulations clearly show the efficacy of eABF combined with the B-S (α_1, α_2) reaction coordinate for effectively sampling pyranose ring pucker, which is not surprising given the excellent convergence properties of $\Delta G(\alpha_1, \alpha_2)$ with eABF as discussed previously. Serving as a negative control, standard (non-biased) triplicate simulations of MeαIdoA starting from the $^1C_4$ chair undergo at most one transition in C-P θ during 200 ns (Figure 5a). Specifically, two of the simulations maintain $\theta \cong 180°$, indicating they are trapped in the initial conformation, while the third transitions at 25 ns to $\theta \cong 90°$ and remains there, indicating it is stuck in the equatorial boat/skew-boat region of puckering space. Therefore, standard sub-microsecond explicit solvent molecular dynamics simulation is inadequate for the task of sampling puckering conformations for pyranoses modeled with the CHARMM force field.

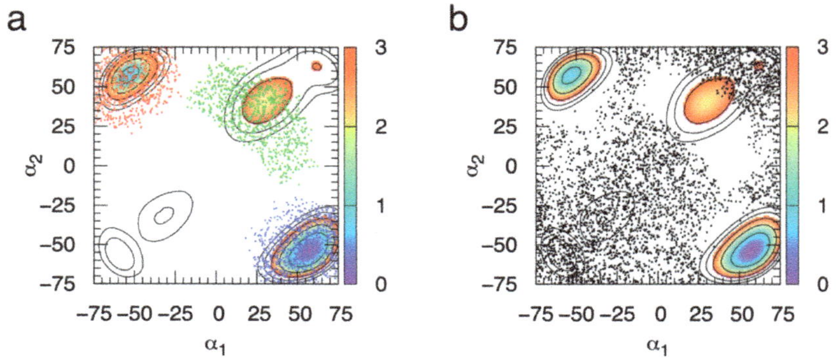

Figure 4. Sampling of specific MeαIdoA ring puckering conformations during eABF simulation with the Babin-Sagui (α_1, α_2) reaction coordinate. Sampled (α_1, α_2) values are separated into those for $^4C_1$, $^1C_4$, and ^2S_O (blue, red, and green dots, respectively, in panel "a") and for all other (black dots, panel "b") puckering conformations. α_1 and α_2 are in degrees. $\Delta G(\alpha_1, \alpha_2)$ is in kcal/mol, with contours drawn every 1 kcal/mol from 0–5 kcal/mol and colored from 0–3 kcal/mol. Puckering data have been aggregated across the triplicate simulations, and $\Delta G(\alpha_1, \alpha_2)$ is from the first simulation in the triplicate.

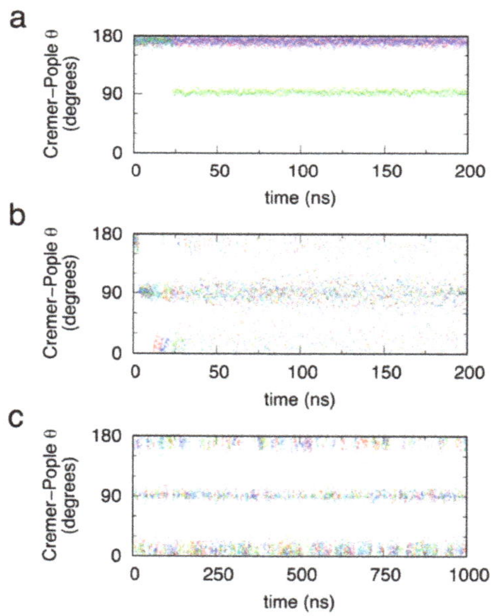

Figure 5. MeαIdoA conformational transitions in standard (non-biased) (**a**), eABF (**b**), and CMAP-biased (**c**) molecular dynamics simulations. eABF and CMAP biased simulations have biasing on the Babin-Sagui (α_1, α_2) reaction coordinate. Data in each panel are from triplicate simulations (blue, red, and green) seeded with different random initial velocities.

In contrast, with eABF sampling, during the first 25 ns, as the time-dependent biasing force becomes a progressively better estimate of the thermodynamic force along (α_1, α_2), transitions in θ start to become induced (Figure 5b). Beyond t = 25 ns, there is rapid transitioning on the nanosecond timescale from the $^4C_1$ chair ($\theta \cong 0°$), through boat/skew-boat conformations ($\theta \cong 90°$), to the $^1C_4$ chair ($\theta \cong 90°$) and back again, indicating sufficient sampling of (α_1, α_2) by eABF to provide an accurate estimate of the thermodynamic force along (α_1, α_2). As a technical point, the eABF approach applies a bias not to (α_1, α_2) directly but to extended degrees of freedom attached to (α_1, α_2), and the thermodynamic force on (α_1, α_2) is recovered from the biasing force applied to these extended degrees of freedom [43,44]. Standard ABF is not possible for sampling (α_1, α_2) since α_1 and α_2 do not meet the required orthogonality condition for standard ABF [66–68] owing to the sharing of atoms carbon 1 and oxygen 5 in both of the dihedral angle definitions. For additional information on this point, we refer interested readers to the cyclohexane data in Figure 2 of reference [44] and the associated discussion therein, which vividly demonstrates errors in estimation of cyclohexane puckering free energy with standard ABF that are corrected with eABF.

As a positive control, and similar to the approach of Babin and Sagui [64], we ran an additional set of simulations that employed CMAP-biased sampling [51,69]. In these simulations, the potential energy was defined by $U_{\text{non-biased}} + U_{\text{CMAP}}$, where $U_{\text{non-biased}}$ is the same CHARMM additive force field function used in the non-biased simulations here and U_{CMAP} is $U_{\text{CMAP}}(\alpha_1, \alpha_2) \cong -0.5 \times \Delta G(\alpha_1, \alpha_2)$. Unlike in the eABF simulations, the bias, in this case from the CMAP term, is fixed. We note that $U_{\text{CMAP}}(\alpha_1, \alpha_2)$ is only approximately equal to $-0.5 \times \Delta G(\alpha_1, \alpha_2)$ since, while $\Delta G(\alpha_1, \alpha_2)$ was computed on a square grid with a grid spacing of 1°, the grid spacing for the CMAP potential is 15°. As expected, there is excellent sampling of C-P θ from the very beginning of the triplicate simulations (Figure 5c). While there is rapid barrier crossing with this approach, there is less uniform sampling across all values of θ as compared to eABF sampling, with a strong tendency to favor sampling of polar and equatorial values of θ as compared to tropical values (Figure 5b vs. Figure 5c). This resulted from the factor of 0.5 used in the definition of $U_{\text{CMAP}}(\alpha_1, \alpha_2)$, and was done to maximize importance sampling of thermodynamically favored regions of (α_1, α_2) space while still lowering barriers sufficiently to achieve ergodic sampling of (α_1, α_2) on the 200-ns time scale of the simulations. As expected, thermodynamically unfavored regions of (α_1, α_2) correspond to tropical values, which in turn are envelope and half-envelope conformations with high degrees of ring strain.

Plotting C-P (θ, ϕ) values sampled during the eABF and the CMAP-biased simulations further validates the degree to which these two biasing methods applied to the (α_1, α_2) reaction coordinate enable sampling of pyranose puckering space. In addition to excellent coverage of the two chair conformations $^4C_1$ and $^1C_4$ located in the polar regions, there is good coverage of the equatorial region for 75° < ϕ < 270°, which includes $^5S_1$, $^{2,5}B$, ^2S_O, $B_{3,O}$, $^1S_3$, $^{1,4}B$, and $^1S_5$, in order of increasing ϕ (Figure 6). That said, there is very limited sampling of equatorial regions outside this range of ϕ values, resulting from the fact that the two-dimensional B-S (α_1, α_2) reaction coordinate is not a perfect replacement for biased sampling of the two-dimensional C-P (θ, ϕ) reaction coordinate. Nonetheless, it is reasonable to assume conformations not sampled are very high in free energy and that the thermodynamically relevant conformations have all been sampled. This latter point is emphasized by comparing these sampling data for eABF versus CMAP biasing. In the case of eABF, as time increases, sampling approaches that for a distribution biased by $-\Delta G(\alpha_1, \alpha_2)$, whereas for CMAP biasing, sampling is that for a distribution biased by $-0.5 \times \Delta G(\alpha_1, \alpha_2)$, as discussed above. As such, eABF provides more complete coverage of (θ, ϕ) space (Figure 6a) as compared to CMAP-biased sampling (Figure 6b).

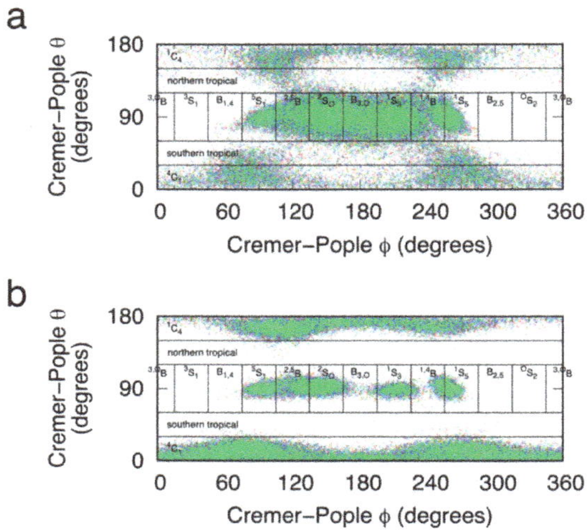

Figure 6. MeαIdoA Cremer-Pople (θ, ϕ) values sampled during eABF (**a**) and CMAP-biased (**b**) molecular dynamics simulations. Pyranose ring puckering regions [56] ("$^4C_1$", "northern tropical", "^2S_O", etc.) are labeled as defined in the Materials and Methods section. Biasing was applied to the Babin-Sagui (α_1, α_2) reaction coordinate. Data in each panel are from triplicate simulations (blue, red, and green) seeded with different random initial velocities. eABF simulations were 200 ns and CMAP-biased simulations were 1000 ns.

2.3. Using eABF-Computed $\Delta G(\alpha_1, \alpha_2)$ to Calculate Specific Ring Puckering Conformation Probabilities

Given that each major thermodynamic minimum for MeαIdoA is populated by a single type of puckering conformation, as shown above, it is possible simply to integrate the probabilities associated with each minimum to determine relative probabilities for specific ring puckering conformations. We operationalized this by converting $\Delta G(\alpha_1, \alpha_2)$ values from eABF simulations to probabilities $p(\alpha_1, \alpha_2)$ using the Boltzmann relationship $p(\alpha_1, \alpha_2) = \exp(\Delta G(\alpha_1, \alpha_2)/-RT)$, where R is the universal gas constant and T is the temperature. We then separated the data based on the (α_1, α_2) quadrant, and summed up all values of p for each (α_1, α_2) bin having an associated value $\Delta G(\alpha_1, \alpha_2) < 3$ kcal/mol within a 20° degree radius of the most favorable thermodynamic minimum in that quadrant. This yields at most one summed probability, P, per quadrant of the (α_1, α_2) coordinate. In the case of MeαIdoA, there are three such values, $P_{+,+}$, $P_{-,+}$, and $P_{+,-}$; the subscript here indicates the quadrant, for example, the quadrant defined by ($\alpha_1 < 0°$, $\alpha_2 > 0°$) for "−, +". As discussed above, for MeαIdoA, the "+, −" minimum corresponds uniquely to the $^4C_1$ ring pucker conformation, "−, +" to $^1C_4$, and "+, +" to ^2S_O (Figure 4), which allows for the assignment of probability values to specific ring pucker conformations based on eABF $\Delta G(\alpha_1, \alpha_2)$ results.

2.4. Ring Puckering Probabilities: Idose and Iduronate

Among the molecules considered in this study (Figure 2), the Ido and IdoA compounds are well known to exhibit significant conformational flexibility with regard to ring pucker. There are recent high-quality experimental results quantifying this, but with variable agreement with prior molecular dynamics simulation studies [46]. Comparison of $^4C_1$:$^1C_4$ ring puckering probability ratios shows good agreement between the present simulation results and these available experimental data (Table 1). In addition to probabilities from the eABF $\Delta G(\alpha_1, \alpha_2)$ results, we have included probabilities computed from the CMAP-biased simulations. These were determined by collecting all $^4C_1$ conformations from a CMAP-biased simulation, assigning a probability $p = \exp(U_{CMAP}/-RT)$ to each conformation to account for the effect of the CMAP bias, and then summing up the p values to get the total probability for the $^4C_1$ pucker. This was likewise carried out for the $^1C_4$ pucker, and the two total probabilities were normalized to sum to 100% (Table 1, "CMAP-biased simulations").

Table 1. $^4C_1$:$^1C_4$ ring puckering probability ratios in idose (Ido) and iduronate (IdoA) compounds.

Compound	eABF Simulations [1]	CMAP-Biased Simulations [1]	Experimental [46]
αIdo	17.6:82.4 (1.8)	15.1:84.9 (1.9)	18:82
MeαIdo	16.1:83.9 (0.7)	18.1:81.9 (2.1)	42:58
βIdo	97.1:2.9 (0.7)	90.7:9.3 (1.7)	82:18
MeβIdo	82.8:17.2 (2.2)	76.6:23.4 (1.5)	74:26
MeαIdoA	82.9:17.1 (1.4)	77.2:22.8 (1.0)	61:39

[1] Data are averages from triplicate simulations with standard error of the mean values in parentheses.

Converting the $^4C_1$:$^1C_4$ ring puckering probability ratios r to free energies using the relationship $\Delta G = -RT\ln(r)$ and plotting these ΔG values further illustrates how well the force field approach treats the close balance between $^4C_1$ and $^1C_4$ ring conformations. These ΔG values for the $^4C_1$ to $^1C_4$ equilibrium from the eABF and from the CMAP-biased simulations are typically within 0.5 kcal/mol of the experimental values (Figure 7a and Figure 7b, respectively). This very small degree of error is excellent for a force field model, and is not much different than what is seen when comparing the results from the two different simulation approaches using the same force field (Figure 7c).

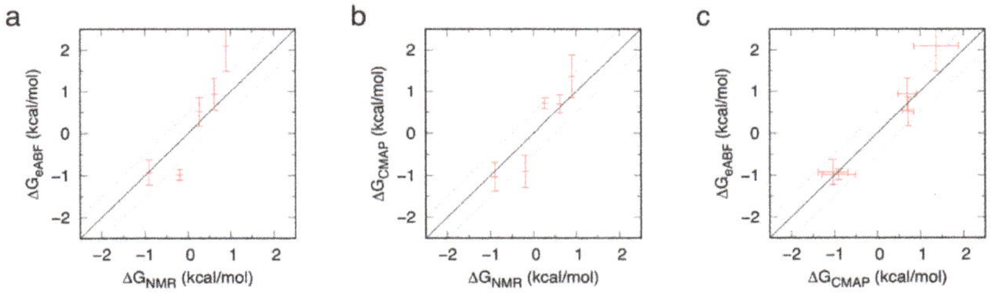

Figure 7. Comparison of ΔG values for the $^4C_1$ to $^1C_4$ equilibrium in Ido and IdoA compounds from eABF simulations, CMAP-biased simulations, and NMR experiments. Data are presented as eABF vs. NMR (**a**), CMAP-biased vs. NMR (**b**), and eABF vs. CMAP-biased (**c**). The specific compounds and the experimental data from NMR experiments are as detailed in Table 1. Simulation data points are averages from triplicate simulations, with error bars representing 95% confidence intervals. The solid diagonal is the line $y = x$, and the dotted diagonal lines are ±0.5 kcal/mol.

2.5. Ring Puckering: $\Delta G(\alpha_1, \alpha_2)$ Minima for All Compounds

Quantitative calculation of pyranose ring puckering probabilities is valuable for comparison to high-quality experimental data for pyranoses with multiple thermodynamically accessible puckering conformations, as in the case of IdoA and Ido. However, such calculation by either integration around eABF $\Delta G(\alpha_1, \alpha_2)$ minima or summing of re-weighted probabilities for individual snapshots from CMAP-biased simulations entails substantial post-simulation effort following the initial computation of $\Delta G(\alpha_1, \alpha_2)$ with eABF. Unlike IdoA and Ido, most of the pyranose monosaccharides considered here are expected to have a single thermodynamically important $\Delta G(\alpha_1, \alpha_2)$ minimum that corresponds to either the $^4C_1$ or $^1C_4$ chair pucker conformation. As such, tabulation of ΔG minima values in the four quadrants of (α_1, α_2) space provides a convenient semi-quantitative means to evaluate the behavior of the force field model for those compounds.

Table 2 lists the ΔG minimum value in each of the four quadrants of (α_1, α_2) space for each of the 45 compounds studied. It also correlates each thermodynamically important minimum (i.e., having a value of <3 kcal/mol) with the puckering conformation associated with the value of (α_1, α_2) for that ΔG minimum. This correlation was carried out using computed Cremer-Pople parameters (detailed in "Materials and Methods: Definition of $^4C_1$, $^1C_4$, ^2S_O, $^OS_2$, and other ring puckering conformations") for trajectory snapshots with (α_1, α_2) values within a 10° radius of the location of the ΔG minimum.

Table 2. Minimum $\Delta G(\alpha_1, \alpha_2)$ values in each of the four quadrants of the (α_1, α_2) reaction coordinate, and the corresponding major ring puckering conformation(s).

Compound	$\Delta G_{+,-}$ [1]	$\Delta G_{-,+}$ [1]	$\Delta G_{-,-}$ [1]	$\Delta G_{+,+}$ [1]	Major Pucker Conformation(s) [2]
αGlc	0	5.47 (0.04)	6.05 (0.04)	8.46 (0.05)	$^4C_1$
MeαGlc	0	6.83 (0.06)	7.06 (0.05)	9.39 (0.06)	$^4C_1$
βGlc	0	8.43 (0.19)	5.44 (0.04)	7.03 (0.01)	$^4C_1$
MeβGlc	0	8.27 (0.08)	5.38 (0.03)	6.91 (0.02)	$^4C_1$
αGlcNAc	0	5.01 (0.05)	6.11 (0.07)	7.19 (0.05)	$^4C_1$
MeαGlcNAc	0	6.19 (0.11)	7.08 (0.04)	8.20 (0.04)	$^4C_1$
βGlcNAc	0	4.95 (0.09)	4.01 (0.02)	6.60 (0.05)	$^4C_1$
MeβGlcNAc	0	4.73 (0.06)	2.83 (0.09)	6.85 (0.04)	$^4C_1 > {}^1S_5$
αGal	0	4.25 (0.06)	6.11 (0.04)	8.73 (0.06)	$^4C_1$
MeαGal	0	5.62 (0.01)	7.35 (0.04)	8.74 (0.03)	$^4C_1$
βGal	0	6.56 (0.09)	5.80 (0.01)	8.35 (0.04)	$^4C_1$
MeβGal	0	7.10 (0.06)	6.63 (0.04)	8.29 (0.04)	$^4C_1$
αGalNAc	0	3.09 (0.09)	7.00 (0.07)	7.72 (0.09)	$^4C_1$
MeαGalNAc	0	4.33 (0.06)	8.21 (0.05)	7.77 (0.06)	$^4C_1$
βGalNAc	0	2.47 (0.05)	3.66 (0.07)	7.03 (0.04)	$^4C_1 > {}^1C_4$
MeβGalNAc	0	2.90 (0.04)	3.58 (0.01)	6.87 (0.05)	$^4C_1 > {}^1C_4$
αMan	0	5.26 (0.6)	6.83 (0.02)	9.99 (0.06)	$^4C_1$
MeαMan	0	5.82 (0.03)	7.54 (0.06)	10.74 (0.03)	$^4C_1$
βMan	0	6.89 (0.05)	7.27 (0.03)	8.98 (0.04)	$^4C_1$
MeβMan	0	6.20 (0.05)	6.24 (0.05)	8.14 (0.01)	$^4C_1$
αXyl	0	2.17 (0.01)	6.03 (0.02)	6.00 (0.05)	$^4C_1 > {}^1C_4$
MeαXyl	0	3.60 (0.02)	6.95 (0.02)	6.73 (0.02)	$^4C_1$
βXyl	0	3.77 (0.05)	5.25 (0.01)	3.24 0.01)	$^4C_1$
MeβXyl	0	4.19 (0.00)	4.91 (0.02)	3.51 (0.01)	$^4C_1$
αFuc	3.87 (0.06)	0	8.11 (0.06)	5.97 (0.02)	$^1C_4$
MeαFuc	5.14 (0.03)	0	8.15 (0.01)	7.10 (0.02)	$^1C_4$
βFuc	6.48 (0.03)	0	8.10 (0.03)	5.73 (0.01)	$^1C_4$
MeβFuc	7.01 (0.04)	0	7.86 (0.02)	6.54 (0.04)	$^1C_4$

Table 2. *Cont.*

Compound	$\Delta G_{+,-}$ [1]	$\Delta G_{-,+}$ [1]	$\Delta G_{-,-}$ [1]	$\Delta G_{+,+}$ [1]	Major Pucker Conformation(s) [2]
αNeu5Ac	2.71 (0.09)	0	2.79 (0.02)	1.42 (0.07)	$^2C_5 > ^3S_O > ^5C_2 \cong ^{4,O}B$
MeαNeu5Ac	4.89 (0.29)	0	6.37 (0.10)	2.71 (0.24)	$^2C_5 > ^3S_O$
βNeu5Ac	6.79 (0.09)	0	7.18 (0.10)	4.01 (0.03)	$^1C_4$
MeβNeu5Ac	8.76 (0.08)	0	8.88 (0.10)	5.93 (0.11)	$^1C_4$
αGlcA	0	4.53 (0.12)	5.64 (0.04)	6.28 (0.05)	$^4C_1$
MeαGlcA	0	5.80 (0.04)	6.75 (0.03)	7.20 (0.03)	$^4C_1$
βGlcA	0	5.96 (0.11)	5.69 (0.01)	4.31 (0.06)	$^4C_1$
MeβGlcA	0	8.30 (0.09)	6.49 (0.02)	6.22 (0.04)	$^4C_1$
αIdoA	0	0.31 (0.09)	3.84 (0.04)	1.74 (0.02)	$^4C_1 \cong ^1C_4 > ^2S_O$
MeαIdoA	0	0.77 (0.05)	3.23 (0.01)	2.04 (0.02)	$^4C_1 > ^1C_4 > ^2S_O$
βIdoA	2.29 (0.03)	0	4.47 (0.08)	3.81 (0.03)	$^1C_4 > ^4C_1$
MeβIdoA	3.29 (0.06)	0	4.31 (0.05)	3.53 (0.06)	$^1C_4$
αIdo	0.73 (0.08)	0	0.88 (0.08)	3.20 (0.06)	$^1C_4 > ^4C_1 \cong ^OS_2$
Meαldo	0.82 (0.04)	0	1.00 (0.02)	2.81 (0.03)	$^1C_4 > ^4C_1 \cong ^OS_2 > ^3S_1$
βIdo	0	2.15 (0.10)	4.17 (0.08)	5.30 (0.09)	$^4C_1 > ^1C_4$
MeβIdo	0	1.12 (0.09)	3.49 (0.04)	5.00 (0.09)	$^4C_1 > ^1C_4$
THP	0	0.03 (0.01)	5.14 (0.01)	5.13 (0.00)	$^4C_1 = ^1C_4$

[1] Data in kcal/mol are averages from triplicate simulations for the minimum $\Delta G(\alpha_1, \alpha_2)$ in that quadrant. For example, "−,+" indicates the quadrant defined by ($\alpha_1 < 0°$, $\alpha_2 > 0°$). Standard error of the mean values are in parentheses. [2] Conformations are listed in order of most likely to least likely. Only conformations corresponding to $\Delta G_{+,-}$, $\Delta G_{-,+}$, $\Delta G_{+,+}$, and/or $\Delta G_{-,-}$ < 3 kcal/mol are listed.

As expected, most of the pyranose monosaccharides have a single major pucker conformation: the $^4C_1$ or the $^1C_4$ chair. Aside from IdoA and Ido, which were discussed in the previous section, exceptions to this are MeβGlcNAc, βGalNAc, MeβGalNAc, αXyl, αNeu5Ac, and MeαNeu5Ac.

MeβGlcNAc, βGalNAc, and MeβGalNAc all have their $\Delta G(\alpha_1, \alpha_2)$ global minimum corresponding to the $^4C_1$ chair conformation, as expected. They each also have a secondary minimum, but in all three cases the associated $\Delta G(\alpha_1, \alpha_2)$ is no less than 2.5 kcal/mol, which translates to a probability of no more than 1.5%. For MeβGlcNAc, the secondary minimum arises from skew-boat puckering, whereas for βGalNAc and MeβGalNAc, the secondary minimum is the $^1C_4$ chair conformation.

αXyl has the $^4C_1$ chair conformation as its global minimum and a secondary $\Delta G(\alpha_1, \alpha_2)$ minimum corresponding to the $^1C_4$ chair and with a value of 2.17 kcal/mol. This compares favorably to the value of 1.65 kcal/mol computed with the GROMOS 56a6CARBO force field (Table 1 in [48]), which is also exactly the value from Angyal's scheme for determining ring puckering free energies [70]. We note that data from Angyal's scheme have been used for quantitative comparison in other force field evaluations for a wide variety of pyranoses. It is worth emphasizing here that the Angyal data, though based in experiment, are indirect and were deemed by Angyal to be "calculated interaction energies". Concerning his "calculated interaction energies", Angyal writes, "an approximate calculation serves as a useful guide and can be readily carried out by adding the values of all of the non-bonded interactions occurring in each conformer, plus the value of the anomeric effect [70]". That is, those Angyal data for the $^4C_1$ to $^1C_4$ equilibrium in D-aldopyranoses listed in Table 1 of [70] are calculated as a simple sum of experimental values from model compounds, in contrast to being directly measured for each monosaccharide, for example, through NMR experiments [46].

Neu5Ac is discussed at length in Appendix A, below. Part of that discussion involves comparison to structures from PDB crystal structures. On the one hand, all simulation data here are for isolated Neu5Ac monosaccharides in liquid water, whereas the PDB data are from crystal environments and typically involve Neu5Ac having non-covalent interactions with other biomolecules and/or being covalently attached to other monosaccharides. On the other hand, there is substantial congruence between the aqueous simulation data and the experimental crystal data for Neu5Ac (Figure A1b,d,f,h in Appendix A). Indeed, a computational study of Neu5Ac ring puckering in vacuum and in explicit water noted that the structure of Neu5Ac bound in influenza neuraminidase belonged to conformations preferentially sampled in the aqueous simulations [71]. And, an analysis of high-resolution PDB data for MeβGlcNAc noted that while nearly 97% of structures in the data set were in the $^4C_1$ chair conformation, 2.6% were boats or skew boats [72], which correlates closely with data from the present work. Therefore, in addition to NMR data from directly analogous experimental systems of monosaccharides in liquid water, PDB data may be useful as benchmarks for the type of force field-based simulations described here.

On a final note, control eABF simulations for THP yield a $\Delta G(\alpha_1, \alpha_2)$ plot that is symmetric about both $\alpha_1 = \alpha_2$ and about $\alpha_1 = -\alpha_2$ (Supplementary Material Figure S12), as expected. There are two equivalent global minima at $^4C_1$ = $^1C_4$, and boat/skew-boat conformations are over 5 kcal/mol higher in free energy. Thus, the exocyclic functional groups in the pyranose monosaccharides considered here can be thought of as introducing two types of perturbations to the THP $\Delta G(\alpha_1, \alpha_2)$: breaking of the symmetry, and altering the balance of chair vs. boat/skew-boat energetics.

3. Materials and Methods

3.1. Force Field

All systems were modeled using the CHARMM all-atom additive force field for carbohydrates [36,38] and the CHARMM-modified TIP3P water parameters [73,74] as contained in the "jul20" release of the CHARMM force field available as "toppar_c36_jul20.tgz" from http://mackerell.umaryland.edu/charmm_ff.shtml (accessed on 3 March 2021). Systems with a carboxylate functional group additionally used sodium ion parameters [75,76] included in the same release. During the course of the present work, we discovered a set of typos in the jul20 parameter file that affect Neu5Ac puckering energetics. Full details are provided in Appendix A. The data presented in this manuscript and the associated Supplementary Material reflect the correct parameters as developed in [38].

3.2. System Construction

Solvated systems were constructed for each monosaccharide in Figure 2 using either the α or the β anomer or one of the corresponding O-methyl glycosides, resulting in four unique systems for each monosaccharide. Monosaccharide coordinates were constructed from default force field internal geometries. The solvent consisted of a cubic box of water molecules at the experimental density of water and having an edge length of the longest dimension of the monosaccharide plus 30 Å; water molecules within 3 Å of the monosaccharide were deleted. In systems with a carboxylate group, a single sodium ion replaced a water molecule randomly selected and at least 6 Å from the monosaccharide. All system construction was carried out using the CHARMM program, v. c45b1 [77]. A single system containing tetrahydropyran (THP) was similarly constructed.

3.3. Molecular Dynamics Simulations

Each system was simulated in triplicate under periodic boundary conditions. Each replicate within the triplicate was assigned random initial velocities using a unique random seed to generate a unique trajectory. Simulations were carried out using the NAMD software, v. 2.13 [78]. Electrostatic and Lennard-Jones interactions employed a 10-Å spherical cutoff. Lennard-Jones interaction energies were smoothly switched to zero in the interval 8–10 Å [79], an isotropic correction was applied for Lennard-Jones interactions beyond the cutoff [80], and the particle-mesh Ewald method with a 1 Å grid spacing accounted for electrostatic interactions beyond the cutoff [81]. After 1000 steps of energy minimization, each system was heated through re-initializing velocities to the target temperature of 298 K every 1000 molecular dynamics steps across 20,000 total steps with an integration timestep of 0.5 fs and positional restraints on solute non-hydrogen atoms. The SHAKE [82] and SETTLE algorithms [83] were respectively used to constrain all bonds involving hydrogen atoms and water geometries to their equilibrium values, and a temperature of 298 K and a pressure of 1 atm were maintained using Langevin thermostatting [84] and Nosé-Hoover Langevin barostatting [85,86]. Following heating, positional restraints were removed and data were collected from 200-ns production simulations (100×10^6 timesteps with an integration timestep of 2.0 fs).

The Extended-System Adaptive Biasing Force (eABF) methodology [43,44] was used to determine the free energy of pyranose ring puckering, $\Delta G(\alpha_1, \alpha_2)$, using reaction coordinates proposed by Babin and Sagui [64], where α_1 is the dihedral angle defined by the atoms O5-C1-C2-C3 and α_2 is the dihedral defined by the atoms C3-C4-C5-O5, except in the case of sialic acid in which these dihedrals are defined by O6-C2-C3-C4 and C4-C5-C6-O6, respectively. $\Delta G(\alpha_1, \alpha_2)$ was computed from the CZAR gradient estimate [43] using a Poisson equation formalism [87] implemented within NAMD via the Colvars software module [88]. eABF parameters included a fictitious particle spring constant of $k_B T$/degree/degree and sampling with a $1^\circ \times 1^\circ$ bin size and restrained with half-harmonic potentials to the range $-75^\circ < \alpha_{1,2} < 75^\circ$. Application of the biasing force in a given bin was scaled by 0 for the first 100 samples and then linearly scaled from 0 to 100% between 100 and 200 samples. Non-biased control simulations followed the same protocol but with no eABF sampling.

Additional CMAP-biased simulations were carried out for iduronate and for idose by applying a fixed bias equal to $-0.5 \times \Delta G(\alpha_1, \alpha_2)$ through the CHARMM force field CMAP term [69]. The representation of this bias using CMAP is not exact relative to the reference values computed by eABF simulation, as CMAP uses a square grid with 15° intervals between grid points and bicubic interpolation approximate $-0.5 \times \Delta G(\alpha_1, \alpha_2)$ for off-grid values of (α_1, α_2). CMAP-biased simulations were run using the OpenMM software, v. 7.5.1 [89] and a molecular dynamics protocol similar to that used for non-biased control NAMD simulations.

3.4. Molecular Dynamics Trajectory Analysis

Molecular dynamics trajectories were analyzed with the CHARMM software, including for the computation of Cremer-Pople ring puckering parameters [55]. VMD [90] was used for visualization and the creation of molecular graphics.

3.5. Definition of $^4C_1$, $^1C_4$, ^2S_O, $^OS_2$, and Other Ring Puckering Conformations

C-P parameters (θ, ϕ) were used to define ring puckering conformations as follows (note: analogous puckers for Neu5Ac compounds have all superscripted/subscripted numbers in puckering conformations incremented by 1 to reflect the different atom numbering in Neu5Ac, as shown in Figure 2):

- $^4C_1$: $0^\circ \leq \theta < 30^\circ$, ϕ = any
- Southern tropical: $30^\circ \leq \theta < 60^\circ$, ϕ = any
- Equatorial: $60^\circ \leq \theta < 120^\circ$, with specific conformations defined by,
    - $^{3,O}B$: $0^\circ \leq \phi < 15^\circ$ or $345^\circ \leq \phi < 360^\circ$
    - $^3S_1$: $15^\circ \leq \phi < 45^\circ$
    - $B_{1,4}$: $45^\circ \leq \phi < 75^\circ$
    - $^5S_1$: $75^\circ \leq \phi < 105^\circ$
    - $^{2,5}B$: $105^\circ \leq \phi < 135^\circ$
    - ^2SO: $135^\circ \leq \phi < 165^\circ$

    - $B_{3,O}$: 165° ≤ ϕ < 195°
    - $^1S_3$: 195° ≤ ϕ < 225°
    - $^{1,4}B$: 225° ≤ ϕ < 255°
    - $^1S_5$: 255° ≤ ϕ < 285°
    - $B_{2,5}$: 285° ≤ ϕ < 315°
    - $^OS_2$: 315° ≤ ϕ < 345°
- Northern tropical: 120° ≤ θ < 150°, ϕ = any
- $^4C_1$: 150° ≤ θ ≤ 180°, ϕ = any

4. Conclusions

The data presented here provide a thorough accounting of the ring puckering free energies for the ten common vertebrate monosaccharides and idose, as represented by the CHARMM force field. In addition to demonstrating that the CHARMM force field reliably models ring puckering across this set diverse of molecules, the results show that doing so is possible with a single set of self-consistent force-field parameters developed using a standardized force field parametrization protocol [36,38]. This, in combination with examples of CHARMM force field studies on glycosidic linkages [91–96], lends confidence to the application of these parameters in the modeling of carbohydrate-containing protein systems, such as glycoproteins and proteoglycans as well as transmembrane proteins in glycolipid-containing bilayers. Accurate simulations for these types of systems can help expand the frontiers of protein structural biology by bridging gaps in experimental approaches for characterizing carbohydrate-containing protein systems.

Supplementary Materials: The following are available online at https://www.mdpi.com/article/10.3390/ijms23010473/s1.

Author Contributions: Conceptualization, O.G.; methodology, O.G. and D.M.; software, O.G. and D.M.; validation, O.G. and D.M.; formal analysis, O.G. and D.M.; resources, O.G.; data curation, O.G.; writing—original draft preparation, O.G.; writing—review and editing, O.G., D.M. and M.G.; visualization, O.G. and M.G.; supervision, O.G.; project administration, O.G.; funding acquisition, O.G. All authors have read and agreed to the published version of the manuscript.

Funding: This research was funded by the National Science Foundation, grant number MCB-1453529 to O.G. The APC was funded by the University of New England School of Pharmacy.

Acknowledgments: O.G. gratefully acknowledges communications with Jérôme Hénin and Anthony Serianni.

Conflicts of Interest: The authors declare no conflict of interest. The funders had no role in the design of the study; in the collection, analyses, or interpretation of data; in the writing of the manuscript; or in the decision to publish the results.

Appendix A

During the course of the present work, we discovered a set of typos in the jul20 parameter file that affect Neu5Ac puckering energetics. These typos affect only Neu5Ac in the present work and will be corrected in a future official update to the CHARMM force field (A. D. MacKerell, Jr., personal communication). For the time being, the jul20 "par_all36_carb.prm" CHARMM parameter file can be manually corrected by adding the following lines to that file and deleting all other lines that refer to these same parameters:

NC2D1	CC3161	CC3161	CC3261	0.20	3	0.0
CC312	CC3163	CC3161	NC2D1	0.20	3	0.0
OC3C61	CC3163	CC3161	NC2D1	0.20	3	0.0

The typos affect two dihedrals in the Neu5Ac ring, with the first parameter affecting rotation about the C4-C5 bond and the second two rotation about the C5-C6 bond. The above three lines revert the parameters to the original values in the publication describing parametrization for Neu5Ac [38]. Figure A1 demonstrates the large qualitative difference between the eABF $\Delta G(\alpha_1, \alpha_2)$ results using the incorrect force field dihedral parameters resulting from the typos and the correct force field dihedral parameters that are the original values from that publication. In Figure A1, (α_1, α_2) values from all instances of Neu5Ac in the PDB are overlaid on top of the $\Delta G(\alpha_1, \alpha_2)$ contour plots, and clearly show the superiority of the correct force field parameters as judged by the overlap of the PDB data with the global minima in the $\Delta G(\alpha_1, \alpha_2)$ contour plots (data were extracted from the PDB on 30 July 2021 by searching with the SMILES string "CC(=O)NC1C(CC(OC1C(C(CO)O)O)(C(=O)O)O)O" and separating hits into either α anomers or β anomers, of which there were 439 and 52, respectively found across a total of 170 PDB entries). In the case of the incorrect parameters, there is poor overlap, while with the correct parameters there is excellent overlap.

For αNeu5Ac and MeαNeu5Ac simulated using the incorrect parameters (Figure A1a and Figure A1c, respectively), the global minimum is in a boat/skew-boat region of (α_1, α_2) space whereas the vast majority of crystallographic structures in the α anomeric form are in the $^2C_5$ chair pucker conformation. However, with the correct force field parameters, the global minimum is in the $^2C_5$ region of (α_1, α_2) space for both αNeu5Ac (Figure A1b) and MeαNeu5Ac (Figure A1d), and the small proportion of α anomeric crystallographic structures outside of this region are located in or near a secondary minimum with favorable free energy (i.e., < 3 kcal/mol).

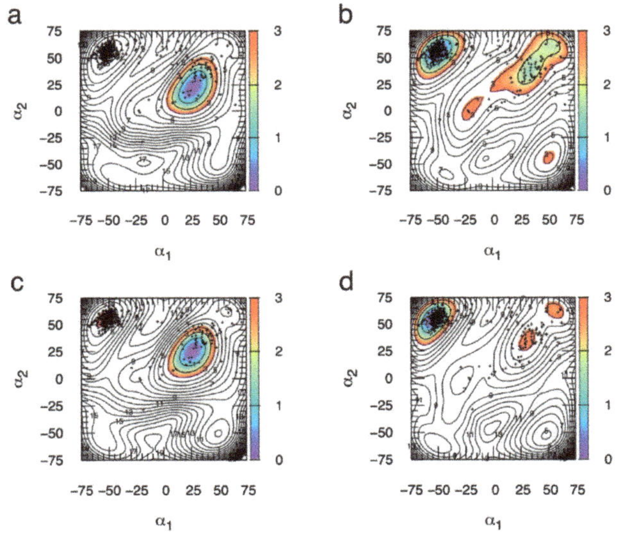

Figure A1. *Cont.*

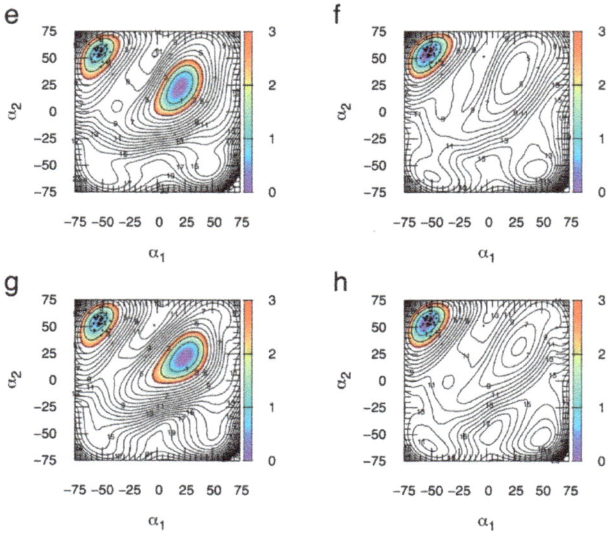

Figure A1. $\Delta G(\alpha_1, \alpha_2)$ from eABF simulations using incorrect versus correct dihedral force field parameters for Neu5Ac along with (α_1, α_2) data from all Neu5Ac structures in the PDB (searched 30 July 2021). $\Delta G(\alpha_1, \alpha_2)$ contour data are for αNeu5Ac with incorrect parameters (**a**), αNeu5Ac with correct parameters (**b**), MeαNeu5Ac with incorrect parameters (**c**), MeαNeu5Ac with correct parameters (**d**), βNeu5Ac with incorrect parameters (**e**), βNeu5Ac with correct parameters (**f**), MeβNeu5Ac with incorrect parameters (**g**), and MeβNeu5Ac with correct parameters (**h**). α_1 and α_2 are in degrees. $\Delta G(\alpha_1, \alpha_2)$ is from the first simulation in the triplicate and is in kcal/mol, with contours drawn every 1 kcal/mol and colored from 0–3 kcal/mol. PDB data were divided into two groups: those from α anomers and those from β anomers. Crystallographic data from the α anomers are displayed as small +'s in (**a**–**d**) and crystallographic data from the β anomers are displayed as small +'s in (**e**–**h**).

For the β anomers simulated using the incorrect parameters (βNeu5Ac (Figure A1e) and MeβNeu5Ac (Figure A1g)), there are no crystallographic Neu5Ac structures in the β anomeric form that coincide with the global minimum. In contrast, with the correct parameters, nearly all of these crystallographic structures in the β anomeric form, which are in the $^2C_5$ chair pucker conformation, coincide with the global $\Delta G(\alpha_1, \alpha_2)$ minimum for both βNeu5Ac (Figure A1f) and MeβNeu5Ac (Figure A1h).

References

1. Schnaar, R.L.; Kinoshita, T. Glycosphingolipids. In *Essentials of Glycobiology*; Varki, A., Cummings, R.D., Esko, J.D., Stanley, P., Hart, G.W., Aebi, M., Darvill, A.G., Kinoshita, T., Packer, N.H., Prestegard, J.H., et al., Eds.; Cold Spring Harbor Laboratory Press: Cold Spring Harbor, NY, USA, 2015; pp. 125–135. [CrossRef]
2. Seeberger, P.H. Monosaccharide Diversity. In *Essentials of Glycobiology*; Varki, A., Cummings, R.D., Esko, J.D., Stanley, P., Hart, G.W., Aebi, M., Darvill, A.G., Kinoshita, T., Packer, N.H., Prestegard, J.H., et al., Eds.; Cold Spring Harbor Laboratory Press: Cold Spring Harbor, NY, USA, 2017; pp. 19–30. [CrossRef]
3. Stanley, P.; Taniguchi, N.; Aebi, M. N-Glycans. In *Essentials of Glycobiology*; Varki, A., Cummings, R.D., Esko, J.D., Stanley, P., Hart, G.W., Aebi, M., Darvill, A.G., Kinoshita, T., Packer, N.H., Prestegard, J.H., et al., Eds.; Cold Spring Harbor Laboratory Press: Cold Spring Harbor, NY, USA, 2015; pp. 99–111. [CrossRef]
4. Brockhausen, I.; Stanley, P. O-GalNAc Glycans. In *Essentials of Glycobiology*; Varki, A., Cummings, R.D., Esko, J.D., Stanley, P., Hart, G.W., Aebi, M., Darvill, A.G., Kinoshita, T., Packer, N.H., Prestegard, J.H., et al., Eds.; Cold Spring Harbor Laboratory Press: Cold Spring Harbor, NY, USA, 2015; pp. 113–123. [CrossRef]
5. Lindahl, U.; Couchman, J.; Kimata, K.; Esko, J.D. Proteoglycans and Sulfated Glycosaminoglycans. In *Essentials of Glycobiology*; Varki, A., Cummings, R.D., Esko, J.D., Stanley, P., Hart, G.W., Aebi, M., Darvill, A.G., Kinoshita, T., Packer, N.H., Prestegard, J.H., et al., Eds.; Cold Spring Harbor Laboratory Press: Cold Spring Harbor, NY, USA, 2015; pp. 207–221. [CrossRef]

6. Varki, A.; Kornfeld, S. Historical Background and Overview. In *Essentials of Glycobiology*; Varki, A., Cummings, R.D., Esko, J.D., Stanley, P., Hart, G.W., Aebi, M., Darvill, A.G., Kinoshita, T., Packer, N.H., Prestegard, J.H., et al., Eds.; Cold Spring Harbor Laboratory Press: Cold Spring Harbor, NY, USA, 2015; pp. 1–18. [CrossRef]
7. Imberty, A.; Prestegard, J.H. Structural Biology of Glycan Recognition. In *Essentials of Glycobiology*; Varki, A., Cummings, R.D., Esko, J.D., Stanley, P., Hart, G.W., Aebi, M., Darvill, A.G., Kinoshita, T., Packer, N.H., Prestegard, J.H., et al., Eds.; Cold Spring Harbor Laboratory Press: Cold Spring Harbor, NY, USA, 2015; pp. 387–400. [CrossRef]
8. Kumar, A.; Narayanan, V.; Sekhar, A. Characterizing Post-Translational Modifications and Their Effects on Protein Conformation Using NMR Spectroscopy. *Biochemistry* **2020**, *59*, 57–73. [CrossRef] [PubMed]
9. Xin, F.; Radivojac, P. Post-translational modifications induce significant yet not extreme changes to protein structure. *Bioinformatics* **2012**, *28*, 2905–2913. [CrossRef] [PubMed]
10. Craveur, P.; Narwani, T.J.; Rebehmed, J.; de Brevern, A.G. Investigation of the impact of PTMs on the protein backbone conformation. *Amino Acids* **2019**, *51*, 1065–1079. [CrossRef]
11. Kermani, A.A. A guide to membrane protein X-ray crystallography. *FEBS J.* **2021**, *288*, 5788–5804. [CrossRef]
12. Woods, R.J. Predicting the Structures of Glycans, Glycoproteins, and Their Complexes. *Chem. Rev.* **2018**, *118*, 8005–8024. [CrossRef] [PubMed]
13. Samsonov, S.A.; Pisabarro, M.T. Computational analysis of interactions in structurally available protein-glycosaminoglycan complexes. *Glycobiology* **2016**, *26*, 850–861. [CrossRef] [PubMed]
14. Whitmore, E.K.; Vesenka, G.; Sihler, H.; Guvench, O. Efficient Construction of Atomic-Resolution Models of Non-Sulfated Chondroitin Glycosaminoglycan Using Molecular Dynamics Data. *Biomolecules* **2020**, *10*, 537. [CrossRef]
15. Whitmore, E.K.; Martin, D.; Guvench, O. Constructing 3-Dimensional Atomic-Resolution Models of Nonsulfated Glycosaminoglycans with Arbitrary Lengths Using Conformations from Molecular Dynamics. *Int. J. Mol. Sci.* **2020**, *21*, 7699. [CrossRef]
16. Bohne-Lang, A.; von der Lieth, C.-W. GlyProt: In silico glycosylation of proteins. *Nucleic Acids Res.* **2005**, *33*, W214–W219. [CrossRef]
17. Singh, A.; Montgomery, D.; Xue, X.; Foley, B.L.; Woods, R.J. GAG Builder: A web-tool for modeling 3D structures of glycosaminoglycans. *Glycobiology* **2019**, *29*, 515–518. [CrossRef]
18. Engelsen, S.B.; Hansen, P.I.; Perez, S. POLYS 2.0: An open source software package for building three-dimensional structures of polysaccharides. *Biopolymers* **2014**, *101*, 733–743. [CrossRef] [PubMed]
19. Kuttel, M.M.; Ståhle, J.; Widmalm, G. CarbBuilder: Software for building molecular models of complex oligo- and polysaccharide structures. *J. Comput. Chem.* **2016**, *37*, 2098–2105. [CrossRef]
20. Clerc, O.; Deniaud, M.; Vallet, S.D.; Naba, A.; Rivet, A.; Perez, S.; Thierry-Mieg, N.; Ricard-Blum, S. MatrixDB: Integration of new data with a focus on glycosaminoglycan interactions. *Nucleic Acids Res.* **2019**, *47*, D376–D381. [CrossRef]
21. Clerc, O.; Mariethoz, J.; Rivet, A.; Lisacek, F.; Pérez, S.; Ricard-Blum, S. A pipeline to translate glycosaminoglycan sequences into 3D models. Application to the exploration of glycosaminoglycan conformational space. *Glycobiology* **2019**, *29*, 36–44. [CrossRef] [PubMed]
22. Park, S.-J.; Lee, J.; Qi, Y.; Kern, N.R.; Lee, H.S.; Jo, S.; Joung, I.; Joo, K.; Lee, J.; Im, W. CHARMM-GUI Glycan Modeler for modeling and simulation of carbohydrates and glycoconjugates. *Glycobiology* **2019**, *29*, 320–331. [CrossRef]
23. Almond, A. Multiscale modeling of glycosaminoglycan structure and dynamics: Current methods and challenges. *Curr. Opin. Struct. Biol.* **2018**, *50*, 58–64. [CrossRef]
24. Sattelle, B.M.; Shakeri, J.; Cliff, M.J.; Almond, A. Proteoglycans and their heterogeneous glycosaminoglycans at the atomic scale. *Biomacromolecules* **2015**, *16*, 951–961. [CrossRef]
25. Frank, M. Conformational analysis of oligosaccharides and polysaccharides using molecular dynamics simulations. *Methods Mol. Biol.* **2015**, *1273*, 359–377. [CrossRef] [PubMed]
26. Widmalm, G. A perspective on the primary and three-dimensional structures of carbohydrates. *Carbohydr. Res.* **2013**, *378*, 123–132. [CrossRef]
27. Kirschner, K.N.; Yongye, A.B.; Tschampel, S.M.; González-Outeiriño, J.; Daniels, C.R.; Foley, B.L.; Woods, R.J. GLYCAM06: A generalizable biomolecular force field. *Carbohydrates. J. Comput. Chem.* **2008**, *29*, 622–655. [CrossRef]
28. Singh, A.; Tessier, M.B.; Pederson, K.; Wang, X.; Venot, A.P.; Boons, G.-J.; Prestegard, J.H.; Woods, R.J. Extension and validation of the GLYCAM force field parameters for modeling glycosaminoglycans. *Can. J. Chem.* **2016**, *94*, 927–935. [CrossRef]
29. Pol-Fachin, L.; Rusu, V.H.; Verli, H.; Lins, R.D. GROMOS 53A6GLYC, an Improved GROMOS Force Field for Hexopyranose-Based Carbohydrates. *J. Chem. Theory Comput.* **2012**, *8*, 4681–4690. [CrossRef] [PubMed]
30. Pol-Fachin, L.; Verli, H.; Lins, R.D. Extension and validation of the GROMOS $53A6_{GLYC}$ parameter set for glycoproteins. *J. Comput. Chem.* **2014**, *35*, 2087–2095. [CrossRef]
31. Hansen, H.S.; Hünenberger, P.H. A reoptimized GROMOS force field for hexopyranose-based carbohydrates accounting for the relative free energies of ring conformers, anomers, epimers, hydroxymethyl rotamers, and glycosidic linkage conformers. *J. Comput. Chem.* **2011**, *32*, 998–1032. [CrossRef]
32. Plazinski, W.; Lonardi, A.; Hünenberger, P.H. Revision of the GROMOS $56A6_{CARBO}$ force field: Improving the description of ring-conformational equilibria in hexopyranose-based carbohydrates chains. *J. Comput. Chem.* **2016**, *37*, 354–365. [CrossRef] [PubMed]

33. Panczyk, K.; Gaweda, K.; Drach, M.; Plazinski, W. Extension of the GROMOS $56a6_{CARBO/CARBO\_R}$ Force Field for Charged, Protonated, and Esterified Uronates. *J. Phys. Chem. B* **2018**, *122*, 3696–3710. [CrossRef]
34. Damm, W.; Frontera, A.; Tirado-Rives, J.; Jorgensen, W.L. OPLS all-atom force field for carbohydrates. *J. Comput. Chem.* **1997**, *18*, 1955–1970. [CrossRef]
35. Kony, D.; Damm, W.; Stoll, S.; van Gunsteren, W.F. An improved OPLS-AA force field for carbohydrates. *J. Comput. Chem.* **2002**, *23*, 1416–1429. [CrossRef] [PubMed]
36. Guvench, O.; Greene, S.N.; Kamath, G.; Brady, J.W.; Venable, R.M.; Pastor, R.W.; Mackerell, A.D., Jr. Additive empirical force field for hexopyranose monosaccharides. *J. Comput. Chem.* **2008**, *29*, 2543–2564. [CrossRef]
37. Guvench, O.; Hatcher, E.R.; Venable, R.M.; Pastor, R.W.; MacKerell, A.D., Jr. CHARMM Additive All-Atom Force Field for Glycosidic Linkages between Hexopyranoses. *J. Chem. Theory Comput.* **2009**, *5*, 2353–2370. [CrossRef]
38. Guvench, O.; Mallajosyula, S.S.; Raman, E.P.; Hatcher, E.; Vanommeslaeghe, K.; Foster, T.J.; Jamison, F.W., II; Mackerell, A.D., Jr. CHARMM additive all-atom force field for carbohydrate derivatives and its utility in polysaccharide and carbohydrate-protein modeling. *J. Chem. Theory Comput.* **2011**, *7*, 3162–3180. [CrossRef] [PubMed]
39. Mallajosyula, S.S.; Guvench, O.; Hatcher, E.; MacKerell, A.D., Jr. CHARMM Additive All-Atom Force Field for Phosphate and Sulfate Linked to Carbohydrates. *J. Chem. Theory Comput.* **2012**, *8*, 759–776. [CrossRef]
40. Sattelle, B.M.; Shakeri, J.; Almond, A. Does Microsecond Sugar Ring Flexing Encode 3D-Shape and Bioactivity in the Heparanome? *Biomacromolecules* **2013**, *14*, 1149–1159. [CrossRef]
41. Sattelle, B.M.; Hansen, S.U.; Gardiner, J.; Almond, A. Free energy landscapes of iduronic acid and related monosaccharides. *J. Am. Chem. Soc.* **2010**, *132*, 13132–13134. [CrossRef] [PubMed]
42. Sattelle, B.M.; Bose-Basu, B.; Tessier, M.; Woods, R.J.; Serianni, A.S.; Almond, A. Dependence of pyranose ring puckering on anomeric configuration: Methyl idopyranosides. *J. Phys. Chem. B* **2012**, *116*, 6380–6386. [CrossRef]
43. Lesage, A.; Lelièvre, T.; Stoltz, G.; Hénin, J. Smoothed Biasing Forces Yield Unbiased Free Energies with the Extended-System Adaptive Biasing Force Method. *J. Phys. Chem. B* **2017**, *121*, 3676–3685. [CrossRef] [PubMed]
44. Fu, H.; Shao, X.; Chipot, C.; Cai, W. Extended Adaptive Biasing Force Algorithm. An On-the-Fly Implementation for Accurate Free-Energy Calculations. *J. Chem. Theory Comput.* **2016**, *12*, 3506–3513. [CrossRef]
45. Plazinski, W.; Plazinska, A. Molecular dynamics simulations of hexopyranose ring distortion in different force fields. *Pure Appl. Chemistry. Chim. Pure Appl.* **2017**, *89*, 1283–1294. [CrossRef]
46. Bose-Basu, B.; Zhang, W.; Kennedy, J.L.W.; Hadad, M.J.; Carmichael, I.; Serianni, A.S. ^{13}C-Labeled Idohexopyranosyl Rings: Effects of Methyl Glycosidation and C6 Oxidation on Ring Conformational Equilibria. *J. Org. Chem.* **2017**, *82*, 1356–1370. [CrossRef]
47. Lins, R.D.; Hünenberger, P.H. A new GROMOS force field for hexopyranose-based carbohydrates. *J. Comput. Chem.* **2005**, *26*, 1400–1412. [CrossRef]
48. Panczyk, K.; Plazinski, W. Pyranose ring puckering in aldopentoses, ketohexoses and deoxyaldohexoses. A molecular dynamics study. *Carbohydr. Res.* **2018**, *455*, 62–70. [CrossRef]
49. Guvench, O.; MacKerell, A.D., Jr. Automated conformational energy fitting for force-field development. *J. Mol. Model.* **2008**, *14*, 667–679. [CrossRef]
50. MacKerell, A.D., Jr.; Bashford, D.; Bellott, M.; Dunbrack, R.L.; Evanseck, J.D.; Field, M.J.; Fischer, S.; Gao, J.; Guo, H.; Ha, S.; et al. All-atom empirical potential for molecular modeling and dynamics studies of proteins. *J. Phys. Chem. B* **1998**, *102*, 3586–3616. [CrossRef]
51. MacKerell, A.D., Jr.; Feig, M.; Brooks, C.L., III. Improved treatment of the protein backbone in empirical force fields. *J. Am. Chem. Soc.* **2004**, *126*, 698–699. [CrossRef] [PubMed]
52. Best, R.B.; Zhu, X.; Shim, J.; Lopes, P.E.; Mittal, J.; Feig, M.; Mackerell, A.D., Jr. Optimization of the additive CHARMM all-atom protein force field targeting improved sampling of the backbone phi, psi and side-chain chi(1) and chi(2) dihedral angles. *J. Chem. Theory Comput.* **2012**, *8*, 3257–3273. [CrossRef]
53. Klauda, J.B.; Venable, R.M.; Freites, J.A.; O'Connor, J.W.; Tobias, D.J.; Mondragon-Ramirez, C.; Vorobyov, I.; MacKerell, A.D., Jr.; Pastor, R.W. Update of the CHARMM all-atom additive force field for lipids: Validation on six lipid types. *J. Phys. Chem. B* **2010**, *114*, 7830–7843. [CrossRef] [PubMed]
54. Klauda, J.B.; Monje, V.; Kim, T.; Im, W. Improving the CHARMM Force Field for Polyunsaturated Fatty Acid Chains. *J. Phys. Chem. B* **2012**, *116*, 9424–9431. [CrossRef]
55. Cremer, D.; Pople, J.A. General definition of ring puckering coordinates. *J. Am. Chem. Soc.* **1975**, *97*, 1354–1358. [CrossRef]
56. Dowd, M.K.; French, A.D.; Reilly, P.J. Modeling of aldopyranosyl ring puckering with MM3 (92). *Carbohydr. Res.* **1994**, *264*, 1–19. [CrossRef]
57. Barducci, A.; Bussi, G.; Parrinello, M. Well-Tempered Metadynamics: A Smoothly Converging and Tunable Free-Energy Method. *Phys. Rev. Lett.* **2008**, *100*, 020603. [CrossRef]
58. Barducci, A.; Bonomi, M.; Parrinello, M. Metadynamics. *WIREs Comput. Mol. Sci.* **2011**, *1*, 826–843. [CrossRef]
59. Autieri, E.; Sega, M.; Pederiva, F.; Guella, G. Puckering free energy of pyranoses: A NMR and metadynamics-umbrella sampling investigation. *J. Chem. Phys.* **2010**, *133*, 095104. [CrossRef]
60. Pickett, H.M.; Strauss, H.L. Conformational structure, energy, and inversion rates of cyclohexane and some related oxanes. *J. Am. Chem. Soc.* **1970**, *92*, 7281–7290. [CrossRef]

61. Hansen, H.S.; Hünenberger, P.H. Using the local elevation method to construct optimized umbrella sampling potentials: Calculation of the relative free energies and interconversion barriers of glucopyranose ring conformers in water. *J. Comput. Chem.* **2010**, *31*, 1–23. [CrossRef]
62. Boeyens, J.C.A.; Evans, D.G. Group theory of ring pucker. *Acta Crystallogr. Sect. B* **1989**, *45*, 577–581. [CrossRef]
63. Sega, M.; Autieri, E.; Pederiva, F. Pickett angles and Cremer–Pople coordinates as collective variables for the enhanced sampling of six-membered ring conformations. *Mol. Phys.* **2011**, *109*, 141–148. [CrossRef]
64. Babin, V.; Sagui, C. Conformational free energies of methyl-alpha-L-iduronic and methyl-beta-D-glucuronic acids in water. *J. Chem. Phys.* **2010**, *132*, 104108. [CrossRef]
65. Alibay, I.; Bryce, R.A. Ring Puckering Landscapes of Glycosaminoglycan-Related Monosaccharides from Molecular Dynamics Simulations. *J. Chem. Inf. Model.* **2019**, *59*, 4729–4741. [CrossRef] [PubMed]
66. Darve, E.; Rodríguez-Gómez, D.; Pohorille, A. Adaptive biasing force method for scalar and vector free energy calculations. *J. Chem. Phys.* **2008**, *128*, 144120. [CrossRef]
67. Hénin, J.; Chipot, C. Overcoming free energy barriers using unconstrained molecular dynamics simulations. *J. Chem. Phys.* **2004**, *121*, 2904–2914. [CrossRef]
68. Hénin, J.; Fiorin, G.; Chipot, C.; Klein, M.L. Exploring Multidimensional Free Energy Landscapes Using Time-Dependent Biases on Collective Variables. *J. Chem. Theory Comput.* **2010**, *6*, 35–47. [CrossRef]
69. MacKerell, A.D., Jr.; Feig, M.; Brooks, C.L., III. Extending the treatment of backbone energetics in protein force fields: Limitations of gas-phase quantum mechanics in reproducing protein conformational distributions in molecular dynamics simulations. *J. Comput. Chem.* **2004**, *25*, 1400–1415. [CrossRef] [PubMed]
70. Angyal, S.J. The Composition and Conformation of Sugars in Solution. *Angew. Chem. Int. Ed. Engl.* **1969**, *8*, 157–166. [CrossRef]
71. Spiwok, V.; Tvaroška, I. Conformational free energy surface of alpha-N-acetylneuraminic acid: An interplay between hydrogen bonding and solvation. *J. Phys. Chem. B* **2009**, *113*, 9589–9594. [CrossRef]
72. Sattelle, B.M.; Almond, A. Is N-acetyl-D-glucosamine a rigid 4C1 chair? *Glycobiology* **2011**, *21*, 1651–1662. [CrossRef]
73. Jorgensen, W.L.; Chandrasekhar, J.; Madura, J.D.; Impey, R.W.; Klein, M.L. Comparison of simple potential functions for simulating liquid water. *J. Chem. Phys.* **1983**, *79*, 926–935. [CrossRef]
74. Durell, S.R.; Brooks, B.R.; Ben-Naim, A. Solvent-induced forces between two hydrophilic groups. *J. Phys. Chem.* **1994**, *98*, 2198–2202. [CrossRef]
75. Beglov, D.; Roux, B. Finite representation of an infinite bulk system: Solvent boundary potential for computer simulations. *J. Chem. Phys.* **1994**, *100*, 9050–9063. [CrossRef]
76. Venable, R.M.; Luo, Y.; Gawrisch, K.; Roux, B.; Pastor, R.W. Simulations of anionic lipid membranes: Development of interaction-specific ion parameters and validation using NMR data. *J. Phys. Chem. B* **2013**, *117*, 10183–10192. [CrossRef] [PubMed]
77. Brooks, B.R.; Brooks, C.L., III; MacKerell, A.D., Jr.; Nilsson, L.; Petrella, R.J.; Roux, B.; Won, Y.; Archontis, G.; Bartels, C.; Boresch, S.; et al. CHARMM: The biomolecular simulation program. *J. Comput. Chem.* **2009**, *30*, 1545–1614. [CrossRef]
78. Phillips, J.C.; Braun, R.; Wang, W.; Gumbart, J.; Tajkhorshid, E.; Villa, E.; Chipot, C.; Skeel, R.D.; Kalé, L.; Schulten, K. Scalable molecular dynamics with NAMD. *J. Comput. Chem.* **2005**, *26*, 1781–1802. [CrossRef]
79. Steinbach, P.J.; Brooks, B.R. New spherical-cutoff methods for long-range forces in macromolecular simulation. *J. Comput. Chem.* **1994**, *15*, 667–683. [CrossRef]
80. Shirts, M.R.; Mobley, D.L.; Chodera, J.D.; Pande, V.S. Accurate and efficient corrections for missing dispersion interactions in molecular simulations. *J. Phys. Chem. B* **2007**, *111*, 13052–13063. [CrossRef] [PubMed]
81. Darden, T.; York, D.; Pedersen, L. Particle mesh Ewald: An N log(N) method for Ewald sums in large systems. *J. Chem. Phys.* **1993**, *98*, 10089–10092. [CrossRef]
82. Ryckaert, J.P.; Ciccotti, G.; Berendsen, H.J.C. Numerical integration of Cartesian equations of motion of a system with constraints: Molecular dynamics of n-alkanes. *J. Comput. Phys.* **1977**, *23*, 327–341. [CrossRef]
83. Miyamoto, S.; Kollman, P.A. Settle: An analytical version of the SHAKE and RATTLE algorithm for rigid water models. *J. Comput. Chem.* **1992**, *13*, 952–962. [CrossRef]
84. Kubo, R.; Toda, M.; Hashitume, N. *Statistical Physics II: Nonequilibrium Statistical Mechanics*, 2nd ed.; Springer: New York, NY, USA, 1991.
85. Martyna, G.J.; Tobias, D.J.; Klein, M.L. Constant pressure molecular dynamics algorithms. *J. Chem. Phys.* **1994**, *101*, 4177–4189. [CrossRef]
86. Feller, S.E.; Zhang, Y.H.; Pastor, R.W.; Brooks, B.R. Constant pressure molecular dynamics simulation: The Langevin piston method. *J. Chem. Phys.* **1995**, *103*, 4613–4621. [CrossRef]
87. Hénin, J. Fast and accurate multidimensional free energy integration. *J. Chem. Theory Comput.* **2021**, *17*, 6789–6798. [CrossRef] [PubMed]
88. Fiorin, G.; Klein, M.L.; Hénin, J. Using collective variables to drive molecular dynamics simulations. *Mol. Phys.* **2013**, *111*, 3345–3362. [CrossRef]
89. Eastman, P.; Swails, J.; Chodera, J.D.; McGibbon, R.T.; Zhao, Y.; Beauchamp, K.A.; Wang, L.P.; Simmonett, A.C.; Harrigan, M.P.; Stern, C.D.; et al. OpenMM 7: Rapid development of high performance algorithms for molecular dynamics. *PLoS Comput. Biol.* **2017**, *13*, e1005659. [CrossRef] [PubMed]
90. Humphrey, W.; Dalke, A.; Schulten, K. VMD: Visual molecular dynamics. *J. Mol. Graph.* **1996**, *14*, 33–38. [CrossRef]

91. Mallajosyula, S.S.; MacKerell, A.D. Influence of Solvent and Intramolecular Hydrogen Bonding on the Conformational Properties of O-Linked Glycopeptides. *J. Phys. Chem. B* **2011**, *115*, 11215–11229. [CrossRef]
92. Faller, C.E.; Guvench, O. Sulfation and cation effects on the conformational properties of the glycan backbone of chondroitin sulfate disaccharides. *J. Phys. Chem. B* **2015**, *119*, 6063–6073. [CrossRef] [PubMed]
93. Yang, M.; Angles d'Ortoli, T.; Säwén, E.; Jana, M.; Widmalm, G.; MacKerell, A.D. Delineating the conformational flexibility of trisaccharides from NMR spectroscopy experiments and computer simulations. *Phys. Chem. Chem. Phys.* **2016**, *18*, 18776–18794. [CrossRef]
94. Ng, C.; Premnath, P.N.; Guvench, O. Rigidity and flexibility in the tetrasaccharide linker of proteoglycans from atomic-resolution molecular simulation. *J. Comput. Chem.* **2017**, *38*, 1438–1446. [CrossRef] [PubMed]
95. Rönnols, J.; Engström, O.; Schnupf, U.; Säwén, E.; Brady, J.W.; Widmalm, G. Inter-residual Hydrogen Bonding in Carbohydrates Unraveled by NMR Spectroscopy and Molecular Dynamics Simulations. *Chem. Bio. Chem.* **2019**, *20*, 2519–2528. [CrossRef]
96. Lutsyk, V.; Plazinski, W. Conformational Properties of Glycosaminoglycan Disaccharides: A Molecular Dynamics Study. *J. Phys. Chem. B* **2021**, *125*, 10900–10916. [CrossRef]

International Journal of *Molecular Sciences*

MDPI

Article

Molecular Dynamics Simulations of Phosphorylated Intrinsically Disordered Proteins: A Force Field Comparison

Ellen Rieloff [1] and Marie Skepö [1,2,*]

[1] Division of Theoretical Chemistry, Lund University, P.O. Box 124, SE-221 00 Lund, Sweden; ellen.rieloff@teokem.lu.se

[2] LINXS—Lund Institute of Advanced Neutron and X-ray Science, Scheelevägen 19, SE-223 70 Lund, Sweden

* Correspondence: marie.skepo@teokem.lu.se

Abstract: Phosphorylation is a common post-translational modification among intrinsically disordered proteins and regions, which helps regulate function by changing the protein conformations, dynamics, and interactions with binding partners. To fully comprehend the effects of phosphorylation, computer simulations are a helpful tool, although they are dependent on the accuracy of the force field used. Here, we compared the conformational ensembles produced by Amber ff99SB-ILDN+TIP4P-D and CHARMM36m, for four phosphorylated disordered peptides ranging in length from 14–43 residues. CHARMM36m consistently produced more compact conformations with a higher content of bends, mainly due to more stable salt bridges. Based on comparisons with experimental size estimates for the shortest and longest peptide, CHARMM36m appeared to overestimate the compactness. The difference between the force fields was largest for the peptide showing the greatest separation between positively charged and phosphorylated residues, in line with the importance of charge distribution. For this peptide, the conformational ensemble did not change significantly upon increasing the ionic strength from 0 mM to 150 mM, despite a reduction of the salt-bridging probability in the CHARMM36m simulations, implying that salt concentration has negligible effects in this study.

Keywords: intrinsically disordered proteins; phosphorylation; force fields

Citation: Rieloff, E.; Skepö, M. Molecular Dynamics Simulations of Phosphorylated Intrinsically Disordered Proteins: A Force Field Comparison. *Int. J. Mol. Sci.* **2021**, *22*, 10174. https://doi.org/10.3390/ijms221810174

Academic Editors: Istvan Simon and Csaba Magyar

Received: 1 September 2021
Accepted: 16 September 2021
Published: 21 September 2021

1. Introduction

Intrinsically disordered proteins (IDPs) are characterized by a lack of a tertiary structure under physiological conditions [1,2], which means that they are better described by an ensemble of different conformations than a single structure. This is reflected in their free energy landscapes, which normally are rather flat without a deep energy minimum as for globular proteins [3]. The flattened energy landscape makes IDPs very sensitive to changes in the environment and post-translational modifications (PTMs) of the sequence. A common type of reversible PTM is phosphorylation, which introduces extra negative charges and the possibility of forming hydrogen bonds and salt bridges [4]. Phosphorylation is commonly employed by cells as a regulatory mechanism, as it can change both the conformational ensemble and the dynamics, as well as the interaction with a binding partner, and therefore affect function. The functional implications of phosphorylation can be drastic, such as for the disordered neuroprotein tau, for which hyperphosphorylation has been related to amyloid fibril formation in Alzheimer's disease [5]. In proteins such as statherin and caseins, the phosphorylated residues are essential for their ability to bind to the tooth surface [6,7] or sequester calcium [8].

Experimental techniques such as small-angle X-ray scattering (SAXS) and fluorescence resonance energy transfer (FRET) have been used to provide information on global conformational changes upon phosphorylation of intrinsically disordered proteins or regions, while circular dichroism spectroscopy and nuclear magnetic resonance (NMR) have detected changes in secondary structure or other local arrangements such as salt bridges [9–14]. However, due to the vast conformational ensembles possessed by IDPs, computer simulations are often a useful complement to obtain more detailed information, though this requires accurate models and force fields. We have previously shown that a coarse-grained "one bead per residue model" has proven to accurately predict average radius of gyration (R_g) and scattering curves for various IDPs, including statherin, although producing overly compact conformations of other more phosphorylated IDPs [15]. The two-site UNRES model has recently been extended with parameters for phosphorylated residues [16] and applied to study phosphorylation-induced folding of an IDP [17]. Although coarse-grained models are more computationally efficient and generally easier to interpret than atomistic models, they can lack in detail. In atomistic modelling, there is continuous development of force fields and water models towards more accurately describing IDPs, and some important adjustment have been the refinement of the backbone dihedral angles and balancing the water–protein and protein–protein interactions; see for example the following reviews and references within [18,19]. However, we recently showed that while the commonly used force fields CHARMM36m and Amber ff99SB-ILDN+TIP4P-D accurately captured the global dimensions of the 15-residue-long N-terminal fragment of Statherin in the nonphosphorylated state, it overestimated the compactness in the phosphorylated state [20]. More recently, overcompaction was also observed for two approximately 80-residue-long phosphorylated IDPs in several force fields, where it was suggested to depend on an overestimation of charge–charge interactions [21], in line with an overstabilization of salt bridges in standard force fields [22]. In this study, we made a further comparison of the two aforementioned force fields, by applying them to four phosphorylated peptides, namely two different fragments from tau, specifically residues 173-183 (Tau1) and 225-246 (Tau2), the first 25 amino acids in the milk protein β-casein (bCPP) and the saliva protein statherin (Stath). For all peptides, CHARMM36m was shown to sample more compact conformations than Amber ff99SB-ILDN+TIP4P-D, associated with a much higher probability for salt bridges. The effect was more pronounced in sequences with large separation between phosphorylated residues and positively charged residues, showing the importance of charge distribution. In bCPP, which showed the largest differences between the force fields, the addition of 150 mM NaCl did not change the average size estimates and shape significantly, despite a significant reduction of salt bridge occurrence in CHARMM36m. This implies that salt bridges are still of importance at 150 mM salt and that we can ignore the effects of salt concentration in this study.

2. Results and Discussion

Four phosphorylated peptides, shown in Table 1, were simulated at physiological pH using two different force fields: Amber ff99SB-ILDN [23] with the TIP4P-D [24] water model and parameters for the phosphorylated residues from Homeyer et al. [25] and Steinbrecher et al. [26] (A99) and CHARMM36m [27] with the CHARMM-modified TIP3P water model [28] (C36). The peptides were chosen based on availability of experimental data to compare with and size considering the computational expense.

Table 1. Full name and sequence of the peptides included in this study. Positively charged residues are marked in blue, negatively charged in red, and phosphorylated residues highlighted with yellow. Note that Tau1 includes three additional residues in accordance with [11], to allow for experimental comparison.

Name	Protein	Sequence
Tau1	$Tau_{173-183}$	CAKTPPAPKTPPAW
Tau2	$Tau_{225-246}$	KVAVVRTPPKSPSSAKSRLQTA
bCPP	β-$casein_{1-25}$	RELEELNVPGEIVESLSSSEESITR
Stath	Statherin	DSSEEKFLRRIGRFGYGYGPYQPVPEQPLYPQPYQPQYQQYTF

2.1. Size and Shape

For all four peptides, the two force fields produced different conformational ensembles, as seen by the distributions of the R_g and the end-to-end distance (R_{ee}) in Figure 1. The C36 distributions were narrower and centered on values lower than the A99 distributions. For Tau2 and bCPP, the R_g distribution had a sharp peak at low values. From the average R_g and R_{ee} presented in Table 2, it is clear that Tau1 showed the smallest differences between the force fields, while bCPP showed the largest differences. The discrepancy was larger for R_{ee} than R_g. For Tau1, Chin et al. [11] determined the average R_{ee} to be ~3.17 nm, based on FRET. To obtain an R_{ee} distance distribution from the FRET data they assumed a semi-flexible polymer model, and the resulting distribution was skewed towards longer distances, with the peak value located at 3.64 nm (Figure 4A in ref. [11]). Comparing A99 and C36 to the experimental average, A99 overestimated it approximately as much as C36 underestimated it. However, the skewed shape and peak position at 3.64 nm produced in A99 was in better experimental agreement than C36, since the distribution in C36 was more symmetrical with multiple peaks and had the main peak located at 3.03 nm.

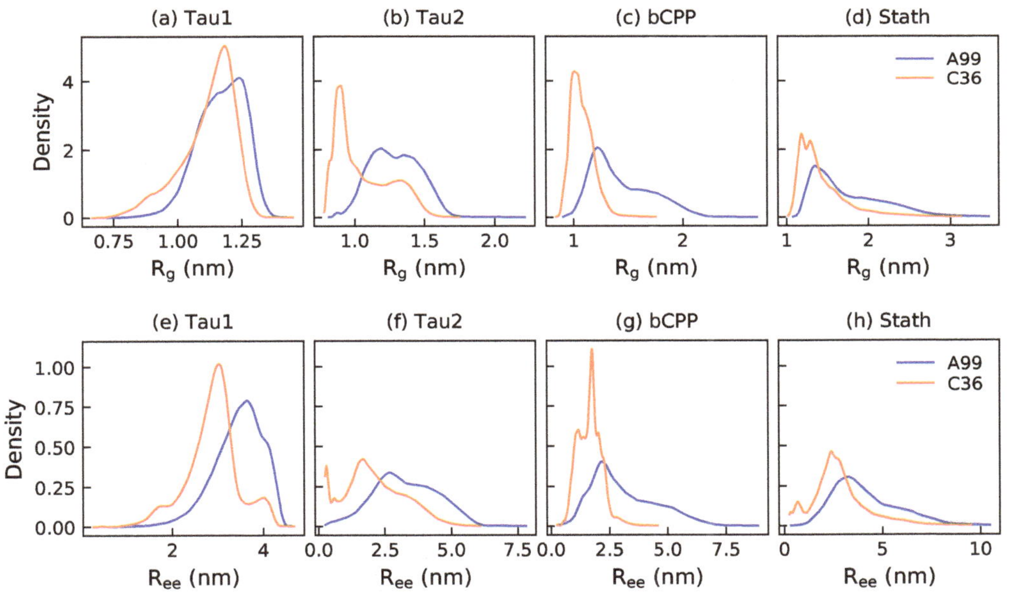

Figure 1. Distribution of the radius of gyration (**top row**) and the end-to-end distance (**bottom row**) of Tau1, Tau2, bCPP, and Stath simulated with Amber ff99SB-ILDN (A99) and CHARMM36m (C36). The legend applies to all panels.

Table 2. Average radius of gyration and end-to-end distance of the peptides simulated with Amber ff99-SB-ILDN (A99) and CHARMM36m (C36). The difference between the force fields is expressed in relation to A99.

Peptide	Radius of Gyration (nm)			End-to-End Distance (nm)		
	A99	C36	Difference (%)	A99	C36	Difference (%)
Tau1	1.17 ± 0.01	1.12 ± 0.01	4	3.44 ± 0.04	2.88 ± 0.07	16
Tau2	1.29 ± 0.03	1.06 ± 0.10	18	3.27 ± 0.17	2.10 ± 0.32	36
bCPP	1.43 ± 0.03	1.08 ± 0.02	24	3.09 ± 0.15	1.65 ± 0.10	47
Stath	1.73 ± 0.09	1.41 ± 0.04	18	4.05 ± 0.17	2.74 ± 0.20	32

For Stath, earlier published SAXS data [15] provided an R_g of 1.93 ± 0.2 nm; hence, R_g was 10% smaller in A99 and 27% smaller in C36. Since R_g determined from SAXS includes a hydration shell, it was expected that R_g calculated from simulations would be slightly smaller, although not to that extent. Since it is not straightforward which contrast to use for the hydration shell in the calculations of scattering curves for IDPs [29], in Supplementary Figure S1 and Table S2, we compared the curves calculated using different contrasts of the hydration shell to the experimental curve for Stath. While the highest contrast used ($0.03\ e/\text{Å}^3$) yielded the best agreement with the scattering curve, it provided the worst agreement with the Kratky plot. Henriques et al. [29] showed that the optimal contrast for IDPs was often between $0.01\ e/\text{Å}^3$ and $0.02\ e/\text{Å}^3$, although varying with both force field and protein. The optimal values for A99 and C36 were suggested to be around $0.0075\ e/\text{Å}^3$ and $0.02\ e/\text{Å}^3$, respectively. While the suggested optimal value gave reasonable agreement with the experimental form factor for A99, this was not the case for C36. For C36, all contrasts > 0 clearly showed larger compaction than the experimental Kratky plot.

Even without experimental scattering curves to compare to, the dimensionless Kratky plot, presented in Figure 2, is a good way of comparing the average shape of the peptides in the two different force fields. The short peptide Tau1 exhibited a more extended shape than the other three peptides, which in A99 were shown to have more of the typical IDP behavior, resembling a Gaussian chain. For all four peptides, the Kratky plot produced in C36 had a lower slope, and for the three longest peptides, the curve started to move towards the bell-shaped curve typical of globular proteins. Hence, this implies that C36 sampled more compact or well-defined conformations than A99, in accordance with the R_g and R_{ee} distributions. Notice also that the Kratky plot of Stath in A99 was in excellent agreement with the experimental data, while the curve corresponding to C36 fell below, as shown in Figure 2d.

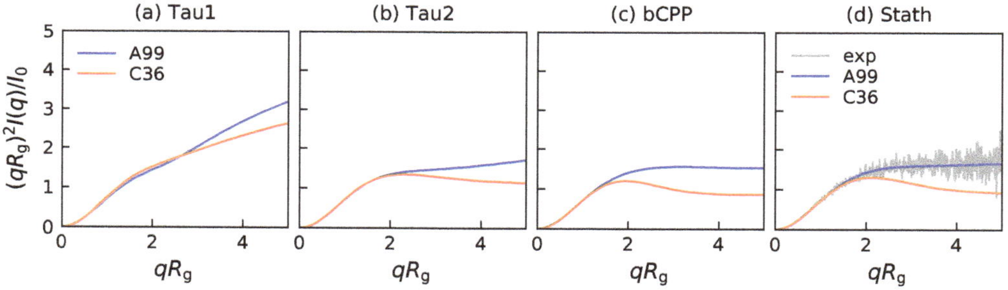

Figure 2. Dimensionless Kratky plot from simulations with Amber ff99SB-ILDN and CHARMM36m for (**a**) Tau1, (**b**) Tau2, (**c**) bCPP, and (**d**) Stath. In Panel (d), experimental data from Cragnell et al. [15] are included for comparison. The legend in Panel (a) is applicable to all panels.

2.2. Salt Bridges and Secondary Structure

Since our previous study [20] suggested that overstabilized salt bridges are the reason why C36 produces more compact conformations than A99, we calculated the occupancy of the possible salt bridge interactions involving the phosphorylated residues. Figure 3 indeed shows that salt bridges were formed much more in C36 than A99, for all the peptides. In Tau2 and bCPP, the strong salt bridges in C36 restricted the conformational ensemble, which explains the smaller and narrower distributions of R_g and R_{ee}. In bCPP, the salt-bridging residues were well separated in the sequence, therefore having a larger effect on the R_g and R_{ee} distributions. In Tau1, the salt bridge interactions almost exclusively appeared between the adjacent residues and between pT175 and the N-terminal.

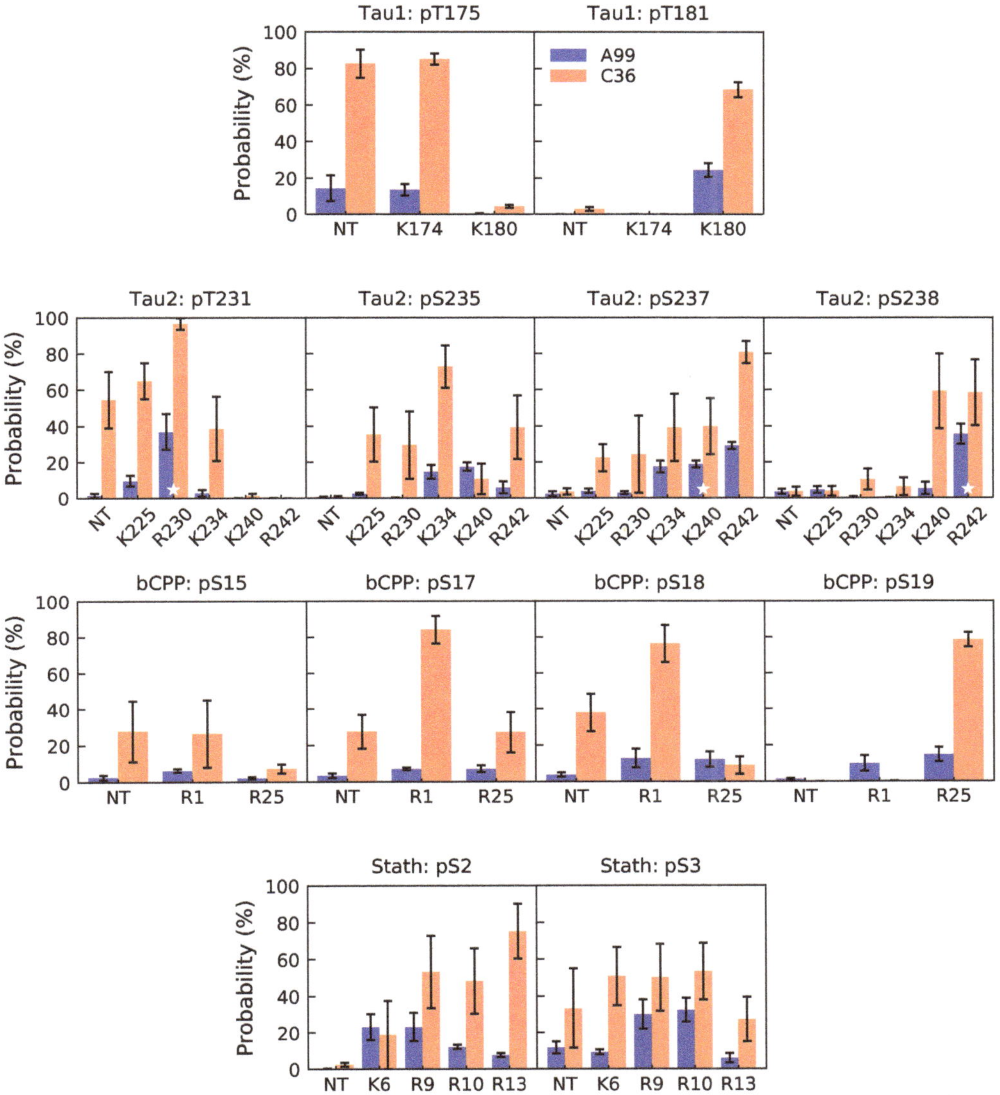

Figure 3. Probability of possible salt bridge interactions for the phosphorylated residues with the N-terminus (NT) and positively charged residues in Tau1 (**first row**), Tau2 (**second row**), bCPP (**third row**), and Stath (**last row**). For Tau2, experimentally established salt bridges [12] are marked with a white star. Error bars correspond to errors calculated by block averaging.

For Tau2, there is experimental evidence of the following salt bridges, detected by NMR experiments: pT231–R230, pS237–K240, and pS238–R242 [12]. pT231–R230 and pS238–R242 are indeed two of the most often occurring salt bridges in A99, while pS237–R242 is more common than pS237–K240. Several other salt bridges are also as frequently present as pS237–K240. In C36, pT231–R230 is the most occurring salt bridge, but both pS327–R242 and pS235–K234 are more probable than pS237–K240. Hence, while both force fields captured the experimentally established salt bridges, they also suggested other salt bridges to be present and some of them to be more common than the experimentally established ones.

Advancing to the secondary structure, Figure 4 shows that the peptides were mainly irregular, although Tau1 contained much of the polyproline type II (PPII) structure as well. In fact, all peptides contained a significant amount of PPII, as well as a significant content of bends. The content of the helical structure (α- and 3_{10}-helix) and β-strands was low in all peptides. Tau1 exhibited the largest differences between the force fields, where A99 produced 16 percentage points more of the PPII structure than C36, which instead contained a more irregular structure. For the other peptides, the differences were smaller. Overall, the peptides only had one significant difference in common, which was a higher content of bends in C36 than A99. Inspecting the content along the sequence, it was evident that it was mostly the same parts of the peptide that were enriched in a certain type of structure in both force fields (see Supplementary Figure S3). However, in C36, the helical content was completely missing from the first ten residues of Stath, which is concerning since the N-terminal region has been shown to possess helical propensity in water, although being mainly disordered [6,30]. Another striking difference between the force fields for Stath is that some residues centered on residues Y21 and Y41 occasionally formed a β-sheet or β-bridge in C36, but not in A99. Notice also that for Tau2, the bend propensity at residues V228–V229 was much higher in C36 than in A99. Since these residues were located right between K225 and pT231, which in C36 formed a stable salt bridge, this suggested that the bend was formed as a result of the salt bridge. Furthermore, for Tau2, NMR data have suggested approximately 40% α-helical propensity in region A15-R18 [12]. Both A99 and C36 sampled the helical structure in this region, however, to a lower extent than what the experimental data suggested.

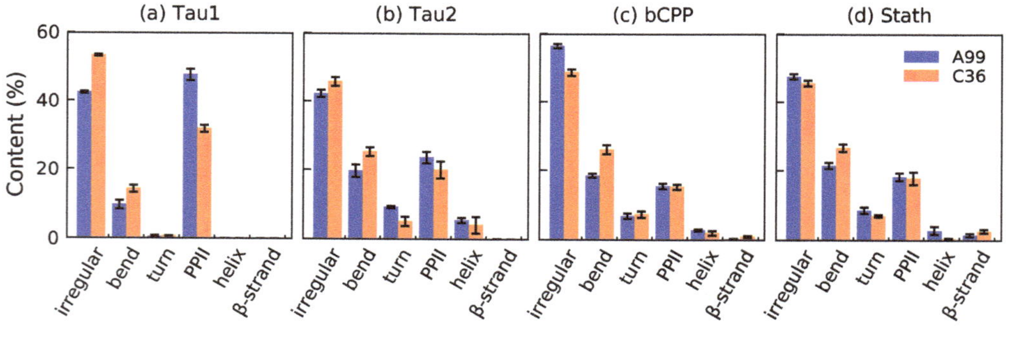

Figure 4. Average content of different types of the secondary structure in (**a**) Tau1, (**b**) Tau2, (**c**) bCPP, and (**d**) Stath simulated with Amber ff99SB-ILDN (A99) and CHARMM36m (C36). The legend applies to all panels. The helix includes the α- 3_{10}- and a negligible content of the π-helix, while the β-strand also includes β-bridge. Error bars correspond to errors calculated by block averaging.

2.3. Energy Landscapes

The differences between the force fields in this study is well summarized by the energy landscapes in Figures 5–8. Tau2, bCPP, and Stath all showed a narrower energy landscape in C36, in line with a more restricted conformational ensemble. Tau1, which is rather short and stiff, actually gained a larger conformational landscape in C36, due to sampling more bent conformations in addition to being more stretched out as in A99; see Figure 5. Notice also that in C36, the global minimum, which was the most populated, contained conformations that were not entirely stretched out. Instead, the N-terminal end was folded over, such that a salt bridge was formed between pT175 and the positively charged N-terminus.

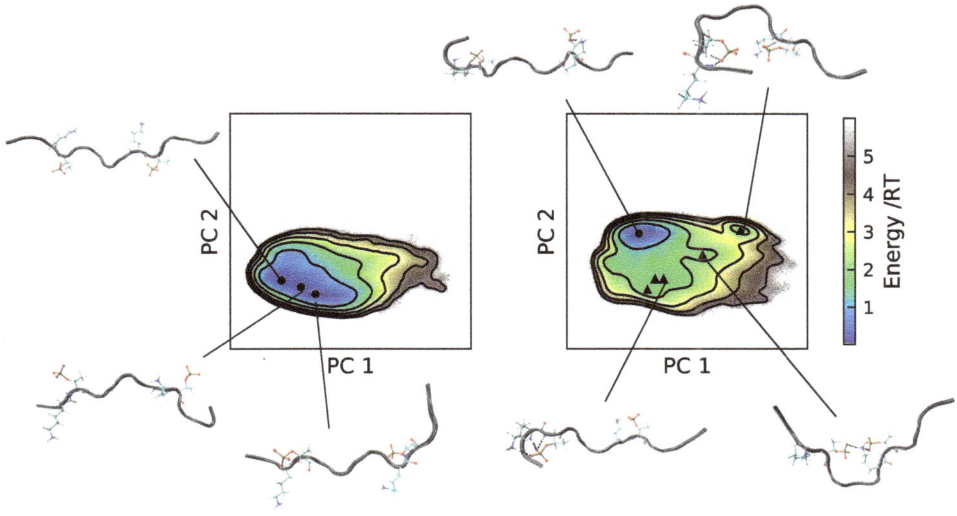

Figure 5. Energy landscapes and conformations in selected minima of Tau1. (**Left**) A99; (**right**) C36. The energy landscapes are constructed using the first two components from principal component analysis, using the same basis set for both force fields, such that they are directly comparable. Contour lines are drawn for integer energy levels in the interval $1 \leq \mathrm{RT} \leq 5$, and the minimum of each basin is represented by a marker: ●: energy $\leq 1RT$, ▲: $\leq 2RT$. In the conformations, the phosphorylated and positively charged residues are shown explicitly. Dashed black lines represent hydrogen bonds. The peptide conformations are color-coded according to the secondary structure determination in VMD, where silver is irregular (coil) and cyan is turns. The N-terminus of each conformation is the leftmost end.

Although the energy landscapes of Tau2 in A99 and C36 were located in almost the same area, the energy levels differed; see Figure 6. The most populated basin in the C36 simulation was a deep and narrow minimum, while the A99 simulation had a larger area of energy ≤1RT, containing several basins, more typical of IDPs. The salt bridges creating more compact conformations were evident in the C36 conformations, while the A99 conformations were more stretched out with fewer salt bridges. Notice that the phosphorylated residues in C36 had a tendency to interact with several positively charged residues simultaneously. In both force fields, a basin minimum with a helical region starting with pS237 and pS238 was found, in line with the secondary structure analysis.

For bCPP, there was indeed many more elongated conformations in the A99 simulation (see Figure 7), and it is clear that what caused the more compact conformations in C36 was the salt bridges between the phosphorylated serines and the arginines. In C36, all depicted conformations contained at least one salt bridge between phosphoserine and arginine, while this was much rarer in A99, explaining why the energy landscapes looked so different. Regarding Stath, comparing the conformations in Figure 8, there were two striking differences. First, there was a higher presence of salt bridges between phosphoserine and positively charged residues in C36, keeping the N-terminal end in a more bent conformation. Secondly, in C36, the β-strand and β-bridge formation between the middle region and C-terminal region detected in Supplementary Figure S3 contributed to making the conformations more compact compared to A99.

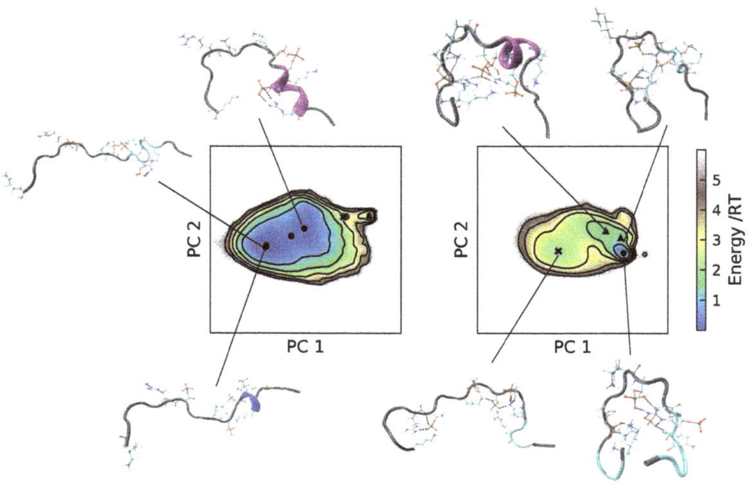

Figure 6. Energy landscapes and conformations in selected minima of Tau2. (**Left**) A99; (**right**) C36. The energy landscapes are constructed using the first two components from principal component analysis, using the same basis set for both force fields, such that they are directly comparable. Contour lines are drawn for integer energy levels in the interval $1 \leq RT \leq 5$, and the minimum of each basin is represented by a marker: ●: energy $\leq 1RT$, ▲: $\leq$2RT, ✖: $\leq$3RT. In the conformations, the phosphorylated and positively charged residues are shown explicitly. Dashed black lines represent hydrogen bonds. The peptide conformations are color-coded according to the secondary structure determination in VMD, where silver is irregular (coil), cyan is turns, magenta is the α-helix, and blue is the 3_{10}-helix. The N-terminus of each conformation is the leftmost end.

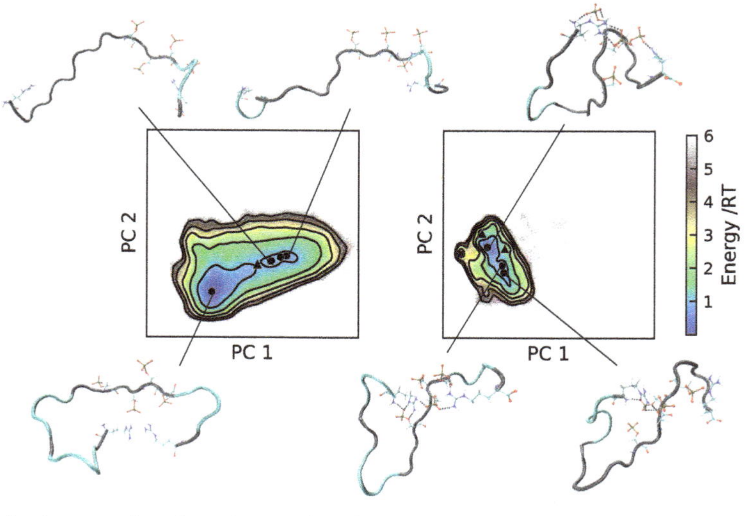

Figure 7. Energy landscapes and conformations in selected minima of bCPP. (**Left**) A99; (**right**) C36. The energy landscapes are constructed using the first two components from principal component analysis, using the same basis set for both force fields, such that they are directly comparable. Contour lines are drawn for integer energy levels in the interval $1 \leq RT \leq 5$, and the minimum of each basin is represented by a marker: ●: energy $\leq$ 1RT, ▲: $\leq$2RT, ✖: $\leq$3RT. In the conformations, the phosphorylated and positively charged residues are shown explicitly. Dashed black lines represent hydrogen bonds. The peptide conformations are color-coded according to the secondary structure determination in VMD, where silver is irregular (coil) and cyan is turns. The N-terminus of each conformation is the leftmost end.

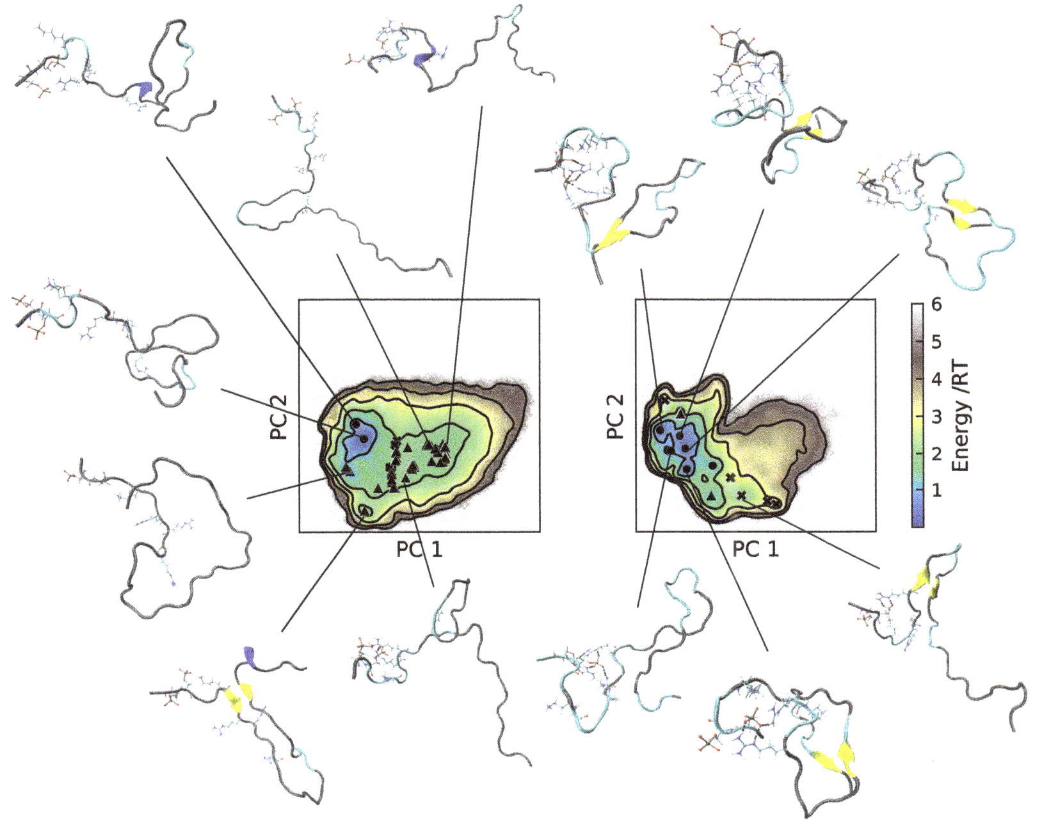

Figure 8. Energy landscapes and conformations in selected minima of Stath. (**Left**) A99; (**right**) C36. The energy landscapes are constructed using the first two components from principal component analysis, using the same basis set for both force fields, such that they are directly comparable. Contour lines are drawn for integer energy levels in the interval $1 \leq RT \leq 5$, and the minimum of each basin is represented by a marker: ●: energy $\leq$ 1RT, ▲: $\leq$2RT, ✖: $\leq$3RT. In the conformations, the phosphorylated and positively charged residues are shown explicitly. Dashed black lines represent hydrogen bonds. The peptide conformations are color-coded according to the secondary structure determination in VMD, where silver is irregular (coil), cyan is turns, blue is the 3_{10}-helix, yellow is the β-sheet, and tan is the β-bridge. The N-terminus of each conformation is the leftmost/topmost end.

2.4. Effect of Salt Concentration

Since the salt bridges formed between phosphorylated and positively charged residues were shown to influence the conformational ensemble, it is of importance to also consider the effect of the screening of the electrostatic interactions. Here, we focused on bCPP, which due to showing the largest differences between force fields and having the highest fraction of charged residues in combination with the largest charge separation (see Supplementary Table S1), was expected to show the largest response to ionic strength. Figure 9 shows that in C36, four of the salt bridges were dramatically reduced upon the addition of 150 mM NaCl; however, the probability of two other salt bridges increased, whereas in A99, only one salt bridge was significantly reduced. At 150 mM salt, the salt-bridging probability was more comparable between A99 and C36, although overall still higher in C36. Supplementary

Figure S3 shows the changes in the contact map upon the addition of 150 mM NaCl for bCPP simulated in A99 and C36. For A99, we clearly saw that the preference for the N-terminal end to be in contact with the phosphorylated and negatively charged region (residues 14–21) diminished. In C36, the strongly conserved R1–pS17 and R1–pS18 contacts were greatly decreased, while the contact of R1 with surrounding residues in the negatively charged region was increased. Hence, this suggested an increased mobility, while still maintaining contact with the negatively charged region. In C36, the cross-diagonal lines also signalized a decrease of the β-sheet; however, the content was relatively low from the beginning.

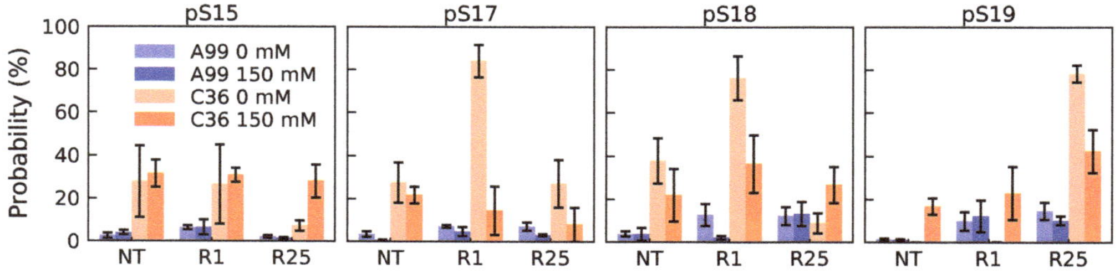

Figure 9. Probability of possible salt bridge interactions for the phosphorylated residues with the N-terminus (NT) and positively charged residues in bCPP, simulated with the two different force fields in the presence of 0 mM or 150 mM NaCl. Error bars corresponds to errors calculated by block averaging.

By comparing the energy landscapes in Figure 10, it is clear that screening of the electrostatic interactions indeed broadened the conformational ensemble, but mainly in C36, which also showed the largest change in salt bridge probability. In C36, the addition of 150 mM NaCl led to the exploration of more stretched out conformations; however, more compact conformations still clearly dominated. A99 also showed an increased probability of visiting more stretched out conformations after the addition of 150 mM NaCl. This shift in the conformational ensemble was also observed in the distributions of R_g and R_{ee} shown in Supplementary Figure S4. However, the changes were actually rather small, such that the average values were indistinguishable. Upon the addition of salt, the R_g changed from 1.43 ± 0.03 nm to 1.45 ± 0.03 nm for A99 and from 1.08 ± 0.02 nm to 1.08 ± 0.03 nm for C36. The changes in R_{ee} were from 3.09 ± 0.15 nm to 3.37 ± 0.13 nm and from 1.65 ± 0.10 nm to 1.67 ± 0.10 nm, respectively. The effect of salt on the calculated scattering curves was also so small that it could be deemed negligible; see Supplementary Figure S5.

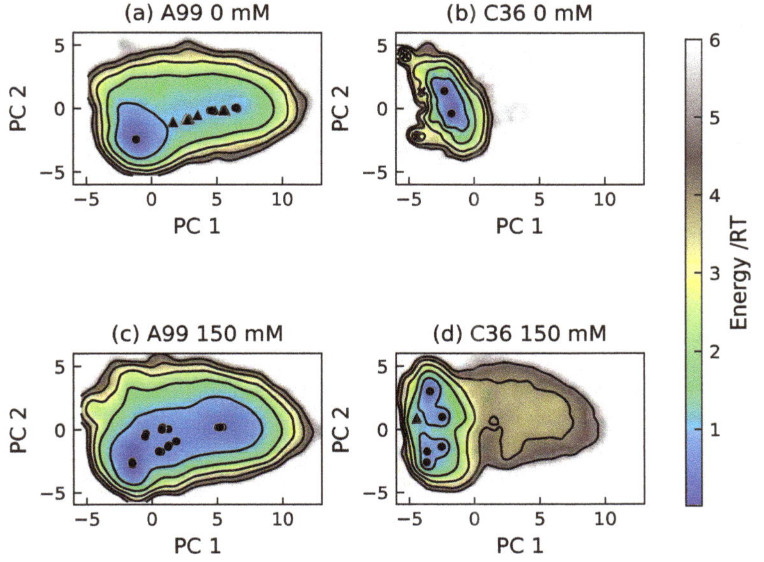

Figure 10. Energy landscapes of bCPP simulated with the two force fields Amber ff99SB-ILDN (A99) and CHARMM36m (C36) in the presence of 0 mM or 150 mM NaCl.

3. Conclusions

C36 produced more compact conformations of all four peptides, which indeed was expected to be caused mainly by salt bridge stability. In Tau1, the salt bridges pT175–K174 and pT181-K180 were formed without much effect on the overall conformation; however, an additional salt bridge between the N-terminus and pT175 decreased R_{ee} and R_g in C36. In Stath, the salt bridges contributed to the discrepancy by restricting the conformation of the first 15 residues, in the same way as previously shown for that fragment studied alone [20]. However, also the β-bridge and β-strand formation between the middle and C-terminal region were shown to contribute to more compact conformations. While C36 produced good results of nonphosphorylated short IDPs, it has been shown to underestimate the size of larger IDPs (>60 residues) [31,32]. Since Stath was 43 residues long, and thus the longest peptide included in this study, it is reasonable to believe that other effects also play a role. That bCPP showed the largest difference between the force fields and Tau1 the smallest implies that the separation between the phosphorylated and positively charged residues controls how much the conformational ensemble is influenced by stable salt bridges. This is in accordance with the importance of considering the level of charge separation for predicting the conformational ensemble of IDPs with a high fraction of charges [33].

When comparing to experimental data, it is important to consider the effect of salt, since most experiments are performed in the presence of buffer and additional salt. In bCPP, the addition of 150 mM NaCl was shown to dramatically reduce the probability of some of the salt bridges in C36, whereas the probability of other salt bridges actually increased. In A99, only one salt bridge was significantly reduced, which suggests that salt bridges still are of importance at 150 mM NaCl. Considering the changes in salt bridge probability for bCPP with salt concentration, it is plausible that the discrepancies between the simulations and experimental reference for Tau2 were caused by nonmatching ionic strength, since the experiments were performed with 50 mM phosphate buffer. At the same time, it can be hard to discern the salt bridges involving close-by residues experimentally, such as for pS237, pS238, K240, and R242.

Despite significant differences in the salt-bridging probability in C36, the effect of salt concentration on the global conformational level, such as R_g and R_{ee}, was small enough to be negligible for both force fields. In fact, the calculated form factor was indistinguishable, implying that comparing simulations performed without salt with experimental SAXS data collected at 150 mM NaCl indeed can be valid. Since bCPP is the peptide for which we expected the largest effects of salt concentration, this further strengthens the comparison with SAXS data for Stath collected at 150 mM NaCl, which showed that A99 was in good agreement, while C36 overestimated the level of compaction. Although the effects of ionic strength seem negligible in this study, this is generally not the case. For example, Jin and Gräter needed 350 mM of salt in simulations with A99 to reach experimental agreement for IDPs that are approximately 80 residues long [21], which suggested that also A99 overestimate the strength of salt bridges. Here, both Tau1 and Stath were compared to experimental size estimates, and only C36 was with certainty shown to underestimate the size. Hence, a possible overestimation of salt bridge stability in A99 is not expected to be a major issue for describing the conformational ensemble of the short IDPs studied in this work. This emphasizes the importance of benchmarking against IDPs of different length and sequence when developing and evaluating force fields. While a reduction of the strength of salt bridges appears to be a crucial step in improving the performance of C36, it appears less critical in A99. However, note that this statement is based only on the global conformational properties and that it might be different for studies of dynamics. Based on observations that many force fields have a tendency to overstabilize salt bridges, which seems to be related to side-chain partial charges [22,34–36], we suggest that readjusting the side-chains' partial charges, especially of the phosphorylated residues, is a way of improving the force fields.

Another area which has not been touched upon in this work is the influence of charge regulation and pH. The simulations have been performed with fixed charges in a state corresponding to physiological pH, where the phosphorylated residues have have a charge of $-2e$. Since the pKa of the phosphorylated residues is around six [37], in reality it can fluctuate between $-1e$ and $-2e$. Recent studies have suggested the importance of the protonation state of phosphorylated residues for molecular interactions [38], hence influencing salt bridge formation and the conformational ensemble. Therefore, this is suggested to be included in future investigations.

Considering the secondary structure, the only general difference between the force fields was a higher content of bends in C36. In Tau2, it was focused on regions between salt-bridging-forming partners, suggesting that highly stable salt bridges can enforce bends depending on the separation between the salt-bridging residues. For Tau2, it was suggested that both force fields underestimated the helical propensity, and in Stath, a lack of helix propensity in the N-terminal regions was concerning for C36. However, to properly assess the performance of force fields regarding the secondary structure, detailed experimental references are important. Hence, we see that NMR experiments of phosphorylated IDPs recording coupling constants, NOEs, and chemical shifts, which capture the effects of both the secondary structure and salt bridges, are an essential part of improving force fields. Since atomistic simulations can be used to carefully detect the secondary structure and salt bridges and their dynamics, it is an important tool in understanding the mechanism behind the regulation of IDP function by phosphorylation, provided that sufficient accuracy of the force fields is achieved.

4. Materials and Methods

Fraction of charged residues and κ, a parameter describing how segregated the charged residues are in the sequence [33] were calculated in CIDER [39], by equalizing the phosphorylated residues to other negatively charged residues. The value of κ is normalized against the most segregated sequence for that sequence composition, therefore adopting a value in the range 0–1, where 1 corresponds to the most segregated sequence possible.

The simulations listed in Supplementary Table S3 were performed in GROMACS 2018.4 [40–44], using two different force fields: Amber ff99SB-ILDN [23] with the TIP4P-D [24] water model and parameters for the phosphorylated residues from Homeyer et al. [25] and Steinbrecher et al. [26], and CHARMM36m [27] with the CHARMM-modified TIP3P water model [28]. Initial configurations of the peptides were constructed from the sequence as linear chains using Avogadro 1.2.0 [45], optimizing the structure with the auto-optimization tool. Each peptide was solvated in a rhombic dodecahedron box, having a minimum distance between the peptide and the box edges of 1 nm. Sodium ions were added to neutralize the system, and two systems were also simulated with sodium and chloride ions in a concentration corresponding to 150 mM. Periodic boundary conditions were employed in all directions. The equations of motion were integrated using the Verlet leapfrog algorithm [46] with a time step of 2 fs. Nonbonded interactions were treated with a Verlet list cutoff scheme. The short-range interactions were calculated using neighbor lists with cutoff 1 nm or 1.2 nm, for the Amber and CHARMM force fields, respectively. For the CHARMM force field, the Lennard–Jones interactions were switched off smoothly (force-switch) between 1 nm and 1.2 nm. Long-range dispersion corrections were applied to energy and pressure in the case of the Amber force field. Long-range electrostatic interactions were treated by particle mesh Ewald [47] with a cubic interpolation and a 1.6 Å grid spacing. The LINCS algorithm [48] was used to constrain all bond lengths in the case of Amber and only bonds with hydrogen atoms in the case of CHARMM. The solute and solvent were separately coupled to temperature baths at 298 K using the velocity rescaling thermostat [49] with a 0.1 ps relaxation time. Parrinello–Raman pressure coupling [50] was used to keep the pressure at 1 bar, using a 2 ps relaxation time and 4.5×10^{-5} bar^{-1} isothermal compressibility.

Energy minimization was performed by the steepest descent algorithm until the system converged within the available machine precision. Initiation of five replicates per system with different starting seeds was performed separately in two steps using position restraints on the peptide. The first step was 500 ps of NVT simulation (constant number of particles, volume, and temperature) performed to stabilize the temperature, followed by the second step of 1000 ps of NPT simulation (constant number of particles, pressure, and temperature) to stabilize the pressure. Production runs of the five replicates per system were performed in the NPT ensemble, for at least 1 µs per replicate. The total simulation time per system is stated in Supplementary Table S3. Energies and coordinates were saved every 10 ps. Supplementary Tables S4 and S5 compile a few differences applied to the salt simulations to reduce the computational time.

Analysis

The convergence and sampling quality were assessed in the following ways. The time evolution of the R_g and the R_{ee} in the simulations were observed for signs of equilibration in the initial stage. Based on this, the first 30 ns were removed from each replicate of bCPP in CHARMM36m and the first 50 ns of each replicate of Tau2 in CHARMM36m before final analysis (see Supplementary Figures S21 and S24). In other systems the equilibration was deemed fast enough to be negligible. The distributions of the R_g and the R_{ee} as well as the energy landscapes were compared between replicates, since similarity indicates sufficient sampling. The autocorrelation function and block average error estimates of the R_g and the R_{ee} in the concatenated simulation were calculated and observed for an estimate of the correlation time and convergence of the error estimates. All this data is presented in the Supplementary Figures S6–S33. Although some systems showed greater dissimilarity between replicates than desired, based on the assessment of the concatenated trajectory, it was deemed sufficiently sampled to allow for a comparison between the force fields.

R_g and R_{ee} were calculated using GROMACS 2018.4 [40–44]. Reported error estimates were calculated using block averaging analysis as implemented in the *gmx analyze* routine in GROMACS. Scattering curves were calculated using CRYSOL Version 2.8.3 [51] with the contrast of the hydration shell being 0.0075 $e/Å^3$ for Amber ff99SB-ILDN+TIP4P-D and

0.02 $e/\text{Å}^3$ for CHARMM36m, as suggested by [29]. The presented curve is the average over 10,000 equally spaced frames. In Supplementary Figure S1 and Table S2, the effect of different contrasts of the hydration shell is shown for Stath. The quality of fit to the experimental curve is computed as:

$$\chi^2(f,c) = N_q^{-1} \sum_{i=1}^{N_q} \left[\frac{I_{\text{ref}}(q_i) - (f I_{\text{obs}}(q_i) + c)}{\sigma_{\text{ref}}(q_i)} \right]^2, \tag{1}$$

where N_q^{-1} is the number of points in the reference curve, I_{ref} and I_{obs} are the reference and observed intensities, respectively, and $\sigma_{\text{ref}}(q_i)$ is the error associated with each data point of the reference curve. The function was minimized using the Nelder–Mead method [52], as implemented in Scipy [53], using linear interpolation to produce I_{obs} at the same q points as the reference [29]. AUTORG in the ATSAS program [54] was used to determine the R_g from Guinier analysis. The secondary structure was determined using the DSSP program Version 2.2.1 [55] with an extension to detect the polyproline type II structure [56,57]. The MDTraj Python library Version 1.9.3 [58] was used to calculate contact probability and analyze salt bridges. Contact between two residues was defined as when the shortest distance between two atoms < 0.4 nm. Since salt bridges are formed as a result of hydrogen bonding and electrostatic interactions, they were assessed by analyzing the presence of hydrogen bonds based on the criterion in [59], as implemented in MDTraj. Energy landscapes were calculated following the Campos and Baptista approach [60], with the differences described by Henriques et al. [61]. In short, principal component analysis was applied to the Cartesian coordinates of the backbone atoms of the protein, obtained after translational and rotational least squares fitting on the central structure of the simulation. The conditional free energy was calculated from the probability density function in the representation space constructed by the first two principal components, obtained by Gaussian kernel density estimation. The basins and minima were assigned as described by Campos and Baptista [60]. It is worth noting that the first two components were shown to account for 46–60% of the variance, hence not providing a complete picture of the conformational classes, but at least an overview sufficient for comparison between the force fields. Snapshots from the simulations were produced using VMD 1.9.3 [62–64].

Supplementary Materials: The following are available online at www.mdpi.com/article/10.3390/ijms221810174/s1.

Author Contributions: Conceptualization, M.S. and E.R.; formal analysis, E.R.; funding acquisition, M.S.; investigation, E.R.; methodology, M.S. and and E.R.; project administration, M.S.; resources, M.S.; supervision, M.S.; validation, M.S. and E.R.; visualization, E.R.; writing, M.S. and E.R. All authors have read and agreed to the published version of the manuscript.

Funding: Financial support was given by the Royal Physiographic Society in Lund and the Bertil Andersson foundation.

Institutional Review Board Statement: Not applicable.

Informed Consent Statement: Not applicable.

Data Availability Statement: Data is included in the article or Supplementary Materials.

Acknowledgments: The simulations were performed on resources provided by the Swedish National Infrastructure for Computing (SNIC) at the Center for Scientific and Technical Computing at Lund University (LUNARC).

Conflicts of Interest: The authors declare no conflict of interest.

Abbreviations

A99	Amber ff99SB-ILDN with TIP4P-D water
C36	CHARMM36m with CHARMM-modified TIP3P water
FRET	Fluorescence resonance energy transfer
IDP	Intrinsically disordered protein
NMR	Nuclear magnetic resonance
PPII	polyproline type II
R_g	Radius of gyration
R_{ee}	End-to-end distance
SAXS	Small-angle X-ray scattering

References

1. Dunker, A.; Lawson, J.; Brown, C.J.; Williams, R.M.; Romero, P.; Oh, J.S.; Oldfield, C.J.; Campen, A.M.; Ratliff, C.M.; Hipps, K.W.; et al. Intrinsically disordered protein. *J. Mol. Graph. Model.* **2001**, *19*, 26–59. [CrossRef]
2. Tompa, P. Intrinsically unstructured proteins. *Trends Biochem. Sci.* **2002**, *27*, 527–533. [CrossRef]
3. Fisher, C.K.; Stultz, C.M. Constructing ensembles for intrinsically disordered proteins. *Curr. Opin. Struct. Biol.* **2011**, *21*, 426–431. [CrossRef] [PubMed]
4. Johnson, L.N.; Lewis, R.J. Structural Basis for Control by Phosphorylation. *Chem. Rev.* **2001**, *101*, 2209–2242. [CrossRef] [PubMed]
5. Gong, C.X.; Iqbal, K. Hyperphosphorylation of microtubule-associated protein tau: a promising therapeutic target for Alzheimer disease. *Curr. Med. Chem.* **2008**, *15*, 2321–2328. [CrossRef]
6. Raj, P.A.; Johnsson, M.; Levine, M.J.; Nancollas, G.H. Salivary statherin. Dependence on sequence, charge, hydrogen bonding potency, and helical conformation for adsorption to hydroxyapatite and inhibition of mineralization. *J. Biol. Chem.* **1992**, *267*, 5968–5976. [CrossRef]
7. Makrodimitris, K.; Masica, D.L.; Kim, E.T.; Gray, J.J. Structure Prediction of Protein–Solid Surface Interactions Reveals a Molecular Recognition Motif of Statherin for Hydroxyapatite. *J. Am. Chem. Soc.* **2007**, *129*, 13713–13722. [CrossRef] [PubMed]
8. De Kruif, C.G.; Holt, C. Casein Micelle Structure, Functions and Interactions. In *Advanced Dairy Chemistry—1 Proteins: Part A/Part B*; Fox, P.F., McSweeney, P.L.H., Eds.; Springer: Boston, MA, USA, 2003; pp. 233–276. [CrossRef]
9. Martin, E.W.; Holehouse, A.S.; Grace, C.R.; Hughes, A.; Pappu, R.V.; Mittag, T. Sequence Determinants of the Conformational Properties of an Intrinsically Disordered Protein Prior to and upon Multisite Phosphorylation. *J. Am. Chem. Soc.* **2016**, *138*, 15323–15335. [CrossRef]
10. Mittag, T.; Marsh, J.; Grishaev, A.; Orlicky, S.; Lin, H.; Sicheri, F.; Tyers, M.; Forman-Kay, J.D. Structure/Function Implications in a Dynamic Complex of the Intrinsically Disordered Sic1 with the Cdc4 Subunit of an SCF Ubiquitin Ligase. *Structure* **2010**, *18*, 494–506. [CrossRef]
11. Chin, A.; Toptygin, D.; Elam, W.; Schrank, T.; Hilser, V. Phosphorylation Increases Persistence Length and End-to-End Distance of a Segment of Tau Protein. *Biophys. J.* **2016**, *110*, 362–371. [CrossRef]
12. Schwalbe, M.; Kadavath, H.; Biernat, J.; Ozenne, V.; Blackledge, M.; Mandelkow, E.; Zweckstetter, M. Structural Impact of Tau Phosphorylation at Threonine 231. *Structure* **2015**, *23*, 1448–1458. [CrossRef]
13. RFarrell, H.; Qi, P.; Wickham, E.; Unruh, J. Secondary Structural Studies of Bovine Caseins: Structure and Temperature Dependence of β-Casein Phosphopeptide (1–25) as Analyzed by Circular Dichroism, FTIR Spectroscopy, and Analytical Ultracentrifugation. *J. Protein Chem.* **2002**, *21*, 307–321. [CrossRef]
14. Brister, M.A.; Pandey, A.K.; Bielska, A.A.; Zondlo, N.J. OGlcNAcylation and Phosphorylation Have Opposing Structural Effects in tau: Phosphothreonine Induces Particular Conformational Order. *J. Am. Chem. Soc.* **2014**, *136*, 3803–3816. [CrossRef] [PubMed]
15. Cragnell, C.; Rieloff, E.; Skepö, M. Utilizing Coarse-Grained Modeling and Monte Carlo Simulations to Evaluate the Conformational Ensemble of Intrinsically Disordered Proteins and Regions. *J. Mol. Biol.* **2018**, *430*, 2478–2492. [CrossRef]
16. Sieradzan, A.K.; Bogunia, M.; Mech, P.; Ganzynkowicz, R.; Giełdoń, A.; Liwo, A.; Makowski, M. Introduction of Phosphorylated Residues into the UNRES Coarse-Grained Model: Toward Modeling of Signaling Processes. *J. Phys. Chem. B* **2019**, *123*, 5721–5729. [CrossRef] [PubMed]
17. Sieradzan, A.K.; Korneev, A.; Begun, A.; Kachlishvili, K.; Scheraga, H.A.; Molochkov, A.; Senet, P.; Niemi, A.J.; Maisuradze, G.G. Investigation of Phosphorylation-Induced Folding of an Intrinsically Disordered Protein by Coarse-Grained Molecular Dynamics. *J. Chem. Theory Comput.* **2021**, *17*, 3203–3220. [CrossRef] [PubMed]
18. Chong, S.H.; Chatterjee, P.; Ham, S. Computer Simulations of Intrinsically Disordered Proteins. *Annu. Rev. Phys. Chem.* **2017**, *68*, 117–134. [CrossRef] [PubMed]
19. Huang, J.; MacKerell, A.D. Force field development and simulations of intrinsically disordered proteins. *Curr. Opin. Struct. Biol.* **2018**, *48*, 40–48. [CrossRef]
20. Rieloff, E.; Skepö, M. Phosphorylation of a Disordered Peptide–Structural Effects and Force Field Inconsistencies. *J. Chem. Theory Comput.* **2020**, *16*, 1924–1935. [CrossRef]
21. Jin, F.; Gräter, F. How multisite phosphorylation impacts the conformations of intrinsically disordered proteins. *PLoS Comput. Biol.* **2021**, *17*, e1008939. [CrossRef]

22. Ahmed, M.C.; Papaleo, E.; Lindorff-Larsen, K. How well do force fields capture the strength of salt bridges in proteins? *PeerJ* **2018**, *6*, e4967. [CrossRef]
23. Lindorff-Larsen, K.; Piana, S.; Palmo, K.; Maragakis, P.; Klepeis, J.L.; Dror, R.O.; Shaw, D.E. Improved side-chain torsion potentials for the Amber ff99SB protein force field. *Proteins* **2010**, *78*, 1950–1958. [CrossRef] [PubMed]
24. Piana, S.; Donchev, A.G.; Robustelli, P.; Shaw, D.E. Water Dispersion Interactions Strongly Influence Simulated Structural Properties of Disordered Protein States. *J. Phys. Chem. B* **2015**, *119*, 5113–5123. [CrossRef]
25. Homeyer, N.; Horn, A.H.C.; Lanig, H.; Sticht, H. AMBER force-field parameters for phosphorylated amino acids in different protonation states: phosphoserine, phosphothreonine, phosphotyrosine, and phosphohistidine. *J. Mol. Model.* **2006**, *12*, 281–289. [CrossRef] [PubMed]
26. Steinbrecher, T.; Latzer, J.; Case, D.A. Revised AMBER Parameters for Bioorganic Phosphates. *J. Chem. Theory Comput.* **2012**, *8*, 4405–4412. [CrossRef]
27. Huang, J.; Rauscher, S.; Nawrocki, G.; Ran, T.; Feig, M.; de Groot, B.L.; Grubmüller, H.; MacKerell, A.D., Jr. CHARMM36m: An improved force field for folded and intrinsically disordered proteins. *Nat. Methods* **2016**, *14*, 71–73.
28. MacKerell, A.D.; Bashford, D.; Bellott, M.; Dunbrack, R.L.; Evanseck, J.D.; Field, M.J.; Fischer, S.; Gao, J.; Guo, H.; Ha, S.; et al. All-Atom Empirical Potential for Molecular Modeling and Dynamics Studies of Proteins. *J. Phys. Chem. B* **1998**, *102*, 3586–3616. [CrossRef]
29. Henriques, J.; Arleth, L.; Lindorff-Larsen, K.; Skepö, M. On the Calculation of SAXS Profiles of Folded and Intrinsically Disordered Proteins from Computer Simulations. *J. Mol. Biol.* **2018**, *430*, 2521–2539. [CrossRef] [PubMed]
30. Naganagowda, G.A.; Gururaja, T.L.; Levine, M.J. Delineation of Conformational Preferences in Human Salivary Statherin by 1H, 31P NMR and CD Studies: Sequential Assignment and Structure-Function Correlations. *J. Biomol. Struct. Dyn.* **1998**, *16*, 91–107. [CrossRef]
31. Robustelli, P.; Piana, S.; Shaw, D.E. Developing a molecular dynamics force field for both folded and disordered protein states. *Proc. Natl. Acad. Sci. USA* **2018**, *115*, E4758–E4766. [CrossRef] [PubMed]
32. Chan-Yao-Chong, M.; Deville, C.; Pinet, L.; van Heijenoort, C.; Durand, D.; Ha-Duong, T. Structural Characterization of N-WASP Domain V Using MD Simulations with NMR and SAXS Data. *Biophys. J.* **2019**, *116*, 1216–1227. [CrossRef]
33. Das, R.K.; Pappu, R.V. Conformations of intrinsically disordered proteins are influenced by linear sequence distributions of oppositely charged residues. *Proc. Natl. Acad. Sci. USA* **2013**, *110*, 13392–13397. [CrossRef] [PubMed]
34. Piana, S.; Lindorff-Larsen, K.; Shaw, D.E. How Robust Are Protein Folding Simulations with Respect to Force Field Parameterization? *Biophys. J.* **2011**, *100*, L47–L49. [CrossRef]
35. Debiec, K.T.; Gronenborn, A.M.; Chong, L.T. Evaluating the Strength of Salt Bridges: A Comparison of Current Biomolecular Force Fields. *J. Phys. Chem. B* **2014**, *118*, 6561–6569. [CrossRef]
36. Best, R.B. Atomistic Force Fields for Proteins. In *Biomolecular Simulations: Methods and Protocols*; Bonomi, M., Camilloni, C., Eds.; Springer: New York, NY, USA, 2019; pp. 3–19. [CrossRef]
37. Bienkiewicz, E.A.; Lumb, K.J. Random-coil chemical shifts of phosphorylated amino acids. *J. Biomol. NMR* **1999**, *15*, 203–206. [CrossRef]
38. Kawade, R.; Kuroda, D.; Tsumoto, K. How the protonation state of a phosphorylated amino acid governs molecular recognition: insights from classical molecular dynamics simulations. *FEBS Lett.* **1999**, *594*, 903–912. [CrossRef] [PubMed]
39. Holehouse, A.S.; Das, R.K.; Ahad, J.N.; Richardson, M.O.G.; Pappu, R.V. CIDER: Resources to Analyze Sequence-Ensemble Relationships of Intrinsically Disordered Proteins. *Biophys. J.* **2017**, *112*, 16–21. [CrossRef] [PubMed]
40. Berendsen, H.; van der Spoel, D.; van Drunen, R. GROMACS: A message-passing parallel molecular dynamics implementation. *Comput. Phys. Commun.* **1995**, *91*, 43–56. [CrossRef]
41. Hess, B.; Kutzner, C.; van der Spoel, D.; Lindahl, E. GROMACS 4: Algorithms for Highly Efficient, Load-Balanced, and Scalable Molecular Simulation. *J. Chem. Theory Comput.* **2008**, *4*, 435–447. [CrossRef]
42. Pronk, S.; Páll, S.; Schulz, R.; Larsson, P.; Bjelkmar, P.; Apostolov, R.; Shirts, M.R.; Smith, J.C.; Kasson, P.M.; van der Spoel, D.; et al. GROMACS 4.5: A high-throughput and highly parallel open source molecular simulation toolkit. *Bioinformatics* **2013**, *29*, 845–854. [CrossRef]
43. Páll, S.; Abraham, M.J.; Kutzner, C.; Hess, B.; Lindahl, E. Tackling Exascale Software Challenges in Molecular Dynamics Simulations with GROMACS. In *Solving Software Challenges for Exascale*; Markidis, S., Laure, E., Eds.; Springer International Publishing: Cham, Switzerland, 2015; pp. 3–27.
44. Abraham, M.J.; Murtola, T.; Schulz, R.; Páll, S.; Smith, J.C.; Hess, B.; Lindahl, E. GROMACS: High performance molecular simulations through multi-level parallelism from laptops to supercomputers. *SoftwareX* **2015**, *1–2*, 19–25. [CrossRef]
45. Hanwell, M.D.; Curtis, D.E.; Lonie, D.C.; Vandermeersch, T.; Zurek, E.; Hutchison, G.R. Avogadro: An advanced semantic chemical editor, visualization, and analysis platform. *J. Cheminform.* **2012**, *4*, 17. [CrossRef] [PubMed]
46. Hockney, R.W.; Eastwood, J.W. *Computer Simulation Using Particles*; McGraw-Hill: New York, NY, USA, 1981.
47. Darden, T.; York, D.; Pedersen, L. Particle mesh Ewald: An N·log(N) method for Ewald sums in large systems. *J. Chem. Phys.* **1993**, *98*, 10089–10092. [CrossRef]
48. Hess, B.; Bekker, H.; Berendsen, H.J.C.; Fraaije, J.G.E.M. LINCS: A linear constraint solver for molecular simulations. *J. Comput. Chem.* **1997**, *18*, 1463–1472. [CrossRef]
49. Bussi, G.; Donadio, D.; Parrinello, M. Canonical sampling through velocity rescaling. *J. Chem. Phys.* **2007**, *126*, 014101. [CrossRef]

50. Parrinello, M.; Rahman, A. Polymorphic transitions in single crystals: A new molecular dynamics method. *J. Appl. Phys.* **1981**, *52*, 7182–7190. [CrossRef]
51. Svergun, D.; Barberato, C.; Koch, M.H.J. *CRYSOL*—A Program to Evaluate X-ray Solution Scattering of Biological Macromolecules from Atomic Coordinates. *J. Appl. Crystallogr.* **1995**, *28*, 768–773. [CrossRef]
52. Nelder, J.A.; Mead, R. A Simplex Method for Function Minimization. *Comput. J.* **1965**, *7*, 308–313. [CrossRef]
53. Virtanen, P.; Gommers, R.; Oliphant, T.E.; Haberland, M.; Reddy, T.; Cournapeau, D.; Burovski, E.; Peterson, P.; Weckesser, W.; Bright, J.; et al. SciPy 1.0: Fundamental Algorithms for Scientific Computing in Python. *Nat. Methods* **2020**, *17*, 261–272. [CrossRef]
54. Franke, D.; Petoukhov, M.V.; Konarev, P.V.; Panjkovich, A.; Tuukkanen, A.; Mertens, H.D.T.; Kikhney, A.G.; Hajizadeh, N.R.; Franklin, J.M.; Jeffries, C.M.; et al. *ATSAS 2.8*: A comprehensive data analysis suite for small-angle scattering from macromolecular solutions. *J. Appl. Crystallogr.* **2017**, *50*, 1212–1225. [CrossRef]
55. Kabsch, W.; Sander, C. Dictionary of protein secondary structure: Pattern recognition of hydrogen-bonded and geometrical features. *Biopolymers* **1983**, *22*, 2577–2637. [CrossRef]
56. Mansiaux, Y.; Joseph, A.P.; Gelly, J.C.; de Brevern, A.G. Assignment of PolyProline II Conformation and Analysis of Sequence—Structure Relationship. *PLoS ONE* **2011**, *6*, e18401. [CrossRef]
57. Chebrek, R.; Leonard, S.; de Brevern, A.G.; Gelly, J.C. PolyprOnline: polyproline helix II and secondary structure assignment database. *Database* **2014**, *2014*, bau102. [CrossRef] [PubMed]
58. McGibbon, R.T.; Beauchamp, K.A.; Harrigan, M.P.; Klein, C.; Swails, J.M.; Hernández, C.X.; Schwantes, C.R.; Wang, L.P.; Lane, T.J.; Pande, V.S. MDTraj: A Modern Open Library for the Analysis of Molecular Dynamics Trajectories. *Biophys. J.* **2015**, *109*, 1528–1532. [CrossRef]
59. Wernet, P.; Nordlund, D.; Bergmann, U.; Cavalleri, M.; Odelius, M.; Ogasawara, H.; Näslund, L.Å.; Hirsch, T.K.; Ojamäe, L.; Glatzel, P.; et al. The Structure of the First Coordination Shell in Liquid Water. *Science* **2004**, *304*, 995–999. [CrossRef] [PubMed]
60. Campos, S.R.R.; Baptista, A.M. Conformational Analysis in a Multidimensional Energy Landscape: Study of an Arginylglutamate Repeat. *J. Phys. Chem. B* **2009**, *113*, 15989–16001. [CrossRef]
61. Henriques, J.; Cragnell, C.; Skepö, M. Molecular Dynamics Simulations of Intrinsically Disordered Proteins: Force Field Evaluation and Comparison with Experiment. *J. Chem. Theory Comput.* **2015**, *11*, 3420–3431. [CrossRef] [PubMed]
62. Humphrey, W.; Dalke, A.; Schulten, K. VMD—Visual Molecular Dynamics. *J. Mol. Graph.* **1996**, *14*, 33–38. [CrossRef]
63. Stone, J. An Efficient Library for Parallel Ray Tracing and Animation. Master's Thesis, Computer Science Department, University of Missouri-Rolla, Rolla, MO, USA, 1998.
64. Frishman, D.; Argos, P. Knowledge-based secondary structure assignment. *Proteins* **1995**, *23*, 566–579. [CrossRef]

International Journal of *Molecular Sciences*

Article

The Effect of Multisite Phosphorylation on the Conformational Properties of Intrinsically Disordered Proteins

Ellen Rieloff [1] and Marie Skepö [1,2,*]

[1] Division of Theoretical Chemistry, Lund University, P.O. Box 124, SE-221 00 Lund, Sweden; ellen.rieloff@teokem.lu.se

[2] LINXS—Lund Institute of Advanced Neutron and X-ray Science, Scheelevägen 19, SE-223 70 Lund, Sweden

* Correspondence: marie.skepo@teokem.lu.se

Abstract: Intrinsically disordered proteins are involved in many biological processes such as signaling, regulation, and recognition. A common strategy to regulate their function is through phosphorylation, as it can induce changes in conformation, dynamics, and interactions with binding partners. Although phosphorylated intrinsically disordered proteins have received increased attention in recent years, a full understanding of the conformational and structural implications of phosphorylation has not yet been achieved. Here, we present all-atom molecular dynamics simulations of five disordered peptides originated from tau, statherin, and β-casein, in both phosphorylated and non-phosphorylated state, to compare changes in global dimensions and structural elements, in an attempt to gain more insight into the controlling factors. The changes are in qualitative agreement with experimental data, and we observe that the net charge is not enough to predict the impact of phosphorylation on the global dimensions. Instead, the distribution of phosphorylated and positively charged residues throughout the sequence has great impact due to the formation of salt bridges. In statherin, a preference for arginine–phosphoserine interaction over arginine–tyrosine accounts for a global expansion, despite a local contraction of the phosphorylated region, which implies that also non-charged residues can influence the effect of phosphorylation.

Keywords: intrinsically disordered proteins; phosphorylation; force fields

Citation: Rieloff, E.; Skepö, M. The Effect of Multisite Phosphorylation on the Conformational Properties of Intrinsically Disordered Proteins. *Int. J. Mol. Sci.* **2021**, *22*, 11058. https://doi.org/10.3390/ijms222011058

Academic Editors: Istvan Simon and Csaba Magyar

Received: 17 September 2021
Accepted: 8 October 2021
Published: 14 October 2021

Publisher's Note: MDPI stays neutral with regard to jurisdictional claims in published maps and institutional affiliations.

1. Introduction

Intrinsically disordered proteins (IDPs) lack tertiary structure under physiological conditions [1,2], such that they adopt a range of different interchanging conformations rather than a single structure. This is reflected in their rather flat free energy landscapes [3], making them sensitive to environmental changes and post-translational modifications (PTMs), which helps to regulate function. Many IDPs also demonstrate the ability to bind to several targets, and adopt different folds depending on the target. These characteristics of IDPs are advantageous in signaling, regulation, and recognition processes, where IDPs are abundantly involved [4,5].

Phosphorylation is a reversible type of PTM, especially prevalent among intrinsically disordered regions and proteins [6–8]. The addition of a bulky phosphoryl group to residues such as serine or threonine adds extra negative charge and enables formation of hydrogen bonds and salt bridges [9], which can induce drastic changes in the conformational ensemble and the dynamics of the IDP. In a simplistic view, assuming electrostatics to be the major determinant of IDP conformation, a net positively charged IDP is expected to contract upon phosphorylation, while a negatively charged or neutral IDP will expand. In a recent atomistic simulation study by Jin and Gräter, this prediction was shown to hold true for multisite phosphorylation of the four peptides studied [10]. Generally, net charge and hydropathy provide good indications of the level of compaction of a protein only in some cases, while many require an additional inspection of the fraction of charged residues and charge pattern, due to their polyampholytic nature [11,12].

In recent years, phosphorylated IDPs have received increased attention [10,13–23]. Changes in global conformation, secondary structure, and local arrangements upon phosphorylation of disordered proteins and regions have been studied experimentally by techniques such as small angle X-ray scattering (SAXS), fluorescence resonance energy transfer, circular dichroism (CD) spectroscopy, and nuclear magnetic resonance (NMR) [13–15,20,24–26]. Due to the vast conformational ensembles possessed by IDPs, a combination of different techniques is required and often well complemented by atomistic simulations, which through detailed information can provide further insight. After many years of important adjustments, such as refinement of backbone dihedral angles and balancing the water–protein and protein–protein interactions, there are several force field and water model combinations that can be applied to IDPs [27,28]. Less attention has been given to charge–charge interactions, although it has been determined that many standard force fields have a tendency to overestimate salt bridges [29,30]. More recently, it has been shown that, for phosphorylated peptides, this can cause serious discrepancies between simulations and experiments [10,20,31].

In our most recent work involving all-atom molecular dynamics simulations of phosphorylated disordered peptides, Amber ff99SB-ILDN in combination with the TIP4P-D water model showed promising results in describing the conformational ensemble of short disordered peptides [20,31]. Here, we have extended the work with simulations of the non-phosphorylated variants of the four peptides in [31], using the aforementioned force field, and additional analyses of a fifth peptide published in [20], to study the conformational and structural effects upon phosphorylation, with the aim of gaining better insight into the controlling factors. By experimental comparison, we also detect limitations of the force field. Two of the peptides are fragments from the neuroprotein tau, involved in stabilizing neuronal microtubules [32]. Phosphorylation of tau regulates its function, and hyperphosphorylation has been implicated to cause pathological effects by involvement in amyloid fibril formation in Alzheimer's disease [33,34]. Another two of the peptides are the saliva protein statherin and its fifteen residue long N-terminal fragment, SN15. Statherin maintains a supersaturated environment of calcium phosphate in the saliva, by preventing spontaneous precipitation and crystal growth [35–37]. This functionality is closely associated with the N-terminal fragment containing the phosphorylated residues [37]. The last peptide is the 25 residue long N-terminal fragment of β-casein, which naturally contains four phosphorylated serines that sequester calcium and promotes the formation of calcium phosphate nanoclusters [38–40].

We observe that, for these peptides, ranging in length from 11 to 43 residues that net charge is not enough to predict the change in global dimensions upon phosphorylation at two to four sites. Instead, salt bridge formation has great impact, depending on the distribution of phosphorylated and positively charged residues throughout the sequence. Furthermore, in statherin, a preference for arginine–phosphoserine interactions over arginine–tyrosine interactions explains the phosphorylation induced changes.

2. Results and Discussion

2.1. Net Charge Is Not Enough to Explain Phosphorylation Induced Changes

Atomistic simulations of five different disordered peptides in both non-phosphorylated and phosphorylated state, shown in Table 1, have been performed at conditions corresponding to physiological pH (approximately pH 7). The peptides were chosen based on the availability of experimental data and their size, considering computational expense.

Table 1. Full name and sequence of the peptides included in this study. Positively charged residues are marked in blue, negatively charged in red, and phosphorylation sites are highlighted with yellow. The number of residues (N_{res}), net charge of the non-phosphorylated variant ($Z_{no\text{-}phos}$), and the phosphorylated variant (Z_{phos}) are also shown.

Name	Peptide	Sequence	N_{res}	$Z_{no\text{-}phos}$	Z_{phos}
Tau1	$Tau_{173-183}$	AKTPPAPKTPP	11	+2	−2
SN15	$Statherin_{1-15}$	DSSEEKFLRRIGRFG	15	+1	−3
Tau2	$Tau_{225-246}$	KVAVVRTPPKSPSSAKSRLQTA	22	+5	−3
bCPP	β-$casein_{1-25}$	RELEELNVPGEIVESLSSSEESITR	25	−5	−13
Stath	Statherin	DSSEEKFLRRIGRFGYGYGPYQPVPEQPLYPQPYQPQYQQYTF	43	0	−4

SN15, Tau2, and bCPP all contract upon phosphorylation, as shown from the peak shift towards lower values of the distributions of radius of gyration (R_g) and end-to-end distance (R_{ee}) in Figure 1, as well as the average values of R_g and R_{ee} presented in Table 2. For SN15 and Tau2, the width of the distribution also decreases, while bCPP keeps the same range, only the shape of the distribution changes. Stath and Tau1 both expand, shown from a peak shift towards larger values in the distributions. For Tau1, the expansion is more clear observing the R_g distribution than the R_{ee} distribution, which only changes shape by the disappearance of a shoulder at lower values. This, however, causes the average R_{ee}, presented in Table 2, to increase. An increase of R_{ee} upon phosphorylation of Tau1 has been detected by fluorescence resonance energy transfer measurements, as reported by Chin et al. [15].

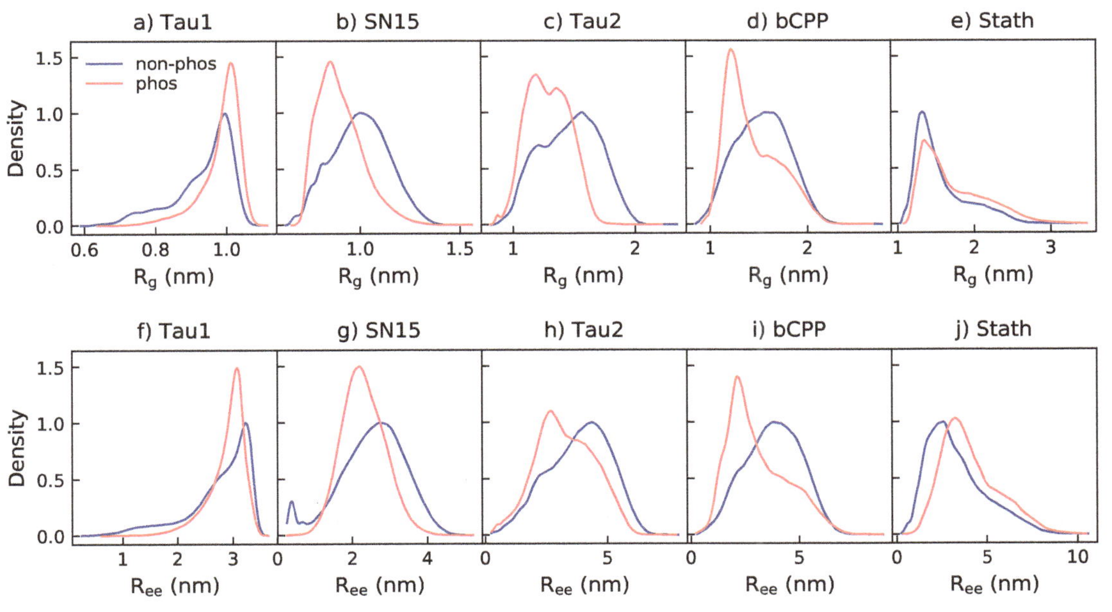

Figure 1. Radius of gyration (R_g) and end-to-end distance (R_{ee}) density distributions of the non-phosphorylated (non-phos) and phosphorylated (phos) variants. The SN15 data are obtained from Ref. [20] (2020 American Chemical Society), and data for the phosphorylated variants of Tau2, bCPP, and Stath from [31].

Table 2. Average radius of gyration (R_g) and end-to-end distance (R_{ee}) of the non-phosphorylated (non-phos) and phosphorylated (phos) variants. Data for SN15 are obtained from [20] and for the phosphorylated peptides of Tau2, bCPP, and Stath from [31].

	Radius of Gyration (nm)		**End-to-End Distance (nm)**	
Peptide	**non-phos**	**phos**	**non-phos**	**phos**
Tau1	0.93 ± 0.01	0.98 ± 0.01	2.74 ± 0.06	2.89 ± 0.02
SN15	1.00 ± 0.01	0.90 ± 0.01	2.54 ± 0.09	2.30 ± 0.03
Tau2	1.46 ± 0.02	1.29 ± 0.03	3.83 ± 0.09	3.27 ± 0.17
bCPP	1.53 ± 0.03	1.43 ± 0.03	3.80 ± 0.08	3.09 ± 0.15
Stath	1.56 ± 0.04	1.73 ± 0.09	3.30 ± 0.24	4.05 ± 0.17

The shape factor, presented in Figure 2, can be used as an estimate of the shape of the peptide. If it behaves as a Gaussian coil, the shape factor is approximately 6, whereas for a stiff rod, it is around 12. SN15, Tau2, and bCPP are shown to behave rather coil-like in non-phosphorylated state, while Tau1 is more stiff, and Stath more contracted. Upon phosphorylation, bCPP becomes more contracted than a Gaussian coil, while Stath expands to become more coil-like.

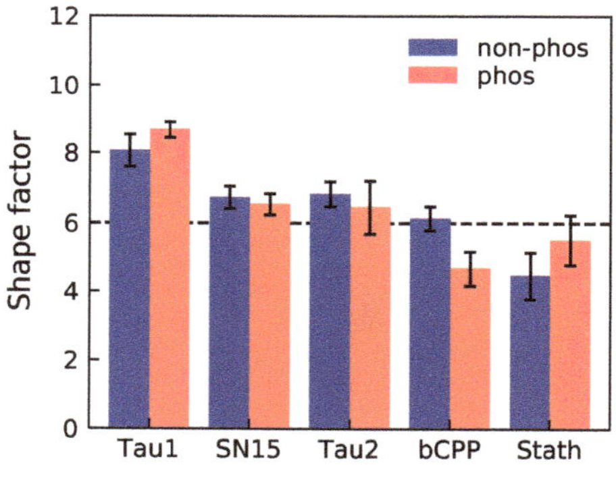

Figure 2. The shape factor of the non-phosphorylated (non-phos) and phosphorylated (phos) variants. The dashed line corresponds to the shape factor of a Gaussian coil. The error bars are based on error propagation of the error estimates determined for radius of gyration and end-to-end distance by block averaging.

Comparing the induced changes of R_g and R_{ee} with the net charge of the non-phosphorylated peptides, it is clear that the prediction of Jin and Gräter, i.e., that net charge controls the effect of phosphorylation [10], only holds for SN15, Tau2, and Stath. bCPP contracts despite having a negative net charge, and Tau1 expands despite the positive net charge. Note that the peptides in this study are distinctly shorter (11–43 residues) compared to the IDPs in the study by Jin and Gräter (approximately 80 residues) [10], hence local interactions are expected to have a more direct effect on the global dimensions. To understand the effect of phosphorylation of these peptides, we therefore need to investigate changes in secondary structure and specific interactions.

2.2. *Phosphorylation of Tau1 Favors Expanded Conformations*

Tau1 is dominated by irregular structure and polyproline type II helix (PPII), as shown in Figure 3a–f. It possesses 46% and 51% PPII in the non-phosphorylated and phosphorylated state, respectively. Elam et al. [41] have predicted close to 50% PPII content in this region of Tau, and CD measurements of this segment indicate an increase of PPII content upon phosphorylation [15]. In Figure 3a–f, it is shown that all structural changes upon phosphorylation at T175 and T181 take place at the C-terminal end of the peptide, from residue 179 and forward. The propensity for bends and turns at residue 179–181 decreases, while the PPII content increases at residues 181–182. There is occasional salt bridge formation between the phosphothreonines and their respective neighboring lysine. Specifically, the probability of salt bridge formation is $7 \pm 2\%$ for pT175–K174 and $9 \pm 2\%$ for pT181–K180. The most occurring salt bridge is, however, formed between pT175 and the N-terminal, with a probability of $49 \pm 9\%$. However, due to the close proximity between the salt bridging residues, the effect on the overall dimensions of the peptide is small. Since Tau1 is a short and rather stiff peptide, as shown by the shape factor in Figure 2, there is limited contact between residues. The change in contact probability upon phosphorylation is also small, according to Figure 3g, which reveals that the main change is a decrease of contact between T181 and the preceding residues A177 and P178, in agreement with the decreased probability of a bend or turn in that region, as shown by Figure 3b,c. The conformational effects of phosphorylation of Tau are well summarized by Figure 3h,i, showing the energy landscape and conformations of non-phosphorylated and phosphorylated Tau1. The energy landscape of non-phosphorylated Tau1 contains several minima, of which the minimum containing expanded conformations dominate, in line with the relatively high shape factor. Other less populated minima contain conformations with a kink in the C-terminal end, originating from a bend or turn. Upon phosphorylation, the minima with kinked conformations disappears, leaving only the minima with expanded conformations. This is in line with decreased contact probability and explains the change in shape of the R_g and R_{ee} distributions, from a peak with a preceding shoulder to a single peak.

2.3. Phosphorylation Increases the Helix Propensity and Induces Salt Bridge Formation in Tau2 and SN15

Tau2 and SN15 are both mainly irregular and report an increase of helicity upon phosphorylation, see Figures 4a–f and 5a–f, respectively. The helical region is identified as "pSpSAKSR" in Tau2 and "pSpSEEKFLR" in SN15, according to Figures 4e and 5e. The sequences, hence, share two characteristics: (1) the helical region starts with two phoshorylation sites, and (2) three or four steps away from the phosphorylation site, a positively charged residue is positioned. Phosphorylation has been shown to stabilize α-helices if the phosphorylation site is located in the N-terminal end of the helix, by electrostatic interaction between phosphorylated serines and the macrodipole of the helix, and by hydrogen bonding with the amide backbone [42]. With a $i, i+4$ spacing between a phosphorylated serine and a lysine, phosphorylation also stabilizes α-helices through salt bridge formation between the side groups [43].

For Tau2, a phosphorylation-induced increase of α-helical structure from 5 to 40% in region A239–R242 has been reported [13]. In these simulations, the main helical increase upon phosphorylation is associated with region S237–K240, where the increase is from 4 to 26%. However, the helical increase is mainly due to 3_{10}-helix, since the increase of α-helix is only from 1 to 5%. Hence, the simulations are in qualitative agreement with the experiments, but the quantitative results should be treated with caution. In addition, in SN15, the larger part of the helical increase is due to 3_{10}-helix, and an increase of α-helix is supported by CD spectroscopy [20], once again giving qualitative support to the findings in this study. Notice also that, while it is hard to make quantitative comparisons with CD data, our study on SN15 suggested that the simulations underestimate the structural content [20], which is the same as observed for Tau2.

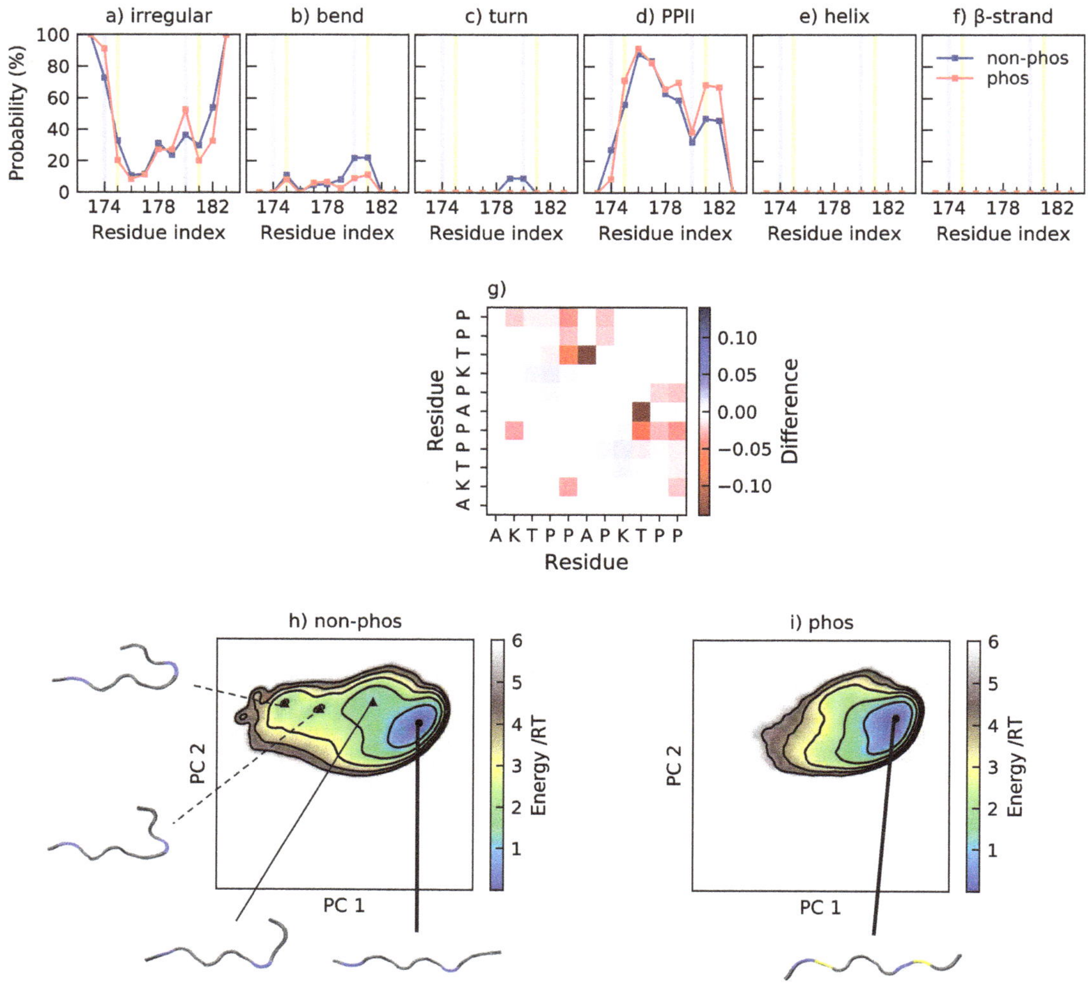

Figure 3. (**a–f**) Secondary structure content along the non-phosphorylated and phosphorylated sequence of Tau1. The helix includes α-helix and 3_{10}-helix. β-strand also includes β-bridge. The positions of phosphorylated and positively charged residues are highlighted in yellow and blue, respectively; (**g**) change in contact probability upon phosphorylation of Tau1; (**h**,**i**) energy landscapes and conformations in minima of non-phosphorylated and phosphorylated Tau1. The energy landscapes are constructed using the first two components from principal component analysis, applying the same basis set for both variants. Hence, they are directly comparable. Contour lines are drawn for integer energy levels in the interval $1 \leq RT \leq 5$ and the minimum of each basin is represented by a marker depending on the energy: ●: $\leq 1RT$, ▲: $\leq 2RT$. A thick line corresponds to the most populated basin, while dashed lines to the least populated basins. In the conformations, positively charged residues are shown in blue, and phosphorylated residues in yellow.

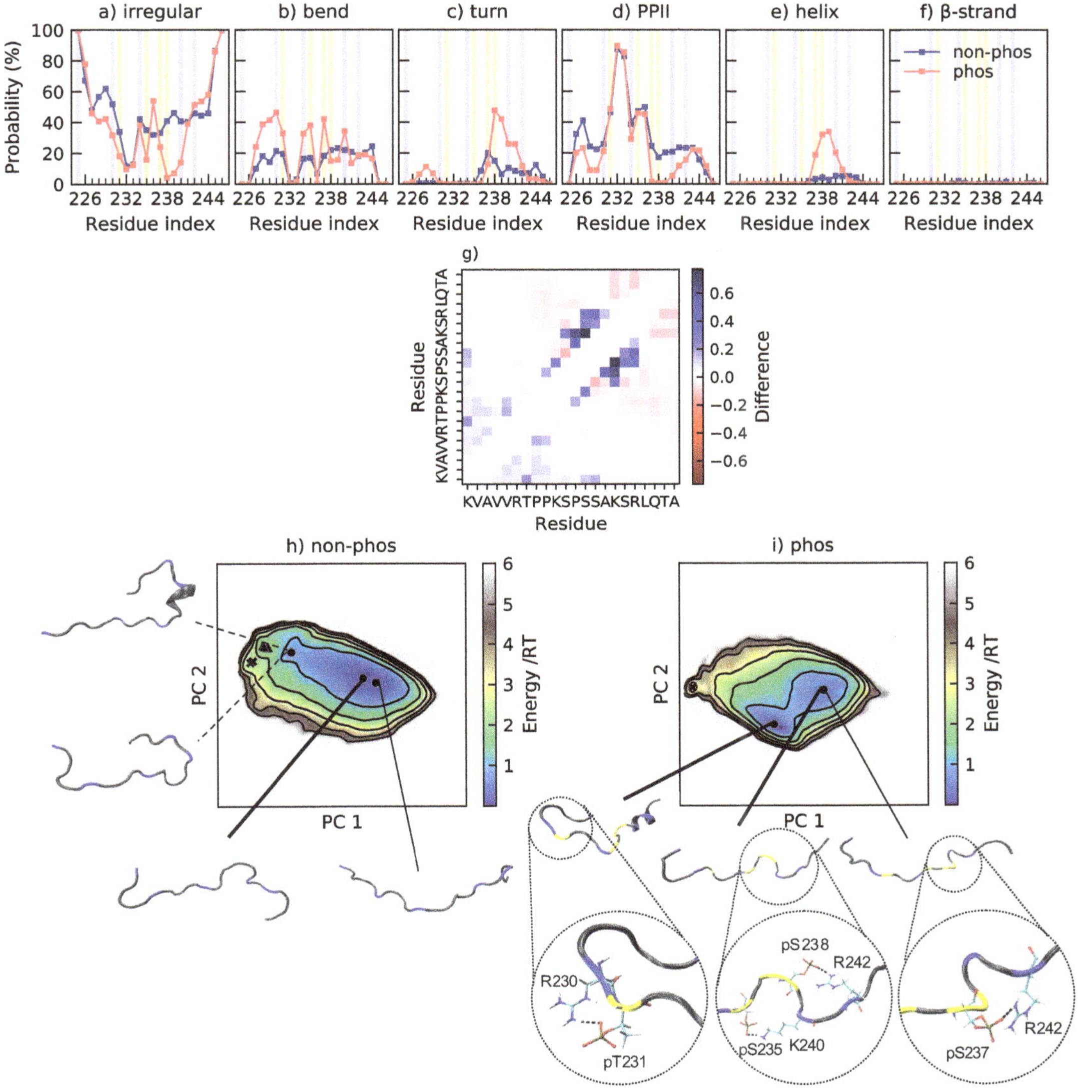

Figure 4. (**a–f**) Secondary structure content along the non-phosphorylated and phosphorylated sequence of Tau2. Helix includes α-helix and 3_{10}-helix. β-strand also includes β-bridge. The data for the phosphorylated peptide are previously published in [31]. The positions of phosphorylated and positively charged residues are highlighted in yellow and blue, respectively; (**g**) change in contact probability upon phosphorylation of Tau2; (**h**,**i**) energy landscapes and conformations in minima of non-phosphorylated and phosphorylated Tau2. The energy landscapes are constructed using the first two components from principal component analysis, using the same basis set for both variants, hence making them directly comparable. Contour lines are drawn for integer energy levels in the interval $1 \leq RT \leq 5$ and the minimum of each basin is represented by a marker depending on the energy: ●: $\leq 1RT$, ▲: $\leq 2RT$, ✖: $\leq 3RT$. Thick lines correspond to the most populated basins, while dashed lines to the least populated basins. In the conformations, positively charged residues are shown in blue and phosphorylated residues in yellow.

While helix formation decreases the R_g and R_{ee}, salt bridge formation can also contribute to the compaction observed upon phosphorylation. In Tau2, several salt bridges have been established from NMR measurements, specifically pT231–R230, pS237–K240, and pS238–R242 [13]. pT231–R230 and pS238–R242 are indeed the two most occurring salt bridges according to Table 3, while pS237–R242 is the third most common. Apart from the increase of helical content related to phosphorylation, Figure 4b reveals an interesting pattern of bends after phosphorylation, where the charged residues R, K, pT, and pS are enriched in bends. The conformations in Figure 4 illustrate how the salt bridges contribute to the formation of bends. Since the probability of a turn at A227–V229 is roughly the same as the probability of the pT231–K225 salt bridge (see Figures 3 and 4c), and V228 is located right between K225 and pT231, we conclude that this turn is also a result of a salt bridge interaction. Hence, this peptide shows that salt bridge formation can induce bends and turns.

Table 3. Probability of salt bridge formation (%) between phosphorylated residues and positively charged residues in Tau2, where NT is the N-terminus. The data are obtained from Ref. [31]. The values printed in bold correspond to the experimentally established salt bridges [13].

Residue	NT	K225	R230	K234	K240	R242
pT231	1 ± 1	10 ± 3	**37 ± 10**	3 ± 2	~ 0	~ 0
pS235	<1	2 ± 1	<1	15 ± 4	17 ± 2	6 ± 3
pS237	2 ± 1	4 ± 3	3 ± 10	17 ± 2	**19 ± 2**	29 ± 2
pS238	4 ± 1	5 ± 2	$3 < 1$	~ 0	5 ± 4	**35 ± 6**

Comparing the energy landscapes of non-phosphorylated and phosphorylated Tau2 in Figure 4h,i, it is shown that, for both peptides, more extended conformations, such as in the minima furthest to the right, are sampled, but to a different extent. These type of conformations are more common in the non-phosphorylated variant, while the most populated basin contains conformations with the N-terminal end folded over, to come closer to the phosphorylated residues. While K225 rarely involves in a proper salt bridge with other residues than pT231, it is still energetically favorable to be in rather close vicinity of the phosphorylated region, considering both the charged side chain and the N-terminus. These types of conformations give rise to an increased contact probability within the N-terminal part of the chain, see Figure 4g. The increased contact probability close to the diagonal in the middle to C-terminal end corresponds to the increase of helical structure and certain salt bridges. Apart from those, there is a decrease of the probability of contacts within the C-terminal end upon phosphorylation. The two minima in the left part of the energy landscape of non-phosphorylated Tau2 in Figure 4h are examples of conformations with a higher level of contact within the C-terminal end. They originate from the electrostatic attraction between the C-terminus and the positively charged residues. In phosphorylated Tau2, that region of the energy landscape is visited much less (see Figure 4i), in agreement with the changes in contact probability. Notice, however, that the probability of conformations with one end folded over is much higher after phosphorylation, which explains the decrease in R_g and R_{ee}. The conformation corresponding to the minimum in the most populated basin for the phosphorylated peptide additionally shows a helix in the C-terminal end, which also contributes to a decreased R_g and R_{ee}.

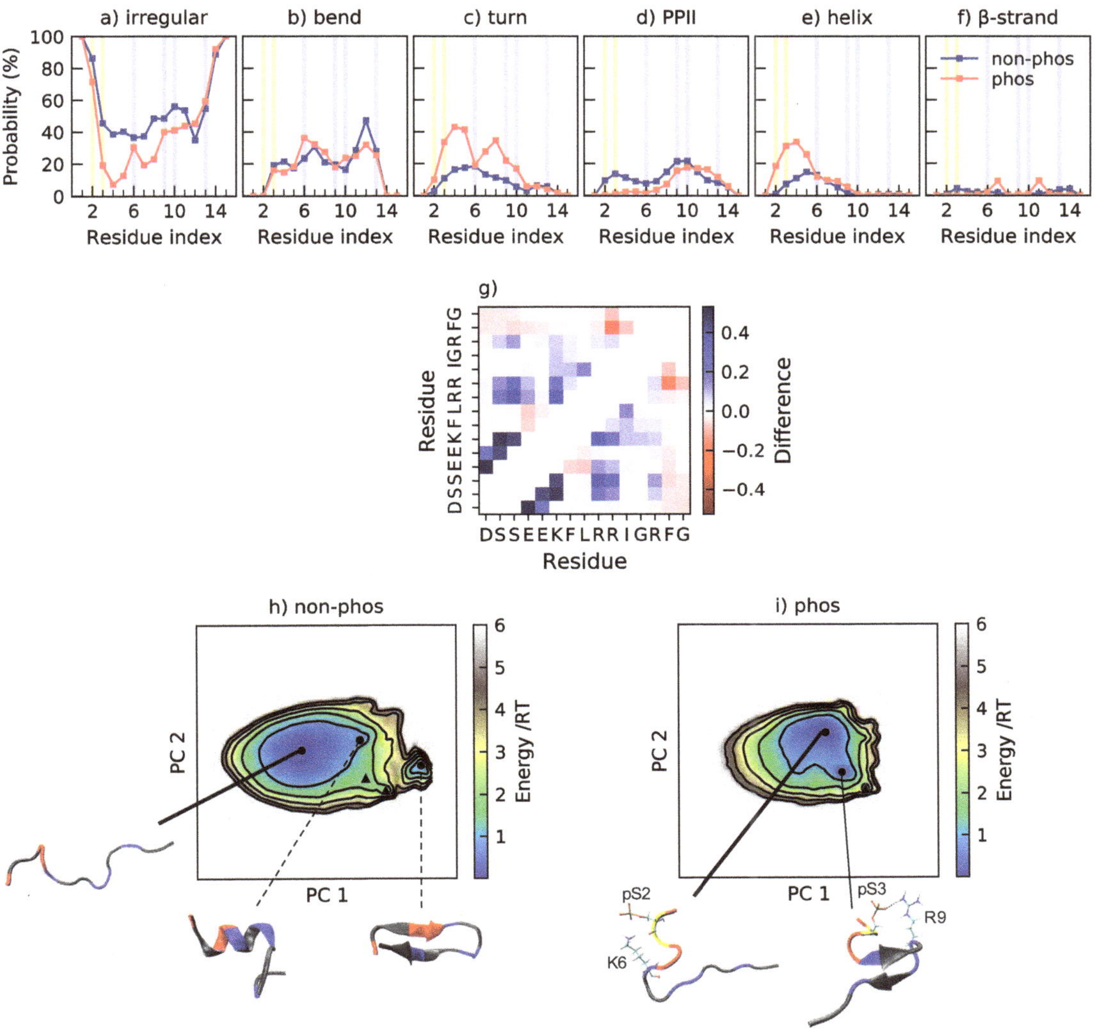

Figure 5. (**a**–**f**) Secondary structure content along the non-phosphorylated and phosphorylated sequence of SN15. Helix includes α-helix and 3_{10}-helix. β-strand includes also β-bridge. These data are obtained from Ref. [20] (2020 American Chemical Society). The positions of phosphorylated and positively charged residues are highlighted in yellow and blue, respectively; (**g**) change in contact probability upon phosphorylation of SN15, based on data from Ref. [20]. (**h**,**i**) Energy landscapes and conformations in minima of non-phosphorylated and phosphorylated SN15. The energy landscapes are constructed using the first two components from principal component analysis, using the same basis set for both variants. Hence, they are directly comparable. Contour lines are drawn for integer energy levels in the interval $1 \leq RT \leq 5$ and the minimum of each basin is represented by a marker depending on the energy: ●: $\leq 1RT$, ▲: $\leq 2RT$. A thick line corresponds to the most populated basin, while dashed lines to the least populated basins. In the conformations, positively charged residues are shown in blue, negatively charged residues in red, and phosphorylated residues in yellow. Phosphorylated and positively charged residues that are close are shown explicitly.

In SN15, the salt bridges pS2–K6, pS3–K6, pS3–R9, and pS3–R10 are the most probable and all form with an approximately 25% occurrence. From the change in contact probability displayed in Figure 5g, it appears that the pS2–K6 and pS3–K6 salt bridges contribute to stabilize the formed helix. The pS3–R9 and pS3–R10 salt bridges are also visible in the contact map and contribute to an increase in the amount of more compact conformations after phosphorylation. In the energy landscape in Figure 5, it is shown that phosphorylation shifts the position of the main minima in the energy landscape, from an area of more coil-like structures to a more compact state. The non-phosphorylated peptide also samples conformations that are more compact with a higher content of secondary structure, but more rarely than the phosphorylated peptide. The conformation corresponding to the minimum in the most populated basin in the phosphorylated peptide has residue pS2 and K6 close enough to be in contact; however, there is no helix, but instead a turn at residues E4–E5. This shows that it is favorable to have pS2 and K6 in contact, but that the interaction does not necessarily imply helix formation. In Figure 5c, it was shown that the turn content in region S3–E5 also increases upon phosphorylation, not only the helix content. There is also an increase of turn content in region F7–R11, which is partly caused by occasional β-strand formation, as shown in the other conformation in Figure 5, and partly by residues pS3 and R9 coming close to form a salt bridge, in line with the turn induced in Tau2. Both of these changes give rise to more compact conformations. We must, however, note that SAXS measurements have indicated that a compaction upon phosphorylation is plausible, but probably smaller than shown in the simulations [20]. While Jin and Gräter found that changes in the hydration shell upon phosphorylation can hide global conformational changes in SAXS measurements, they also concluded that the force field used in this study overestimates the charge effect, thus providing two different explanations of the deviations between the simulations and experiments [10]. Note also that the contact map reports a decrease of contact between R10 and F14, a contact probably formed due to cation–π interaction, which will be discussed further in the section regarding Stath.

2.4. Salt Bridge Formation Shifts the Conformational Ensemble of bCPP

For bCPP, the secondary structure content is dominated by an irregular structure and is highly similar in phosphorylated and non-phosphorylated states, as shown by Figure 6a–f, in agreement with CD spectroscopy results by Farrell et al. [25]. The small difference that occurs upon phosphorylation at S14, S17, S18, and S19 is a change from helix and turn to irregular structure in region E14–S17. The vanishing of helical content is in agreement with the conclusion of Andrew et al. that phosphorylation of a residue in the interior of a helix, without a positively charged residue within suitable distance, destabilizes the helix [42]. Since disruption of a short helix would not cause a contraction of the peptide, the conformational changes in bCPP upon phosphorylation are not explained by secondary structure. Instead, the contraction is due to electrostatic attraction including salt bridge formation between the positively charged end residues and the phosphorylated residues, as seen in Table 4. Although both end residues are arginines, there is a preference of R1 to interact with the phosphorylated region over R25, due to the respective charges of the termini. This is evident from the fact that the N-terminus is also involved in salt bridges with the phosphorylated residues, and further shown by the difference in contact probability in Figure 6g. When R1 interacts with the phosphorylated residues, it causes the peptide to fold over, reducing R_g and R_{ee} substantially. From the energy landscapes in Figure 6h,i, it is shown that before phosphorylation the minima with lowest energy contain more extended conformations, while after phosphorylation the minima with lowest energy instead showcase the N-terminal part being folded over.

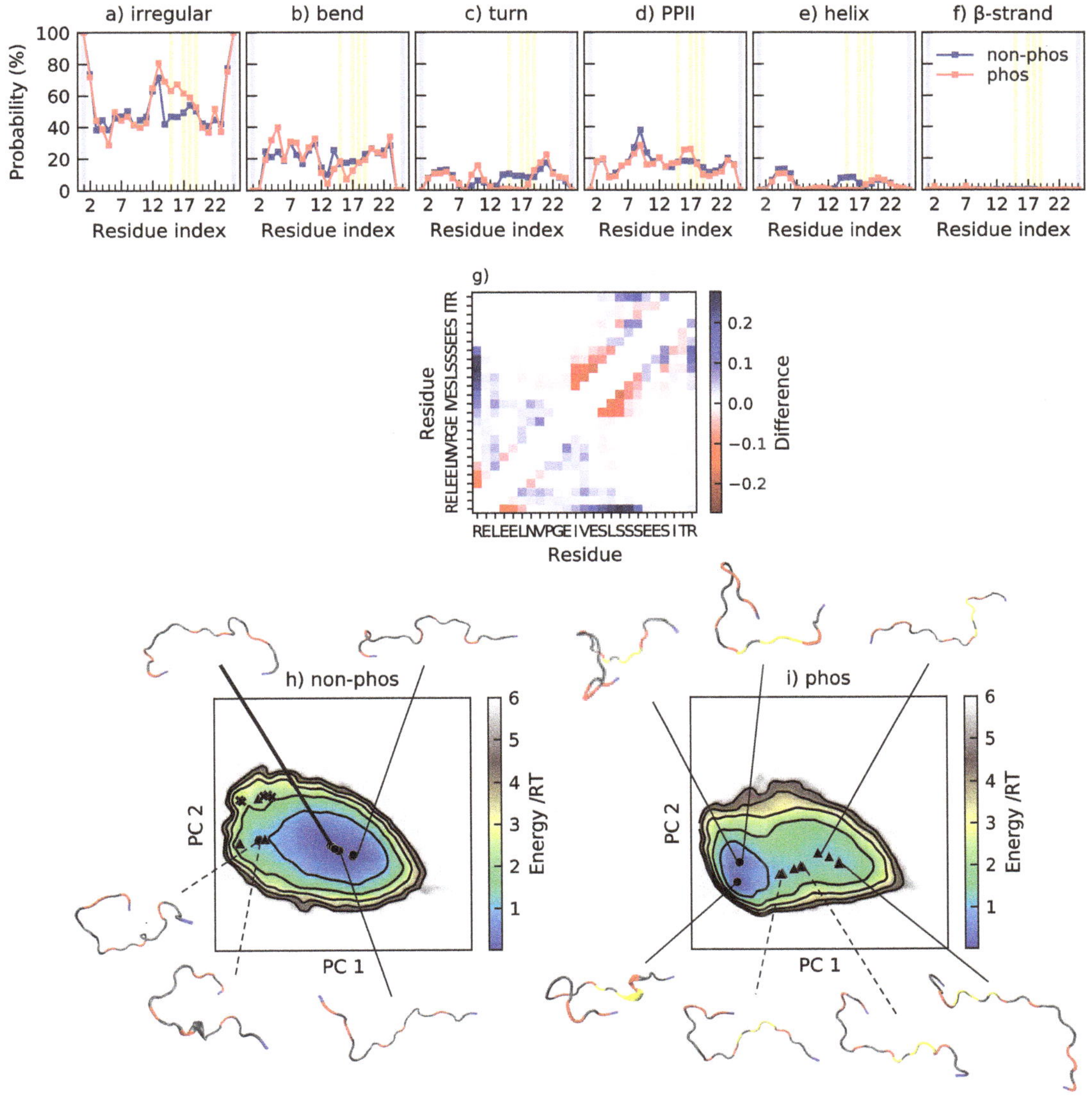

Figure 6. (**a**–**f**) Secondary structure content along the non-phosphorylated and phosphorylated sequence of bCPP. Helix includes α-helix and 3_{10}-helix. β-strand includes also β-bridge. The data for the phosphorylated peptide are previously published in [31]. The positions of phosphorylated and positively charged residues are highlighted in yellow and blue, respectively; (**g**) change in contact probability upon phosphorylation of bCPP; (**h**,**i**) energy landscapes and conformations in minima of non-phosphorylated and phosphorylated bCPP. The energy landscapes are constructed using the first two components from principal component analysis, using the same basis set for both variants. Hence, they are directly comparable. Contour lines are drawn for integer energy levels in the interval $1 \leq RT \leq 5$ and the minimum of each basin is represented by a marker depending on the energy: ●: $\leq 1RT$, ▲: $\leq 2RT$, ✖: $\leq 3RT$. A thick line corresponds to the most populated basin, while dashed lines to the least populated basins. In the conformations, positively charged residues are shown in blue, negatively charged residues in red and phosphorylated residues in yellow.

Table 4. Probability of salt bridge formation (%) between phosphorylated residues and positively charged residues in bCPP, where NT is the N-terminus. The data are obtained from Ref. [31].

Residue	NT	R1	R25
pS15	2 ± 1	6 ± 1	2 ± 1
pS17	3 ± 1	7 ± 1	7 ± 2
pS18	4 ± 1	13 ± 4	12 ± 4
pS19	1 ± 1	10 ± 4	15 ± 4

Based only on the net charge of non-phosphorylated bCPP, it was expected that it would expand upon phosphorylation. Considering only region E13–E21, which contains the four phosphorylation sites, this effect was noticed. The average distance between the C_α atoms of residue 13 and 21 increases from 1.91 ± 0.03 nm to 2.12 ± 0.03 nm upon phosphorylation. However, due to the strong electrostatic interaction between the arginines and the phosphorylated region that are far apart in the sequence, the global result is compaction. Hence, the relative position of charged residues is very important to consider for the effects of phosphorylation on the overall dimensions of the peptide.

We previously showed that the addition of 150 mM NaCl had negligible effects on the salt bridges and global conformational properties of phosphorylated bCPP [31]. The same applies to non-phosphorylated bCPP, as presented in Supplementary File S1, Figures S1 and S2. However, although the average values of R_g at 0 and 150 mM are within error, there is a slight increase in the phosphorylated variant and decrease in the non-phosphorylated variant, see Table 5. Hence, at 150 mM NaCl, the difference observed in R_g between the two variants vanishes, considering the associated error. Note, however, that the distributions still have distinctly different shapes, hence we argue that the conformational ensembles are still different. The same trend is observed in the average R_{ee} values, although a difference with respect to phosphorylation state still remains at 150 mM NaCl, see Table 5. In addition, in the calculated scattering curve (Supplementary File S1, Figure S2), the effect of salt is smaller than the effect of phosphorylation. The difference between the form factor of non-phosphorylated and phosphorylated bCPP is, however, still rather small, so we suspect that it can be hard to detect experimentally with SAXS. Based on the fraction of charged residues and level of charge separation, we expect the other peptides in this study to show smaller effects in regard to salt concentration than bCPP. Hence, we expect the results observed here to be also valid at 150 mM NaCl.

Table 5. Average radius of gyration and end-to-end distance of the non-phosphorylated (non-phos) and phosphorylated (phos) bCPP in the presence of 0 and 150 mM NaCl. The data for the phosphorylated peptide are previously published in [31].

	Radius of Gyration (nm)		End-to-End Distance (nm)	
	0 mM	150 mM	0 mM	150 mM
non-phos	1.53 ± 0.03	1.48 ± 0.02	3.80 ± 0.08	3.64 ± 0.09
phos	1.43 ± 0.03	1.45 ± 0.03	3.09 ± 0.15	3.37 ± 0.13

2.5. Arginine—Phosphoserine Interactions Outshines Arginine—Tyrosine Interactions in Stath

Upon phosphorylation of Stath, the three largest changes in secondary structure are a decrease of β-strand structure, an increase of helical structure, and an increase of turns, according to Figure 7a–f. The increase of helical structure is in the same region as observed for the N-terminal fragment SN15. Figure 7f implies that residues R10, Y18, Y21, and Y41 are of extra importance for the formation of β-sheet. The cation–π interaction that can occur between aromatic residues, such as tyrosine, and cationic residues, such as arginine, have been shown to be common in proteins [44]. A correlation between β-strands and cation–

π interactions have also been established [45]. Table 6 show that the cation–π interaction indeed is more occurring in non-phosphorylated Stath than in phosphorylated Stath, suggesting that it drives the formation of β-strands. The conformations in Figure 7I–III show examples of the cation–π interaction in non-phosphorylated Stath. Although the aromatic-cation interactions are more common in non-phosphorylated Stath, they still occur in phosphorylated Stath, as exemplified by Figure 7. Upon phosphorylation, the occurrence of cation–π interaction decreases substantially, while salt bridge formation appears according to Table 7. Notice that R10, which was shown to interact with tyrosines, is involved in one of the most probable salt bridges, pS3–R10. Hence, the arginine–phosphoserine interaction is deemed stronger than the arginine–tyrosine interaction. The replacement of arginine–tyrosine interaction with arginine–phosphoserine causes the β-strands to vanish, which explains the observed expansion.

Table 6. Probability of cation–π interaction (%) for certain pairs of residues in non-phosphorylated (non-phos) and phosphorylated (phos) Stath.

Residues	non-phos	phos
R10–Y18	13.8 ± 6.3	1.6 ± 0.9
R10–Y21	32.0 ± 8.6	3.9 ± 0.7
R10–Y41	9.2 ± 4.3	0.4 ± 0.2

Table 7. Probability of salt bridge formation (%) between phosphorylated residues and positively charged residues in Stath, where NT is the N-terminus. The data are obtained from Ref. [31].

Residue	NT	K6	R9	R10	R13
pS2	<1	23 ± 7	23 ± 8	12 ± 1	8 ± 1
pS3	12 ± 3	9 ± 1	30 ± 8	32 ± 7	6 ± 3

As presented above, SN15, which is the first fifteen residues of Stath, contracts upon phosphorylation, which was explained by the increased helicity and formation of salt bridges. Supplementary File S1, Figure S3 shows that, in phosphorylated Stath, the global dimensions of the first fifteen residues, $Stath_{1-15}$ agree with those of the fragment (SN15). In the non-phosphorylated variant, the distributions are also rather similar, except for a sharp peak in both the R_g and R_{ee} distributions, which corresponds to a basin in the energy landscape with the conformation shown in Supplementary File S1, Figure S3c. Regarding the secondary structure, according to Supplementary File S1, Figure S4, the largest difference between SN15 and $Stath_{1-15}$ is caused by β-strand not forming in SN15, due to lacking its partner further on in the sequence. There are also some differences in bends and turns, but the increase of helical propensity is similar. Hence, overall, the first fifteen residues of Stath behave rather similarly in the full peptide and as a standalone fragment, although especially the presence of the rest of the sequence induces β-strand formation. Despite this discrepancy, we can conclude that phosphorylation of Stath causes a contraction of the first fifteen residues, but an expansion of the full peptide, due to disruption of β-sheets.

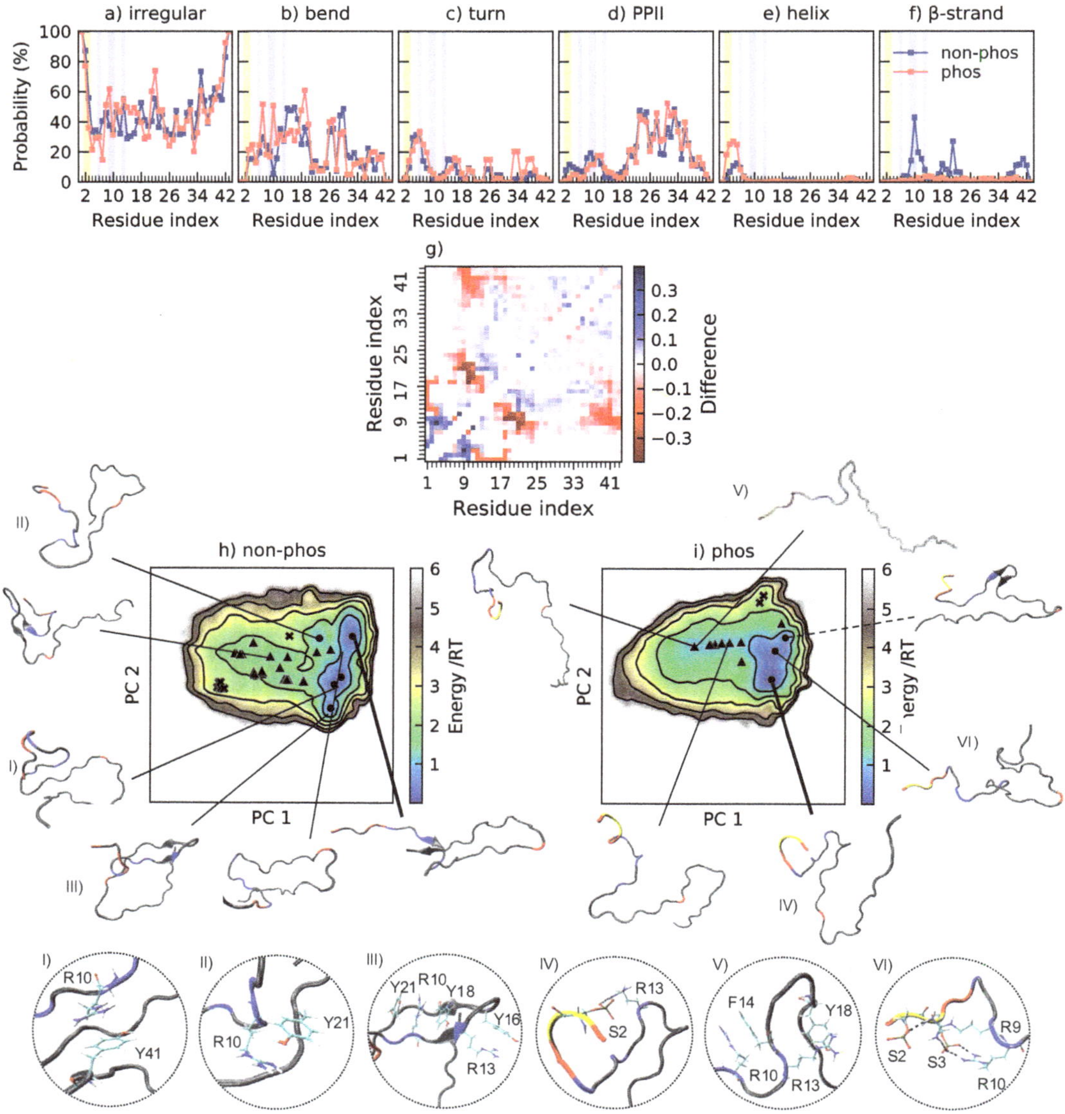

Figure 7. (**a**–**f**) Secondary structure content along the non-phosphorylated and phosphorylated sequence of Stath. Helix includes α-helix and 3_{10}-helix. β-strand includes also β-bridge. The data for the phosphorylated peptide are previously published in [31]. The positions of phosphorylated and positively charged residues are highlighted in yellow and blue, respectively; (**g**) change in contact probability upon phosphorylation of Stath; (**h**,**i**) energy landscapes and conformations in minima of non-phosphorylated and phosphorylated Stath. The energy landscapes are constructed using the first two components from principal component analysis, using the same basis set for both variants, hence making them directly comparable. Contour lines are drawn for integer energy levels in the interval $1 \leq RT \leq 5$ and the minimum of each basin is represented by a marker depending on the energy: ●: $\leq 1RT$, ▲: $\leq 2RT$, ✖: $\leq 3RT$. A thick line corresponds to the most populated basin, while a dashed line to the least populated basin. In the conformations, positively charged residues are shown in blue, negatively charged residues in red, and phosphorylated residues in yellow. The circles show specific interactions within the peptide in the conformations corresponding to the Roman numerals.

3. Conclusions

Some of the peptides in this study contracted upon phosphorylation, while others became more expanded. However, the net charge was not enough to predict the effect in these short peptides. Instead, we have identified factors that appeared to be of greater importance, of which the first is the distribution of charged residues, in line with the influence of linear charge distribution on the conformational ensemble of IDPs [46]. Especially the relative position of phosphorylated and positively charged residues mattered, considering that salt bridges formed between residues far from each other in the sequence had the largest effect on the overall dimensions of the peptide. Regarding salt bridges, Kumar et al. have shown that phosphorylation can re-wire salt bridges by competing with already present E–R salt bridges [47], but no such tendencies were observed for these peptides. Here, the possible salt bridges in the non-phosphorylated peptides were either low in probability or did not change much upon phosphorylation. In Stath, competitive interactions between positively charged residues, aromatic residues, and phosphorylated residues accounted for the changes upon phosphorylation. This shows that, for peptides which include arginine, it can be of importance to also consider aromatic residues. In both bCPP and Stath, phosphorylation induced the opposite effect on the local and global dimensions, hence, to understand the purpose/implications of the phosphorylated residues, both length-scales should be studied. This is especially important dealing with longer IDPs where local/non-local effects can have larger compensatory effect than observed for short peptides [14].

Regarding secondary structure, the separation between phosphorylated and positively charged residues was shown to control the helix propensity, and salt bridges additionally induced changes in the amount of bends and turns. Comparison with experimental data on secondary structure for SN15 and Tau2 indicates that the simulations underestimate the structural content. For these peptides, a preference of 3_{10}- over α-helix was also observed, while the experimental data only considered α-helix. Hence, the simulations were better at indicating trends than producing exact measurements of secondary structure. Overall, the simulation results were often in qualitative agreement with available experimental data, suggesting that, despite the deficiency related to secondary structure and the reported tendency of the force field to overestimate charge–charge interactions, simulations with this force field can still contribute to an increased understanding of the implications of phosphorylation.

As a final note, this study shows that there are several factors contributing to the outcome of phosphorylation, and that they are of varying importance in different peptides. This shows that phosphorylation indeed is complex; however, it is still possible to obtain a better understanding of these factors individually. Therefore, we have an ongoing project in which the number of phosphorylated residues and their positions are varied in a controlled manner, to investigate the effects of those factors systematically.

4. Materials and Methods

All-atom molecular dynamics simulations of the systems shown in Table 8 were performed using GROMACS version 2018.4 (version 4.6.7 for simulation of Stathn) [48–52] with the AMBER ff99SB-ILDN [53] force field and the TIP4P-D [54] water model. Parameters for phosphorylated residues were derived from Homeyer et al. [55] and Steinbrecher et al. [56]. Please note that some of the data sets are previously published and only reanalyzed for this study.

Table 8. Details of the simulations included in this work. The suffix n stands for non-phosphorylated peptide, while the suffix p stands for phosphorylated.

Peptide	Box Volume (nm^3)	Number of Solvent Molecules	Number of Sodium Ions	Number of Chloride Ions	Total Simulation Length (μs)
Tau1n	157.63	5130	0	2	10.0
Tau1p	140.55	4594	2	0	5.0
Tau2n	724.974	23862	0	5	6.0
SN15n [a]	272.13	8839	0	1	14.4
SN15p [a]	294.52	9703	3	0	22.0
Tau2p [b]	722.941	23816	3	0	11.0
bCPPn	1009.24	32975	5	0	5.0
bCPPn, 150 mM	1009.24	32793	96	91	5.0
bCPPp [b]	1002.41	32815	13	0	6.0
bCPPp, 150 mM [b]	1002.41	32633	104	91	7.0
Stathn [c]	930.47	30651	0	0	17.0
Stathp [b]	942.11	30942	4	0	12.0

[a] Previously published [20]. [b] Previously published [31]. [c] Using GROMACS version 4.6.7.

Initial configurations of the peptides were constructed from the sequence as linear chains using Avogadro 1.2.0 [57], optimizing the structure with the auto-optimization tool. SN15n and Stathn were constructed as linear chains in PyMOL [58]. Each peptide was placed in a rhombic dodecahedron box with a minimum distance between the peptide and the box edges of 1 nm, and solvated. The number of water molecules is specified in Table 8, alongside the number of chloride and sodium ions that were added to neutralize the system and in two cases obtain a salt concentration of 150 mM. Periodic boundary conditions were employed in all directions. The equations of motion were integrated using the Verlet leapfrog algorithm [59] with a time step of 2 fs. Non-bonded interactions were treated with a Verlet list cutoff scheme. The short-ranged interactions were calculated using neighbor lists with a cutoff of 1 nm. Long-ranged dispersion corrections were applied to energy and pressure and long-ranged electrostatic interactions were treated by Particle Mesh Ewald [60] with a cubic interpolation and 0.16 nm grid spacing. All bond lengths were constrained using the LINCS algorithm [61]. Solute and solvent were separately coupled to temperature baths at 298 K using the velocity rescaling thermostat [62] with a 0.1 ps relaxation time. Parrinello–Raman pressure coupling [63] was used to keep the pressure at 1 bar, using a 2 ps relaxation time and $4.5 \cdot 10^{-5}$ bar^{-1} isothermal compressibility.

Energy minimization was performed by the steepest descent algorithm until the system was converged within the available machine precision. Initiation of five replicates per system with different starting seeds were performed separately in two steps using position restraints on the peptide. The first step was 500 ps of NVT simulation (constant number of particles, volume, and temperature) performed to stabilize the temperature, followed by the second step of 1000 ps of NPT simulation (constant number of particles, pressure, and temperature) to stabilize the pressure. Production runs of the five replicates per system were performed in the NPT ensemble, for at least 1 μs per replicate. bCPPp with 150 mM salt was simulated in 10 replicates for 0.7 μs each. The total simulation time per system is stated in Table 8. Energies and coordinates were saved every 10 ps, except for in the simulations with 150 mM NaCl. The saving frequency there was every 50 or 40 ps, for bCPPn and bCPPp, respectively.

Analysis

R_g and R_{ee} were calculated using GROMACS 2018.4 and the *gmx analyze* routine was used to obtain averages and error estimates from block averaging analysis. Distributions were obtained by Gaussian kernel estimation using the SciPy package version 1.5.4 [64]. The shape factor, r_s, was calculated from the average values of R_g and R_{ee} according to:

$$r_s = \frac{R_{ee}^2}{R_g^2}. \quad (1)$$

Secondary structure was determined using the DSSP program version 2.2.1 [65] with an extension to detect polyproline type II structure [66,67], on 10,000 equally spaced frames from the combined trajectory. The MDTraj Python library version 1.9.3 [68] was used to obtain contact maps, analyze salt bridges, and cation–π interactions. For the contact maps, contact was defined as when two atoms of different residues were within 0.4 nm of each other. Since salt bridges are formed as a result of hydrogen bonding and electrostatic interactions, they have been assessed by analyzing the presence of hydrogen bonds based on the criterion in reference [69], as implemented in MDTraj. Cation–π interactions were analyzed based on the position of the NZ atom in arginine and CG and CZ in tyrosine. Interaction was defined to occur when both the distances R:NZ–Y:CG and R:NZ–Y:CZ were ≤0.6 nm [44]. The energy landscapes were calculated using principal component analysis following the approach described by Campos and Baptista [70], with the differences described by Henriques et al. [71]. In short, principal component analysis was applied to the Cartesian coordinates of the backbone atoms of the protein, obtained after translational and rotational least square fitting on the central structure of the simulation. The conditional free energy was calculated from the probability density function in the representation space constructed by the first two principal components, obtained by Gaussian kernel density estimation. Snapshots from the simulations were produced using VMD 1.9.3 [72–74]. Data were plotted using a Jupyter Notebook [75] with Python version 3.6.4 and packages NumPy version 1.19.5 [76] and Matplotlib version 2.1.2 [77].

Convergence and sampling quality were assessed by comparing the R_g and R_{ee} distributions, and energy landscapes, between the replicates, as well as by observing the auto-correlation function and convergence of the block average error estimate of R_g and R_{ee} in the concatenated simulation. These data are available in Supplementary File S2.

Supplementary Materials: The following are available online at https://www.mdpi.com/article/10.3390/ijms222011058/s1.

Author Contributions: Conceptualization, M.S. and E.R.; data curation, E.R.; formal analysis, M.S. and E.R.; funding acquisition, M.S.; investigation, E.R.; methodology, M.S. and E.R.; project administration, M.S. and E.R.; resources, M.S.; supervision, M.S.; validation, M.S. and E.R.; visualization, M.S. and E.R.; writing—original draft, E.R.; writing—review and editing, M.S. and E.R. All authors have read and agreed to the published version of the manuscript.

Funding: Financial support has been given by the Royal Physiographic Society in Lund and the Bertil Andersson foundation.

Institutional Review Board Statement: Not applicable.

Informed Consent Statement: Not applicable.

Data Availability Statement: Data are included in the article or Supplementary Material.

Acknowledgments: The simulations were performed on resources provided by the Swedish National Infrastructure for Computing (SNIC) at the Center for Scientific and Technical Computing at Lund University (LUNARC).

Conflicts of Interest: The authors declare no conflict of interest.

Abbreviations

The following abbreviations are used in this manuscript:

IDP	Intrinsically disordered protein
CD	Circular dichroism
NMR	Nuclear magnetic resonance
R_g	Radius of gyration
R_{ee}	End-to-end distance
SAXS	Small-angle X-ray scattering

References

1. Dunker, A.; Lawson, J.; Brown, C.J.; Williams, R.M.; Romero, P.; Oh, J.S.; Oldfield, C.J.; Campen, A.M.; Ratliff, C.M.; Hipps, K.W.; et al. Intrinsically disordered protein. *J. Mol. Graph. Model.* **2001**, *19*, 26–59. [CrossRef]
2. Tompa, P. Intrinsically unstructured proteins. *Trends Biochem. Sci.* **2002**, *27*, 527–533. [CrossRef]
3. Fisher, C.K.; Stultz, C.M. Constructing ensembles for intrinsically disordered proteins. *Curr. Opin. Struct. Biol.* **2011**, *21*, 426–431. [CrossRef] [PubMed]
4. Uversky, V.N. Wrecked regulation of intrinsically disordered proteins in diseases: pathogenicity of deregulated regulators. *Front. Mol. Biosci.* **2014**, *1*, 6. [CrossRef] [PubMed]
5. Babu, M.M.; van der Lee, R.; de Groot, N.S.; Gsponer, J. Intrinsically disordered proteins: regulation and disease. *Curr. Opin. Struct. Biol.* **2011**, *21*, 432–440. [CrossRef] [PubMed]
6. Dunker, A.K.; Brown, C.J.; Lawson, J.D.; Iakoucheva, L.M.; Obradovićá, Z. Intrinsic Disorder and Protein Function. *Biochemistry* **2002**, *41*, 6573–6582. [CrossRef]
7. Iakoucheva, L.M.; Radivojac, P.; Brown, C.J.; O'Connor, T.R.; Sikes, J.G.; Obradovic, Z.; Dunker, A.K. The importance of intrinsic disorder for protein phosphorylation. *Nucleic Acids Res.* **2004**, *32*, 1037–1049. [CrossRef]
8. Gao, J.; Xu, D. Correlation Between Posttranslational Modification and Intrinsic Disorder in Protein. In *Biocomputing 2012*; Altman, R.B., Dunker, A.K., Hunter, L., Murray, T.A., Klein, T.E., Eds.; World Scientific Publishing Co. Pte. Ltd.: Singapore; pp. 94–103. [CrossRef]
9. Johnson, L.N.; Lewis, R.J. Structural Basis for Control by Phosphorylation. *Chem. Rev.* **2001**, *101*, 2209–2242. [CrossRef]
10. Jin, F.; Gräter, F. How multisite phosphorylation impacts the conformations of intrinsically disordered proteins. *PLoS Comput. Biol.* **2021**, *17*, e1008939. [CrossRef] [PubMed]
11. Firman, T.; Ghosh, K. Sequence charge decoration dictates coil-globule transition in intrinsically disordered proteins. *J. Chem. Phys.* **2018**, *148*, 123305. [CrossRef]
12. Uversky, V.N. Intrinsically Disordered Proteins and Their "Mysterious" (Meta)Physics. *Front. Phys.* **2019**, *7*, 10. [CrossRef]
13. Schwalbe, M.; Kadavath, H.; Biernat, J.; Ozenne, V.; Blackledge, M.; Mandelkow, E.; Zweckstetter, M. Structural Impact of Tau Phosphorylation at Threonine 231. *Structure* **2015**, *23*, 1448–1458. [CrossRef] [PubMed]
14. Martin, E.W.; Holehouse, A.S.; Grace, C.R.; Hughes, A.; Pappu, R.V.; Mittag, T. Sequence Determinants of the Conformational Properties of an Intrinsically Disordered Protein Prior to and upon Multisite Phosphorylation. *J. Am. Chem. Soc.* **2016**, *138*, 15323–15335. [CrossRef] [PubMed]
15. Chin, A.F.; Toptygin, D.; Elam, W.A.; Schrank, T.P.; Hilser, V.J. Phosphorylation Increases Persistence Length and End-to-End Distance of a Segment of Tau Protein. *Biophys. J.* **2016**, *110*, 362–371. [CrossRef] [PubMed]
16. Kulkarni, P.; Jolly, M.K.; Jia, D.; Mooney, S.M.; Bhargava, A.; Kagohara, L.T.; Chen, Y.; Hao, P.; He, Y.; Veltri, R.W.; et al. Phosphorylation-induced conformational dynamics in an intrinsically disordered protein and potential role in phenotypic heterogeneity. *Proc. Natl. Acad. Sci. USA* **2017**, *114*, E2644–E2653. [CrossRef]
17. Wang, K.; Ning, S.; Guo, Y.; Duan, M.; Yang, M. The regulation mechanism of phosphorylation and mutations in intrinsically disordered protein 4E-BP2. *Phys. Chem. Chem. Phys.* **2020**, *22*, 2938–2948. [CrossRef]
18. Rani, L.; Mittal, J.; Mallajosyula, S.S. Effect of Phosphorylation and O-GlcNAcylation on Proline-Rich Domains of Tau. *J. Phys. Chem. B* **2020**, *124*, 1909–1918. [CrossRef]
19. Liu, N.; Guo, Y.; Ning, S.; Duan, M. Phosphorylation regulates the binding of intrinsically disordered proteins via a flexible conformation selection mechanism. *Commun. Chem.* **2020**, *3*, 123. [CrossRef]
20. Rieloff, E.; Skepö, M. Phosphorylation of a Disordered Peptide—Structural Effects and Force Field Inconsistencies. *J. Chem. Theory Comput.* **2020**, *16*, 1924–1935. [CrossRef]
21. Willet, A.H.; Igarashi, M.G.; Chen, J.S.; Bhattacharjee, R.; Ren, L.; Cullati, S.N.; Elmore, Z.C.; Roberts-Galbraith, R.H.; Johnson, A.E.; Beckley, J.R.; et al. Phosphorylation in the intrinsically disordered region of F-BAR protein Imp2 regulates its contractile ring recruitment. *J. Cell Sci.* **2021**, *134*, jcs258645. [CrossRef] [PubMed]
22. Papamokos, G.V.; Tziatzos, G.; Papageorgiou, D.G.; Georgatos, S.; Kaxiras, E.; Politou, A.S. Progressive Phosphorylation Modulates the Self-Association of a Variably Modified Histone H3 Peptide. *Front. Mol. Biosci.* **2021**, *8*, 558. [CrossRef] [PubMed]
23. Nicolaou, S.T.; Hebditch, M.; Jonathan, O.J.; Verma, C.S.; Warwicker, J. PhosIDP: A web tool to visualize the location of phosphorylation sites in disordered regions. *Sci. Rep.* **2021**, *11*, 9930. [CrossRef] [PubMed]

24. Mittag, T.; Marsh, J.; Grishaev, A.; Orlicky, S.; Lin, H.; Sicheri, F.; Tyers, M.; Forman-Kay, J.D. Structure/Function Implications in a Dynamic Complex of the Intrinsically Disordered Sic1 with the Cdc4 Subunit of an SCF Ubiquitin Ligase. *Structure* **2010**, *18*, 494–506. [CrossRef] [PubMed]
25. Farrell, H.; Qi, P.; Wickham, E.; Unruh, J. Secondary Structural Studies of Bovine Caseins: Structure and Temperature Dependence of β-Casein Phosphopeptide (1–25) as Analyzed by Circular Dichroism, FTIR Spectroscopy, and Analytical Ultracentrifugation. *J. Protein Chem.* **2002**, *21*, 307–321. [CrossRef] [PubMed]
26. Brister, M.A.; Pandey, A.K.; Bielska, A.A.; Zondlo, N.J. OGlcNAcylation and Phosphorylation Have Opposing Structural Effects in tau: Phosphothreonine Induces Particular Conformational Order. *J. Am. Chem. Soc.* **2014**, *136*, 3803–3816. [CrossRef] [PubMed]
27. Chong, S.H.; Chatterjee, P.; Ham, S. Computer Simulations of Intrinsically Disordered Proteins. *Annu. Rev. Phys. Chem.* **2017**, *68*, 117–134. [CrossRef]
28. Huang, J.; MacKerell, A.D., Jr. Force field development and simulations of intrinsically disordered proteins. *Curr. Opin. Struct. Biol.* **2018**, *48*, 40–48. [CrossRef]
29. Ahmed, M.C.; Papaleo, E.; Lindorff-Larsen, K. How well do force fields capture the strength of salt bridges in proteins? *PeerJ* **2018**, *6*, e4967. [CrossRef]
30. Piana, S.; Lindorff-Larsen, K.; Shaw, D.E. How Robust Are Protein Folding Simulations with Respect to Force Field Parameterization? *Biophys. J.* **2011**, *100*, L47–L49. [CrossRef]
31. Rieloff, E.; Skepö, M. Molecular Dynamics Simulations of Phosphorylated Intrinsically Disordered Proteins: A Force Field Comparison. *Int. J. Mol. Sci.* **2021**, *22*. [CrossRef]
32. Cleveland, D.W.; Hwo, S.Y.; Kirschner, M.W. Purification of tau, a microtubule-associated protein that induces assembly of microtubules from purified tubulin. *J. Mol. Biol.* **1977**, *116*, 207–225. [CrossRef]
33. Buée, L.; Bussière, T.; Buée-Scherrer, V.; Delacourte, A.; Hof, P.R. Tau protein isoforms, phosphorylation and role in neurodegenerative disorders11These authors contributed equally to this work. *Brain Res. Rev.* **2000**, *33*, 95–130. [CrossRef]
34. Gong, C.X.; Iqbal, K. Hyperphosphorylation of microtubule-associated protein tau: A promising therapeutic target for Alzheimer disease. *Curr. Med. Chem.* **2008**, *15*, 2321–2328. [CrossRef]
35. Hay, D.; Smith, D.; Schluckebier, S.; Moreno, E. Basic Biological Sciences Relationship between Concentration of Human Salivary Statherin and Inhibition of Calcium Phosphate Precipitation in Stimulated Human Parotid Saliva. *J. Dent. Res.* **1984**, *63*, 857–863. [CrossRef]
36. Moreno, E.; Zahradnik, R. Demineralization and Remineralization of Dental Enamel. *J. Dent. Res.* **1979**, *58*, 896–903. [CrossRef] [PubMed]
37. Raj, P.A.; Johnsson, M.; Levine, M.J.; Nancollas, G.H. Salivary statherin. Dependence on sequence, charge, hydrogen bonding potency, and helical conformation for adsorption to hydroxyapatite and inhibition of mineralization. *J. Biol. Chem.* **1992**, *267*, 5968–5976. [CrossRef]
38. Holt, C.; Timmins, P.A.; Errington, N.; Leaver, J. A core-shell model of calcium phosphate nanoclusters stabilized by β-casein phosphopeptides, derived from sedimentation equilibrium and small-angle X-ray and neutron-scattering measurements. *Eur. J. Biochem.* **1998**, *252*, 73–78. [CrossRef]
39. Little, E.M.; Holt, C. An equilibrium thermodynamic model of the sequestration of calcium phosphate by casein phosphopeptides. *Eur. Biophys. J.* **2004**, *33*, 435–447. [CrossRef]
40. Ferraretto, A.; Gravaghi, C.; Fiorilli, A.; Tettamanti, G. Casein-derived bioactive phosphopeptides: role of phosphorylation and primary structure in promoting calcium uptake by HT-29 tumor cells. *FEBS Lett.* **2003**, *551*, 92–98. [CrossRef]
41. Austin Elam, W.; Schrank, T.P.; Campagnolo, A.J.; Hilser, V.J. Evolutionary conservation of the polyproline II conformation surrounding intrinsically disordered phosphorylation sites. *Protein Sci.* **2013**, *22*, 405–417. [CrossRef]
42. Andrew, C.D.; Warwicker, J.; Jones, G.R.; Doig, A.J. Effect of Phosphorylation on α-Helix Stability as a Function of Position. *Biochemistry* **2002**, *41*, 1897–1905. [CrossRef]
43. Errington, N.; Doig, A.J. A Phosphoserine-Lysine Salt Bridge within an α-Helical Peptide, the Strongest α-Helix Side-Chain Interaction Measured to Date. *Biochemistry* **2005**, *44*, 7553–7558. [CrossRef]
44. Gallivan, J.P.; Dougherty, D.A. Cation-πinteractions in structural biology. *Proc. Natl. Acad. Sci. USA* **1999**, *96*, 9459–9464. [CrossRef]
45. Tayubi, I.; Sethumadhavan, R. Nature of cation-πinteractions and their role in structural stability of immunoglobulin proteins. *Biochemistry* **2010**, *75*, 912–918. [CrossRef]
46. Das, R.K.; Pappu, R.V. Conformations of intrinsically disordered proteins are influenced by linear sequence distributions of oppositely charged residues. *Proc. Natl. Acad. Sci. USA* **2013**, *110*, 13392–13397. [CrossRef]
47. Kumar, P.; Chimenti, M.S.; Pemble, H.; Schönichen, A.; Thompson, O.; Jacobson, M.P.; Wittmann, T. Multisite Phosphorylation Disrupts Arginine-Glutamate Salt Bridge Networks Required for Binding of Cytoplasmic Linker-associated Protein 2 (CLASP2) to End-binding Protein 1 (EB1) *. *J. Biol. Chem.* **2012**, *287*, 17050–17064. [CrossRef] [PubMed]
48. Berendsen, H.; van der Spoel, D.; van Drunen, R. GROMACS: A message-passing parallel molecular dynamics implementation. *Comput. Phys. Commun.* **1995**, *91*, 43–56. [CrossRef]
49. Hess, B.; Kutzner, C.; van der Spoel, D.; Lindahl, E. GROMACS 4: Algorithms for Highly Efficient, Load-Balanced, and Scalable Molecular Simulation. *J. Chem. Theory Comput.* **2008**, *4*, 435–447. [CrossRef]

50. Pronk, S.; Páll, S.; Schulz, R.; Larsson, P.; Bjelkmar, P.; Apostolov, R.; Shirts, M.R.; Smith, J.C.; Kasson, P.M.; van der Spoel, D.; et al. GROMACS 4.5: A high-throughput and highly parallel open source molecular simulation toolkit. *Bioinformatics* **2013**, *29*, 845–854. [CrossRef] [PubMed]
51. Páll, S.; Abraham, M.J.; Kutzner, C.; Hess, B.; Lindahl, E. Tackling Exascale Software Challenges in Molecular Dynamics Simulations with GROMACS. In *Solving Software Challenges for Exascale*; Markidis, S., Laure, E., Eds.; Springer International Publishing: Cham, Switzerland, 2015; pp. 3–27.
52. Abraham, M.J.; Murtola, T.; Schulz, R.; Páll, S.; Smith, J.C.; Hess, B.; Lindahl, E. GROMACS: High performance molecular simulations through multi-level parallelism from laptops to supercomputers. *SoftwareX* **2015**, *1–2*, 19–25. [CrossRef]
53. Lindorff-Larsen, K.; Piana, S.; Palmo, K.; Maragakis, P.; Klepeis, J.L.; Dror, R.O.; Shaw, D.E. Improved side-chain torsion potentials for the Amber ff99SB protein force field. *Proteins* **2010**, *78*, 1950–1958. [CrossRef]
54. Piana, S.; Donchev, A.G.; Robustelli, P.; Shaw, D.E. Water Dispersion Interactions Strongly Influence Simulated Structural Properties of Disordered Protein States. *J. Phys. Chem. B* **2015**, *119*, 5113–5123. [CrossRef]
55. Homeyer, N.; Horn, A.H.C.; Lanig, H.; Sticht, H. AMBER force-field parameters for phosphorylated amino acids in different protonation states: Phosphoserine, phosphothreonine, phosphotyrosine, and phosphohistidine. *J. Mol. Model.* **2006**, *12*, 281–289. [CrossRef]
56. Steinbrecher, T.; Latzer, J.; Case, D.A. Revised AMBER Parameters for Bioorganic Phosphates. *J. Chem. Theory Comput.* **2012**, *8*, 4405–4412. [CrossRef]
57. Hanwell, M.D.; Curtis, D.E.; Lonie, D.C.; Vandermeersch, T.; Zurek, E.; Hutchison, G.R. Avogadro: An advanced semantic chemical editor, visualization, and analysis platform. *J. Cheminform.* **2012**, *4*, 17. [CrossRef] [PubMed]
58. Schrödinger, L.L.C. *The PyMOL Molecular Graphics System*; Version 1.2r1. 2009. Available online: https://pymol.org/2/ (accessed on 10 September 2021).
59. Hockney, R.W.; Eastwood, J.W. *Computer Simulation Using Particles*; McGraw-Hill: New York, NY, USA, 1981.
60. Darden, T.; York, D.; Pedersen, L. Particle mesh Ewald: An N·log(N) method for Ewald sums in large systems. *J. Chem. Phys.* **1993**, *98*, 10089–10092. [CrossRef]
61. Hess, B.; Bekker, H.; Berendsen, H.J.C.; Fraaije, J.G.E.M. LINCS: A linear constraint solver for molecular simulations. *J. Comput. Chem.* **1997**, *18*, 1463–1472. [CrossRef]
62. Bussi, G.; Donadio, D.; Parrinello, M. Canonical sampling through velocity rescaling. *J. Chem. Phys.* **2007**, *126*, 014101. [CrossRef] [PubMed]
63. Parrinello, M.; Rahman, A. Polymorphic transitions in single crystals: A new molecular dynamics method. *J. Appl. Phys.* **1981**, *52*, 7182–7190. [CrossRef]
64. Virtanen, P.; Gommers, R.; Oliphant, T.E.; Haberland, M.; Reddy, T.; Cournapeau, D.; Burovski, E.; Peterson, P.; Weckesser, W.; Bright, J.; et al. SciPy 1.0: Fundamental Algorithms for Scientific Computing in Python. *Nat. Methods* **2020**, *17*, 261–272. [CrossRef]
65. Kabsch, W.; Sander, C. Dictionary of protein secondary structure: Pattern recognition of hydrogen-bonded and geometrical features. *Biopolymers* **1983**, *22*, 2577–2637. [CrossRef] [PubMed]
66. Mansiaux, Y.; Joseph, A.P.; Gelly, J.C.; de Brevern, A.G. Assignment of PolyProline II Conformation and Analysis of Sequence—Structure Relationship. *PLoS ONE* **2011**, *6*, e18401. [CrossRef] [PubMed]
67. Chebrek, R.; Leonard, S.; de Brevern, A.G.; Gelly, J.C. PolyprOnline: polyproline helix II and secondary structure assignment database. *Database* **2014**, *2014*, bau102. [CrossRef] [PubMed]
68. McGibbon, R.T.; Beauchamp, K.A.; Harrigan, M.P.; Klein, C.; Swails, J.M.; Hernández, C.X.; Schwantes, C.R.; Wang, L.P.; Lane, T.J.; Pande, V.S. MDTraj: A Modern Open Library for the Analysis of Molecular Dynamics Trajectories. *Biophys. J.* **2015**, *109*, 1528–1532. [CrossRef] [PubMed]
69. Wernet, P.; Nordlund, D.; Bergmann, U.; Cavalleri, M.; Odelius, M.; Ogasawara, H.; Näslund, L.Å.; Hirsch, T.K.; Ojamäe, L.; Glatzel, P.; et al. The Structure of the First Coordination Shell in Liquid Water. *Science* **2004**, *304*, 995–999. [CrossRef]
70. Campos, S.R.R.; Baptista, A.M. Conformational Analysis in a Multidimensional Energy Landscape: Study of an Arginylglutamate Repeat. *J. Phys. Chem. B* **2009**, *113*, 15989–16001. [CrossRef] [PubMed]
71. Henriques, J.; Cragnell, C.; Skepö, M. Molecular Dynamics Simulations of Intrinsically Disordered Proteins: Force Field Evaluation and Comparison with Experiment. *J. Chem. Theory Comput.* **2015**, *11*, 3420–3431. [CrossRef] [PubMed]
72. Humphrey, W.; Dalke, A.; Schulten, K. VMD—Visual Molecular Dynamics. *J. Mol. Graph.* **1996**, *14*, 33–38. [CrossRef]
73. Stone, J. An Efficient Library for Parallel Ray Tracing and Animation. Master's Thesis, Computer Science Department, University of Missouri-Rolla, Rolla, MO, USA, 1998.
74. Frishman, D.; Argos, P. Knowledge-based secondary structure assignment. *Proteins* **1995**, *23*, 566–579. [CrossRef]
75. Kluyver, T.; Ragan-Kelley, B.; Pérez, F.; Granger, B.; Bussonnier, M.; Frederic, J.; Kelley, K.; Hamrick, J.; Grout, J.; Corlay, S.; et al. Jupyter Notebooks—A Publishing Format for Reproducible Computational Workflows. In *Positioning and Power in Academic Publishing: Players, Agents and Agendas*; Loizides, F., Schmidt, B., Eds.; IOS Press: Amsterdam, The Netherlands 2016; pp. 87–90.
76. Harris, C.R.; Millman, K.J.; van der Walt, S.J.; Gommers, R.; Virtanen, P.; Cournapeau, D.; Wieser, E.; Taylor, J.; Berg, S.; Smith, N.J.; et al. Array programming with NumPy. *Nature* **2020**, *585*, 357–362. [CrossRef]
77. Hunter, J.D. Matplotlib: A 2D Graphics Environment. *Comput. Sci. Eng.* **2007**, *9*, 90–95. [CrossRef]

International Journal of *Molecular Sciences*

MDPI

Article

Saturation Mutagenesis of the Transmembrane Region of HokC in *Escherichia coli* Reveals Its High Tolerance to Mutations

Maria Teresa Lara Ortiz, Victor Martinell García and Gabriel Del Rio *

Department of Biochemistry and Structural Biology, Institute of Cellular Physiology at UNAM, Mexico City 04510, Mexico; mlara@ifc.unam.mx (M.T.L.O.); vmartinell@tutamail.com (V.M.G.)
* Correspondence: gdelrio@ifc.unam.mx

Abstract: Cells adapt to different stress conditions, such as the antibiotics presence. This adaptation sometimes is achieved by changing relevant protein positions, of which the mutability is limited by structural constrains. Understanding the basis of these constrains represent an important challenge for both basic science and potential biotechnological applications. To study these constraints, we performed a systematic saturation mutagenesis of the transmembrane region of HokC, a toxin used by Escherichia coli to control its own population, and observed that 92% of single-point mutations are tolerated and that all the non-tolerated mutations have compensatory mutations that reverse their effect. We provide experimental evidence that HokC accumulates multiple compensatory mutations that are found as correlated mutations in the HokC family multiple sequence alignment. In agreement with these observations, transmembrane proteins show higher probability to present correlated mutations and are less densely packed locally than globular proteins; previous mutagenesis results on transmembrane proteins further support our observations on the high tolerability to mutations of transmembrane regions of proteins. Thus, our experimental results reveal the HokC transmembrane region high tolerance to loss-of-function mutations that is associated with low sequence conservation and high rate of correlated mutations in the HokC family sequences alignment, which are features shared with other transmembrane proteins.

Keywords: transmembrane proteins; saturation mutagenesis; deep sequencing; residue packing

Citation: Lara Ortiz, M.T.; Martinell García, V.; Del Rio, G. Saturation Mutagenesis of the Transmembrane Region of HokC in *Escherichia coli* Reveals Its High Tolerance to Mutations. *Int. J. Mol. Sci.* **2021**, *22*, 10359. https://doi.org/10.3390/ijms221910359

Academic Editor: Csaba Magyar

Received: 3 September 2021
Accepted: 22 September 2021
Published: 26 September 2021

1. Introduction

Understanding the structure–function relationship of proteins represents a challenge to design effective pharmacological compounds [1,2]. Transmembrane (TM) proteins represent 30% of all proteins and less than 3% of these proteins have their three-dimensional atomic (3D) structures solved [3]. Most TM proteins are targets for pharmacologic intervention given their role in transport and signaling [4], thus anticipating the ability of TM proteins to adapt their sequence without affecting their activity has both basic and applied motivations. A common way to study the structure–function relationship of proteins involves the prediction of residues important for protein function based on the 3D structure of TM proteins, which are seldom available. In the absence of a 3D structure, critical residues for protein function may be predicted based on multiple sequence alignments (MSA) of similar proteins; MSA are built based on substitution matrices that, until recently, have been developed specific for TM proteins [5,6]. In either case, the precise identification of critical residues for protein function is accomplished by saturation mutagenesis of proteins, which up to date have been performed mostly on globular proteins [7–24]; a recent report on the rat neurotensin 1 D03 receptor showed that TM regions allowed for more diverse mutations than the globular regions [25].

Critical residues for protein function are commonly considered positions in a protein that upon mutation affect the folding, stability, binding, and/or catalytic activity of proteins; note that performing single-point mutations may identify loss-of-function mutations, which are mutations that eliminate protein function. We have previously reviewed the different experimental criteria used to define what a critical residue is and proposed a quantitative measurement, Criticality Index (CI), that efficiently relates protein mutations with their functional effect [26]. Several approaches have been described to predict these critical residues [27–34] and all failed to identify several known critical residues [35]. These non-predicted critical residues may be either false-positives or truly hard to predict critical residues. To filter out false-positives, especially on large-scale mutagenesis experiments of proteins, we have reported a combined experimental and computational method that CHecks for Incorrect Sequence-Phenotype Assignments, or CHISPAs [36]. ISPAs (i.e., false positives) are those protein mutants observed with both wild type and mutant phenotypes at a frequency equal or smaller than the expected experimental error introduced to generate/discover mutations. In the present study, we will use this method to study the structure–function relationship of a bitopic protein.

Bitopic proteins (i.e., having a single helical TM region) constitute a convenient model to study the structure–function relationship of TM proteins; besides having a single helical TM region, the activity of these proteins usually is associated to their lateral dimerization in cell membranes [37]; thus, bitopic proteins represent the minimum protein unit that crosses biological membranes. In the present study, we performed both experimental and computational analyses of a bitopic TM helical polypeptide, HokC. This peptide is a toxin that kills *Escherichia coli* cells that express it [38], constituting a convenient system to identify critical residues for its toxic function (e.g., loss-of-function mutations will allow cells to growth). The size of this toxin is also convenient to identify single and multiple mutations, since the sequence of the whole gene may be obtained in a single read by any DNA deep sequencing technology available. We provide experimental evidence that HokC accumulates multiple compensatory mutations that are found as correlated mutations in the HokC family multiple sequences alignment. These correlated mutations are twice as much frequently found in transmembrane proteins than in the globular ones, which is accompanied by a lower local density of residue packing in transmembrane proteins compared with globular proteins. Our results together with previous experimental results support the idea that transmembrane proteins are more tolerant to loss-of-function mutations.

2. Results

2.1. Sensitivity of Experimental Screening

Under growing conditions, *E. coli* cells repress HokC expression to prevent cell death. To disrupt this cellular control, the HokC gene was cloned in the pEXT22 plasmid under the tac promoter; the plasmid also harbors the $lacI^Q$ repressor, to ensure maximal repression of the tac promoter. Hence, this expression system guarantees no transcription leakiness of the gene under the tac promoter, which is important to study the effect of this gene expression on cell survival. To derepress the tac promoter from the $lacI^Q$ repressor, isopropyl-beta-D-thiogalactoside (IPTG) is commonly used. The chromosomal copy of hokC has 3 ATG codons; we noticed that over-expression of the ORF including the 3 ATG codons did not kill all cells; on the contrary, the *hokC* gene expressed from the second ATG had more toxic effect on *E. coli* cells (data not shown); hence, we used that short version of *hokC* in our experiments. To determine how much IPTG is required to activate the expression of HokC, we used a range of IPTG concentrations (see Methods) and a dilution factor of 0.25×10^{-2}; hence, if no colonies were detected it meant that the IPTG was preventing the growth of at least 25 times the initial cells exposed to IPTG. We observed than in all, but one, tested IPTG concentrations, *E. coli* cells did not grow (see Supplementary Materials Table S1). Since we did not observe any difference in cell viability at different levels of IPTG induction, we assumed that for a mutant to be detected in our system, this has to reproduce the effect of having HokC expression repressed, i.e., we would mainly detect loss-of-function mutations. The mutants that reduced up to 25 times the toxicity of HokC would be detected as wild type.

2.2. Mutagenesis of HokC

To reduce the size of the screening, the 23 amino acid residues of the TM region of HokC was mutagenized in regions. For instance, a 3-residue region will generate 30 single point mutations to 1000 multiple mutations (we mutated each position for 10 other residues, see Methods) that will likely be identified by screening 1000 clones or more. Therefore, we selected an average of 1000 isolated colonies for each of the seven mutated regions of the TM region of HokC and classify their phenotypes (see Figure 1). We defined as a wild-type phenotype those cells that upon expression of a HokC mutation no cell colony was observed and, a mutant phenotype corresponds with cells expressing a HokC mutant that upon expression allow the growth of cell colonies (see Methods). The number of colonies analyzed for each of the seven mutated regions and the observed phenotypes are presented in Supplementary Materials Table S2. Note that from this first line of results, we may anticipate that regions II (residues 7–9) and VI (residues 19–21) are less likely to contain loss-of-function mutations than the other regions.

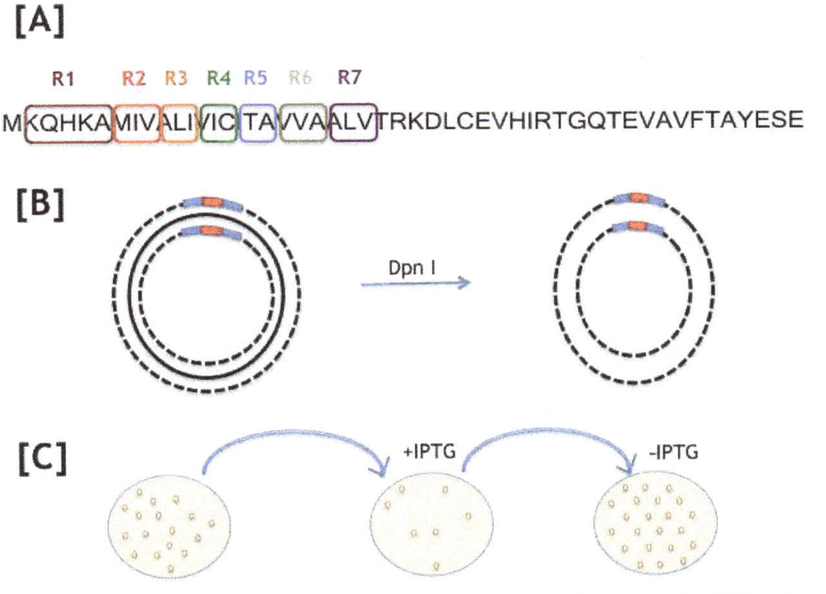

Figure 1. Mutagenesis strategy. (**A**) Seven regions were selected to mutate the ORF coding for HokC; the full sequence of HokC is shown and the regions are marked. (**B**) Oligonucleotides (blue bars) were designed to introduce mutants (red bars) using a QuickChange strategy; the plasmid harboring the wild-type sequence for HokC was amplified (indicated by a punctuated line) and the original plasmid was eliminated by digestion with Dpn I (see Methods for details). (**C**) The plasmids harboring the desired mutations were transferred to competent *E. coli* cells and each colony obtained was replicated into two plates, one with (+IPTG) and another without IPTG (−IPTG), the inducer of HokC expression; cells growing in IPTG harbor a mutation that inactivated the HokC activity and those not growing harbored a mutation that did not affect HokC activity.

After isolating and pooling the DNA from these colonies, we obtained 2,266,368 DNA reads with mutant phenotype and 1,881,708 DNA reads with wild-type phenotype. The sequencing procedure identified mutations beyond the targeted TM region of the protein (all single-residue mutations found in this study are presented in Supplementary Materials Table S3A,B). Yet, the occurrences of mutations beyond position 24 (105,301 sequences contained mutations above this position), where the TM region ends, are rare and, consequently, were not taken into account in our analysis (see Supplementary Materials

Figure S1). The incorrect sequence-phenotype assignments were identified following the CHISPAs procedure using a rate of experimental error of 4% (see Methods). Supplementary Materials Table S4 summarizes all significant single mutants for HokC that rendered a mutant and wild-type phenotype. Two quantitative traits are expected for every position: the number of mutations that rendered a wild-type phenotype (tolerance) and the number of mutations rendering a mutant phenotype (intolerance). We defined as a critical residue any position in the protein sequence for which the ratio of intolerant over tolerant mutations was larger than 1. Our results indicate that none of the residues in the TM region of HokC are critical for its function, yet 19 single-point mutations at 13 different residues eliminate its function (see Supplementary Materials Table S4); these are referred to as deleterious or loss-of-function mutations.

We observed that any amino acid substitution (e.g., Ala for Val or Ile for Trp or any other substitution at any given position) in the HokC rendering a mutant phenotype was also found to render a wild type phenotype (see Supplementary Materials Figure S2). These observations indicate that the position where the substitution takes place is relevant (an Ala for Ile mutation at i-position in the transmembrane region of HokC will not have the same effect if it occurs at j-position) and/or that HokC is able to tolerate many of these mutations. In fact, 4 out of the 13 single-residue substitutions identified to render loss-of-function mutations were found as substitutions in the multiple sequence alignment in the HokC family, suggesting that such natural variants included in the HokC family should have tolerated the mutation if the toxic activity was conserved. We will next explore this idea.

Our experimental design allowed us to identify multiple mutations: HokC variants that include more than one point mutation (see Methods). Among these multiple mutations, we detected compensatory mutations, indicating any combination (double, triple, and so on) of single loss-of-function mutations that showed a wild-type phenotype. Table 1 shows the most frequently observed compensatory mutations in our study; for a full list of these compensatory mutations, see Supplementary Materials Table S5. It is noticeable that residue 7 is the only residue in region II that presented deleterious mutations and was the position most frequently observed among compensatory mutations (see Table 1); this result explains the observation about region II (residues 7–9) presenting most of the wild type phenotypes (see Supplementary Materials Table S2). All 19 mutations rendering a mutant phenotype (see Table 1) may be compensated (see Supplementary Materials Table S5), providing an explanation for the high tolerance of the TM region of HokC to maintain the toxic function of this peptide.

Table 1. Compensatory mutations in the TM region of HokC.

Combined Mutations (Experimental)	Counts	Combined Mutations (MSA)	Counts
M7W, I12S	613	V13I, A6T	3
I12S, I14S	317	V19L, A6T	8
L11P, I12S	276	A22T, V19L	81
M7W, I12C	221	A6T, K2M	1
M7W, I14S	220	A22S, V19L	2
M7W, L11P	184	A22T, V13I	1
I12S, V19G	145	V19L, V13I	11
I12S, A22T	136	A21T, V19L	5

The observed combinations of deleterious single mutations (see Supplementary Materials Table S4) that occurred in our experimental set up rendering a wild-type phenotype that were considered compensatory mutations. The table only shows compensatory mutations that are present more than 100 times in our experimental setup. Please note that these compensatory mutations may be present in combination with other tolerated mutations (see Methods); for the list of all compensatory mutations see Supplementary Materials

Table S5. For a full list of single-point mutations observed in compensatory mutations, see Supplementary Materials Table S6.

Interestingly, residue Cys15 tolerated every mutation. Since the previously reported Cysteine to Serine tolerated mutation at that position is conservative and several of the mutations identified at this position were not conservative, we performed a site-directed mutagenesis of this Cys15 residue by three different residues (Cys15Ser, Cys15Glu and Cys15Ala) to validate the tolerance for HokC toxic function of these mutations; our site-directed mutagenesis validated the saturation mutagenesis observations at this position (data not shown).

The orientation of HokC in the TM region is important for its activity. To test for the orientation of the TM region of single (Met7Trp or Ile12Ser) and multiple (Met7Trp-Ile12Ser) mutations of HokC that rendered mutant and wild-type phenotypes, respectively, we fused GFP or phoA to the C-terminus of these mutants. Such constructs have been previously reported to assess the orientation of both N- and C-terminus of TM regions of *E. coli* proteins [39]. As control, we fused GFP or phoA to the wild-type sequence of HokC. Our results showed that the GFP fusions (to wild type or any of the mutants) eliminated the toxic activity of HokC upon induction (see Supplementary Materials Figure S3A). Alternatively, phoA fusions kept the activity of wild type and every mutant tested (see Supplementary Materials Figure S3B). Accordingly, phoA and not GFP fusions, displayed enzymatic activity (see Supplementary Materials Figure S4). These results indicated that HokC has its C-terminus oriented towards the periplasmic space and that the mutants kept this orientation and the level of expression of the wild type sequence.

In summary, our experimental results revealed that HokC tolerates all single point mutations by accumulating multiple compensatory mutations. This result suggested that: (i) sequence conservation analysis may show low correlation with deleterious mutations, and (ii) TM regions have structural features that allow for accommodating multiple compensatory mutations. To test these hypotheses, we next performed a computational analysis of the HokC protein family and on TM proteins in general.

2.3. Are Critical Residues in the TM Region of HokC Conserved?

Using a sequence alignment reported for the HokC family derived from PFAM (see Methods), only one residue (Cys15) identified in the TM region of HokC was invariant (data not shown). To test if this lack of relationship between critical residues and invariant character of residues is the consequence of using an alignment not optimized for TM proteins, we generated a multiple sequence alignment (MSA) with the 148 protein sequences of the PFAM family PF01848 using TM-COFFEE (see Supplementary Materials Table S7). Our results indicate that only residue Thr17 was invariant and Val24 presented some degree of conservation, yet none of these positions are critical for protein function. This MSA was also analyzed to compute conservation scores based on the rate4site algorithm (see Methods). According to this analysis (see Supplementary Materials Table S8), residues 1, 15, and 17 show the lowest mutability (conservation score $\geq$8) in the TM region of HokC; furthermore, modifying the parameters of rate4site, it was noted that some correlation between experiments and conservation could be found (data not shown). We explored a third method, PROVEAN (see Methods), which predicted positions 1, 12, and 13 to include deleterious mutations (see Supplementary Materials Table S9). Interestingly, position 13 presented substitutions in the MSA that rendered a deleterious effect in our experimental screening (see Table 1). These results confirmed the expected poor correlation between sequence conservation and the loss-of-function mutations in HokC.

One possible mechanism to maintain function without conserving amino acids is by compensatory mutations, i.e., multiple mutations that compensate the deleterious effect of individual mutations. Hence, it is expected that natural variants of HokC may have accumulated compensatory mutations if they were to keep the biological function of HokC. To test this idea, we compared the mutability of each position in the HokC family alignment with that observed in our mutagenesis experiment. As shown in Supplementary Materials

Table S4, the MSA included residue substitutions at positions K2, V6, A13, I14, V19, A21, and A22 that, in our experimental, data rendered a mutant phenotype (deleterious mutations in Supplementary Materials Table S4). This result supports the notion that these loss-of-function mutations must have been compensated if the homologous proteins of HokC should keep their toxic function. To test this idea, we identified all the multiple mutations in the MSA for the HokC family that harbored deleterious mutations for HokC and observed that 91 out of 148 protein sequences included this class of multiple mutations (see Supplementary Materials Table S10). Thus, correlated mutations in the HokC family correspond with compensatory mutations identified in our screening. To study whether this is a particular property of HokC or a general trend of TM proteins, we decided to extend our analysis to other TM proteins.

2.4. *Compensatory Mutations Correlate to High Order Residue Contacts in HokC*

According to the expected helical structure of the TM region of HokC, residues that are closer than four residues apart in the sequence may be close in the three-dimensional structure; hence these may be suitable to accommodate compensatory mutations. In agreement with this idea, we observed compensatory mutations in residues that are close at the sequence level (see Table 1). Furthermore, it has been shown that the TM region of HokC may be engaged in the formation of a homodimer as inferred from the mutagenesis of Cys15 for Serine [40]. Our results revealed compensatory mutations between residues far away in the TM region (e.g., positions 6 and 7 with positions 13 and 12, respectively), suggesting that these residues may interact when these are at different monomers; otherwise, an unusual bend on the helix has to be assumed for these residues to interact within the same monomer, which may prevent this region to fully traverse the membrane. The recent prediction reported for the HokC monomer by AlphaFold software version 2, indicates that this TM region does not present an unusual bend in the helix [41], in agreement with the idea that positions 6 and 7 with positions 13 and 12 in HokC monomers participate in the dimerization.

Thus, compensatory mutations in HokC are in agreement with the helical structure of this TM peptide and revealed some other residues that may participate in the dimerization of HokC.

2.5. *Implications for TM Proteins*

Our results indicate that compensatory mutations accumulate among the HokC family of toxins. It has been shown that the loss-of-function single-point mutations may be reverted by combining these with other deleterious mutations [42]. Such mutations are referred to as compensatory mutations that usually correspond with residues close in the 3D structure of proteins [43]. Based on these observations, we wondered whether these mutations accumulated among residues close in the 3D structure of TM proteins (these proteins are structurally classified as mainly alpha or mainly beta) and compared these with globular proteins that presented these same structural classes (see Methods). Our results indicate that TM proteins tend to favor, at least twice as much, the presence of multiple mutations between nearby residues in the 3D structure of proteins (see Figure 2).

To evaluate if the observed increased rate of compensatory mutations is associated with the difference in compactness of TM versus globular proteins, we carried out an analysis of the residue contacts in these two groups of proteins. We observed that as proteins (both globular and TM proteins) change in size, the number of three-dimensional contacts among residues increases proportionally (see Figure 3). This indicates that both globular and TM proteins present a constant packing density, with similar average number of contacts per residue for globular (5.4) and TM proteins (5.4).

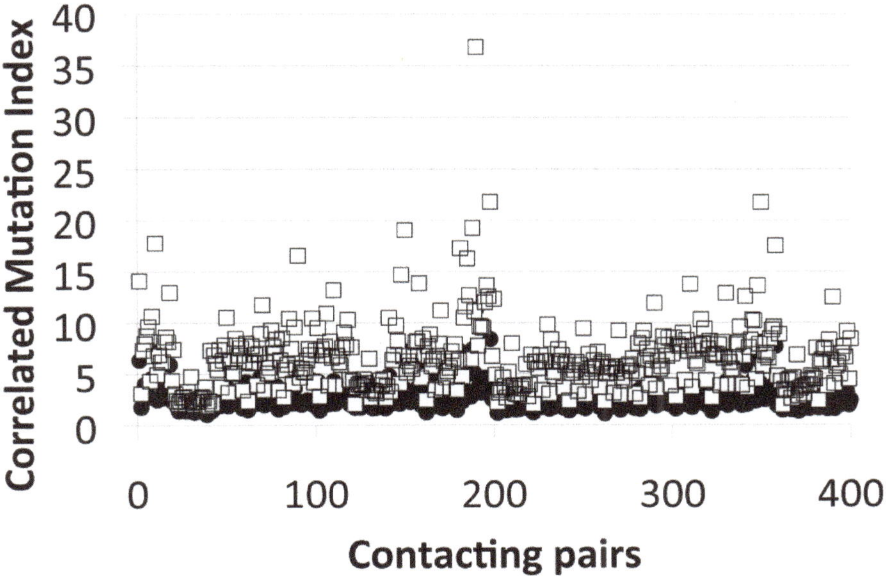

Figure 2. Correlated mutation index of globular and transmembrane proteins. The normalized frequency for all 400 residue pairs at distance of 5 Å in the three-dimensional protein structure (represented in x-axis) that were simultaneously mutated as observed in multiple sequence alignments for their corresponding protein families (correlated mutation index) is presented for both, globular (black circles) and transmembrane (white squares) proteins.

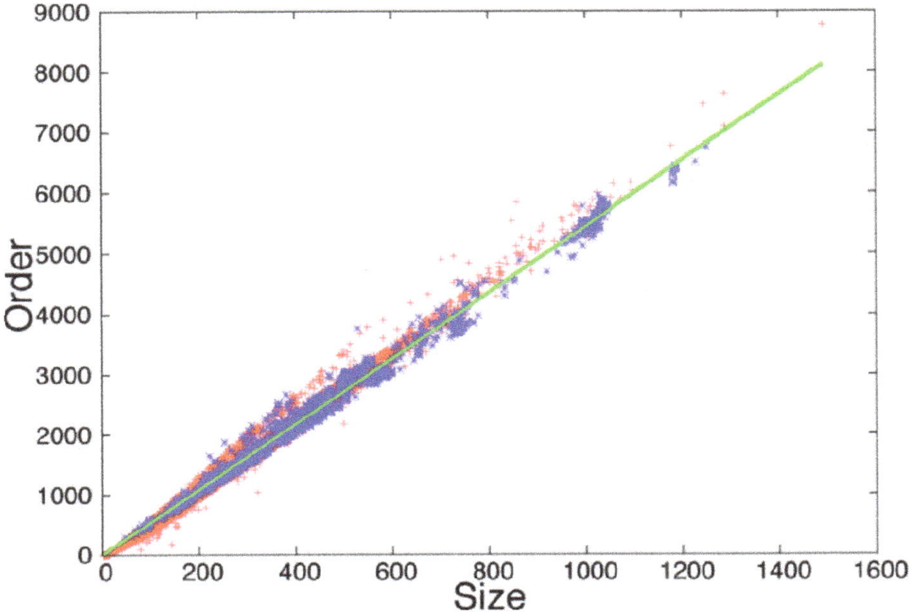

Figure 3. Density of residue contacts for globular and transmembrane proteins. Protein structures were transformed into contact maps at 5 Å to obtain the number of residues (Size) and the total number of reside contacts (Order) for each protein analyzed (see Methods). Size and Order are plotted for both globular (+) and transmembrane (+) proteins. The green line represents the best linear adjustment to both data sets and has a slope of 5.4. The plot was generated using gnuplot.

In an attempt to identify local differences in packing between these classes of proteins, we looked for maximal cliques in their residue contact maps. Maximal cliques are those cliques (group of residues that are all in contact in the 3D space) that are not part of any larger clique, hence correspond with the densest regions within proteins. We observed that TM proteins accumulated small maximal cliques (size 3) more than globular proteins (see Figure 4). Thus, the set of TM proteins analyzed are less densely packed than the globular proteins as a consequence of reducing the number of large maximal cliques.

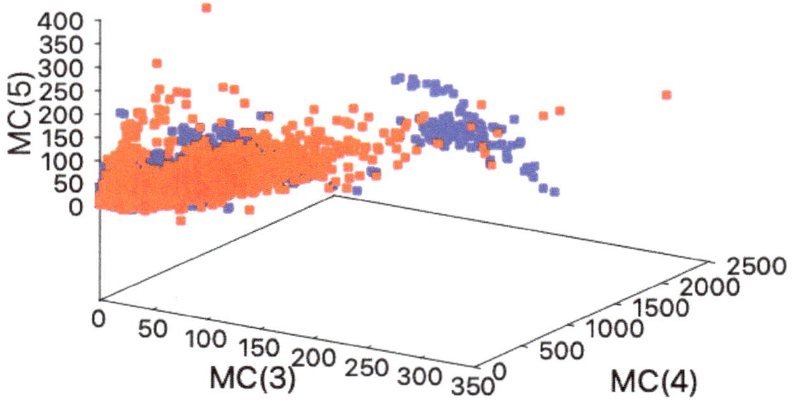

Figure 4. Maximal cliques observed in globular and transmembrane proteins. Protein structures were transformed into contact maps at 5 Å to identify the maximal cliques including 3, 4, or 5 residues using Tomita algorithm (see Methods); maximal cliques correspond with the protein regions where residues are highly packed. Maximal cliques occurrences of size 3, 4, and 5 (axis labeled MC(3), MC(4), and MC(5), respectively), are presented for both globular (■) and transmembrane (■) proteins. Please note the cumulus of blue squares on the right side of the image, which include the maximal cliques of size 3 that are accumulated in transmembrane proteins.

Finally, we analyzed the spherical angles between contacting hydrophobic residues (see Methods) to test if this difference in packing may be associated with differences in the arrangement of contacting hydrophobic residues, i.e., we aimed to compare the core of TM proteins with those of globular proteins that belong to the same structural class. To quantify this, we used the Haussdorff distance that estimates the overall difference of two sets of vectors; in this case, hydrophobic residues that are close in distance were represented in vectors, each element in the vector include the angle between the hydrophobic pair of residues. We observed that the orientation of contacting hydrophobic residues of TM proteins and globular proteins differs; particularly, globular proteins (see Figure 5A) tend to have on average smaller Haussdorff distances among their hydrophobic contacting residues compared with TM proteins (see Figure 5B), yet with larger dispersion. Besides the trend presented in Figure 5A,B, we also noticed that 57% of every pair of globular protein analyzed had identical orientation between contacting hydrophobic residues while this occurred in only 18% of the TM proteins. Despite these differences, we observed a group of globular and TM proteins with mainly alpha helical compositions (with the same CATH classification) that showed a very similar contacting geometry (see Figure 5C; for instance structural class 1.10.405.10 or 1.20.5.110). These results indicate that while there is a trend to maintain the geometrical arrangement of hydrophobic residues in globular proteins more than in TM proteins, there are some exceptions to this trend.

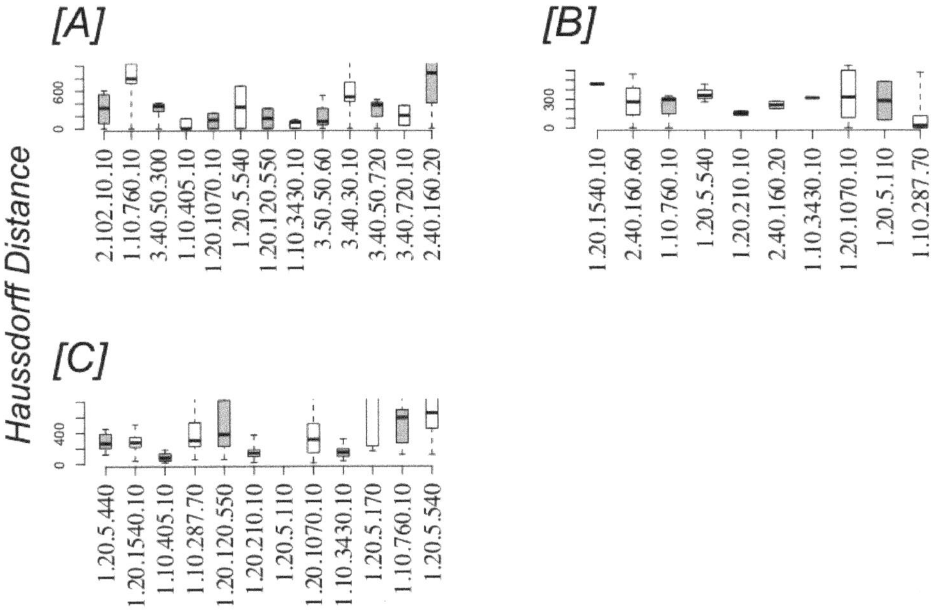

Figure 5. Geometrical differences between globular and transmembrane proteins. The Haussdorff distance (see Methods) was calculated for each protein structure present in the indicated CATH classes on the x-axis for globular (**A**) and transmembrane (**B**). This comparison was conducted also for pairs of globular and transmembrane proteins with the same CATH class with alpha helical structure (**C**). The differences are plotted as boxes, where the median is presented as a horizontal line within the box and the horizontal lines away from the box denote the minimum and maximum values of these distances per CATH class. To facilitate the visualization of these trends, the y-axis value range was ≤700.

3. Discussion

Experimental data derived from saturation mutagenesis of proteins indicates that both TM and globular proteins are more tolerant to mutations than expected from phylogenies; however, these previous studies have not addressed the difference in the tolerance to mutations between these two classes of proteins, if any. The relevance of this comparison is that it may help anticipate which of these proteins may adapt more easily to drugs used to control cell fate or to reveal possible reservoirs for new protein sequences and functions, among others. From sequence analysis, it has been observed that TM proteins tend to present a lower degree of sequence conservation than globular proteins [44,45], yet this observation may be the consequence of the method used to align these sequences rather than a property of these proteins. The pioneer work by Bowie's group showed that the TM regions were as tolerant to mutations as the globular parts of the diacylglycerol kinase from *Escherichia coli*, despite the fact that most critical active-site residues reside in the cytoplasmic domain [46]; yet the coverage of mutations in this experiment was reduced, preventing to fully identify critical residues or compensatory mutations. More recently, it has been shown for the rat neurotensin 1 D03 GPCR that TM regions accepted more diverse mutations than its globular regions [25]; hence, the authors suggested that TM regions are more tolerant to mutations than globular regions. Whether this applies to all TM proteins requires further investigation. To contribute to address this idea, in the present work, we explored the sequence–phenotype space of a TM protein, HokC from *E. coli*. It is relevant to note that the toxic function of HokC depends on its homodimerization and that while we could infer some aspects of this dimerization, our experimental assay cannot discriminate functional defects as a consequence of the monomer or dimer inactivation.

Our experimental results show that 97% (233 out of 240) of all single mutations expected were detected in our screening and only 19 mutations (8%) of these rendered an inactive (non-toxic) HokC peptide (see Supplementary Materials Table S4). Hence, the TM region of HokC tolerates most (92%) single-point mutations. For comparison, the C-terminal domain that lays at the periplasmic space of *E. coli* has been proposed to encode for the toxic domain based on two results: (i) the absence of mutations that alter protein function at the N-terminus and (ii) the substitution of the C-terminal region by the phoA resulted in a non-toxic protein [40]. Here, we show that the TM region actually encodes for positions that, upon mutation, alter protein function and that fusing HokC variants to GFP renders an inactive protein. Hence, our results show that HokC toxicity depends on the N-terminal domain and that such domain is more tolerant to mutations than those previously reported for the C-terminus domain. Furthermore, this rate of tolerance for the TM region of HokC is larger than in previous experimental reports showing that globular proteins only tolerate 30–40% of all possible single point mutations [47]. To evaluate whether this is a property of HokC or if this is a general property of TM regions, we performed complementary computational analysis.

In this regard, it has been noted that the core of globular proteins and that of the TM regions are mainly composed of hydrophobic residues, yet different forces drive this similarity in composition. Particularly, globular proteins are subjected to the hydrophobic collapse [48] while the folding of TM regions is commonly assisted laying the hydrophobic residues inside the lipid membrane [49]. This difference suggests that TM regions may tolerate any hydrophobic mutations, yet our results indicate that not every hydrophobic residue is tolerated in the TM region (see Supplementary Materials Table S4). This indicates that more complex rules for protein folding take place at the TM region.

This high tolerance to mutations is accompanied with a low degree of sequence conservation observed in the TM region the of HokC family (see Supplementary Materials Table S7). Our results indicate that in the case of the HokC family, this sequence diversity is the consequence of combining multiple mutations that harbor deleterious single amino acid mutations (see Table 1, Supplementary Materials Tables S3 and S10). Such multiple mutations may reduce the conservation of many positions in the HokC family and consequently, methods based on sequence-conservation scores fail to properly identify deleterious mutations in this family of toxins. To study the nature of this capacity of TM proteins to accumulate compensatory mutations, we compared the correlated mutations observed between globular and TM proteins and observed that TM proteins tend to accommodate twice as much correlated mutations as globular proteins (see Figure 2). This observation was then compared with the protein packing properties of TM and globular proteins. It has been shown that globular proteins have a constant atomic density [50], i.e., globular proteins with different folds and different sizes all have a similar average number of atoms per volume within a crystal. We have previously reported that the number of contacting residues in the 3D structure of proteins reproduces this phenomenon [51]. Here, we extend these observations to TM proteins and observed that TM proteins have a similar linear trend in the number of contacting residues than globular proteins (see Figure 3). Yet, we observed local differences in the packing of TM and globular proteins, where globular proteins tend to accommodate more residues per unit volume (see Figure 4). This trend is consistent with the observation that TM proteins tend to incorporate voids within their core to fulfill their biological function (e.g., channels [52]) while voids in globular proteins are destabilizing [53] and, consequently, tend to be avoided. Alternatively, voids in any protein have been proposed to locate where proteins are more flexible [54]. From that perspective, our results may be interpreted as TM proteins being more flexible. Thus, our computational analysis shows that TM proteins are locally less densely packed than globular proteins.

In agreement with this concept, we observed that the more dense packing in globular proteins is related to their regular orientation of contacting residues (see Figure 5A,B). In contrast with these observations, geometrical similarities of contacting helix–helix pairs in globular and TM proteins have been reported [55]; here, we show that the density of contacting residues among proteins in the mainly alpha-helical family of proteins are consistent with these previous observations (see Figure 5C). These results indicate that while there are similarities between alpha-helical TM and globular proteins, overall globular proteins tend to vary less the packing in their core than TM proteins. Relevant to these observations is the idea that proteins fold to a minimum energy accessible by densely packing their residues [56]. A solution to this packing problem may be the regular packing proposed by Kepler in the XVII century [57]. Our results provide evidence that globular proteins packed their residues in a more regular way than TM proteins, suggesting that these may approach Kepler's conjecture. In agreement with these observations, a recent study observed that globular proteins seem to follow Kepler's arrangement [58]. Thus, these observations indicate that globular proteins tend to maintain a regular packing to comply with the hydrophobic collapse during protein folding. On the contrary, TM proteins allow for more compensatory mutations and have less regular packing than globular proteins; whether this packing affects the mutability of TM proteins deserves further investigation.

Finally, our results complement previous observations about the prevalence of compensatory mutations at sectors in protein structures [59]. Sectors are the regions where compensatory mutations lay in the protein structure that are linked to protein function, with different sectors controlling different biochemical properties of proteins. More recently, it has been noted that in many cases, proteins tend to have a single sector that is dominated by sequence conservation; thus, the relevance of correlated mutations is diminished in those protein regions [60]. Here, we found that the TM region of a toxin that binds to another TM region (homodimerizes) has one sector (TM region accumulates large number of compensatory mutations) with low sequence conservation (see Table 1 and Supplementary Materials Tables S4–S7). These results suggest that sectors in TM proteins may have different properties than those in globular proteins; this deserves to be further explored.

In summary, we presented a systematic mutagenesis and deep sequencing of the TM region of a bitopic protein, the toxin HokC, to explore its structure–function relationship. We observed that most mutations are tolerated, in agreement with the low degree of sequence conservation of this family of toxins. This poor sequence conservation has an impact on the reliability of prediction methods aimed to identify critical residues. We observed that this family of toxins, and TM proteins in general, tend to accumulate mutations among contacting residues more than globular proteins do. The density of packing between globular and TM proteins may be associated with this trend, by revealing that contacts between residues within membranes follow rules different from those observed in globular proteins. Future mutagenesis of TM proteins may help reveal such rules.

4. Materials and Methods

4.1. Strains and Reagents

The bacterial strains used in our studies were *Escherichia coli* MC4100 Δ(argF-lac)U169 araD139 rpsL150 relA1 flbB5301 deoC1 ptsF25 rbsR; *E. coli* XL1-Blue supE44 hsdR17 recA1 endA1 gyrA96 thi-1 relA1 lac-; *E. coli* DH5α supE44 ΔlacU169 (φ80 lacZ DM15) hsdR17 recA1 endA1 gyrA96 thi-1 relA1. The alkaline phosphatase activity assay was performed in the CC118 strain and the GFP activity on the BL21(DE3)pLysS strain.

The plasmid pEXT22/frg-hokC containing the gene hokC starting at the second ATG was used as template for both PCR random mutagenesis and for the site-directed mutagenesis. The plasmids for the expression of HokC fused to GFP or phoA were pGFPe and pHA1-yedZ, respectively.

4.2. Mutagenesis

Site-directed mutagenesis on the coding region of HokC trans-membrane region was performed using the QuikChange Site-Directed Mutagenesis Kit (Agilent Stratagene, Santa Clara, CA, USA). To that end, we designed a strategy to mutate the TM region of HokC at 7 different groups of neighbor residues as summarized in Supplementary Materials Table S2. The following libraries of oligonucleotides were used for this goal:

- Region 1

R1 Forward 5′ GGA GAA GAG AGC AAT G NNS NNS NNS NNS NNS ATG ATT GTC GCC C 3′

R1 Reverse 5′ GGG CGA CAA TCA T NNS NNS NNS NNS NNS CAT TGC TCT CTT CTC C 3′

- Region 2

R2 Forward 5′ GCA GCA TAA GGC G NNS NNS NNS GC CCT GAT CGT CAT C 3′

R2 Reverse 5′ GAT GAC GAT CAG GGC SNN SNN SNN CGC CTT ATG CTG C 3′

- Region 3

R3 Forward 5′ GGC GAT GAT TGT C NNS NNS NNS GTC ATC TGT ATC ACC G 3′

R3 Reverse 5′ CGG TGA TAC AGA TGA C SNN SNN SNN GAC AAT CAT CGC C 3′

- Region 4

R4 Forward 5′ GTC GCC CTG ATC NNS NNS NNS ATC ACC GCC GTA GTG 3′

R4 Reverse 5′ CAC TAC GGC GGT GAT SNN SNN SNN GAT CAG GGC GAC 3′

- Region 6

R6 Forward 5′ CTG TAT CAC CGC C NNS NNS NNS GCG CTG GTA ACG 3′

R6 Reverse 5′ CGT TAC CAG CGC SNN SNN SNN GGC GGT GAT ACA G 3′

- Region 7

R7 Forward 5′ CGC CGT AGT GGC G NNS NNS NNS ACG AGA AAA GAC CTC TG 3′

R7 Reverse 5′ CAG AGG TCT TTT CTC GT SNN SNN SNN CGC CAC TAC GGC G 3′

where S stand for G or C nucleotides and N for any of the four nucleotides. Note that these oligonucleotides will generate mutant codons with SNS composition coding for 10 (L, P, H, Q, R, V, A, D, E, G) out of the 20 conventional amino acid residues. In this way, the number of variants to be screened is reduced and at the same time keeping the diversity of physicochemical properties of the amino acid residues. Please note that each pair of oligonucleotides will hybridize at the corresponding regions that are targeted in the mutagenesis experiment. For instance, the oligonucleotides for region 1 include a 5′ tail (GGA GAA GAG AGC AAT G) required for hybridization that includes the first coding codon (ATG) of the gene followed by 5 codons that are mutated by SNS and followed by a tail in the 3′ end (ATG ATT GTC GCC C) for hybridization purposes. For the site-directed mutagenesis reactions we followed the instructions of the manufacturer: 50 ng of plasmid (pEXT22/frg-hokC), a pair of mutagenic oligonucleotides (125 ng), 1 μL dNTP mix, 5 μL of 10× reaction buffer and 2.5 U of Pfu Turbo DNA Polymerase (Agilent Technologies, Santa Clara, CA, USA) in a 50 μL total volume.

To obtain the HokC mutants Met7Trp, Ile12Ser, and double mutants Met7Trp and Ile12Ser, the QuikChange Lightning site-directed mutagenesis Kit (Agilent Technologies, Santa Clara, CA, USA) was used. The following oligonucleotides were used for this goal:

7MxWForw:5′GCAGCATAAGGCGTGGATTGTCGCCCTGATCG 3′

7MxWRev:5′CGATCAGGGCGACAATCCACGCCTTATGCTGC3′

12IxSForw:5′CGATGATTGTCGCCCTGAGCGTCATCTGTATCACC3′

12IxSRev:5′GGTGATACAGATGACGCTCAGGGCGACAATCATCG3′

For GFP fusions, both plasmid pGFPe and PCR products were digested and ligated using XhoI and BamHI restriction enzyme sites. For phoA fusions, the PCR product and plasmid pHA1-yedZ were digested ligated with XhoI and KpnI.

4.3. Selection of Clones

To select the hokC variants with wild-type and mutant phenotypes, we performed the following procedure. *E. coli* cells were grown in Luria broth with kanamycin to select for those carrying the plasmid expressing hokC mutations. The plasmid, pEXT22, includes a non-leaky promoter induced by IPTG. The over-expression of hokC was achieved by adding IPTG to the media; this would kill cells expressing a wild-type-like HokC activity. However, cells expressing a mutation critical for HokC activity will grow. All our mutagenesis experiments were performed on a short version of hokC starting from the second ATG codon. To select colonies for sequencing, we looked for isolated colonies; for that end, we used Corning square BioAssay dishes (245 mm × 245 mm of area) (Merck, Kenilworth, NJ, USA).

4.4. Sensitivity of Screening

The expression system is reported not to leak transcripts of the genes cloned into the system. To test this and to evaluate how much transcription of the hokc gene was required to kill cells, we conducted a dose–response experiment, where IPTG was added to the media in different concentrations: 0.01, 0.05, 0.1, 0.2, 0.4, and 0.8 mM. *E. coli* DH5a cells were grown overnight to reach a cell density measured at 600 nm of 0.65 measured with a spectrophotomer Genesys 10S UV-Vis (Thermo Scientific, Waltham, MA, USA). These cells were diluted by a factor of 0.25×10^{-4} and 100 mL of this dilution were plated on Petri dishes with LB + Kan 10 mg/mL with or without IPTG at different concentrations: 0.01 mM, 0.05 mM, 0.1 mM, 0.2 mM, 0.4 mM, 0.6 mM, and 0.8 mM. These cells were grown for 19 h at 37 °C and the number of colonies that grew in these conditions were counted on a Freedom EVO 150 robotic station using the Pickolo software version 3.5 (SciRobotics, Kfar Saba, Israel.

4.5. Sequencing

To sequence mutants in the trans-membrane coding region of hokC, we implemented the following procedure. Colonies with wild-type or mutant phenotypes were picked and grown overnight in 3 mL of LB media with kanamycin 10 mg/mL (Sigma-Aldrich, Estado de Mexico, Mexico). These colonies were pooled in 2 groups according to their origin: cells with a wild-type and mutant phenotypes. From these pools, DNA was extracted. Thus, two pools of plasmids were obtained: from wild-type and mutant phenotype colonies. From these DNA molecules, the mutated hokC region was amplified by PCR to generate the amplicons used for sequencing; the final size of the PCR products was 450 bp. This sample was mixed at equimolar ratios and sequenced at the "Unidad Universitaria de Secuenciación Masiva de DNA-UNAM" using MySeq from Illumina company, with the MySeq reagent kit (Illumina, San Diego, CA, USA) version 2 for 500 cycles, 250 nt each read. Note that the hokC gene is smaller than the reads, thus we will be able to identify the full-length gene sequence of every mutant. TrueSeq DNA PCR-free sample preparation Kit (Illumina, San Diego, CA, USA) was used to add the adapters to our amplicons, without fragmenting the amplicons. Since this sequencer has the capacity to generate 107 DNA reads and the number of bacterial colonies to be sequenced is substantially smaller than this number (103), the experiment could generate thousands of clusters with exactly the same sequence. However, only 80% of the amplicons may have the same sequence and thus, we mixed our amplicons with sequences provided by the "Unidad Universitaria de Secuenciación Masiva de DNA-UNAM".

4.6. Activity of PhoA Fusion Proteins

Strains expressing phoA fusions were grown overnight and inoculated into 50-mL cultures of Luria broth with antibiotic (50 µg/mL ampicillin) at 37 °C to reach an OD at

600 nm of 0.4; then, cells were induced with arabinose (final concentration of 0.2%) and grown for 1 h. The activity assay was carried out as described before [39]. Briefly:

1. Centrifuge 1.2 mL of the bacterial culture in Eppendorf tube.
2. Wash cells in cold WB and resuspend pellet in 1.2 mL cold PM1 buffer.
3. To permeabilize the cells, add 100 μL chloroform and 100 μL 0.05% SDS to 1 mL of the washed cells, vortex for 10 s, and incubate for 5 min at 37 °C. Then place tubes on ice for 5 min. After the chloroform has settled, transfer 100 μL of the upper phase of the bacterial suspension to a 96 plate well.
4. To start the reaction, add 50 μL of the pNPP solution (0.15% in 1 M Tris–HCl, pH 8.0) to the bacterial suspension and incubate at RT until yellow color develops. Add 50 μL 2N NaOH to stop the reaction. Record incubation time and OD at 405 nm for each sample.
5. Calculate enzymatic activity in relative units (A) according to the following formula:

A = 1000 × (OD405sample − OD405control well)/(OD595 sample − OD595control well)/t (min) of incubation

4.7. Sequence Data Analysis

DNA reads were trimmed using the Phred algorithm implemented in seqtk (seqtk trimfq option); this process eliminated low quality bases from both ends of the DNA sequences. Then, these fastq files were transformed to fasta files using seqtk (seqtk seq –a option).

The relative frequency of each mutation ($F(mut_i)$) was quantified by the following formula:

$$F(mut_i) = 100 \times (WT_i - MUT_i)/(WT_i + MUT_i) \quad (1)$$

where WT_i corresponds to the number of times the i-mutation (mut_i) was found with a wild-type phenotype and MUT_i is the number of times the i-mutation (mut_i) was found with a mutant phenotype. Then, an ISPA was identified if $|F(mut_i)| \leq$ Experimental errors. Note that $F(mut_i)$ may be positive or negative, indicating whether the mutant is over-represented in mutant phenotypes or wild-type phenotypes, respectively.

4.8. Sequence Alignment

PFAM alignments were obtained from the PFAM web site. By counting the number of sequences that maintain the same residue than the reference sequence (HOKC_ECOLI) the residue conservation score was derived. The same set of sequences was used to align them using TM-COFFEE, an optimized algorithm and substitution matrix for TM proteins [61].

The identification of conserved and critical residues was performed using the Multiple Sequence Alignment generated for the HokC family and the conservation scores were computed based on the rate4site algorithm as implemented in the ConSurf server [62]. Alternatively, PROVEAN was used as an alternative method to identify functionally relevant substitutions [63].

4.9. Correlated Mutations Index

Two data sets were used for this analysis: (i) Globular set: 150 globular proteins including different folds and PFAM domain families [64] (see Supplementary Materials Table S11A) and (ii) TM set: 593 TM proteins from TOPDB [65] (see Supplementary Materials Table S11B). For each entry in each data set, a multiple sequence alignment (MSA) was obtained from the HSSP database [66]. Additionally, every contacting residue was identified using a 5Å distance criterion as we have previously described elsewhere [67]. Finally, every combined mutation for every contacting residue was identified from the MSA. In this case, each of the 400 possible amino acid pairs were identified and normalized according to number of residue pairs of each kind observed for each protein. For instance, if protein P presented 30 times the pair Ala–Ala and this Ala–Ala pair was mutated 15 imes in the MSA, the normalized frequency of correlated mutations for the Ala–Ala pair in protein P is 50% or 0.5. This is the value reported as the correlated mutation index of a protein. The codes to compute this mutation index and datasets are available at [68].

4.10. Analysis of Contacts in Proteins

To compare the degree of compactness between globular and TM proteins, we used two larger sets for globular and TM proteins, LG (see Supplementary Materials Table S12A) and LTM (see Supplementary Materials Table S12B) sets, respectively. For each set, we computed the size and order of the contact map derived by identifying as contacting residues those closer than 5 Å in at least one pair of atoms as described above. Then, we adjusted the size (number of residues in a given protein) versus the order (number of contacts between residues in a given three-dimensional structure of a protein) to a linear equation using the gnuplot function fit [69]. The difference on the slopes of these two data sets represents the level of difference in packing between these classes of proteins. The size and order for each chain of PDB entries in each dataset and codes are available at [68].

To determine the type of arrangement these proteins adopt upon folded, we compared the spherical angles of clusters of residues. Briefly, every amino acid in a protein and their contacting residues were identified; then, the angles between the central residue and its neighbors were calculated. The angle values obtained for each set were compared using the Haussdorff distance as implemented by Java Topology Suite [70]; to compute the minimum Haussdorff distance for every pair of proteins, we used a simulated annealing algorithm. The codes to compute the spherical coordinates and the minimum Haussdorff distances and associated datasets are available at [68]. Only proteins from the same CATH class with a difference in length no bigger than 20 residues were used for our analysis.

Finally, the number of residue cluster classes (RCCs) of size 3, 4, and 5 were computed as previously described by our group (software version 1 to generate RCCs is available at [71]) and accumulated. Briefly, residue-contacts at 5Å apart were identified and the maximal cliques of size 3, 4, and 5 were quantified.

Supplementary Materials: Supplementary Materials can be found at https://www.mdpi.com/article/10.3390/ijms221910359/s1. Table S1: Colonies counted under different IPTG concentrations; Table S2: Observed phenotypes per mutated region of HokC; Table S3: Single-residue mutations found in every mutant of HokC with wild-type (A) and mutant (B) phenotypes; Table S4: Single-residue mutations on the TM region of HokC; Table S5: Multiple mutations in HokC with wild type phenotype; Table S6: Occurrence of deleterious single-point mutations in compensatory mutations; Table S7: Multiple sequence alignment of 148 protein sequences from PF01848 family obtained with TM-COFFEE; Table S8: HokC ConSurf Results; Table S9: Predictions of deleterious mutations in HokC family by PROVEAN; Table S10: Multiple deleterious mutations found in sequences of the HokC family; Table S11A: Globular proteins used to compute the correlated mutation index; Table S11B: Transmembrane proteins used to compute the correlated mutation index; Table S12A: Globular proteins used to compute maximal cliques, size–order, and spherical coordinates; Table S12B: Transmembrane proteins used to compute maximal cliques, size-order and spherical coordinates; Figure S1: Distribution of identified mutants over HokC sequence; Figure S2: Heat map for HokC substitutions; Figure S3: Cell growth of HokC fusions; Figure S4: Phosphatase activity measured on cells expressing HokC or variants.

Author Contributions: Conceptualization, G.D.R.; methodology, M.T.L.O., G.D.R., and V.M.G.; software, G.D.R. and V.M.G.; validation, M.T.L.O. and V.M.G.; resources, G.D.R.; data curation, V.M.G.; writing—original draft preparation, G.D.R.; writing—review and editing, M.T.L.O. and V.M.G.; visualization, M.T.L.O. and V.M.G.; supervision, G.D.R. All authors have read and agreed to the published version of the manuscript.

Funding: This research was funded in part by Instituto de Fisiologia Celular and grants FOINS-219 from CONACYT, IN208014, IN205911 and IT200320 from the Programa de Apoyo a Proyectos de Investigacion e Innovacion Tecnologica at UNAM to GDR. VMG was supported in part by scholarship number 207415 from the Programa de Apoyo a Proyectos de Investigacion e Innovacion Tecnologica at UNAM.

Institutional Review Board Statement: Not applicable.

Informed Consent Statement: Not applicable.

Data Availability Statement: All the newly generated data is available as supplemental data in this publication and/or is available as indicated in reference [68]. The previously published data and software used in this study are included as references [41,65,66,69–71].

Acknowledgments: To the Unidad de servicios de cómputo and taller de mantenimiento at the Instituto de fisiología celular, UNAM México.

Conflicts of Interest: The authors declare no conflict of interest. The funders had no role in the design of the study; in the collection, analyses, or interpretation of data; in the writing of the manuscript, or in the decision to publish the results.

Abbreviations

TM	Transmembrane
IPTG	isopropyl-beta-D-thiogalactoside
MSA	Multiple Sequence Alignment
3D	Three-dimensional

References

1. Maggiora, G.M. The reductionist paradox: Are the laws of chemistry and physics sufficient for the discovery of new drugs? *J. Comput. Mol. Des.* **2011**, *25*, 699–708. [CrossRef]
2. Besnard, J.; Ruda, G.F.; Setola, V.; Abecassis, K.; Rodriguiz, R.M.; Huang, X.-P.; Norval, S.; Sassano, M.F.; Shin, A.I.; Webster, L.A.; et al. Automated design of ligands to polypharmacological profiles. *Nat. Cell Biol.* **2012**, *492*, 215–220. [CrossRef]
3. Kozma, D.; Simon, I.; Tusnády, G.E. PDBTM: Protein data bank of transmembrane proteins after 8 years. *Nucleic Acids Res.* **2012**, *41*, D524–D529. [CrossRef] [PubMed]
4. Arinaminpathy, Y.; Khurana, E.; Engelman, D.M.; Gerstein, M.B. Computational analysis of membrane proteins: The largest class of drug targets. *Drug Discov. Today* **2009**, *14*, 1130–1135. [CrossRef]
5. Forrest, L.; Tang, C.L.; Honig, B. On the accuracy of homology modeling and sequence alignment methods applied to membrane proteins. *Biophys. J.* **2006**, *91*, 508–517. [CrossRef] [PubMed]
6. Ng, P.C.; Henikoff, J.G.; Henikoff, S. PHAT: A transmembrane-specific substitution matrix. *Bioinformatics* **2000**, *16*, 760–766. [CrossRef] [PubMed]
7. Loeb, D.D.; Swanstrom, R.; Everitt, L.; Manchester, M.; Stamper, S.E.; Hutchison, C.A. Complete mutagenesis of the HIV-1 protease. *Nat. Cell Biol.* **1989**, *340*, 397–400. [CrossRef] [PubMed]
8. Rennell, D.; Bouvier, S.E.; Hardy, L.W.; Poteete, A.R. Systematic mutation of bacteriophage T4 lysozyme. *J. Mol. Biol.* **1991**, *222*, 67–88. [CrossRef]
9. Suckow, J.; Markiewicz, P.; Kleina, L.G.; Miller, J.; Kisters-Woike, B.; Müller-Hill, B. Genetic studies of the Lac repressor. XV: 4000 single amino acid substitutions and analysis of the resulting phenotypes on the basis of the protein structure. *J. Mol. Biol.* **1996**, *261*, 509–523. [CrossRef]
10. Huang, W.; Petrosino, J.; Hirsch, M.; Shenkin, P.S.; Palzkill, T. Amino acid sequence determinants of beta-lactamase structure and activity. *J. Mol. Biol.* **1996**, *258*, 688–703. [CrossRef]
11. Guo, H.H.; Choe, J.; Loeb, L.A. Protein tolerance to random amino acid change. *Proc. Natl. Acad. Sci. USA* **2004**, *101*, 9205–9210. [CrossRef] [PubMed]
12. Fowler, D.M.; Araya, C.L.; Fleishman, S.J.; Kellogg, E.H.; Stephany, J.J.; Baker, D.; Fields, S. High-resolution mapping of protein sequence-function relationships. *Nat. Methods.* **2010**, *7*, 741–746. [CrossRef] [PubMed]
13. Ernst, A.; Gfeller, D.; Kan, Z.; Seshagiri, S.; Kim, P.M.; Bader, G.; Sidhu, S.S. Coevolution of PDZ domain–ligand interactions analyzed by high-throughput phage display and deep sequencing. *Mol. BioSyst.* **2010**, *6*, 1782–1790. [CrossRef] [PubMed]
14. Hietpas, R.T.; Jensen, J.; Bolon, D.N.A. Experimental illumination of a fitness landscape. *Proc. Natl. Acad. Sci. USA* **2011**, *108*, 7896–7901. [CrossRef]
15. Jr, R.N.M.; Poelwijk, F.J.; Raman, A.; Gosal, W.S.; Ranganathan, R. The spatial architecture of protein function and adaptation. *Nat. Cell Biol.* **2012**, *491*, 138–142. [CrossRef]
16. Deng, Z.; Huang, W.; Bakkalbasi, E.; Brown, N.G.; Adamski, C.J.; Rice, K.; Muzny, D.; Gibbs, R.A.; Palzkill, T. Deep sequencing of systematic combinatorial libraries reveals β-lactamase sequence constraints at high resolution. *J. Mol. Biol.* **2012**, *424*, 150–167. [CrossRef]
17. Adkar, B.; Tripathi, A.; Sahoo, A.; Bajaj, K.; Goswami, D.; Chakrabarti, P.; Swarnkar, M.K.; Gokhale, R.S.; Varadarajan, R. Protein model discrimination using mutational sensitivity derived from deep sequencing. *Structures* **2012**, *20*, 371–381. [CrossRef]
18. Traxlmayr, M.W.; Hasenhindl, C.; Hackl, M.; Stadlmayr, G.; Rybka, J.D.; Borth, N.; Grillari, J.; Rüker, F.; Obinger, C. Construction of a stability landscape of the CH3 domain of human IgG1 by combining directed evolution with high throughput sequencing. *J. Mol. Biol.* **2012**, *423*, 397–412. [CrossRef]
19. Araya, C.; Fowler, D.M.; Chen, W.; Muniez, I.; Kelly, J.W.; Fields, S. A fundamental protein property, thermodynamic stability, revealed solely from large-scale measurements of protein function. *Proc. Natl. Acad. Sci. USA* **2012**, *109*, 16858–16863. [CrossRef]

20. Wu, N.C.; Young, A.P.; Dandekar, S.; Wijersuriya, H.; Al-Mawsawi, L.Q.; Wu, T.-T.; Sun, R. Systematic identification of H274Y compensatory mutations in influenza A virus neuraminidase by high-throughput screening. *J. Virol.* **2013**, *87*, 1193–1199. [CrossRef]
21. Melamed, D.; Young, D.L.; Gamble, C.E.; Miller, C.R.; Fields, S. Deep mutational scanning of an RRM domain of the Saccharomyces cerevisiae poly(A)-binding protein. *RNA* **2013**, *19*, 1537–1551. [CrossRef] [PubMed]
22. Roscoe, B.P.; Thayer, K.M.; Zeldovich, K.B.; Fushman, D.; Bolon, D.N. Analyses of the effects of all ubiquitin point mutants on yeast growth rate. *J. Mol. Biol.* **2013**, *425*, 1363–1377. [CrossRef] [PubMed]
23. Starita, L.M.; Pruneda, J.; Lo, R.S.; Fowler, D.M.; Kim, H.J.; Hiatt, J.B.; Shendure, J.; Brzovic, P.S.; Fields, S.; Klevit, R.E. Activity-enhancing mutations in an E3 ubiquitin ligase identified by high-throughput mutagenesis. *Proc. Natl. Acad. Sci. USA* **2013**, *110*, E1263–E1272. [CrossRef] [PubMed]
24. Shin, H.; Cho, Y.; Choe, D.; Jeong, Y.; Cho, S.; Kim, S.C.; Cho, B.-K. Exploring the functional residues in a flavin-binding fluorescent protein using deep mutational scanning. *PLoS ONE* **2014**, *9*, e97817. [CrossRef] [PubMed]
25. Schlinkmann, K.M.; Honegger, A.; Tureci, E.; Robison, K.E.; Lipovsek, D.; Plückthun, A. Critical features for biosynthesis, stability, and functionality of a G protein-coupled receptor uncovered by all-versus-all mutations. *Proc. Natl. Acad. Sci. USA* **2012**, *109*, 9810–9815. [CrossRef]
26. Corral-Corral, R.; Beltrán, J.A.; Brizuela, C.A.; Del Rio, G. Systematic identification of machine-learning models aimed to classify critical residues for protein function from protein structure. *Molecules* **2017**, *22*, 1673. [CrossRef]
27. Studer, R.A.; Dessailly, B.H.; Orengo, C.A. Residue mutations and their impact on protein structure and function: Detecting beneficial and pathogenic changes. *Biochem. J.* **2013**, *449*, 581–594. [CrossRef] [PubMed]
28. Taylor, N.R. Small world network strategies for studying protein structures and binding. *Comput. Struct. Biotechnol. J.* **2013**, *5*, e201302006. [CrossRef]
29. Fajardo, J.E.; Fiser, A. Protein structure based prediction of catalytic residues. *BMC Bioinform.* **2013**, *14*, 63. [CrossRef]
30. Cusack, M.P.; Thibert, B.; Bredesen, D.E.; Del Río, G. Efficient identification of critical residues based only on protein structure by network analysis. *PLoS ONE* **2007**, *2*, e421. [CrossRef]
31. Göbel, U.; Sander, C.; Schneider, R.; Valencia, A. Correlated mutations and residue contacts in proteins. *Proteins Struct. Funct. Bioinform.* **1994**, *18*, 309–317. [CrossRef]
32. Fodor, A.A.; Aldrich, R.W. Influence of conservation on calculations of amino acid covariance in multiple sequence alignments. *Proteins Struct. Funct. Bioinform.* **2004**, *56*, 211–221. [CrossRef] [PubMed]
33. Kowarsch, A.; Fuchs, A.; Frishman, D.; Pagel, P. Correlated mutations: A hallmark of phenotypic amino acid substitutions. *PLoS Comput. Biol.* **2010**, *6*, e1000923. [CrossRef] [PubMed]
34. Thibert, B.; Bredesen, D.E.; Del Rio, G. Improved prediction of critical residues for protein function based on network and phylogenetic analyses. *BMC Bioinform.* **2005**, *6*, 213. [CrossRef]
35. MacArthur, D.G.; Manolio, T.A.; Dimmock, D.; Rehm, H.L.; Shendure, J.; Abecasis, G.R.; Adams, D.R.; Altman, R.; Antonarakis, S.; Ashley, E.A.; et al. Guidelines for investigating causality of sequence variants in human disease. *Nat. Cell Biol.* **2014**, *508*, 469–476. [CrossRef]
36. Ortiz, M.T.L.; Rosario, P.B.L.; Luna-Nevárez, P.; Gamez, A.S.; Campo, A.M.-D.; Del Río, G. Quality control test for sequence-phenotype assignments. *PLoS ONE* **2015**, *10*, e0118288. [CrossRef]
37. Bocharov, E.V.; Volynsky, P.E.; Pavlov, K.V.; Efremov, R.G.; Arseniev, A.S. Structure elucidation of dimeric transmembrane domains of bitopic proteins. *Cell Adh. Migr.* **2010**, *4*, 284–298. [CrossRef]
38. Poulsen, L.K.; Larsen, N.W.; Molin, S.; Andersson, P. A family of genes encoding a cell-killing function may be conserved in all Gram-negative bacteria. *Mol. Microbiol.* **1989**, *3*, 1463–1472. [CrossRef]
39. Rapp, M.; Drew, D.; Daley, D.O.; Nilsson, J.; Carvalho, T.; Melén, K.; De Gier, J.-W.; Von Heijne, G. Experimentally based topology models for *E. coli* inner membrane proteins. *Protein Sci.* **2004**, *13*, 937–945. [CrossRef]
40. Poulsen, L.K.; Refn, A.; Molin, S.; Andersson, P. Topographic analysis of the toxic Gef protein from *Escherichia coli*. *Mol. Microbiol.* **1991**, *5*, 1627–1637. [CrossRef] [PubMed]
41. AlphaFold v2 Server. Available online: https://alphafold.ebi.ac.uk/entry/P0ACG4 (accessed on 21 September 2021).
42. Davis, B.H.; Poon, A.; Whitlock, M. Compensatory mutations are repeatable and clustered within proteins. *Proc. R. Soc. B Boil. Sci.* **2009**, *276*, 1823–1827. [CrossRef] [PubMed]
43. Bhattacherjee, A.; Mallik, S.; Kundu, S. Compensatory mutations occur within the electrostatic interaction range of deleterious mutations in protein structure. *J. Mol. Evol.* **2014**, *80*, 10–12. [CrossRef] [PubMed]
44. Julenius, K.; Pedersen, A.G. Protein evolution is faster outside the cell. *Mol. Biol. Evol.* **2006**, *23*, 2039–2048. [CrossRef] [PubMed]
45. Spielman, S.J.; Wilke, C. Membrane environment imposes unique selection pressures on transmembrane domains of G Protein-coupled receptors. *J. Mol. Evol.* **2013**, *76*, 172–182. [CrossRef] [PubMed]
46. Wen, J.; Chen, X.; Bowie, J.U. Exploring the allowed sequence space of a membrane protein. *Nat. Genet.* **1996**, *3*, 141–148. [CrossRef]
47. Rockah-Shmuel, L.; Tóth-Petróczy, Á.; Tawfik, D.S. Systematic mapping of protein mutational space by prolonged drift reveals the deleterious effects of seemingly neutral mutations. *PLoS Comput. Biol.* **2015**, *11*, e1004421. [CrossRef]
48. Haran, G. How, when and why proteins collapse: The relation to folding. *Curr. Opin. Struct. Biol.* **2012**, *22*, 14–20. [CrossRef]
49. Popot, J.-L.; Engelman, D.M. Membranes do not tell proteins how to fold. *Biochemistry* **2015**, *55*, 5–18. [CrossRef]

50. Fischer, H.; Polikarpov, I.; Craievich, A.F. Average protein density is a molecular-weight-dependent function. *Protein Sci.* **2009**, *13*, 2825–2828. [CrossRef]
51. Corral, R.C.; Chavez, E.; Del Rio, G. Machine learnable fold space representation based on residue cluster classes. *Comput. Biol. Chem.* **2015**, *59*, 1–7. [CrossRef]
52. Pellegrini-Calace, M.; Maiwald, T.; Thornton, J.M. PoreWalker: A novel tool for the identification and characteri-zation of channels in transmembrane proteins from their three-dimensional structure. *PLoS Comput. Biol.* **2009**, *5*, e1000440. [CrossRef]
53. Eriksson, A.E.; Baase, W.A.; Zhang, X.J.; Heinz, D.W.; Blaber, M.; Baldwin, E.; Matthews, B.W. Response of a protein structure to cavity-creating mutations and its relation to the hydrophobic effect. *Science* **1992**, *255*, 178–183. [CrossRef]
54. Halle, B. Flexibility and packing in proteins. *Proc. Natl. Acad. Sci. USA* **2002**, *99*, 1274–1279. [CrossRef]
55. Gimpelev, M.; Forrest, L.; Murray, D.; Honig, B. Helical packing patterns in membrane and soluble proteins. *Biophys. J.* **2004**, *87*, 4075–4086. [CrossRef]
56. Istrail, S.; Lam, F. Combinatorial algorithms for protein folding in lattice models: A survey of mathematical results. *Commun. Inf. Syst.* **2009**, *9*, 303–346. [CrossRef]
57. Hales, T. A proof of the Kepler conjecture. *Ann. Math.* **2005**, *162*, 1065–1185. [CrossRef]
58. Bagci, Z.; Jernigan, R.L.; Bahar, I. Residue coordination in proteins conforms to the closest packing of spheres. *Polymers* **2002**, *43*, 451–459. [CrossRef]
59. Halabi, N.; Rivoire, O.; Leibler, S.; Ranganathan, R. Protein sectors: Evolutionary units of three-dimensional structure. *Cell* **2009**, *138*, 774–786. [CrossRef] [PubMed]
60. Teşileanu, T.; Colwell, L.J.; Leibler, S. Protein sectors: Statistical coupling analysis versus conservation. *PLoS Comput. Biol.* **2015**, *11*, e1004091. [CrossRef]
61. Chang, J.-M.; Di Tommaso, P.; Taly, J.-F.; Notredame, C. Accurate multiple sequence alignment of transmembrane proteins with PSI-Coffee. *BMC Bioinform.* **2012**, *13*, S1. [CrossRef]
62. Ashkenazy, H.; Erez, E.; Martz, E.; Pupko, T.; Ben-Tal, N. ConSurf 2010: Calculating evolutionary conservation in sequence and structure of proteins and nucleic acids. *Nucleic Acids Res.* **2010**, *38*, W529–W533. [CrossRef]
63. Choi, Y.; Chan, A. PROVEAN web server: A tool to predict the functional effect of amino acid substitutions and indels. *Bioinformatics* **2015**, *31*, 2745–2747. [CrossRef] [PubMed]
64. Kosciolek, T.; Jones, D.T. De novo structure prediction of globular proteins aided by sequence variation-derived contacts. *PLoS ONE* **2014**, *9*, e92197. [CrossRef] [PubMed]
65. TopDB Web Server. Available online: http://topdb.enzim.hu/ (accessed on 21 September 2021).
66. HSSP Database. Available online: http://swift.cmbi.ru.nl/gv/hssp/ (accessed on 21 September 2021).
67. Fontove, F.; Del Rio, G. Residue cluster classes: A unified protein representation for efficient structural and functional classification. *Entropy* **2020**, *22*, 472. [CrossRef] [PubMed]
68. Supplementary Data. Available online: https://github.com/gdelrioifc/MutagenesisHokC (accessed on 21 September 2021).
69. GnuPlot Software. Available online: http://www.gnuplot.info (accessed on 21 September 2021).
70. Java Topology Suite. Available online: http://tsusiatsoftware.net/jts/main.html (accessed on 21 September 2021).
71. RCC Software. Available online: https://github.com/C3-Consensus/RCC (accessed on 21 September 2021).

International Journal of *Molecular Sciences*

MDPI

Commentary

Oxidative Folding of Proteins: The "Smoking Gun" of Glutathione

Alessio Bocedi [1], Giada Cattani [1], Giorgia Gambardella [1], Linda Schulte [2], Harald Schwalbe [2] and Giorgio Ricci [1,*]

[1] Department of Chemical Sciences and Technologies, University of Rome "Tor Vergata", 00133 Rome, Italy; bcdlss01@uniroma2.it (A.B.); giada.cattani@gmail.com (G.C.); giorgia.gambardella@gmail.com (G.G.)

[2] Center for Biomolecular Magnetic Resonance (BMRZ), Institute for Organic Chemistry and Chemical Biology, Goethe-University Frankfurt, 60438 Frankfurt, Germany; schulte@nmr.uni-frankfurt.de (L.S.); Schwalbe@nmr.uni-frankfurt.de (H.S.)

* Correspondence: riccig@uniroma2.it; Tel.: +39-0672594353

Abstract: Glutathione has long been suspected to be the primary low molecular weight compound present in all cells promoting the oxidative protein folding, but twenty years ago it was found "not guilty". Now, new surprising evidence repeats its request to be the "smoking gun" which reopens the criminal trial revealing the crucial involvement of this tripeptide.

Keywords: oxidative folding; glutathionylation; nitrosylation; cysteine reactivity; ribosomal exit tunnel; transient complex; glutathione

Citation: Bocedi, A.; Cattani, G.; Gambardella, G.; Schulte, L.; Schwalbe, H.; Ricci, G. Oxidative Folding of Proteins: The "Smoking Gun" of Glutathione. *Int. J. Mol. Sci.* **2021**, *22*, 10148. https://doi.org/10.3390/ijms221810148

Academic Editors: Istvan Simon and Csaba Magyar

Received: 20 August 2021
Accepted: 16 September 2021
Published: 20 September 2021

1. Introduction

For many years the oxidized form of glutathione (GSSG) was considered the main culprit for the oxidative folding of many proteins. Indeed, GSSG displays an unusually high concentration in the endoplasmic reticulum. Further, its role in establishing the cellular redox potential is undisputed. In addition, a few disulfide containing proteins, when reduced and incubated with a GSH/GSSG mixture in a ratio similar to the one found in this cell compartment, refolded, correctly forming native disulfides. However, twenty years ago Cuozzo and Kaiser [1] claimed that GSSG cannot be considered the culprit because, when the cell is deprived of this compound, oxidative folding still occurs. At this stage, ER oxidoreductin 1 (Ero1) and the protein disulfide isomerase (PDI) were indicated as the main responsible for protein folding [1]. This hypothesis was rapidly accepted by the scientific community although conflicting evidence emerged from Kaiser's own study. In fact, why does the disulfide bond formation still occur in cells that are simultaneously defective in both glutathione biosynthesis and Ero1 function? Bardwell and co-workers, in an interesting comment on these results, postulated the existence of a second, yet-to-be discovered oxidizing pathway [2]. They concluded that the ultimate source of oxidizing equivalents for the protein disulfide formation still has to be identified and that it remains "a complete mystery" [2].

In this context, other comments were also instructive. By considering that the rate-limiting steps for native disulfide bond formation in vivo are the late, complex, isomerization steps, whereas oxidation is much more rapid [3], Freedman and co-workers concluded that "*there is no reason to exclude the possibility that GSSG is on the normal oxidative pathway for secretory proteins, since in the absence of GSSG a normally minor direct oxidative pathway may become the major pathway. In such a case, the overall rate of production of native proteins would not be compromised by the change in oxidation pathway as the oxidative steps are not themselves rate-limiting*" [4]. Despite these counterarguments, no striking evidence was able to reverse the Cuozzo and Kaiser dogma. As a consequence, in almost all recent reviews about oxidative folding, glutathione was only related as a redox regulating agent for PDI and no direct interaction of this compound with the nascent reduced protein was considered [5–7].

Now we can show surprising findings that could light up the crime scene, at least in its early phase, and that can reverse the previous sentence.

Recently, we found that a few cysteines in the fully reduced albumin, adopting a molten globule-like conformation, showed unusual hyper-reactivity toward GSSG and various thiol reagents [8]. In particular, a single cysteine, identified as Cys75, displayed a second-order kinetic constant > 250 M^{-1} s^{-1} which corresponds to more than one thousand times higher reactivity toward GSSG than the one of an unperturbed protein cysteine (k = 0.2 M^{-1} s^{-1}) (Figure 1) [9]. At first, we considered this surprising reactivity as a specific feature of a single protein. However, soon after this first observation, we discovered a similar, but even more striking hyper-reactivity in a cysteine (Cys94) of the reduced lysozyme in its unfolded state [10]. In this case, the reactivity toward GSSG was found to be more than 3000 times higher than that of a normal amino acid cysteine. We then hypothesized a possible function of this hyper-reactivity: when lysozyme lacks its four disulfides it rapidly collapses into irreversible and insoluble aggregates. The very fast reaction of Cys94 with GSSG inhibits instantaneously the aggregation [10]. This evidence gathered for a second protein represented a strong indication that this phenomenon was not a specific feature of albumin, as we initially thought, but could be a more general mechanism linked to protein folding.

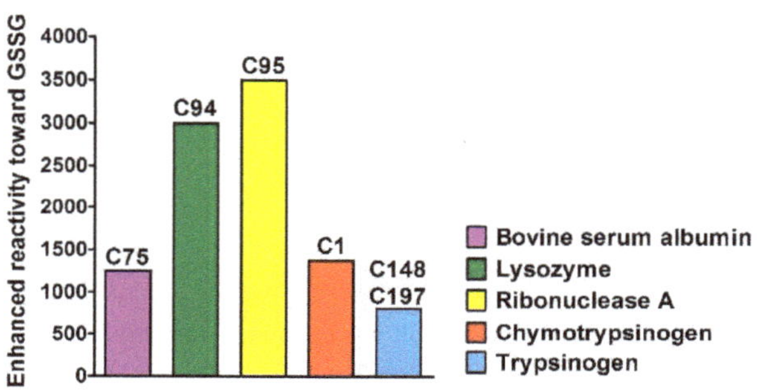

Figure 1. Hyper-reactivity of structural cysteines in five different proteins. Hyper-reactivity of Cys75, Cys94, Cys95, Cys1, Cys148 and Cys197 toward GSSG found in albumin, lysozyme, ribonuclease A, chymotrypsinogen, and trypsinogen, respectively. Pseudo first-order kinetic constants were normalized to that of an unperturbed protein cysteine.

Motivated by this observation, we searched for other hyper-reactive cysteines. We found a thousand times increased reactivity toward GSSG for Cys95, Cys1 and for both Cys148 and Cys197 in the reduced molten globule conformations of ribonuclease [11], chymotrypsinogen [12] and trypsinogen [13], respectively. In all these proteins the occurrence of a transient protein-GSSG complex was demonstrated on the basis of the quenching of intrinsic fluorescence occurring before the glutathionylation event in ribonuclease [11], lysozyme [10], chymotrypsinogen [12], and trypsinogen [13] (Table 1). The transient complex represents the origin of this unknown kinetic property.

Table 1. Values of K_D for Protein-GSSG complex.

Proteins	K_D (mM)
Lysozyme [a]	0.3 ± 0.1
Ribonuclease [a]	0.12 ± 0.05
Chymotrypsinogen [b]	1.5 ± 0.1
Trypsinogen [b]	0.4 ± 0.1

[a] Values obtained at pH 7.4 from Refs. [10,11]; [b] values obtained at pH 5.0 from Refs. [12,13].

A possible role in this phenomenon of a lowered pK_a of the sulfhydryl group was also considered but a recent investigation [9] likely demonstrated that a low pK_a cannot produce more than three times increased reactivity toward GSSG (Figure 2).

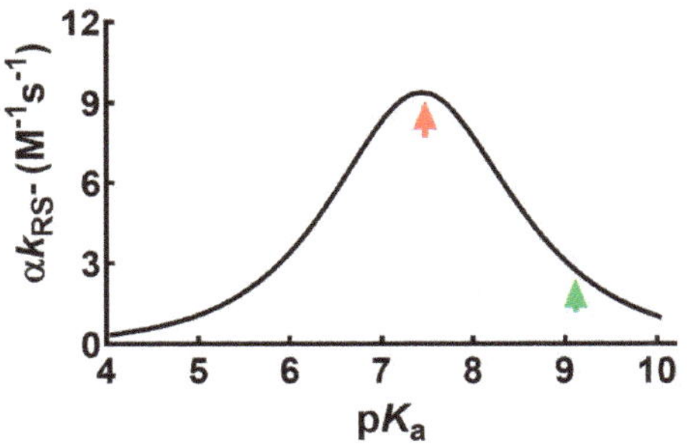

Figure 2. Dependence of the second-order kinetic constants (αk_{RS^-}) on pK_a for the reaction of several thiols with different pK_a with different disulfides at pH 7.4 (modified from Ref. [9]). The red arrow marks the maximum value of the bell-shaped graph. The pK_a of the unperturbed protein cysteine is labelled with the green arrow. The maximum implement of reaction rate due to a lowered pK_a was found to be 3 times.

As a further important discovery is that scarce or no hyper-reactivity was observed toward other natural disulfides such as cystine, homocystine, and cystamine, confirming an almost exclusive specificity of interaction toward GSSG (Figure 3) [10–13].

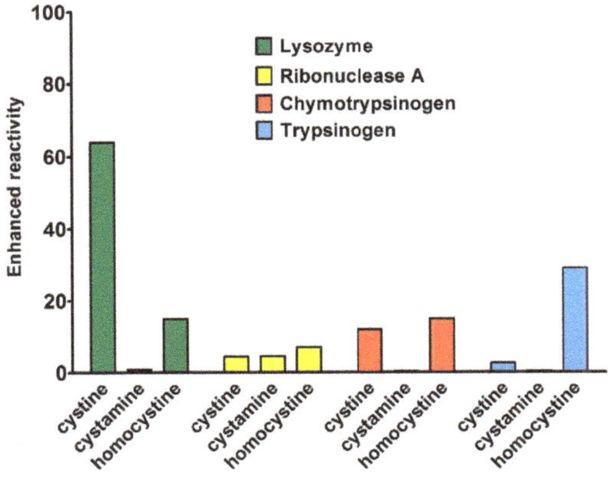

Figure 3. Reactivity of protein cysteines toward natural disulfides. The enhanced reactivity represents the second-order kinetic constants normalized to that of GSH. All proteins did not show any evident hyper-reactivity except the small enhanced reactivity found in Lysozyme toward cystine (65 times) which is small compared to the one of Cys94 toward GSSG (about 3000 times).

Of particular interest is also the observation that similar hyper-reactivity is saved during a divergent evolution, as observed for chymotrypsinogen and trypsinogen, both coming from a common ancestral peptidase. This preservation during evolution was again a relevant clue for the implication of glutathione in the folding process.

These results demonstrate that the reduced molten globule conformations of all these proteins display a sophisticated propensity to interact with GSSG, a property typically unknown to the biochemist community. While this supports an early participation of glutathione in the folding pathway, it cannot be considered as final proof of it. A recent study, based on earlier studies [14] could, however, represent a decisive turn of this investigation [15]. It was in fact demonstrated that a nascent protein, the bovine γB-crystallin, could interact with glutathione in the ribosomal exit tunnel. Such protein, in fact, displays one of its seven cysteines (Cys18) either as a mixed disulfide with GSH or nitrosylated (Figure 4).

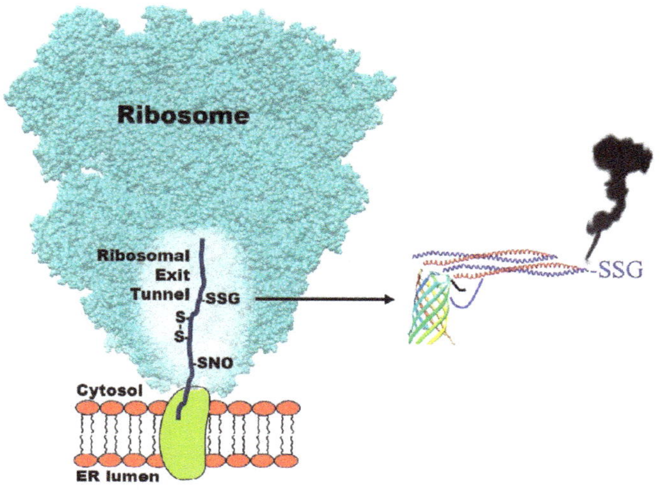

Figure 4. Visualization of ribosomal 50S subunit at the interface of endoplasmic reticulum with a nascent polypeptide chain. Modified cysteines of the bovine γB-crystallin found in the ribosomal exit tunnel during its nascent phase, as demonstrated in Ref. [15]. On the right, an *"imaginary joke structure"* of the glutathionylated protein, which represents the "smoking gun" for glutathione in the early scenario of the oxidative folding (the β-barrel structure represents the revolver grip, while the coiled coil is the revolver barrel).

More surprisingly, detectable amounts of other cysteines have already been found in the form of disulfides (Cys15-Cys32; Cys22-Cys32; Cys32-Cys41; Cys15-Cys32) [15]. This finding provides strong evidence for the involvement of glutathione in the oxidative folding scenario. Apart from its presence as a mixed disulfide with Cys18, all the early protein disulfides found in this compartment are reasonably formed after a first glutathionylation or nitrosylation step caused by GSSG or S-nitrosoglutathione (GSNO) as represented in Figure 5.

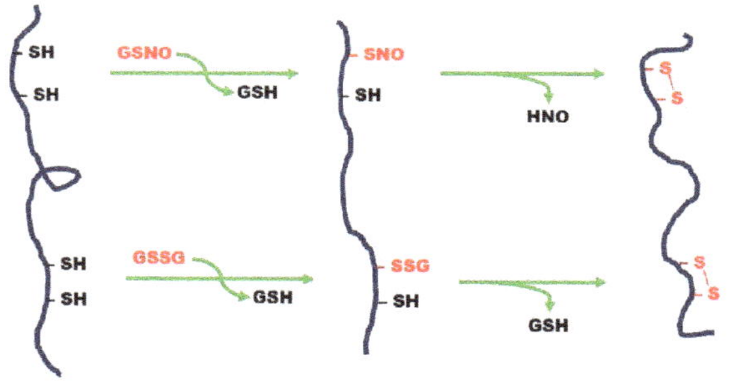

Figure 5. Schematization of modified cysteines on a polypeptide. The upper side represents the nitrosylation due to GSNO. The lower side represents the glutathionylation due to GSSG.

In all these oxidative events no involvement can be evoked for PDI or Ero1: both these enzymes have not been found inside the tunnel and, more importantly, they cannot enter in this narrow ribosomal compartment having much more steric hindrance (diameter 50–60 Å for PDI and 40–60 Å for Ero1 as calculated from crystal structures) [16,17] compared to the one of the exit tunnel (diameter 10–20 Å) [15]. There is no reasonable objection that a similar phenomenon may occur in the nascent phase for many other disulfide containing proteins and this will be verified in the future.

A further interesting observation: the ribosomal synthesis of all proteins proceeds at about 20 amino acids/s and the synthesis of full-length γB-crystallin, made up of 174 amino acids, requires around 9 s. However, the tunnel contains only 34 residues [15] so the permanence of Cys18 as well as of the other cysteines in this compartment cannot exceed 1.5–2 s. Assuming that the Cys18 modification occurs only during its path through the ribosomal exit tunnel, we can consider 20 s to be a reasonable $t_{1/2}$ for the nitrosylation and glutathionylation events. This value is easily estimated taking into account that 20% of Cys18 is found as a modified residue by NMR spectroscopy [15]. This putative $t_{1/2}$ can be compared to the one resulting from the known kinetic constants for the reaction of GSSG and GSNO with a free cysteine (i.e., 0.7 M^{-1} s^{-1} [10] and 60 M^{-1} s^{-1} [18], respectively) and from the intracellular levels of these two compounds (0.4 mM for GSSG in the endoplasmic reticulum [5] and micromolar level for GSNO [19]). From these values, we can estimate much slower kinetics for both reactions ($t_{1/2} \approx$ 1–2 h). These data suggest a strong hyper-reactivity of Cys18 and other cysteines whose cause remains a fascinating enigma to be solved in the future. This property resembles the recently discovered hyper-reactivity toward GSSG of specific cysteines in the molten globular structures of albumin, lysozyme, ribonuclease trypsinogen, and chymotrypsinogen [8–12] but its origin is likely different. In fact, in the exit tunnel no globular structure of the protein can exist, thus no active-site-like cavity may be able to bind GSSG as it occurs in the molten globules of the above cited proteins. We can speculate that the internal membrane of the tunnel behaves like a proper surface able to catalyze the interaction of a few cysteines with GSSG and GSNO.

2. Conclusions

In conclusion, after twenty years from the first judgment, the criminal trial can be reopened to assess possible responsibility of glutathione at least in the early phase of the oxidative folding of several proteins. This does not exonerate PDI and Ero1 from any complicity in this scenario, but their involvement could be confined in a second phase after an initial very fast glutathionylation or nitrosylation step of a single or a few hyper-reactive cysteines triggered by GSSG or GSNO inside the ribosomal exit tunnel or in the endoplasmic reticulum as soon as the molten globule is formed.

Author Contributions: Conceptualization, G.R.; writing—original draft and preparation, A.B., H.S., L.S. and G.R.; writing—review and editing, G.C. and G.G.; supervision, G.R. All authors have read and agreed to the published version of the manuscript.

Funding: This research received no external funding.

Acknowledgments: The authors would like to thank L. Stella for helpful discussion and Sara Notari for editorial assistance.

Conflicts of Interest: The authors declare no conflict of interest.

References

1. Cuozzo, J.W.; Kaiser, C.A. Competition between glutathione and protein thiols for disulphide-bond formation. *Nat. Cell Biol.* **1999**, *1*, 130–135. [CrossRef] [PubMed]
2. Bader, M.; Winther, J.R.; Bardwell, J.C.A. Protein oxidation: Prime suspect found "not guilty". *Nat. Cell Biol.* **1999**, *1*, E57–E58. [CrossRef] [PubMed]
3. Molinari, M.; Helenius, A. Glycoproteins form mixed disulphides with oxidoreductases during folding in living cells. *Nature* **1999**, *402*, 90–93. [CrossRef] [PubMed]

4. Bass, R.; Ruddock, L.W.; Klappa, P.; Freedman, R.B. A major fraction of endoplasmic reticulum-located glutathione is present as mixed disulfides with protein. *J. Biol. Chem.* **2004**, *279*, 5257–5262. [CrossRef] [PubMed]
5. Chakravarthi, S.; Jessop, C.E.; Bulleid, N.J. The role of glutathione in disulphide bond formation and endoplasmic-reticulum-generated oxidative stress. *EMBO Rep.* **2006**, *7*, 271–275. [CrossRef] [PubMed]
6. Delaunay-Moisan, A.; Ponsero, A.; Toledano, M.B. Reexamining the function of glutathione in oxidative protein folding and secretion. *Antioxd. Redox Signal.* **2017**, *27*, 1178–1199. [CrossRef] [PubMed]
7. Wang, L.; Yu, J.; Wang, C.-C. Protein disulfide isomerase is regulated in multiple ways: Consequences for conformation, activities, and pathophysiological functions. *BioEssay* **2020**, *43*, e2000147. [CrossRef] [PubMed]
8. Bocedi, A.; Fabrini, R.; Pedersen, J.Z.; Federici, G.; Iavarone, F.; Martelli, C.; Castagnola, M.; Ricci, G. The extreme hyper-reactivity of selected cysteines drives hierarchical disulfide bond formation in serum albumin. *FEBS J.* **2016**, *283*, 4113–4127. [CrossRef] [PubMed]
9. Gambardella, G.; Cattani, G.; Bocedi, A.; Ricci, G. New Factors Enhancing the Reactivity of Cysteines in Molten Globule Like Structures. *Int. J. Mol. Sci.* **2020**, *21*, 6949. [CrossRef] [PubMed]
10. Bocedi, A.; Cattani, G.; Martelli, C.; Cozzolino, F.; Castagnola, M.; Pucci, P.; Ricci, G. The extreme hyper-reactivity of Cys94 in lysozyme avoids its amorphous aggregation. *Sci. Rep.* **2018**, *8*, 16050. [CrossRef] [PubMed]
11. Bocedi, A.; Cattani, G.; Gambardella, G.; Ticconi, S.; Cozzolino, F.; Di Fusco, O.; Pucci, P.; Ricci, G. Ultra-Rapid Glutathionylation of Ribonuclease: Is this the Real Incipit of its Oxidative Folding? *Int. J. Mol. Sci.* **2019**, *20*, 5440. [CrossRef] [PubMed]
12. Bocedi, A.; Gambardella, G.; Cattani, G.; Bartolucci, S.; Limauro, D.; Pedone, E.; Iavarone, F.; Castagnola, M.; Ricci, G. Ultra-rapid glutathionylation of chymotrypsinogen in its molten globule-like conformation: A comparison to archaeal proteins. *Sci. Rep.* **2020**, *10*, 8943. [CrossRef] [PubMed]
13. Cattani, G.; Bocedi, A.; Gambardella, G.; Iavarone, F.; Boroumand, M.; Castagnola, M.; Ricci, G. Trypsinogen and chymotrypsinogen: The mysterious hyper-reactivity of selected cysteines is still present after their divergent evolution. *FEBS J.* **2021**. [CrossRef] [PubMed]
14. Buhr, F.; Jha, S.; Thommen, M.; Mittelstaet, J.; Kutz, F.; Schwalbe, H.; Rodnina, M.; Komar, A. Synonymous codons direct co-translational folding towards different protein conformations. *Mol. Cell* **2016**, *61*, 341–351. [CrossRef] [PubMed]
15. Schulte, L.; Mao, J.; Reitz, J.; Sreeramulu, S.; Kudlinzki, D.; Hodirnau, V.V.; Meier-Credo, J.; Saxena, K.; Buhr, F.; Langer, J.D.; et al. Cysteine oxidation and disulfide formation in the ribosomal exit tunnel. *Nat. Commun.* **2020**, *11*, 5569. [CrossRef] [PubMed]
16. Wang, C.; Li, W.; Ren, J.; Fang, J.; Ke, H.; Gong, W.; Feng, W.; Wang, C.C. Structural insights into the redox-regulated dynamic conformations of human protein disulfide isomerase. *Antioxid. Redox Signal.* **2013**, *19*, 36–45. [CrossRef] [PubMed]
17. Inaba, K.; Masui, S.; Iida, H.; Vavassori, S.; Sitia, R.; Suzuki, M. Crystal structures of human Ero1α reveal the mechanisms of regulated and targeted oxidation of PDI. *EMBO J.* **2010**, *29*, 3330–3343. [CrossRef] [PubMed]
18. Meyer, D.J.; Kramer, H.; Ozer, N.; Coles, B.; Ketterer, B. Kinetics and equilibria of S-nitrosothiol-thiol exchange between glutathione, cysteine, penicillamines and serum albumin. *FEBS Lett.* **1994**, *345*, 177–180. [CrossRef]
19. Gaston, B. Nitric oxide and thiol groups. *Biochim. Biophys. Acta* **1999**, *1411*, 323–333. [CrossRef]

International Journal of *Molecular Sciences*

MDPI

Article

Fenton-Chemistry-Based Oxidative Modification of Proteins Reflects Their Conformation

Thomas Nehls †, Tim Heymann †, Christian Meyners, Felix Hausch and Frederik Lermyte *

Clemens-Schöpf-Institute, Department of Chemistry, Technical University of Darmstadt, Alarich-Weiss-Straße 4, 64287 Darmstadt, Germany; thomas.nehls@tu-darmstadt.de (T.N.); tim.heymann@tu-darmstadt.de (T.H.); christian_stephan.meyners@tu-darmstadt.de (C.M.); felix.hausch@tu-darmstadt.de (F.H.)

* Correspondence: frederik.lermyte@tu-darmstadt.de

† These authors contributed equally to this work.

Abstract: In order to understand protein structure to a sufficient extent for, e.g., drug discovery, no single technique can provide satisfactory information on both the lowest-energy conformation and on dynamic changes over time (the 'four-dimensional' protein structure). Instead, a combination of complementary techniques is required. Mass spectrometry methods have shown promise in addressing protein dynamics, but often rely on the use of high-end commercial or custom instruments. Here, we apply well-established chemistry to conformation-sensitive oxidative protein labelling on a timescale of a few seconds, followed by analysis through a routine protein analysis workflow. For a set of model proteins, we show that site selectivity of labelling can indeed be rationalised in terms of known structural information, and that conformational changes induced by ligand binding are reflected in the modification pattern. In addition to conventional bottom-up analysis, further insights are obtained from intact mass measurement and native mass spectrometry. We believe that this method will provide a valuable and robust addition to the 'toolbox' of mass spectrometry researchers studying higher-order protein structure.

Keywords: mass spectrometry; protein folding; protein–ligand interactions; protein dynamics; FK506-binding protein; FKBP12; FKBP51

Citation: Nehls, T.; Heymann, T.; Meyners, C.; Hausch, F.; Lermyte, F. Fenton-Chemistry-Based Oxidative Modification of Proteins Reflects Their Conformation. *Int. J. Mol. Sci.* **2021**, *22*, 9927. https://doi.org/10.3390/ijms22189927

Academic Editors: Istvan Simon and Csaba Magyar

Received: 8 August 2021
Accepted: 11 September 2021
Published: 14 September 2021

1. Introduction

Protein structure is inherently a four-dimensional phenomenon, and the dynamic aspects of a protein can be just as important as the mostly static structures typically associated with conventional high-resolution structural biology methods, such as X-ray crystallography. In recent years, gas-phase methods, such as native mass spectrometry and ion mobility spectrometry, have been developed that lack the high resolution of crystallography, nuclear magnetic resonance, or cryo-electron microscopy, but can complement these methods such that the combination of different techniques yields a better understanding of the structural ensemble [1–7]. A benefit of MS-based methods is their near-universal applicability and relatively high throughput. While it is now generally accepted that the gas-phase structure of proteins in native MS reflects important aspects of the solution structure, the question remains to what extent subtle interactions are preserved [3].

One way to combine the analytical benefits of mass spectrometry with the ability to confidently probe protein structure as it appears in solution is to use chemical labelling [8–10]. This can be performed in a way that is sensitive to protein conformation in solution, and subsequent MS analysis allows identification of the modification sites, which facilitates correlation of primary and higher-order structure. Such methods provide valuable complementary information to conventional structural biology, for example, allowing the convenient probing of membrane proteins, or monitoring the response to a thermal or chemical perturbation of the structure on a (sub)millisecond timescale [11–15]. These methods also have the benefit that their experimental feasibility typically only has a limited dependence on protein mass [16].

Arguably, the most common labelling method in use today is hydrogen–deuterium exchange (HDX), which allows modification of backbone amide groups in a way that is sensitive to both solvent accessibility and hydrogen bonding, i.e., secondary structure [17–20]. While this is a powerful method, the kinetics of the exchange reaction are highly sensitive to experimental factors, such as temperature and pH, and, therefore, considerable expertise is required for a successful HDX experiment, which, as a result, is still hardly a routine approach. Furthermore, analysis is typically performed with enzymatic digestion at low pH and temperature, followed by chromatographic separation and MS analysis of peptides. Due to the reversible nature of the labelling reaction, there is always a degree of back-exchange during this step of the experiment, and this has to be carefully controlled. For this reason, the best results are often obtained using automated sample preparation and handling, which allows better precision and accuracy than human-level control over the experiment. Finally, due to the mobilisation of protons (or deuterons) in gas-phase peptides under high-energy conditions, the most common fragmentation technique in mass spectrometry—collision-induced dissociation—causes randomisation ('scrambling') of labelling sites, which limits the resolution of the method to the peptide level [21,22].

For these reasons, irreversible covalent labelling in MS-based structural biology can offer valuable additional information. Different chemistries have been used for this over the years, ranging from conventional substitutions to radical chemistry, for example, with carbenes [10,23–25]. Great strides have been made in the use of hydroxyl radicals for footprinting after homolytic dissociation of hydrogen peroxide upon ultraviolet irradiation, typically employing an excimer laser. This method is known as fast photochemical oxidation of proteins (FPOP) [26–28]. Due to the very short lifetime of the radicals (particularly as a scavenger is typically added to the solution) it is possible to probe the evolution of protein structure on a microsecond timescale, although it should be noted that Vahidi and Konermann have shown evidence that it can take up to milliseconds for all metastable secondary radicals to be destroyed [29]. The use of hydroxyl radicals is particularly attractive, as they are essentially the same size as water molecules and labelling, therefore, captures the biologically relevant solvent-accessible surface. Before the development of FPOP, Chance and colleagues demonstrated the use of synchrotron radiation to form hydroxyl radicals from water, also leading to selective labelling of the exposed surface of a protein [30,31]. While powerful, these methods rely on access to very specialised equipment and, as such, are out of reach for most researchers. Therefore, there is a need for robust conformation-sensitive labelling approaches that can be easily implemented in more routine settings.

Hydroxyl radicals can of course be conveniently produced in solution in a number of ways, most famously by Fenton chemistry. In this case, hydroxyl radicals are produced by reaction of hydrogen peroxide with Fe(II), yielding Fe(III), OH^-, and $OH^\bullet$, the latter of which can react with amino acid side chains. For a residue-specific overview of possible modifications, we refer the reader to the excellent recent review by Gross and colleagues [10]. Pioneering work by Tullius and Dombrowski demonstrated the use of an elegant Fenton system for probing protein–DNA interactions [32]. In this work, radical production was achieved by a redox cycle involving Fe(II)–EDTA, hydrogen peroxide, and ascorbate to regenerate the Fe(II). Attack by the hydroxyl radical then led to cleavage of exposed parts of the DNA backbone. Subsequent analysis of DNA fragments was performed by gel electrophoresis; however, possible modifications to the protein interaction partner were largely ignored. Interestingly, despite continued use of Fenton chemistry for characterising the structure of nucleic acids bound to protein, there has been limited interest in its use for oxidative footprinting of proteins compared to radiolysis [10,33–36]. Several reasons have been reported for this; for example, (i) the Fenton reaction cannot be initiated and quenched on the same timescale as FPOP; (ii) there is the risk that iron, EDTA, or other components in solution might interact with the protein of interest and induce a conformational change; and (iii) letting the process continue for too long could permit undesired secondary reactions [37].

Despite these concerns, here we explored whether Fenton chemistry can be combined with routine proteomics sample preparation and LC-MS to reveal insight into protein structure and dynamics. The overall workflow is represented schematically in Figure 1. Using a set of model proteins with masses between 10 and 150 kDa, we found that we were indeed able to selectively label the exposed protein surface. For the FK1 domain of the immunophilin FKBP51—currently an important drug target due to its relevance in mood disorders [38,39]—differences in the modification pattern between the ligand-free and -bound form were consistent with the known binding site of two different ligands, based on crystallography [40,41]. We were able to apply the same method to complexes of the homologue FKBP12 without any particular difficulty. Finally, in the case of myoglobin, we unexpectedly found that the iron centre in the noncovalently bound haem group was also able to participate in the Fenton reaction. As a result, oxidative footprinting for this protein reflected not only surface exposure in a way consistent with the known structure and labelling experiments from the FPOP literature, but also reflected the binding mode of oxygen to this prosthetic group.

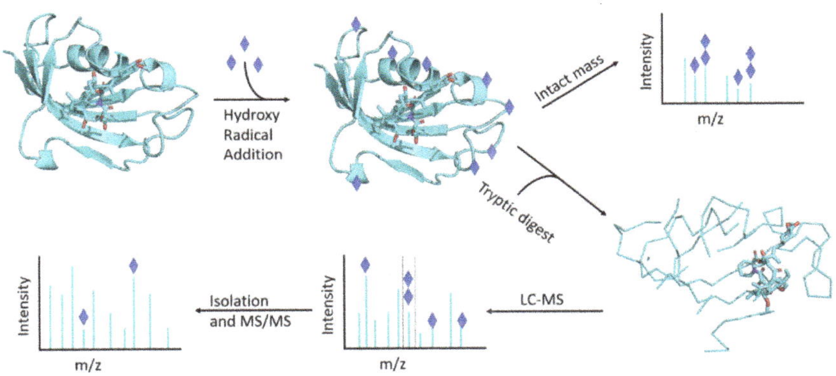

Figure 1. Schematic representation of the oxidative footprinting workflow.

2. Results

We used four model proteins to benchmark our method in this study: one large, noncovalent complex in which sequence regions involved in protein–protein binding interfaces constitute a clearly defined protected 'core', a smaller protein with a noncovalent haem group, and two small drug targets. Interestingly, each of these model systems highlighted a different aspect of the use of Fenton chemistry for oxidative footprinting, and the presentation of the results is, therefore, organised by protein substrate.

2.1. *Myoglobin: Intrinsic Fenton Reactivity Reflects Co-Ordination Mode of the Haem Group*

As a first model system, we decided to use myoglobin, as this 17.6 kDa protein (including a haem group with a mass of 616 Da) has been well characterised by various techniques, including several reports in the early FPOP literature [26,27]. As such, we were eager to determine how the oxidation pattern in Fenton-chemistry-based footprinting would compare to those published results. Furthermore, it is well known that loss of the prosthetic haem group causes significant structural destabilisation in this protein, especially in the C-terminal half of the sequence (helices F, G, and H) [26,27,42]. Therefore, we anticipated that comparing results for *holo*- and *apo*-myoglobin would provide a clue about the ability of our method to distinguish between different conformational states. After quenching the footprinting reaction, an aliquot was taken for intact mass measurement, which was performed under denaturing conditions. Results are shown in Figure 2. In agreement with the FPOP literature, a fairly similar overall level of oxidation was observed for *apo*- and *holo*-myoglobin, with the latter appearing slightly more oxidation-sensitive [26,27].

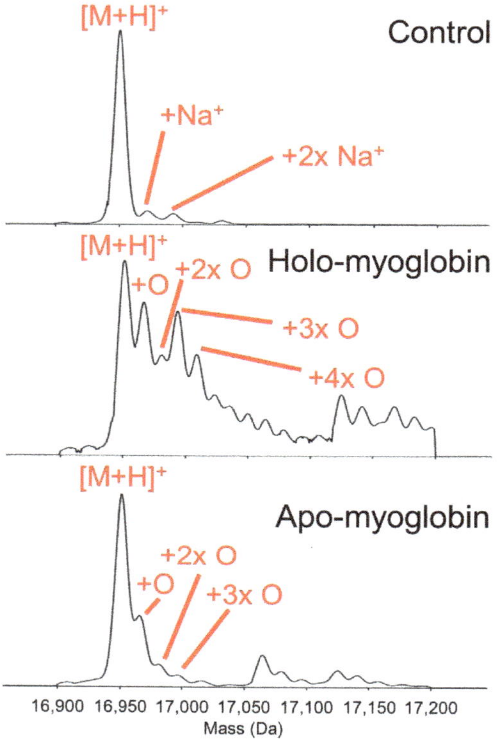

Figure 2. Deconvoluted mass spectra of *apo*-myoglobin and *holo*-myoglobin before ('control' experiment in the top panel) or after oxidative footprinting. Oxidative footprinting was carried out under native-like conditions, but samples were chemically denatured prior to MS measurement, which is why the haem group is not retained in *holo*-myoglobin. Only one spectrum is shown in the top panel as *apo*- and *holo*-myoglobin are identical after chemical denaturation.

Subsequently, oxidation site determination was performed through bottom-up analysis, where differences between both states of the protein were observed. Overall, our data were consistent with those in the early FPOP work, where a good agreement with the exposed surface was already established based on the known crystal structure (see Figure 3) [26,27]. Importantly, as described by Gross and colleagues, His93, which directly co-ordinates the iron centre in the haem group, was efficiently oxidised in *apo*-myoglobin, but was protected from oxidation in *holo*-myoglobin [27]. The main discrepancy between our results and those in previous reports was significant oxidation in *holo*-myoglobin of residues Leu32, Lys42, Phe43, Val68, Leu72, Ile99, Tyr103, and Phe138 in the binding pocket of the haem group (coloured orange in Figure 3). Note that, as described in Materials and Methods, oxidised residues were identified using tandem MS—for example, fragments b_{10}, b_{11}, y_4, and y_6 from the oxidised peptide L(32)FTGHPETLEKFDK(45) bracketed the Lys42 and Phe43 residues, allowing us to say that these were both oxidised, rather than the mass shift being a result of, for example, double oxidation of phenylalanine.

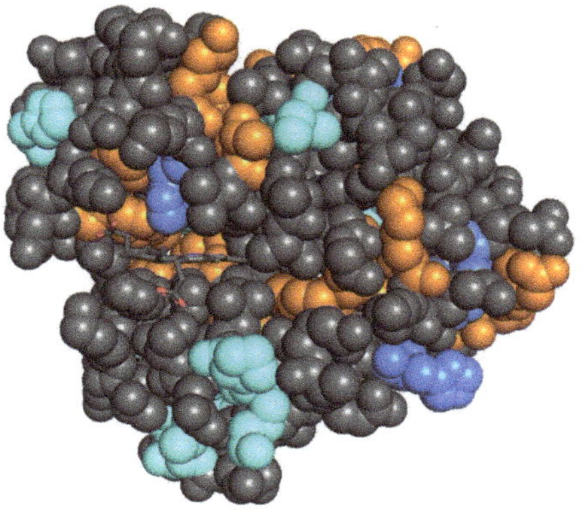

Figure 3. Crystal structure of *holo*-myoglobin (Protein Data Bank accession code 1YMB), with important residues for oxidative footprinting highlighted. Residues that were oxidised in our work that were previously identified as oxidation-sensitive in the FPOP literature are coloured blue. New modifications identified in our work (mostly near the haem group) are coloured orange. Modification sites that were identified in the FPOP literature but not in our experiments are coloured cyan.

As we used the same commercial supplier as Gross and colleagues, and the control experiment shown in Figure 2 confirmed the molecular mass and lack of modifications, we assumed the aforementioned discrepancy was due to a difference in reaction conditions rather than a difference in protein structure. Given the different method for generating hydroxyl radicals (flash photolysis vs. Fe/ascorbate-driven redox chemistry), we hypothesised that the iron centre within the haem group was involved in Fenton-like redox cycling in the presence of ascorbate, and that radicals were generated in close proximity to the haem group in this process. To test this, we repeated the experiment, but did not add any extrinsic iron in the form of Fe(II)–EDTA to the reaction mixture (i.e., only ascorbate and hydrogen peroxide). As before, the reaction was quenched after 15 s.

Under these conditions, we observed very little oxidation at surface regions remote from the haem group, while residues near this group were, again, oxidised extensively, in good agreement with our hypothesis. Interestingly, His93 and other residues on its side of the plane of the haem group were largely protected, and oxidation was mainly observed in residues located on the opposite side in the native structure, including His64, which also co-ordinates the iron centre. This opposite side is where oxygen binds to iron, displacing His64 in the process (see Figure 4); hence, this oxidation pattern seems to reflect the increased local conformational flexibility and space required to allow the protein to perform its oxygen transport function in vivo. Metal-catalysed oxidation of amino acid residues in the vicinity of biologically relevant metal ions has been observed before [43]; however, here, the oxidation pattern not only provided information about the binding region, but also reflected the solution-phase dynamics of ligand binding to and release from the metal.

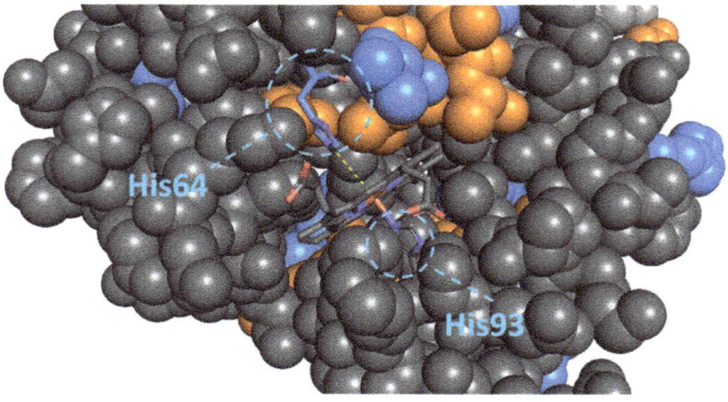

Figure 4. Crystal structure of *holo*-myoglobin, focussed on the binding pocket of the haem group. Highlighted residues were oxidised in the presence of ascorbic acid and hydrogen peroxide, without addition of Fe(II–EDTA. Residues in blue were identified in two independent ways with MaxQuant (either through fragmentation of two overlapping peptides, or from two different oxidative chemical modifications in the same peptide) [44,45]; those in orange were detected once (see Section 4 for details).

2.2. *ADH: Highly Reactive Sulphur-Containing Side Chains and Surface-Selective Labelling*

While our results for myoglobin—including the redox activity of the haem group—were intriguing, we wanted to further investigate the ability of our Fenton-chemistry-based method to selectively modify the solvent-exposed surface of a protein. For this reason, we wanted to use a model system without any redox-active metals. We selected alcohol dehydrogenase (ADH), which forms a 148 kDa tetramer in solution. Bottom-up analysis is not particularly limited by protein mass, and due to the large size of ADH, there is a clear 'core' region which is effectively shielded from the solvent. ADH contains two Zn^{2+} ions per monomer (so eight in total for the tetramer); however, this does not interfere with the labelling reaction, as Zn is redox-inactive under these conditions.

Overall, less extensive modification was observed in ADH compared to myoglobin, possibly reflecting the larger size of the tetramer (see Figure 5). In total, eleven modification sites were identified (note that each chain comprises 347 residues). Comparing the identified modification sites to the accessible surface based on the crystal structure [46], most of the oxidation sites were indeed solvent-exposed, which supports our hypothesis of selective labelling without major structure disruption on the timescale we used. Two exceptions, for which oxidation was observed despite the residues not being classified as accessible, were Met270 and Cys277. For Met270, we found that significant oxidation occurred even during the analysis of a sample where no oxidative footprinting was performed, indicating that this was likely an artefact that occurred on the peptide level during sample preparation or the electrospray process.

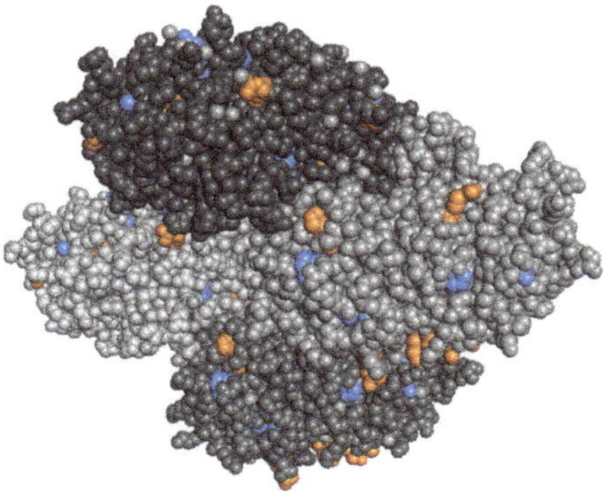

Figure 5. Labelling sites in ADH, indicated in the crystal structure (PDB accession code: 4W6Z) with the same colour code as in Figure 4.

In contrast, Cys277 was not oxidised in a control sample that was not exposed to hydroxyl radicals, indicating that this was indeed a result of the footprinting reaction. This was surprising, as our calculations based on the crystal structure indicated that this residue was not solvent-accessible; however, it should be noted that this residue is located in a cleft facing the solvent, rather than being involved in the protein–protein interface. The nearby residues Ala272 and Gly273 are classified as solvent-accessible in the crystal structure, and the residues in this region have relatively high crystallographic B-factors in the 38–46 $Å^2$ range, indicating significant local conformational flexibility. As such, it is plausible that Cys277 occasionally comes into contact with the solvent during the normal 'breathing' of the protein structure and, given the high intrinsic reactivity of cysteine toward hydroxyl radicals, this could account for the oxidation of this residue that we observed. In this way, it is plausible that our method indirectly provides insights into transient states that are normally 'invisible'. Overall, our results for ADH support the notion that oxidative labelling under the conditions used by us, indeed, selectively modifies the solvent-accessible surface.

2.3. FKBP51 and FKBP12: Key Interactions Drive Remarkable Structural Stabilisation

The immunophilin FKBP51 (see Figure 6A) belongs to the class of FK506-binding proteins and is a potentially important drug target in the context of depression, obesity-related diabetes, and chronic pain [38,39]. Drug development is hindered by the presence of homologues in the human body, including FKBP52 and FKBP12, which poses a challenge for designing selective ligands that avoid off-target effects [40]. One of the most promising ways for selective inhibition of FKBP51 is the targeting of minor conformational states, which exhibit a greater structural difference between homologues than the most abundant conformation [47]. Understanding such differences in the dynamics of protein structure and identifying possible transient binding sites is a major challenge for structural biology and difficult to achieve with conventional methods [48]. Given the importance of this protein family for human health, we decided to investigate two homologues with our footprinting method—FKBP12 (11.8 kDa, Figure 6B) and FKBP51. For the latter, rather than the full-length protein, a 14.0 kDa construct was used consisting of the FKBP-type peptidyl-prolyl cis-trans isomerase (PPIase) domain (called the FK1 domain), and this construct will be referred to as FKBP51FK1 in the rest of this work. Both proteins were analysed in their free state, as well as bound to two different ligands: SAFit1 and FK[4.3.1]-16h [40,41]. Binding affinity (K_i) to FKBP12 is approximately 163 nM for SAFit1 and 1.8 nM for FK[4.3.1]-16h. Binding affinity to FKBP51 is approximately 4 nM for SAFit1 and 57 nM for FK[4.3.1]-16h [41,49].

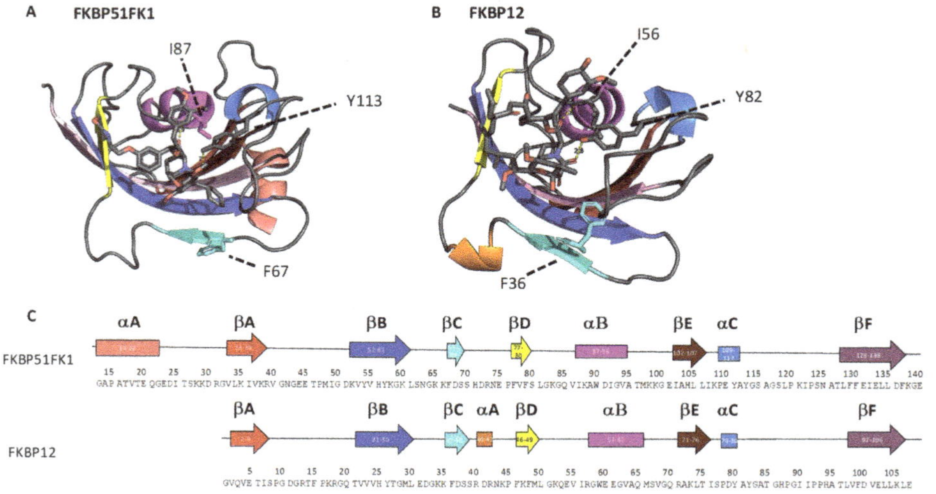

Figure 6. (**A**) Crystal structure of FKBP51FK1 in complex with the SAFit analogue iFit4 (Protein Data Bank accession code 4TW7), with key interacting amino acid residues labelled with a dashed line. (**B**) Crystal structure of FKBP12 in complex with FK506 (accession code 1FKJ), with key interacting amino acid residues labelled with a dashed line. (**C**) Sequence overview for FKBP51FK1 and FKBP12, where homologous sequence regions are aligned. Secondary structure elements are colour-coded and correspond to the same colour in the crystal structures.

In our initial experiments with the FK506-binding proteins, we found that reproducibility between replicate experiments was somewhat poor, leading to wide confidence intervals on the degree of oxidation, particularly for the complex with FK[4.3.1]-16h (data not shown). Manual inspection of the spectra revealed elution of a significant amount of intact protein near the end of the LC gradient in samples where ligand was present, indicating incomplete digestion. We hypothesised that this was due to the protein structure being sufficiently stabilised by interactions with the ligand to resist unfolding under our standard denaturing conditions (incubation with 6 M urea at 28 °C for one hour). This apparently led to inefficient digestion of largely folded protein by trypsin, somewhat similar to a limited proteolysis experiment [50,51].

Given the suboptimal reproducibility in these initial experiments, we discarded the results from bottom-up analysis and repeated the experiments with more aggressive denaturation conditions (vide infra); however, we did wish to further test the hypothesis of ligand-induced stabilisation toward chemical denaturation. For this, we performed MS of intact FKBP12 in the presence of FK[4.3.1]-16h (the sample that showed the highest abundance of remaining intact protein after digestion) with different concentrations of acetonitrile (data not shown). Under native-like conditions with no organic solvent, the protein was mostly in its ligand-bound form and was observed at low charge states, as commonly observed in native MS. Intriguingly, we found that a significant amount of low-charge-state (likely compact) protein was observed until 45% acetonitrile was added, and even a non-negligible amount of protein–ligand complex was still present under these conditions. In contrast, for the ligand-free FKBP12, a 'steady-state' of mostly high-charge-state (likely unfolded) protein was observed at 35% organic solvent, and this did not change until >50%, at which point precipitation of the protein occurred. This observation supports the notion that ligand binding stabilised the protein toward denaturation and subsequent enzymatic digestion. For comparison, the presence of 30% acetonitrile was sufficient to cause myoglobin to mostly lose its haem group. To avoid incomplete protein digestion and ensure reproducibility, the experiments were repeated, with the denaturation step being extended to six hours. Three independent samples were prepared for each condition (two proteins, each in their free state and bound to both ligands), and each sample was injected onto the column twice (i.e., a total of 36 injections were performed).

As before, aliquots were taken and intact mass measurements performed immediately after the oxidative footprinting reaction (Figure 7). This revealed strongly reduced reactivity toward oxidation upon ligand binding, consistent with ligand-induced protection. In addition to the insight into the global labelling extent, a further benefit of this intact mass measurement was that it demonstrated that reaction between FKBP12 and hydroxyl radicals mostly occurred through 'simple' oxidation rather than side reactions that have been reported in the literature [10]. This was revealed through the observation of a pattern of mass increases in steps of 16 Da, up to an addition of 48 Da (more extensively oxidised protein was visible in the spectrum, but at lower abundance). With this knowledge, we were able to significantly speed up our data analysis (a necessity, given the sizeable data set) by focussing on peptides with these modifications. In practice, addition of a single oxygen atom was by far the most common modification at the peptide level, which is consistent with global addition of only a few oxygen atoms to the entire protein.

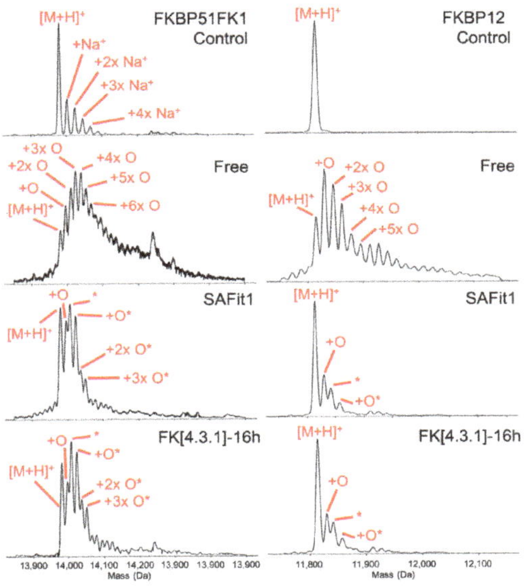

Figure 7. Intact mass measurement of FKBP51FK1 and FKBP12 (deconvoluted spectra shown) under denaturing conditions before (control; top panel) and after oxidative footprinting, either in the absence (second row; 'free' protein) or presence of ligands SAFit1 and FK[4.3.1]-16h. Peaks labelled with an asterisk carry an additional modification of 27.01 Da.

Peptides that were detected with sufficient signal-to-noise for quantification covered 92% of the sequence of FKBP12 and 88% of the sequence of FKBP51FK1. Oxidation sites were identified qualitatively with single-residue specificity through tandem MS; however, signal-to-noise in fragment spectra was insufficient to determine site-specific changes in oxidation level with statistical significance; therefore, quantitative analysis was limited to the peptide level. Results of this analysis are summarised in Figure 8. The fact that three samples were prepared for each condition allowed us to evaluate the reproducibility of our method in this case. Visually, it is apparent that most error bars are small; more quantitatively, the median coefficient of variation for the fraction of oxidised peptides for FKBP51FK1 was 10.2%. A similar value (10.5%) was observed for FKBP12 peptides.

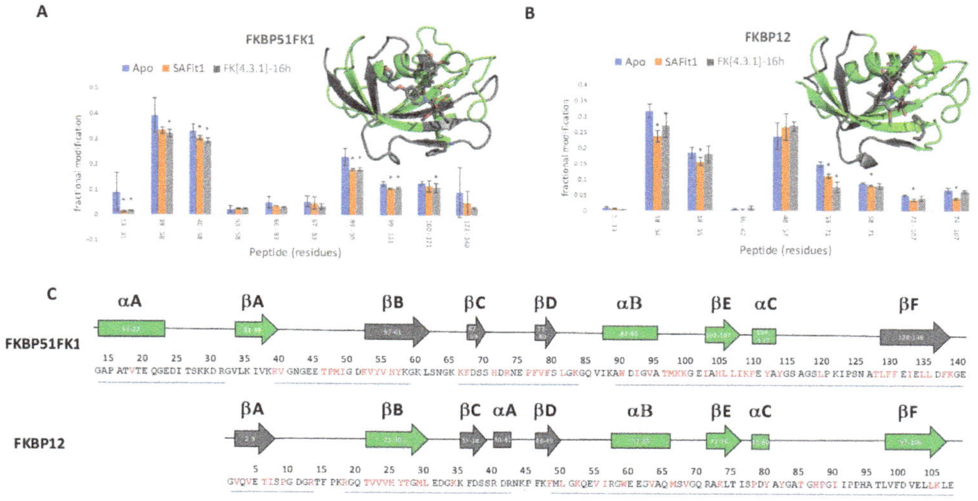

Figure 8. Results of the peptide-level analysis of the FK506-binding proteins: (**A**) fractional oxidative modification of the peptides of FKBP51FK1 for the *apo* protein as well as SAFit1- and FK[4.3.1]-16h-bound protein, in which statistically significant ($p < 0.05$) differences between ligand-bound and -free states are indicated with an asterisk (*). Insets show crystal structures with regions that show reduced oxidation after binding of either ligand in green. (**B**) Fractional modification of the peptides of FKBP12 for the *apo* protein, as well as ligand-bound states. (**C**) Sequence overview for FKBP51FK1 and FKBP12, with sequence regions covered by the observed peptides underlined. Secondary structure elements are labelled as in Figure 6. Elements that show reduced oxidation after ligand binding are coloured green in Panel (**C**), and unaffected elements are in grey (same colour code as the insets in Panel (**A**); no regions were observed where ligand binding led to increased oxidative labelling). Detected oxidative labelling sites are coloured red in the sequence.

The first thing that stands out from these results is the overall reduction in degree of oxidation upon ligand binding for both proteins. This is consistent with the intact mass measurements and could be partly due to direct shielding of reactive residues by the ligands, but also supports the hypothesis that overall structural compaction led to the incomplete digestion of ligand-bound protein that we observed in our initial attempts. Analysing the results in more detail, some interesting differences between both ligands, and between both proteins become apparent. In this discussion, the homology between both proteins is important; specifically, it should be noted that residues 1–106 of FKBP12 show a striking similarity to residues 32–137 of FKBP51FK1. Despite this, even in the ligand-free form, some differences are apparent. Most striking is the very limited degree of oxidation in peptides spanning residues 53–83 in FKBP51FK1, corresponding to the region 22–52 in FKBP12, where significant oxidation was observed. We attribute this to the presence of the N-terminal extension of 31 residues in FKBP51FK1, which appears to shield part of the main β-sheet region of the protein from the solvent in the crystal structure (see Figure 6A). Perhaps due to this limited initial degree of oxidation in FKBP51FK1, no statistically significant reduction was observed in this region after ligand binding, while binding of SAFit1, but not FK[4.3.1]-16h, did lead to a significant protective effect in the 18–34 region of FKBP12, possibly reflecting a greater degree of direct shielding by the bulkier SAFit1.

An alternative, more intriguing explanation for this behaviour than a simple steric effect involves the fact that binding of SAFit1 to FKBP51FK1 requires the side chain of residue Phe67 to be displaced, and this alternative conformation is at the core of the ability of this type of ligand to distinguish between homologues [40]. The binding affinity of SAFit1 to FKBP12, where Phe36 is the counterpart to Phe67 in FKBP51FK1 (see Figure 6), is an order of magnitude lower than to FKBP51FK1 [41,49]. Interestingly, in both cases the key phenylalanine residue is part of a β-sheet, with the adjacent strand (labelled as 'βB' in Figures 6 and 8) composed of residues 21–30 in FKBP12 and residues 52–61 in FKBP51FK1. In this context, the increased protection observed in the βB region of FKBP12 after binding of SAFit1 compared to FK[4.3.1]-16h (which binds to an *apo*-like conformation), and the lack of such protection in the corresponding region of FKBP51FK1 upon binding of either ligand, could indicate that a more significant structural rearrangement is required for FKBP12 than for FKBP51 to adopt the conformation that can efficiently bind SAFit1. It is plausible that this need to undergo a more significant rearrangement makes the SAFit1-binding conformation less favourable for FKBP12, which might contribute to the previously established lower binding affinity.

The N-terminal extension itself also exhibits significant protection upon ligand binding to FKBP51FK1, as reflected by the decrease in oxidation in the peptide spanning residues 13–33. This was surprising as, in the crystal structure, this region is fairly distant from the ligand binding site. This effect may be a result of an indirect stabilization caused by the adjacent beta strands that are protected upon ligand binding. In both FKBP12 and FKBP51FK1, the C-terminal portion of the protein showed significant protection after ligand binding. This is unsurprising and can largely be attributed to steric effects, as this region contains many residues that are either part of, or close to, the binding site. Of note is the protection of the peptide with residues 89–98 in FKBP51FK1 and, similarly, that with residues 53–71 in FKBP12. These span an α-helix (αB) containing, or being close to, a key interacting amino acid residue (Ile56 in FKBP12 and Ile87 in FKBP51FK1) that forms a strong hydrogen bond with both ligands through its amide nitrogen atom [49,52–54]. Furthermore, a very reactive tryptophan (based on both the inherent reactivity of the side chain toward oxidative labelling, and the direct observation, as shown in Figure 8C) that is located within the binding pocket and directly exposed to solvent in the absence of a ligand is located in this region. The shielding of this reactive tryptophan (Trp59 in FKBP12; Trp90 in FKBP51FK1), combined with the strong interaction of the isoleucine with the ligand, causes one of the most significant protections of the protein.

3. Discussion

We have shown that oxidative labelling through Fenton chemistry can be employed for structural characterisation of a set of model proteins, with a reaction time of only a few seconds, and that conformational changes are reflected in the modification pattern. For the smaller (11–18 kDa) proteins we studied, extensive oxidation was observed, and it was demonstrated how the oxidation pattern correlated to protein structure, including dynamic aspects. Information was sparser on the 148 kDa ADH tetramer, where fewer oxidation sites were identified. A plausible explanation for this is that the reaction rate was limited by the concentration of Fenton reagents (specifically Fe(II)–EDTA at 94 μM) under these conditions, leading to approximately the same number of oxidation events being distributed over a much larger number of reactive residues, resulting in the observed greater selectivity for highly reactive sites. A possible way to address this in the future and obtain a consistent degree of oxidation across a range of protein masses would be to use a consistent mass-based concentration for proteins in the labelling solution, rather than consistent molarity.

For the ADH tetramer, we showed that oxidation occurs primarily at the exposed protein surface, in agreement with other hydroxyl radical footprinting techniques. In the cases of myoglobin and the two FK506-binding proteins we tested, there were clear differences in the oxidation pattern between the ligand-bound and ligand-free state of the protein. For myoglobin specifically, a key histidine residue that binds to the native iron centre was highly reactive in the *apo* state, and protected in the *holo* state. Furthermore, oxidation was observed of residues within the binding pocket of the haem group, but exclusively on the side of the plane of this prosthetic group at which biologically relevant ligands, such as oxygen, bind. It is reasonable to assume that this reflects increased conformational flexibility on this side of the haem group, which is necessary to accommodate the exchange and transport of gas molecules by myoglobin. This supports the hypothesis that our method is able to inform on dynamic aspects of protein conformation, rather than just a static lowest-energy structure.

In addition to these strengths, we also identified several practical limitations to the method presented in this work. The reaction time for our oxidative footprinting method is limited in practice to several seconds, which leads to a degree of ensemble averaging and precludes the probing of protein structure on a microsecond timescale, as is possible in labelling methods based on photolysis. The use of microfluidics in future studies could significantly reduce the reaction time [55,56], but—even assuming the extent of the labelling reaction on such a short timescale would be sufficient to obtain structural information—this approach would still be orders of magnitude slower than FPOP.

Another potential concern is the effect of sulphur-containing residues. While this did not pose an issue for most methionine- or cysteine-containing peptides in our hands, we did find that, in the case of ADH, the residue Met270 was consistently and spontaneously oxidised—possibly during sample handling or the electrospray process—even in control samples. This needs to be carefully controlled and could pose a challenge for the analysis of methionine- or cysteine-rich proteins by oxidative footprinting, regardless of the exact chemistry used to generate hydroxyl radicals. The main bottleneck we identified in implementing this type of experiment was data analysis. Given that most amino acid residues are at least somewhat reactive toward hydroxyl radicals, and that many residue types are able to undergo several competing reactions under these conditions, the number of (modified) peptides that need to be matched to an experimental data set, even for a known protein sequence, quickly becomes very large. Even using a high-end desktop PC, searching for all possible reaction products is not feasible, or is at least sufficiently time-consuming to be impractical, and an optimal trade-off between 'complete' data analysis and processing time needs to be determined empirically in the absence of access to high-performance computing. For quantitative analysis of oxidised peptides from FK506-binding proteins, we found that a targeted software package (pepFoot) provided good performance while requiring far less computational power than MaxQuant [44,45,57]. In future work, we will further optimise the data processing workflow, as well as extend the method to a greater set of protein–ligand systems, and compare it to other, more conventional labelling techniques.

Combining bottom-up proteomics analysis with intact mass measurement and/or top-down fragmentation can be helpful for optimising the analytical workflow, as it provides a clue regarding the overall extent of modification and possibly some of the labelling sites, which can inform the subsequent more in-depth bottom-up data analysis. Similarly, especially when studying ligand binding, native mass spectrometry can provide important complementary insight to oxidative footprinting. Finally, incomplete protein digestion complicates the data analysis and potentially leads to poor reproducibility, but, at the same time, can be indicative of high structural stability, similar to limited proteolysis approaches developed in recent years. Combining the insights from all the aforementioned data points—and with insights from conventional structural biology methods—leads to an improved understanding of the 'four-dimensional' structure of a protein in solution. We believe that the underexplored labelling method used in this work shows sufficient promise to be further developed in the future as a technique for hydroxyl radical footprinting with low barriers to entry compared to radiolysis. As such, this will potentially provide a valuable addition to the toolset of researchers interested in MS-based conformational protein analysis.

4. Materials and Methods

4.1. Proteins, Reagents and Solvents

Most materials were acquired from commercial suppliers: HPLC-grade acetonitrile (Roth, Karlsruhe, Germany, HN44.2); LC-MS grade acetonitrile (Supelco LiChrosolv, Darmstadt, Germany, 1.00029.2500); alcohol dehydrogenase (Sigma-Aldrich, St. Louis, MI, USA, A3263); ammonia, 30% *w*/*w* (Sigma-Aldrich, St. Louis, MI, USA, 221228); ammonium acetate, 7.5 M (Sigma-Aldrich, St. Louis, MI, USA, A2706); *apo*-myoglobin (Sigma-Aldrich, St. Louis, MI, USA, A8673); L-ascorbic acid (Sigma-Aldrich, St. Louis, MI, USA, 255564); Discovery® DSC-18 SPE Tubes (Sigma-Aldrich, St. Louis, MI, USA, 62602-U); 1,4-dithiothreitol (Roth, Karlsruhe, Germany, 6909.1); H_4EDTA (Sigma-Aldrich, St. Louis, MI, USA, 431788); formic acid (Fisher Chemical, Waltham, MA, USA, A117-50); *holo*-myoglobin (Sigma-Aldrich, St. Louis, MI, USA, M0630); hydrogen peroxide, 30% *w*/*w* (Sigma-Aldrich, St. Louis, MI, USA, 95321); iodoacetamide (Sigma-Aldrich, St. Louis, MI, USA, I6125); iron(II)-chloride tetrahydrate (Sigma-Aldrich, St. Louis, MI, USA, 44939); Pierce™ Trypsin Protease (Thermo Fisher, Waltham, MA, USA, 90058); triethylammonium bicarbonate buffer (Fluka, St. Louis, MI, USA, 17902); trifluoroacetic acid (Roth, Karlsruhe, Germany, P088.3); urea (Roth, Karlsruhe, Germany, 2317.1). FK506-binding proteins were expressed based on previously described methods [58,59].

4.2. Oxidative Footprinting

Protein stock solutions of FKBP12 and FKBP51FK1 in 20 mM HEPES, pH 8.5, and 150 mM NaCl were diluted in 200 mM ammonium acetate to 40.7 µM. Ligand stock solutions were prepared in DMSO at 250 times the final concentration and prediluted to 50 times the final concentration in acetonitrile. A total of 49.2 µL of the protein solution and 0.8 µL of the ligand solution or 200 mM ammonium acetate were mixed to achieve a final protein concentration of 40 µM and ligand concentration of 80 µM, and incubated for 15 min at room temperature, after which the oxidative footprinting reaction was initiated.

For each reaction, fresh solutions were prepared of 0.3 M hydrogen peroxide, 37.5 mM L-ascorbic acid, 187.5 mM thiourea, and iron(II)–EDTA solution. The 0.3 M hydrogen peroxide solution was prepared by diluting a 30% *w*/*w* stock solution with milliQ water. The ascorbic acid solution was prepared by dissolving 6.6 mg of ascorbic acid in 1 mL of milliQ water and neutralising with 2.3 µL of 30% *w*/*w* ammonia. Note that ascorbic acid is oxidation-sensitive in air and that this solution was stable for approximately one hour at room temperature. For a 187.5 mM thiourea solution, we dissolved 14.27 mg of thiourea in 1 mL of 200 mM ammonium acetate solution. The iron(II)–EDTA solution was prepared using a stock solution of 3 mM H_4EDTA with 12 mM ammonia. A 1.5 mM iron(II)-chloride solution in milliQ water was made fresh for the reaction. Equal volumes of the EDTA stock solution and the iron(II)-chloride solution were mixed in a reaction tube to obtain the iron(II)–EDTA solution.

For the oxidative footprinting reaction, 50 µL of a 40 µM protein solution (in 200 mM ammonium acetate) was pipetted into a 500 µL reaction tube. Next, 10 µL of the iron(II)–EDTA solution was added, followed by 10 µL of the L-ascorbic acid solution. Immediately after subsequently adding 10 µL of the 0.3 M hydrogen peroxide solution, the reaction tube was vortexed and the reaction was allowed to proceed for 15 s. After 15 s, the reaction was quenched by adding 20 µL of the thiourea solution into the reaction tube and vortexing. The final concentration of the protein after the reaction was 5 to 20 µM with 140 mM ammonium acetate.

4.3. Tryptic Digest

For digestion, 300 µL of an 8 M urea solution in 50 mM TEAB, pH 8.5, with 100 mM NaCl, as well as 8 µL of a 0.5 M solution of DTT were added to the oxidative footprinting reaction mixture. After incubating the mixture for one hour at 28 °C, 1 mL of 50 mM TEAB, pH 8.5, and 1.2 µL of 1 mg/mL trypsin in 50 mM acetic acid were added in succession. The digest was incubated overnight at 37 °C. After the samples cooled down to room temperature, trifluoroacetic acid was added until the solution was at a pH value of 2. A 100 mg C18-SPE cartridge was conditioned with 1 mL HPLC-grade acetonitrile and 1 mL 0.6% *v*/*v* TFA solution with milliQ water. Next, the sample was loaded on the cartridge and washed with 1 mL 0.6% *v*/*v* TFA solution in milliQ water. Elution was performed with 1 mL of an 80% *v*/*v* acetonitrile solution. The eluate was dried in a vacuum centrifuge

(UniVapo 150H; UniEquip, Planegg, Germany) and redissolved in 100 µL of a 5% *v/v* acetonitrile solution containing 0.1% *v/v* formic acid.

4.4. LC-MS/MS Analysis

LC-MS/MS analysis was performed with an LTQ Orbitrap XL (Thermo Fisher Scientific, Waltham, MA, USA) controlled by Xcalibur 2.1 and a micro-LC system consisting of a Micro Pro syringe pump (Eldex Laboratories, Napa, CA, USA), and an Endurance autosampler (Spark Holland, Emmen, The Netherlands) controlled by the Endurance software. Acquisitions were started upon injection by contact closure.

Samples (5 µL) were injected with a flushed loop injection and peptides were separated on a ZORBAX StableBond C18, 0.3 × 150 mm, 3.5 µm column (Agilent, Santa Clara, CA, USA) at a flow rate of 5 µL/min using the following gradient: linear gradient from 5% B to 60% B in 60 min, 10 min linear gradient to 100% B, 10 min at 100% B isocratic, followed by re-equilibration at 5% B for 15 min, with solvent A being water with 0.1% formic acid and solvent B being acetonitrile with 0.1% formic acid.

The mass spectrometer was operated in a data-dependent mode with a precursor scan in the Orbitrap with a resolution of 60,000 at m/z 400, followed by fragmentation of peptide ions with a charge state of 2 or higher, giving rise to the four most intense signals in the ion trap using CID with a normalized collision energy of 25. Dynamic exclusion was enabled and set to a repeat count of 2 with a repeat duration of 30 s, the exclusion list size was 200, and the exclusion duration was 50 s. The ESI source was operated with 10 units of sheath gas flow rate, a spray voltage of 4 kV, a capillary temperature of 300 °C, a capillary voltage of 3 V, and tube lens set to 30.

For intact mass measurements and for native MS, including the experiments with different concentrations of acetonitrile, 10 µL of sample was loaded into a glass needle that was pulled to a tip of ca. 1-µm orifice diameter with a P97 Flaming/Brown type micropipette puller (Sutter Instrument Co., Novato, CA, USA), starting from 1.2-mm thin-walled glass capillaries (World Precision Instruments, Friedberg, Germany). Ionisation was then performed using a home-built nano-electrospray source that was coupled to the LTQ Orbitrap XL instrument. Intact protein spectra were deconvoluted with UniDec [60].

4.5. Data Analysis Using MaxQuant

The MaxQuant calculations were separated into two parts and, in all cases, a precursor mass accuracy of 4.5 ppm was used. In the first part with one calculation run, the unmodified peptides were identified. Only the fasta file of the target protein was used to search against. The default settings were used, with the following exceptions: no fractions—yes; min. peptide length—5; max. peptide mass (Da)—4800; min. score for modified peptides—0; second peptides—off; unknown MS/MS match tolerance and unit—0.5 Da; unknown MS/MS de novo tolerance and unit—0.25 Da; unknown deisotoping—off. For every identified peptide, a separate fasta file was then created for use in the second step. This step comprised three calculation runs to determine the modifications. The first calculation run included +15.995 Da (+O) for M, C, W, Y, F, K, R, Q, D, T, S, A, E, L, I, K, H, N, V and +31.990 Da (+2xO) for M, C, W, Y, F, both as variable modifications additional to carbamidomethyl- and acetyl-(N-term) modifications. The second calculation run included +15.995 Da (+O) for W, C, M, Y, F, H; +47.985 Da (+3xO) for C, W, Y, F; −43.053 Da (+O -5xH -3xN -C) for R; −32.008 Da (+O -S -4xH -C) for M; −30.011 Da (-2xH -C -O) for E, D; +13,9792645 (+O -2H) for L, I, V, P, R, K, E, Q; and −2.016 Da (-2xH) for T, S, all as variable modifications additional to carbamidomethyl- and acetyl-(N-term) modifications. The last calculation run included +15.995 Da (+O) for W, C, M, Y, F, H; −10.032 Da (+2xO -2xH -2xN -C) for H; −4.979 Da (+2xO -H -C -N) for H; −22.032 Da (+2xO -2xH -2xC -2xN) for H; and −23.016 Da (+O -H -N -2xC) for H, all as variable modifications additional to carbamidomethyl- and acetyl-(N-term) modifications. The default settings were used, with the following exceptions: no fractions—yes; digestion mode—no digestion; include contaminates—off; min. peptide length—5; max. peptide mass (Da)—4800; min. score for modified peptides—0; second peptides—off; unknown MS/MS match tolerance and unit—0.5 Da; unknown MS/MS de novo tolerance and unit—0.25 Da; unknown deisotoping—off. For positive identification of an oxidatively modified residue, while avoiding false positive results, we typically required that a +15.995 Da (+O) modification was detected two times in different modified peptides, or, alternatively, that a product from a side reaction of oxidative footprinting was found in addition to a +15.995 Da (+O) modification. Finally, an additional search was run against the entire UniProt database to ensure that peptides identified as oxidised were genuine and not false positives due to overlap with peptides from protein contaminants (note that such overlap would need to occur at both the MS and MS/MS level and is, therefore, very unlikely). Other than the trypsin used for proteolysis and a low-level contamination of keratin in a handful of samples, no other contaminants were found, which confirms sample purity and rules out false positives.

4.6. *Quantifying Peptides Using pepFoot*

Raw files were converted to the mz5 format using MSConvert and then processed in the pepFoot software [23,24,57] using the following parameters: modifications—carbamidomethyl, variable modifications—oxidation (+ oxygen), digestion—trypsin, peptide length—5–40, peptide charge—1–6, # missed cleavages—2, MS tolerance—20 ppm. Extracted ion chromatograms of identified peptides by MaxQuant in acceptable abundance (S/N > 9:1) were then integrated for the modified and unmodified peptide, and the degree of modification was calculated by the software for *apo* and *holo* proteins.

Author Contributions: F.L., T.N. and T.H. designed the study. T.N. and T.H. performed oxidative footprinting and LC-MS experiments, as well as data analysis and visualisation. T.N. performed native MS and intact mass measurements. C.M. performed expression of FK506-binding proteins. F.H. and F.L. supervised the project. F.L. wrote the first draft of the manuscript. All authors have read and agreed to the published version of the manuscript.

Funding: Funding by the LOEWE project TRABITA funded by the Hessian Ministry of Higher Education, Research, and the Arts (HMWK) is gratefully acknowledged. T.H. was supported by the Kooperative Promotionsplattform, financed by the HMWK. We acknowledge support by the Deutsche Forschungsgemeinschaft (DFG—German Research Foundation) and the Open Access Publishing Fund of Technical University of Darmstadt.

Institutional Review Board Statement: Not applicable.

Informed Consent Statement: Not applicable.

Data Availability Statement: The data and materials underlying this article will be shared on request to the corresponding author.

Acknowledgments: We are grateful to Tanja Habeck (TU Darmstadt) for help with the initial optimisation of LC-MS experiments, and to Kevin Kretschmer (TU Darmstadt) for assistance with the initial work with the nano-ESI source. We thank Günes Barka from SunChrom GmbH for supporting us with the micro-LC. Alexander Schießer and the MS facility of the Department of Chemistry of TU Darmstadt are gratefully acknowledged for access to the Orbitrap instrument. We are grateful to the reviewers for their insightful comments.

Conflicts of Interest: The authors declare no conflict of interest.

References

1. Konijnenberg, A.; Butterer, A.; Sobott, F. Native ion mobility-mass spectrometry and related methods in structural biology. *Biochim. Biophys. Acta* **2013**, *1834*, 1239–1256. [CrossRef]
2. Politis, A.; Stengel, F.; Hall, Z.; Hernandez, H.; Leitner, A.; Walzthoeni, T.; Robinson, C.V.; Aebersold, R. A mass spectrometry-based hybrid method for structural modeling of protein complexes. *Nat. Methods* **2014**, *11*, 403–406. [CrossRef]
3. Leney, A.C.; Heck, A.J. Native Mass Spectrometry: What is in the Name? *J. Am. Soc. Mass Spectrom.* **2017**, *28*, 5–13. [CrossRef]
4. Robinson, C.V. Mass spectrometry: From plasma proteins to mitochondrial membranes. *Proc. Natl. Acad. Sci. USA* **2019**, *116*, 2814–2820. [CrossRef] [PubMed]

5. Allison, T.M.; Barran, P.; Cianferani, S.; Degiacomi, M.T.; Gabelica, V.; Grandori, R.; Marklund, E.G.; Menneteau, T.; Migas, L.G.; Politis, A.; et al. Computational Strategies and Challenges for Using Native Ion Mobility Mass Spectrometry in Biophysics and Structural Biology. *Anal. Chem.* **2020**, *92*, 10872–10880. [CrossRef] [PubMed]
6. Zhou, M.; Lantz, C.; Brown, K.A.; Ge, Y.; Pasa Tolic, L.; Loo, J.A.; Lermyte, F. Higher-order structural characterisation of native proteins and complexes by top-down mass spectrometry. *Chem. Sci.* **2020**, *11*, 12918–12936. [CrossRef]
7. Groves, K.; Ashcroft, A.E.; Cryar, A.; Sula, A.; Wallace, B.A.; Stocks, B.B.; Burns, C.; Cooper-Shepherd, D.; De Lorenzi, E.; Rodriguez, E.; et al. Reference Protocol to Assess Analytical Performance of Higher Order Structural Analysis Measurements: Results from an Interlaboratory Comparison. *Anal. Chem.* **2021**, *93*, 9041–9048. [CrossRef]
8. Wang, L.; Chance, M.R. Protein Footprinting Comes of Age: Mass Spectrometry for Biophysical Structure Assessment. *Mol. Cell Proteom.* **2017**, *16*, 706–716. [CrossRef] [PubMed]
9. Kaur, U.; Johnson, D.T.; Chea, E.E.; Deredge, D.J.; Espino, J.A.; Jones, L.M. Evolution of Structural Biology through the Lens of Mass Spectrometry. *Anal. Chem.* **2019**, *91*, 142–155. [CrossRef] [PubMed]
10. Liu, X.R.; Zhang, M.M.; Gross, M.L. Mass Spectrometry-Based Protein Footprinting for Higher-Order Structure Analysis: Fundamentals and Applications. *Chem. Rev.* **2020**, *120*, 4355–4454. [CrossRef]
11. Pan, Y.; Konermann, L. Membrane protein structural insights from chemical labeling and mass spectrometry. *Analyst* **2010**, *135*, 1191–1200. [CrossRef] [PubMed]
12. Konermann, L.; Pan, Y.; Stocks, B.B. Protein folding mechanisms studied by pulsed oxidative labeling and mass spectrometry. *Curr. Opin. Struct. Biol.* **2011**, *21*, 634–640. [CrossRef] [PubMed]
13. Khanal, A.; Pan, Y.; Brown, L.S.; Konermann, L. Pulsed hydrogen/deuterium exchange mass spectrometry for time-resolved membrane protein folding studies. *J. Mass Spectrom.* **2012**, *47*, 1620–1626. [CrossRef] [PubMed]
14. Pan, Y.; Piyadasa, H.; O'Neil, J.D.; Konermann, L. Conformational dynamics of a membrane transport protein probed by H/D exchange and covalent labeling: The glycerol facilitator. *J. Mol. Biol.* **2012**, *416*, 400–413. [CrossRef]
15. Pan, Y.; Brown, L.; Konermann, L. Hydrogen exchange mass spectrometry of bacteriorhodopsin reveals light-induced changes in the structural dynamics of a biomolecular machine. *J. Am. Chem. Soc.* **2011**, *133*, 20237–20244. [CrossRef]
16. Benesch, J.L.; Ruotolo, B.T. Mass spectrometry: Come of age for structural and dynamical biology. *Curr. Opin. Struct. Biol.* **2011**, *21*, 641–649. [CrossRef]
17. Konermann, L.; Pan, J.; Liu, Y.H. Hydrogen exchange mass spectrometry for studying protein structure and dynamics. *Chem. Soc. Rev.* **2011**, *40*, 1224–1234. [CrossRef]
18. Karch, K.R.; Coradin, M.; Zandarashvili, L.; Kan, Z.Y.; Gerace, M.; Englander, S.W.; Black, B.E.; Garcia, B.A. Hydrogen-Deuterium Exchange Coupled to Top- and Middle-Down Mass Spectrometry Reveals Histone Tail Dynamics before and after Nucleosome Assembly. *Structure* **2018**, *26*, 1651–1663.e1653. [CrossRef]
19. Masson, G.R.; Burke, J.E.; Ahn, N.G.; Anand, G.S.; Borchers, C.; Brier, S.; Bou-Assaf, G.M.; Engen, J.R.; Englander, S.W.; Faber, J.; et al. Recommendations for performing, interpreting and reporting hydrogen deuterium exchange mass spectrometry (HDX-MS) experiments. *Nat. Methods* **2019**, *16*, 595–602. [CrossRef] [PubMed]
20. Zheng, J.; Strutzenberg, T.; Pascal, B.D.; Griffin, P.R. Protein dynamics and conformational changes explored by hydrogen/deuterium exchange mass spectrometry. *Curr. Opin. Struct. Biol.* **2019**, *58*, 305–313. [CrossRef]
21. Zehl, M.; Rand, K.D.; Jensen, O.N.; Jorgensen, T.J. Electron transfer dissociation facilitates the measurement of deuterium incorporation into selectively labeled peptides with single residue resolution. *J. Am. Chem. Soc.* **2008**, *130*, 17453–17459. [CrossRef]
22. Wang, Q.; Borotto, N.B.; Hakansson, K. Gas-Phase Hydrogen/Deuterium Scrambling in Negative-Ion Mode Tandem Mass Spectrometry. *J. Am. Soc. Mass Spectrom.* **2019**, *30*, 855–863. [CrossRef] [PubMed]
23. Manzi, L.; Barrow, A.S.; Scott, D.; Layfield, R.; Wright, T.G.; Moses, J.E.; Oldham, N.J. Carbene footprinting accurately maps binding sites in protein-ligand and protein-protein interactions. *Nat. Commun.* **2016**, *7*, 13288. [CrossRef] [PubMed]
24. Manzi, L.; Barrow, A.S.; Hopper, J.T.S.; Kaminska, R.; Kleanthous, C.; Robinson, C.V.; Moses, J.E.; Oldham, N.J. Carbene Footprinting Reveals Binding Interfaces of a Multimeric Membrane-Spanning Protein. *Angew. Chem. Int. Ed. Engl.* **2017**, *56*, 14873–14877. [CrossRef]
25. Barth, M.; Bender, J.; Kundlacz, T.; Schmidt, C. Evaluation of NHS-Acetate and DEPC labelling for determination of solvent accessible amino acid residues in protein complexes. *J. Proteom.* **2020**, *222*, 103793. [CrossRef]
26. Hambly, D.M.; Gross, M.L. Laser flash photolysis of hydrogen peroxide to oxidize protein solvent-accessible residues on the microsecond timescale. *J. Am. Soc. Mass Spectrom.* **2005**, *16*, 2057–2063. [CrossRef] [PubMed]
27. Hambly, D.; Gross, M.L. Laser flash photochemical oxidation to locate heme binding and conformational changes in myoglobin. *Int. J. Mass Spectrom.* **2007**, *259*, 124–129. [CrossRef]
28. Li, K.S.; Shi, L.; Gross, M.L. Mass Spectrometry-Based Fast Photochemical Oxidation of Proteins (FPOP) for Higher Order Structure Characterization. *Acc. Chem. Res.* **2018**, *51*, 736–744. [CrossRef]
29. Vahidi, S.; Konermann, L. Probing the Time Scale of FPOP (Fast Photochemical Oxidation of Proteins): Radical Reactions Extend Over Tens of Milliseconds. *J. Am. Soc. Mass Spectrom.* **2016**, *27*, 1156–1164. [CrossRef]
30. Maleknia, S.D.; Brenowitz, M.; Chance, M.R. Millisecond radiolytic modification of peptides by synchrotron X-rays identified by mass spectrometry. *Anal. Chem.* **1999**, *71*, 3965–3973. [CrossRef]
31. Xu, G.; Chance, M.R. Hydroxyl radical-mediated modification of proteins as probes for structural proteomics. *Chem. Rev.* **2007**, *107*, 3514–3543. [CrossRef] [PubMed]

32. Tullius, T.D.; Dombroski, B.A. Hydroxyl radical "footprinting": High-resolution information about DNA-protein contacts and application to lambda repressor and Cro protein. *Proc. Natl. Acad. Sci. USA* **1986**, *83*, 5469–5473. [CrossRef] [PubMed]
33. Jain, S.S.; Tullius, T.D. Footprinting protein-DNA complexes using the hydroxyl radical. *Nat. Protoc.* **2008**, *3*, 1092–1100. [CrossRef] [PubMed]
34. Zhu, Y.; Guo, T.; Park, J.E.; Li, X.; Meng, W.; Datta, A.; Bern, M.; Lim, S.K.; Sze, S.K. Elucidating in vivo structural dynamics in integral membrane protein by hydroxyl radical footprinting. *Mol. Cell Proteom.* **2009**, *8*, 1999–2010. [CrossRef] [PubMed]
35. Leser, M.; Chapman, J.R.; Khine, M.; Pegan, J.; Law, M.; Makkaoui, M.E.; Ueberheide, B.M.; Brenowitz, M. Chemical Generation of Hydroxyl Radical for Oxidative 'Footprinting'. *Protein Pept. Lett.* **2019**, *26*, 61–69. [CrossRef]
36. Johnson, D.T.; Di Stefano, L.H.; Jones, L.M. Fast photochemical oxidation of proteins (FPOP): A powerful mass spectrometry-based structural proteomics tool. *J. Biol. Chem.* **2019**, *294*, 11969–11979. [CrossRef]
37. Wang, L.; Chance, M.R. Structural mass spectrometry of proteins using hydroxyl radical based protein footprinting. *Anal. Chem.* **2011**, *83*, 7234–7241. [CrossRef]
38. Kolos, J.M.; Voll, A.M.; Bauder, M.; Hausch, F. FKBP Ligands-Where We Are and Where to Go? *Front. Pharmacol.* **2018**, *9*, 1425. [CrossRef]
39. Hahle, A.; Merz, S.; Meyners, C.; Hausch, F. The Many Faces of FKBP51. *Biomolecules* **2019**, *9*, 35. [CrossRef]
40. Gaali, S.; Kirschner, A.; Cuboni, S.; Hartmann, J.; Kozany, C.; Balsevich, G.; Namendorf, C.; Fernandez-Vizarra, P.; Sippel, C.; Zannas, A.S.; et al. Selective inhibitors of the FK506-binding protein 51 by induced fit. *Nat. Chem. Biol.* **2015**, *11*, 33–37. [CrossRef]
41. Pomplun, S.; Sippel, C.; Hahle, A.; Tay, D.; Shima, K.; Klages, A.; Unal, C.M.; Riess, B.; Toh, H.T.; Hansen, G.; et al. Chemogenomic Profiling of Human and Microbial FK506-Binding Proteins. *J. Med. Chem.* **2018**, *61*, 3660–3673. [CrossRef] [PubMed]
42. Pan, J.; Han, J.; Borchers, C.H.; Konermann, L. Hydrogen/deuterium exchange mass spectrometry with top-down electron capture dissociation for characterizing structural transitions of a 17 kDa protein. *J. Am. Chem. Soc.* **2009**, *131*, 12801–12808. [CrossRef]
43. Lim, J.; Vachet, R.W. Development of a methodology based on metal-catalyzed oxidation reactions and mass spectrometry to determine the metal binding sites in copper metalloproteins. *Anal. Chem.* **2003**, *75*, 1164–1172. [CrossRef]
44. Cox, J.; Mann, M. MaxQuant enables high peptide identification rates, individualized p.p.b.-range mass accuracies and proteome-wide protein quantification. *Nat. Biotechnol.* **2008**, *26*, 1367–1372. [CrossRef] [PubMed]
45. Tyanova, S.; Temu, T.; Cox, J. The MaxQuant computational platform for mass spectrometry-based shotgun proteomics. *Nat. Protoc.* **2016**, *11*, 2301–2319. [CrossRef] [PubMed]
46. Fraczkiewicz, R.; Braun, W. Exact and Efficient Analytical Calculation of the Accessible Surface Areas and Their Gradients for Macromolecules. *J. Comp. Chem.* **1998**, *19*, 319–333. [CrossRef]
47. Jagtap, P.K.A.; Asami, S.; Sippel, C.; Kaila, V.R.I.; Hausch, F.; Sattler, M. Selective Inhibitors of FKBP51 Employ Conformational Selection of Dynamic Invisible States. *Angew. Chem. Int. Ed. Engl.* **2019**, *58*, 9429–9433. [CrossRef] [PubMed]
48. Parker, B.W.; Goncz, E.J.; Krist, D.T.; Statsyuk, A.V.; Nesvizhskii, A.I.; Weiss, E.L. Mapping low-affinity/high-specificity peptide-protein interactions using ligand-footprinting mass spectrometry. *Proc. Natl. Acad. Sci. USA* **2019**, *116*, 21001–21011. [CrossRef]
49. Voll, A.M.; Meyners, C.; Taubert, M.C.; Bajaj, T.; Heymann, T.; Merz, S.; Charalampidou, A.; Kolos, J.; Purder, P.L.; Geiger, T.M.; et al. Macrocyclic FKBP51 Ligands Define a Transient Binding Mode with Enhanced Selectivity. *Angew. Chem. Int. Ed. Engl.* **2021**, *60*, 13257–13263. [CrossRef]
50. Schopper, S.; Kahraman, A.; Leuenberger, P.; Feng, Y.; Piazza, I.; Muller, O.; Boersema, P.J.; Picotti, P. Measuring protein structural changes on a proteome-wide scale using limited proteolysis-coupled mass spectrometry. *Nat. Protoc.* **2017**, *12*, 2391–2410. [CrossRef]
51. Leuenberger, P.; Ganscha, S.; Kahraman, A.; Cappelletti, V.; Boersema, P.J.; von Mering, C.; Claassen, M.; Picotti, P. Cell-wide analysis of protein thermal unfolding reveals determinants of thermostability. *Science* **2017**, *355*, eaai7825. [CrossRef]
52. Wilson, K.P.; Yamashita, M.M.; Sintchak, M.D.; Rotstein, S.H.; Murcko, M.A.; Boger, J.; Thomson, J.A.; Fitzgibbon, M.J.; Black, J.R.; Navia, M.A. Comparative X-ray structures of the major binding protein for the immunosuppressant FK506 (tacrolimus) in unliganded form and in complex with FK506 and rapamycin. *Acta Crystallogr. D Biol. Crystallogr.* **1995**, *51*, 511–521. [CrossRef] [PubMed]
53. Gaali, S.; Feng, X.; Hahle, A.; Sippel, C.; Bracher, A.; Hausch, F. Rapid, Structure-Based Exploration of Pipecolic Acid Amides as Novel Selective Antagonists of the FK506-Binding Protein 51. *J. Med. Chem.* **2016**, *59*, 2410–2422. [CrossRef]
54. Bauder, M.; Meyners, C.; Purder, P.L.; Merz, S.; Sugiarto, W.O.; Voll, A.M.; Heymann, T.; Hausch, F. Structure-Based Design of High-Affinity Macrocyclic FKBP51 Inhibitors. *J. Med. Chem.* **2021**, *64*, 3320–3349. [CrossRef] [PubMed]
55. Resetca, D.; Wilson, D.J. Characterizing rapid, activity-linked conformational transitions in proteins via sub-second hydrogen deuterium exchange mass spectrometry. *FEBS J.* **2013**, *280*, 5616–5625. [CrossRef]
56. Deng, B.; Zhu, S.; Macklin, A.M.; Xu, J.; Lento, C.; Sljoka, A.; Wilson, D.J. Suppressing allostery in epitope mapping experiments using millisecond hydrogen/deuterium exchange mass spectrometry. *MAbs* **2017**, *9*, 1327–1336. [CrossRef]
57. Bellamy-Carter, J.; Oldham, N.J. PepFoot: A Software Package for Semiautomated Processing of Protein Footprinting Data. *J. Proteome Res.* **2019**, *18*, 2925–2930. [CrossRef] [PubMed]
58. Kozany, C.; Marz, A.; Kress, C.; Hausch, F. Fluorescent probes to characterise FK506-binding proteins. *Chembiochem* **2009**, *10*, 1402–1410. [CrossRef]

59. Bracher, A.; Kozany, C.; Thost, A.K.; Hausch, F. Structural characterization of the PPIase domain of FKBP51, a cochaperone of human Hsp90. *Acta Crystallogr. D Biol. Crystallogr.* **2011**, *67*, 549–559. [CrossRef] [PubMed]
60. Marty, M.T.; Baldwin, A.J.; Marklund, E.G.; Hochberg, G.K.; Benesch, J.L.; Robinson, C.V. Bayesian deconvolution of mass and ion mobility spectra: From binary interactions to polydisperse ensembles. *Anal. Chem.* **2015**, *87*, 4370–4376. [CrossRef]

International Journal of *Molecular Sciences*

Article

An In Vitro Comparative Study of the Effects of Tetrabromobisphenol A and Tetrabromobisphenol S on Human Erythrocyte Membranes—Changes in ATP Level, Perturbations in Membrane Fluidity, Alterations in Conformational State and Damage to Proteins

Monika Jarosiewicz [1,2,*], Piotr Duchnowicz [1], Paweł Jarosiewicz [3], Bogumiła Huras [4] and Bożena Bukowska [1]

1 Department of Biophysics of Environmental Pollution, Faculty of Biology and Environmental Protection, University of Lodz, Pomorska 141/143, 90-236 Lodz, Poland; piotr.duchnowicz@biol.uni.lodz.pl (P.D.); bozena.bukowska@biol.uni.lodz.pl (B.B.)
2 Department of Cytobiochemistry, Faculty of Biology and Environmental Protection, University of Lodz, Pomorska 141/143, 90-236 Lodz, Poland
3 European Regional Centre for Ecohydrology of the Polish Academy of Sciences, Tylna 3, 90-364 Lodz, Poland; p.jarosiewicz@erce.unesco.lodz.pl
4 Łukasiewicz Research Network, Institute of Industrial Organic Chemistry, Annopol 6 Str, 03-236 Warsaw, Poland; bogumila.huras@ipo.lukasiewicz.gov.pl
* Correspondence: monika.jarosiewicz@biol.uni.lodz.pl

Citation: Jarosiewicz, M.; Duchnowicz, P.; Jarosiewicz, P.; Huras, B.; Bukowska, B. An In Vitro Comparative Study of the Effects of Tetrabromobisphenol A and Tetrabromobisphenol S on Human Erythrocyte Membranes—Changes in ATP Level, Perturbations in Membrane Fluidity, Alterations in Conformational State and Damage to Proteins. *Int. J. Mol. Sci.* **2021**, *22*, 9443. https://doi.org/10.3390/ijms22179443

Received: 2 August 2021
Accepted: 27 August 2021
Published: 31 August 2021

Publisher's Note: MDPI stays neutral with regard to jurisdictional claims in published maps and institutional affiliations.

Abstract: Brominated flame retardants (BFRs) are substances used to reduce the flammability of plastics. Among this group, tetrabormobisphenol A (TBBPA) is currently produced and used on the greatest scale, but due to the emerging reports on its potential toxicity, tetrabromobisphenol S (TBBPS)—a compound with a very similar structure—is used as an alternative. Due to the fact that the compounds in question are found in the environment and in biological samples from living organisms, including humans, and due to the insufficient toxicological knowledge about them, it is necessary to assess their impacts on living organisms and verify the validity of TBBPA replacement by TBBPS. The RBC membrane was chosen as the research model. This is a widely accepted research model for assessing the toxicity of xenobiotics, and it is the first barrier to compounds entering circulation. It was found that TBBPA and TBBPS caused increases in the fluidity of the erythrocyte membrane in their hydrophilic layer, and conformational changes to membrane proteins. They also caused thiol group elevation, an increase in lipid peroxidation (TBBPS only) and decreases in the level of ATP in cells. They also caused changes in the size and shape of RBCs. TBBPA caused changes in the erythrocyte membrane at lower concentrations compared to TBBPS at an occupational exposure level.

Keywords: tetrabromobisphenol A; tetrabromobisphenol S; erythrocyte membrane; retardants; erythrocytes

1. Introduction

Tetrabromobisphenol A (TBBPA) is a compound belonging to the group of brominated flame retardants (BFRs). These compounds have been used since the 1970s to reduce the flammability of plastics in many consumer products, such as household articles, furniture, mattresses (including products for babies), textiles, insulation and electronic equipment housings [1]. Currently, TBBPA, produced in the amount of over 200 thousand tons per year, is the most important among that group of compounds [2]. It is worth noting that the production of TBBPA accounts for approximately 60% of all BFRs used, and it is not subject to monitoring or restrictions. This is largely due to the fact that most TBBPA (approximately

90%) is used as a reactive compound, i.e., covalently bonded to a polymer matrix, which limits the possibility of its migration to the environment, but the rest is used in an additive form, which can be released from the product much more easily [3].

TBBPA's widespread use has contributed to the contamination of the environment. TBBPA was found in environmental samples such as soil, water and air [4–7]. It was also found in the air and dust of residential interiors and offices, which significantly contributes to human exposure [8,9]. TBBPA, due to its wide presence in the natural environment and direct human environment, may pose a significant risk to health. Due to the high hydrophobicity of the compound, it can bioaccumulate in living organisms, including humans [10]. Numerous studies document the presence of TBBPA in adipose tissue, milk and serum of mothers, and also in neonatal serum, which is particularly disturbing [10,11]. Currently, more and more publications are appearing indicating the potential toxicity of high doses of TBBPA for mammals. It was found that this compound may be involved in the development of many diseases, such as diabetes, or participate in the neoplastic process [12–14].

Tetrabromobisphenol S (TBBPS) is increasingly being used as an alternative to TBBPA. These compounds have very similar molecular structures, with one difference: in TBBPS, sulfone groups are present instead of the methyl groups of TBBPA. The use of this compound is supported by the presence of a sulfone group in the molecule, which may limit its toxicity, and the fact that this compound is characterized by better flame retardancy [15]. Additionally, apart from the fact that TBBPS is used as a flame retardant additive, it is also used as a herbicide, to control selected weeds in the cultivation of cucumbers, tomatoes and white radish [16]. The increasing use of this compound and its derivatives is resulting in greater presence in the ecosystem and the exposure of living organisms to it. The compound was found, among others, in aquatic organisms and blood serum samples from pregnant women in China [17,18]. So far, there are few toxicological studies that could answer the question of the safety of TBBPS as an alternative. However, due to the structural similarity of the two compounds discussed, it seems reasonable to assume that they may have similar potentially adverse effects on living organisms. Therefore, it is necessary to determine and compare the toxicities of TBBPA and TBBPS.

The erythrocyte membrane is a widely accepted research model in the assessment of xenobiotic toxicity. The changes within it correlate with changes in the membranes of other cell types. Damage to the cell membrane may affect its function and contribute to cell death [19]. Changes in the properties of cell membrane may also be involved in the development of many diseases, including anemia, diabetes, heart diseases and cancer. It has been found that TBBPA may contribute to the induction of cancer and may be involved in the development of type II diabetes and obesity [12–14], therefore, the mechanisms of action of TBBPA and TBBPS in the model cell membrane should be investigated. Therefore, the aim of the study was to evaluate the influences of common BFRs, i.e., TBBPA and TBBPS, on the properties of the erythrocyte membrane.

2. Results

2.1. Membrane Fluidity

In RBCs, an increase in fluidity of the hydrophilic layer of the membrane was observed (an increase in the order parameter S). No statistically significant changes were observed in fluidity of deeper regions of the lipid bilayer.

Both compounds caused an increase in the value of the S parameter proportional to the concentration (5-DSA labeling). There were slight statistically significant differences with the lowest tested concentration of TBBPA (1 μg/mL) and with 10 μg/mL of TBBPS. Changes in relation to the control, both of approximately 105%, were recorded for TBBPA at the concentration of 25 μg/mL, and for TBBPS at twice that high concentration. However, no significant changes were observed in the correlation time coefficients τB and τC (16-DSA labeling) (Table 1).

Table 1. Changes in parameter S and correlation times of τB and τC in human control erythrocytes and the erythrocytes incubated with TBBPA at 1 to 25 μg/mL and TBBPS at 1 to 100 μg/mL for 48 h. (*) Significantly different from control ($p < 0.05$).

Compound	Concentration [μg/mL]	Order Parameter S [%]	Correlation Time τ_B [%]	Correlation Time τ_C [%]
TBBPA	0	100.00 ± 0.009	100.00 ± 0.444	100.00 ± 0.476
	1	101.86 ± 0.003 *	103.75 ± 0.691	104.92 ± 0.925
	10	103.01 ± 0.002 *	104.79 ± 0.823	103.95 ± 1.021
	15	103.95 ± 0.001 *	105.24 ± 0.727	103.70 ± 1.387
	25	105.45 ± 0.009 *	105.05 ± 0.740	104.32 ± 0.887
TBBPS	0	100.00 ± 0.009	100.00 ± 1.080	100.00 ± 1.305
	1	102.57 ± 0.010	100.47 ± 0.231	102.37 ± 0.925
	10	104.45 ± 0.009 *	100.27 ± 0.433	101.32 ± 0.774
	50	105.99 ± 0.008 *	104.18 ± 0.735	104.40 ± 0.455
	100	108.33 ± 0.021 *	106.93 ± 0.903	108.38 ± 1.330

2.2. W/S Ratio

It was found that TBBPA caused a statistically significant decrease in mobility of the attached marker at the lowest concentration; changes ranged from 92.7 to 84.1% (for 1 μg/mL and 25 μg/mL, respectively) compared to the control (100%) (Figure 1). On the other hand, TBBPS caused statistically significant increases in the mobility of the attached marker at the concentrations of 50 and 100 μg/mL (112.52 and 116.53%, respectively) (Figure 2).

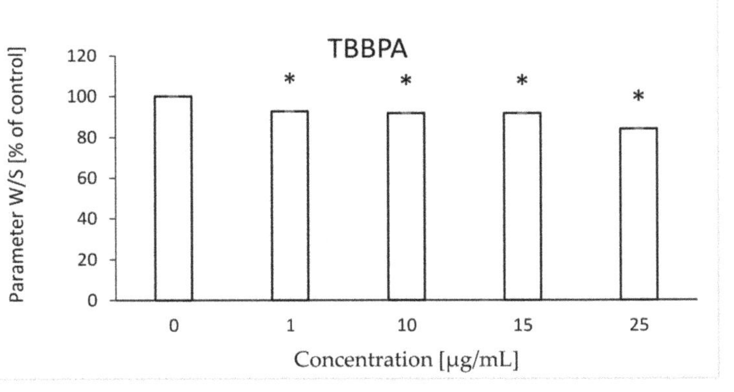

Figure 1. Changes in W/S in human control erythrocytes and the erythrocytes incubated with TBBPA at 1 to 25 μg/mL for 48 h. (*) Significantly different from control ($p < 0.05$).

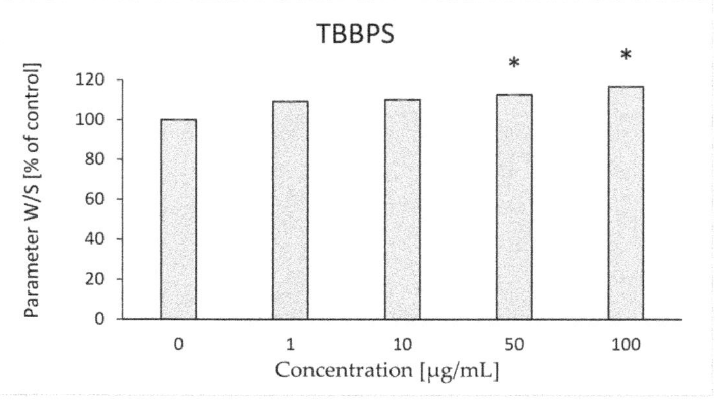

Figure 2. Changes in W/S in human control erythrocytes and the erythrocytes incubated with TBBPS at 1 to 100 μg/mL for 48 h. (*) Significantly different from control ($p < 0.05$).

2.3. Internal Viscosity of Erythrocytes

It was observed that the compounds slightly increased the intrinsic viscosity of RBCs, but these changes were statistically insignificant (Table 2).

Table 2. Internal viscosity of human control erythrocytes and the erythrocytes incubated with TBBPA at 1 to 25 μg/mL incubated and TBBPS at 1 to 100 μg/mL for 48 h. (*) Significantly different from control ($p < 0.05$).

Compound	Concentration [μg/mL]	Internal Viscosity [%]
TBBPA	0	100.00 ± 0.004
	1	103.32 ± 0.004
	10	104.22 ± 0.005
	15	105.19 ± 0.004
	25	106.40 ± 0.004
TBBPS	0	100.00 ± 0.003
	1	99.42 ± 0.004
	10	99.53 ± 0.007
	50	103.37 ± 0.006
	100	107.50 ± 0.007

2.4. Thiol Groups

The level of thiol groups in erythrocyte membrane was assessed after 12 h of incubation with the analyzed compounds. It was found that both TBBPA and TBBPS caused statistically significant increases in that parameter. TBBPA at the concentration of 25 μg/mL increased the level of thiol groups by 8% compared to the control (Figure 3). Statistically significant changes for TBBPS were observed at concentrations of 50 and 100 μg/mL (by 5 and 12% compared to the control, respectively) (Figure 4).

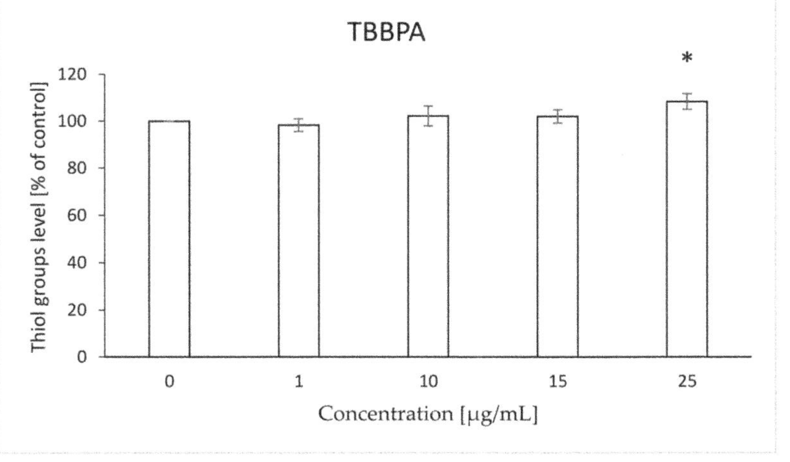

Figure 3. Changes in the thiol groups level in human control erythrocytes and the erythrocytes incubated with TBBPA at 1 to 25 μg/mL for 12 h. (*) Significantly different from control ($p < 0.05$).

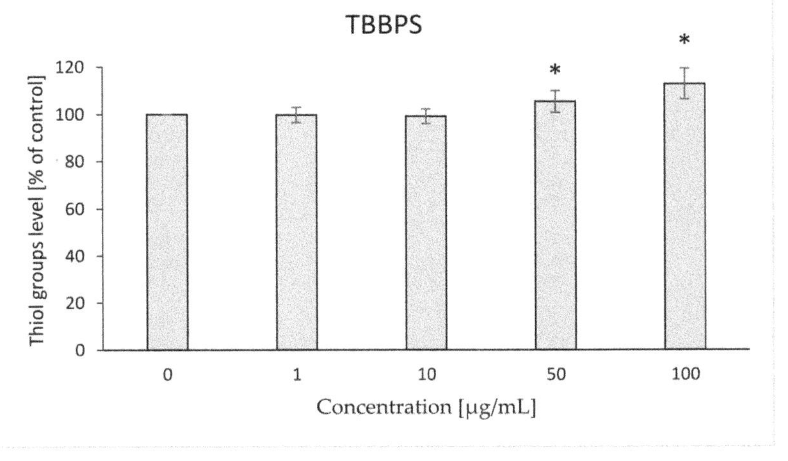

Figure 4. Changes in the thiol groups level in human control erythrocytes and the erythrocytes incubated with TBBPS at 1 to 100 μg/mL for 12 h. (*) Significantly different from control ($p < 0.05$).

2.5. ATP Level

It was found that the tested bromobisphenols reduced the level of intracellular ATP. TBBPA caused statistically significant decreases in ATP level compared to the control at concentrations of 15 and 25 μg/mL—78 and 42%, respectively (Figure 5). In the case of TBBPS, statistically significant decreases in the discussed parameter in relation to the control were observed at the concentrations of 50 and 100 μg/mL—60 and 58%, respectively (Figure 6).

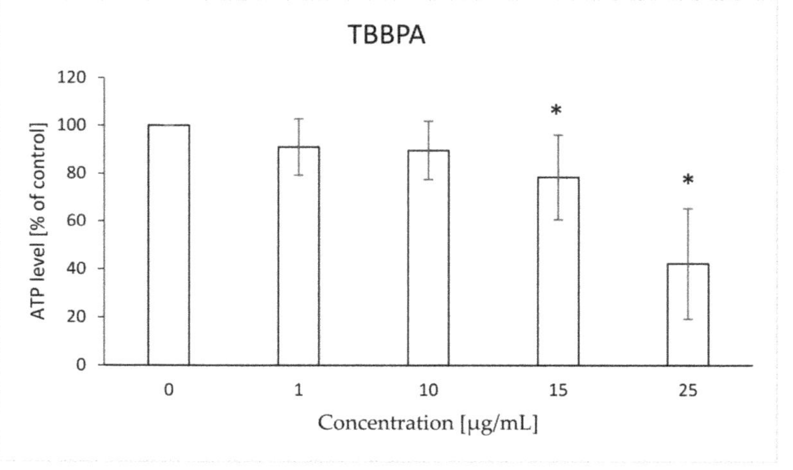

Figure 5. Changes in ATP level in human control erythrocytes and the erythrocytes incubated with TBBPA at 1 to 25 μg/mL for 12 h. (*) Significantly different from control ($p < 0.05$).

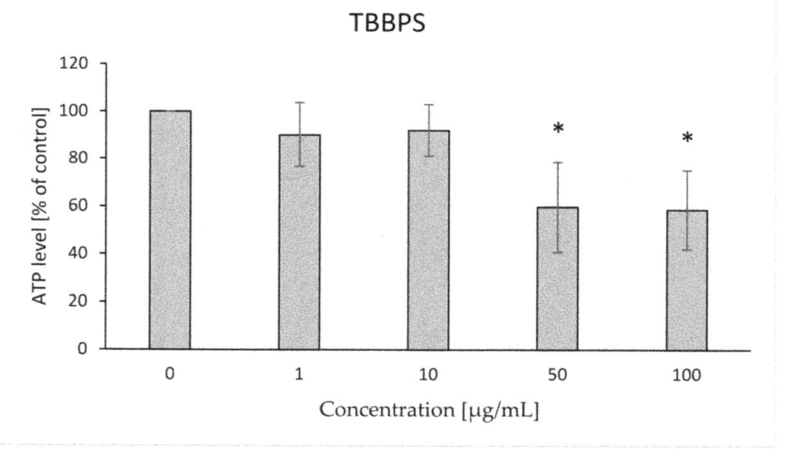

Figure 6. Changes in ATP level in human control erythrocytes and the erythrocytes incubated with TBBPS at 1 to 100 μg/mL for 12 h. (*) Significantly different from control ($p < 0.05$).

2.6. Lipid Peroxidation

After 48 h of incubation of RBCs with the analyzed compoundsin the case of TBBPA, no statistically significant changes were found (Figure 7) within the range of pre-hemolytic concentrations. Statistically significant increases in lipid peroxidation under the influence of TBBPS were observed for concentrations of 50, 100 and 250 μg/mL (up to 120, 135, 145%, respectively) (Figure 8).

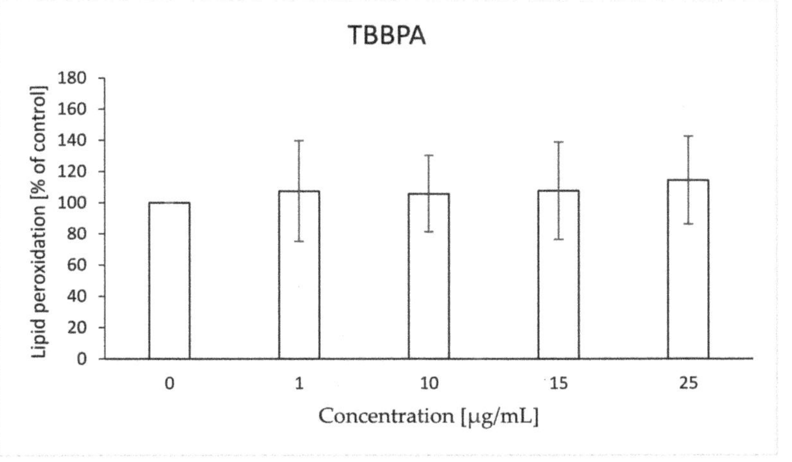

Figure 7. Lipid peroxidation in human control erythrocytes and the erythrocytes incubated with TBBPA at 1 to 25 μg/mL for 48 h.

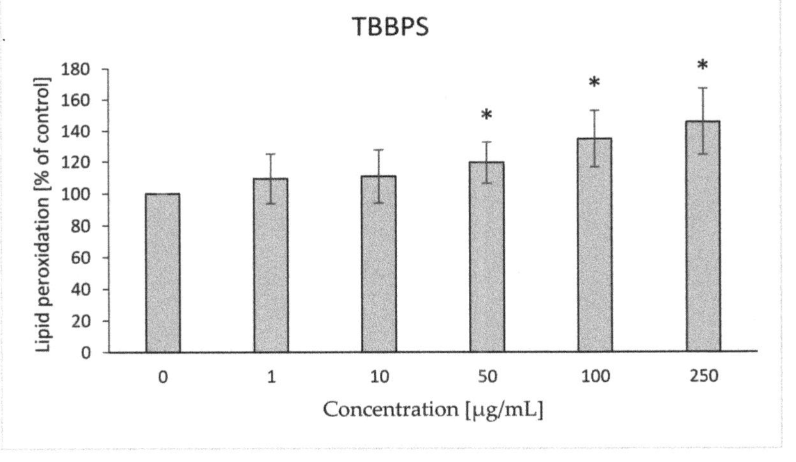

Figure 8. Lipid peroxidation in human control erythrocytes and the erythrocytes incubated with TBBPS at 1 to 250 μg/mL for 48 h. (*) Significantly different from control ($p < 0.05$).

2.7. Osmotic Fragility

It was observed that TBBPA at the lowest concentration (1 μg/mL) caused a slight increase in osmotic resistance, and the highest increases of this parameter were found for the concentration of 15 and 25 μg/mL. At the concentration of NaCl equal to 0.52%, decreases in RBC hemolysis were found to be 6.09 and 5.57%, with the control value of 18.18% (Figure 9). It was shown that in the presence of TBBPA there was a statistically significant decrease in the IC_{50} value for NaCl (Figure 10). On the other hand, there was no effect of TBBPS on the RBC osmotic resistance under the influence of various concentrations of NaCl, nor were there changes in the IC_{50} value (Figures 11 and 12).

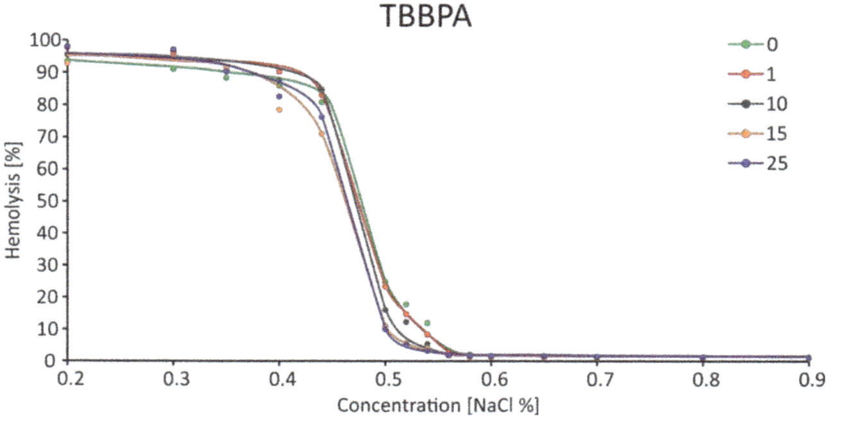

Figure 9. Changes in osmotic resistance of human erythrocytes incubated with TBBPA at 1 to 25 μg/mL for 3 h.

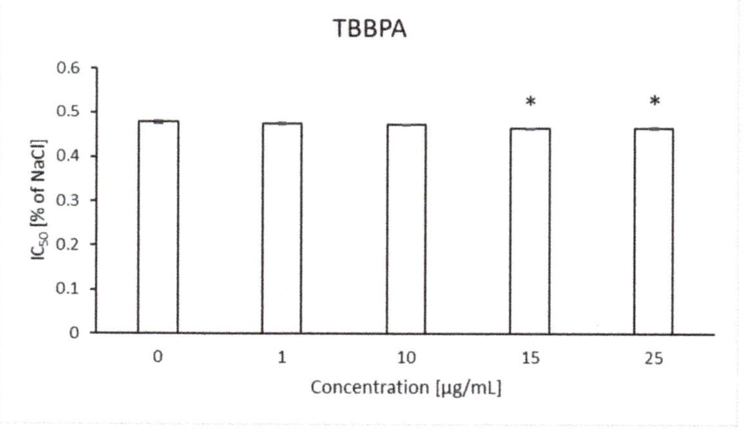

Figure 10. IC_{50} parameter for control erythrocytes and erythroctes incubated with TBBPA at 1–25 μg/mL for 3 h. (*) Significantly different from control ($p < 0.05$).

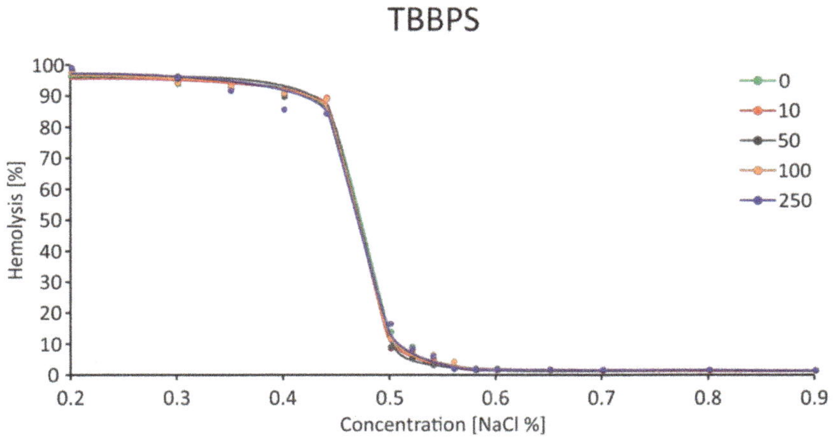

Figure 11. Changes in osmotic resistance of human erythrocytes incubated with TBBPS at 1 to 250 μg/mL for 3 h.

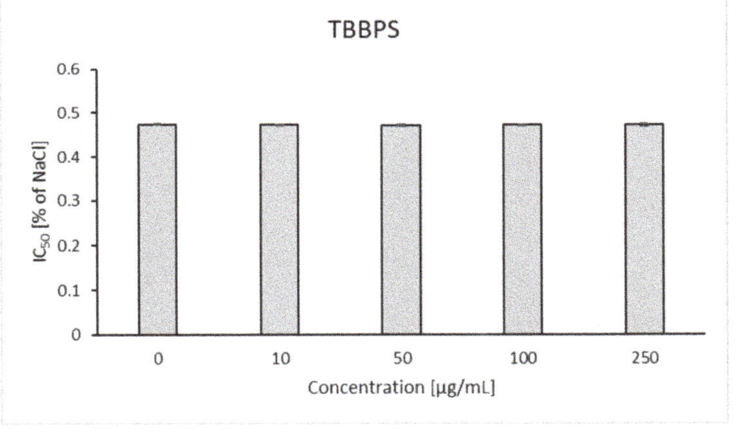

Figure 12. IC_{50} parameter for control erythrocytes and erythroctes incubated with TBBPS at 10–250 μg/mL for 3 h.

2.8. Morphological Changes of Erythrocytes, FSC and SSC Parameter

The analyzed compounds after 48 h of incubation with RBCs caused changes in the FSC and SSC parameters that can be used to assess the size and shape of the cell. After 48 h of RBC incubation with bromobisphenols both compounds increased the FSC parameter, compared to control erythrocytes. There was a statistically significant increase in FSC caused by TBBPA at 25 μg/mL (122%) (Figure 13) and TBBPS at 50 and 100 μg/mL (106, 108%) (Figure 14).

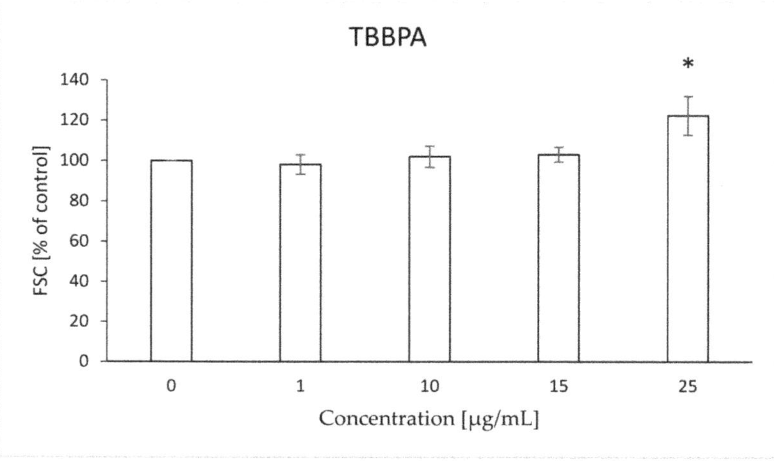

Figure 13. Changes in the FSC parameter in human control erythrocytes and the erythrocytes incubated with TBBPA at 1 to 25 μg/mL for 48 h. (*) Significantly different from control ($p < 0.05$).

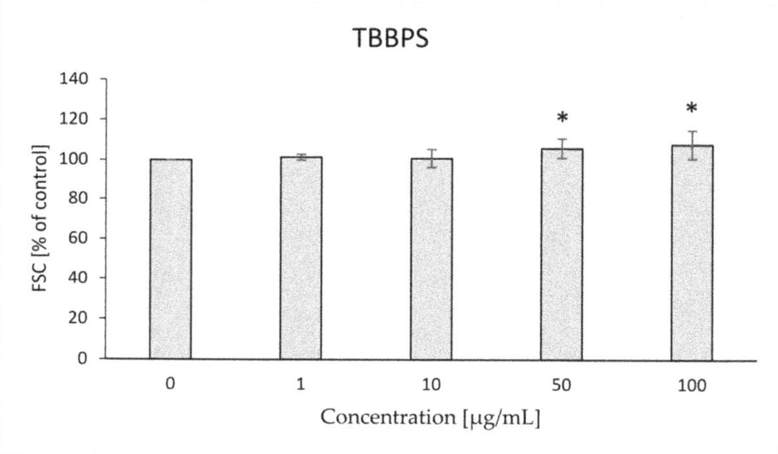

Figure 14. Changes in the FSC parameter in human control erythrocytes and the erythrocytes incubated with TBBPS at 1 to 100 μg/mL for 48 h. (*) Significantly different from control ($p < 0.05$).

In the case of SSC, it was found that TBBPA caused a slight statistically significant increase of the parameter at concentrations of 10 and 15 μg/mL (103, 104%, respectively), while at the concentration of 25 μg/mL the compound caused a significant decrease in relation to the control (71%) (Figure 15). In the case of TBBPS, no statistically significant changes were found (Figure 16) within the range of presented concentrations. The histograms and SSC/FSC dot plots of control and TBBPA and TBBPS in the final concentration were found in Figure 17.

2.9. Microscopic Analysis

Microscopic analysis confirmed that 48 h incubation of TBBPA and TBBPS induced morphological changes within cells. Pictures were taken for TBBPA at concentrations of

25 µg/mL, and for TBBPS at concentrations of 100 µg/mL. The discussed compounds induced the formation of echinocytes (Figure 18).

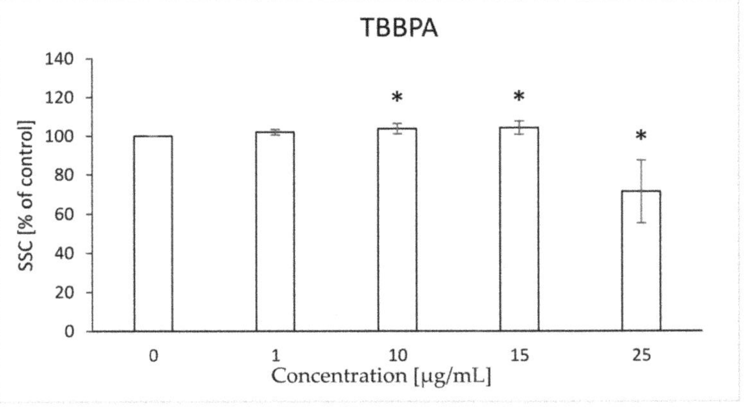

Figure 15. Changes in the SSC parameter in human control erythrocytes and the erythrocytes incubated with TBBPA at 1 to 25 µg/mL for 48 h. (*) Significantly different from control ($p < 0.05$).

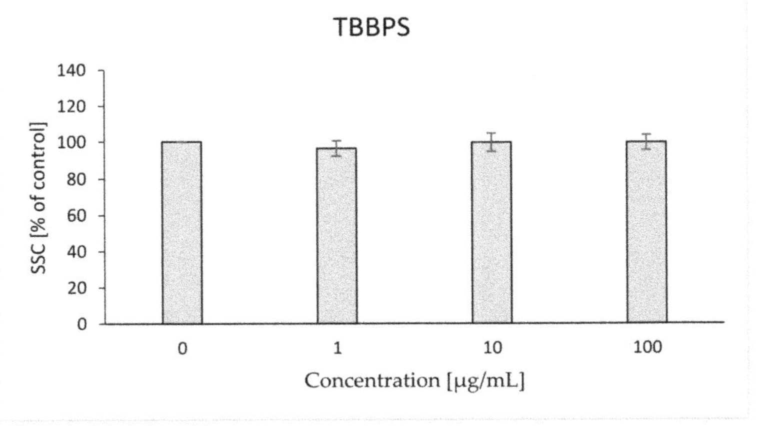

Figure 16. Changes in the SSC parameter in human control erythrocytes and the erythrocytes incubated with TBBPS at 1 to 100 µg/mL for 48 h.

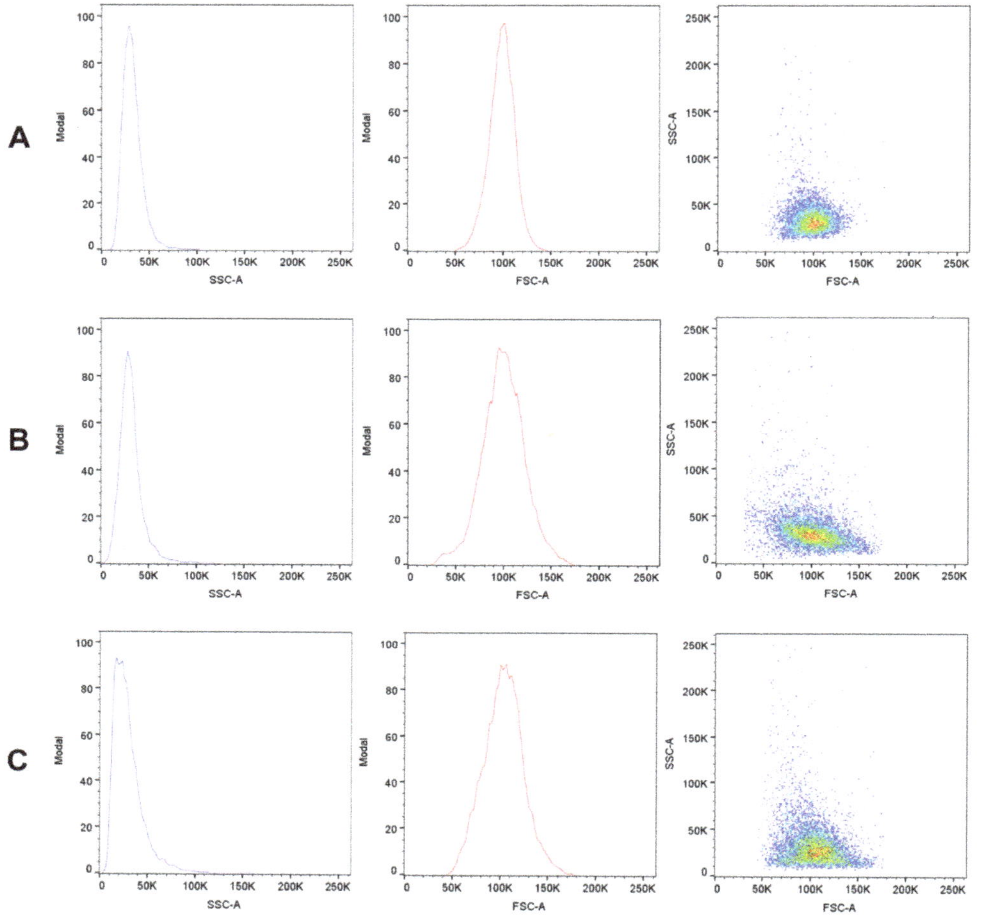

Figure 17. Scattering diagrams of human control erythrocytes (**A**), erythrocytes incubated with TBBPS at 100 μg/mL (**B**) and erythrocytes incubated with TBBPA at 25 μg/mL (**C**) for 48 h. The FSC-A diagrams represent the light scattered near the forward direction (proportional to the value of the particles). The SSC-A diagrams represent scattering at a right angle (depended on cell shape and internal properties). The FSC-A/SSC-A diagram is a dual parameter contour plot proportional to the total cell diversity.

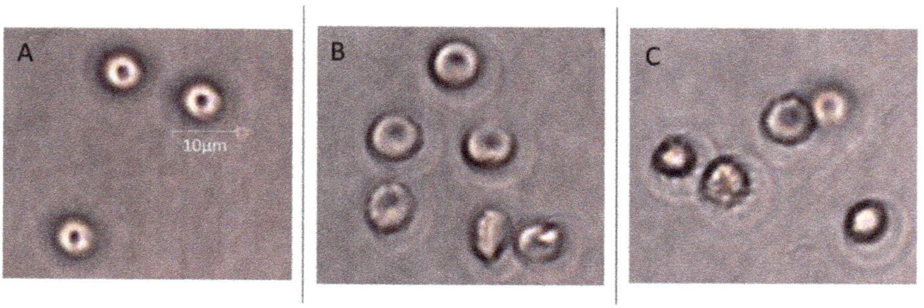

Figure 18. Micrographs of human control erythrocytes (**A**) and the erythrocytes incubated with TBBPA at 25 μg/mL (**B**) and TBBPS at 100 μg/mL (**C**) for 48 h.

3. Discussion

The protein and lipid components of the RBC membrane create a complex structure of interrelationships enabling the maintenance of the proper shape and physiology. Disturbances in these membrane components may cause changes in the shape and deformation capacity of erythrocytes, which may contribute to disturbances in their function and shorten their lifespan [19]. In this study, we assessed the effects of TBBPA and TBBPS on the erythrocyte membrane, which is the first barrier to xenobiotics entering circulation.

One of the properties of a biological membrane, resulting from its structure and interactions between its components, is fluidity [20]. Using the electron paramagnetic resonance (EPR) method, the placement of 5-DSA and 16-DSA probes in the erythrocyte membrane exposed to TBBPA and TBBPS was assessed. These probes diffuse into the environment of the lipid bilayer of biological membranes: 5-DSA is located with the hydrocarbon chains, and 16-DSA is located deeper, in the middle of the lipid bilayer. Under the influence of TBBPA or TBBPS, a statistically significant increase in fluidity of the hydrophilic layer of the RBC membrane was observed (as evidenced by an increased order parameter S). An upward trend was also observed in the deeper regions of the lipid bilayer, but it was statistically insignificant (Table 1). It can be assumed that the analyzed compounds, due to the presence of bromine atoms in them (large in size), initially localized in the shallower regions of the lipid bilayer. The inverse ability to locate in the membrane was demonstrated by Maćczak et al. (2017) in their studies on the effects of bisphenol A and its analogues on RBCs [21]. These authors found that bisphenols did not affect the S parameter, and therefore, did not localize in the hydrophilic layer, but changed the relaxation times τB and τC, which indicates penetration of compounds into the hydrophobic layer, which is explained by the high hydrophobicity of the analyzed compounds [21]. Perhaps the additional presence of bromine atoms in these compounds makes it difficult for them to penetrate deeper regions to the level of carbon 16, where bisphenols devoid of bromine atoms were located.

The W/S parameter determines the state of the internal conformation of membrane proteins, so it is a very sensitive parameter determining changes in their properties. This parameter was also tested with the use of the EPR method by using the MSL spin marker covalently binding to the sulfhydryl groups (-SH) of cytoskeleton proteins, mainly spectrin and actin. It was found that TBBPA, starting from the concentration of 1 μg/mL, caused a decrease in the mobility of the attached marker, and TBBPS caused an increase starting from the concentration of 50 μg/mL (Figures 1 and 2). An increase in the W/S ratio may indicate conformational changes in the structures of membrane proteins that lead to higher exposure of thiol groups to chemical reactions, and/or may contribute to disulfide bond breakage. On the other hand, a decreased W/S ratio may indicate oxidation of thiol groups by reactive oxygen species (ROS), which reduces their availability for the marker. The increase in ROS level under the influence of TBBPA, even at very low concentrations (0.001 μg/mL), was confirmed in another paper [22]. A decrease in W/S may also indicate formation of protein aggregates, which may also be caused by ROS [23,24]. Moreover, we found that the analyzed compounds contributed to increases in the level of thiol groups (TBBPA from 25 μg/mL, and TBBPS from 50 μg/mL) (Figures 3 and 4), which may partially confirm the influences of TBBPA and TBBPS on the conformation of membrane proteins caused by interactions with thiol groups. Pocernich et al. (2001) also found that lipid peroxidation products could covalently bind to cysteine, lysine or histidine, which may also result in conformational changes of membrane proteins [25]. In the case of our research, a statistically significant increase in lipid peroxidation was found only after incubation of RBCs with TBBPS at the highest concentration (Figure 8). Moreover, in the previous paper [26] we found that TBBPA (at the concentration of 25 μg/mL) and TBBPS (at the concentration of 250 μg/mL) caused decreases in tryptophan fluorescence in erythrocyte membranes, which also confirms that these compounds may damage membrane proteins via ROS.

Changes in the viscosity of the interior of RBCs may be associated with changes in their shape. Although we observed upward trends in the assessment of intrinsic viscosity of erythrocytes, these changes were not statistically significant (Table 2). On the other hand, it was found that the analyzed compounds induced changes in the forward scatter channel (FSC) and side scatter channel (SSC) parameters obtained from flow cytometric analysis (Figures 13–17), which may reflect changes in the size, shape and external structure of the membranes of the investigated cells. These changes are also visible in the pictures achieved from a phase-contrast microscopic examination (Figure 18). An increase in the size of erythrocytes and changes in their shape are usually associated with damage done to the cell membrane, and may result from influx of water into the cell and/or incorporation of compounds into the structure of the membrane. It is also known that cytoskeleton proteins and integral membrane proteins are responsible for maintaining the shape of RBCs. Various xenobiotics, including phenols, can damage erythrocyte proteins [27,28], resulting in conversion of a normal discocyte into an echinocyte or a stomatocyte [29]. In the case of compounds analyzed by us, formation of echinocytes may be conditioned by incorporation of TBBPA or TBBPS into the hydrophilic region of the membrane, as indicated by changes in the S parameter. A decrease in the ATP level in the cell may also contribute to changes in the shape of the cell, and thus to a decrease in its survival [30]. We observed that both TBBPA and TBBPS significantly decreased cellular ATP levels—the highest concentrations by nearly 60% with TBBPA (25 μg/mL), and more than 40% when incubated with TBBPS at 100 μg/mL.

Shape changes can also be associated with water loss, increased intracellular viscosity and decreased osmotic resistance in RBCs. In the case of our research, slight increases in osmotic resistance were observed when RBCs were incubated with TBBPA at concentrations of 15 and 25 μg/mL (Figure 9), which could also have been related to incorporation of the compound into the membrane and its partial stiffening, which would result in reduced susceptibility to hemolysis [29,31]. Moreover, in the case of this compound, the IC_{50} value decreased as the concentration increased, which may confirm the above observations.

4. Materials and Methods

4.1. Chemicals

TBBPA (purity 99%, 2,6-dibromo-4-[2-(3,5-dibromo-4-hydroxyphenyl)propan-2-yl]phenol)) was purchased from LGC Standards (Wesel, Germany). Tetrabromobisphenol S (purity 98.8%, 2,6-dibromo-4-(3,5-dibromo-4-hydroxyphenyl)sulfonylphenol) was synthetized in the Institute of Industrial Organic Chemistry in Warsaw, Poland. DMSO (99.5%), 16-doxylstearic acid (16-DSA), 4-N-maleimide-2,2,6,6-tetramethylopiperidine-1-oxyl (MSL), 2,2,6,6-tetramethyl piperidine-Noxyl-4-amine (TEMPAMINE) and ouabain were bought from Sigma-Aldrich (Merck, Kenilworth, NJ, USA). 5-Doxylstearic acid (5-DSA) was bought from Santa Cruz Biotechnology (Dallas, TX, USA). ATP Determination Kit was purchased from Thermo Fisher Scientific (Waltham, MA, USA). Ethylenediaminetetraacetic acid tetrasodium salt (EDTA), tris (hydroxymethyl)aminomethane (Tris), 5,5-dithiobis-2-nitrobenzoic acid (DTNB), sodium dodecyl sulfate (SDS), phenylmethylsulfonyl fluoride (PMSF) and other chemicals were obtained from Carl Roth (Roth, Germany), POCh, (Gliwice, Poland) or Alfachem (Lublin, Poland).

4.2. Erythrocyte and Erythrocyte's Membranes Isolation

RBCs were isolated from leukocyte-buffy coat separated from blood from healthy donors from the Regional Centre of Blood Donation and Blood Treatment (Lodz, Poland).

The RBCs' isolation and treatment procedure was previously described by Jarosiewicz et al. (2017) [32]. RBCs with a hematocrit of 5% (about 630 mln cells x mL^{-1}) were incubated with the analyzed compounds at concentrations ranging from 1 to 25 μg/mL for TBBPA and 1–250 μg/mL for TBBPS, at 37 °C for 48, 12 or 3 h, depending on the experiment. Differences in the concentrations of the compounds studied were dictated by hemolytic properties of BFRs tested. Compounds were dissolved in DMSO (to final concentration of 0.4%). Concentrations of the compounds were selected on the basis of their hemolytic abilities described in the previous paper [32]. Moreover, in the case of the 48 h incubation period, an additional antibiotic was used (0.2% streptomycin and penicillin). Appropriate controls were performed to exclude the effect of antibiotic and DMSO on RBCs. The exact conditions of incubation are described in the article by Jarosiewicz et al., 2020 [26].

Isolation of RBCs membranes followed the incubation of the RBCs with the analyzed compounds. Isolation of RBC membranes was carried out using the Dodge et al. (1963) method with some modifications [33]. The exact isolation procedure was described in the previous paper by Jarosiewicz et al., 2020 [26].

The research was approved by the Bioethics Committee of the University of Lodz No. 7/KBBN-UŁ/II/2015.

4.3. Membrane Fluidity

The RBC's membrane fluidity was analyzed by electron paramagnetic resonance (EPR) spectroscopy (Brucker 300 Spectrometer, Ettlingen, Germany) using spin labeled fatty acids: 5-doxylstearic acid (5-DSA) and 16-doxylstearic acid (16-DSA). From the EPR spectra obtained for the 5-DSA spin label, the ordering parameter S was calculated, and the correlation times τB and τC were calculated for the 16-DSA spin label. Order parameter S and the correlation times τB and τC were calculated as described in the study of Koter et al. (2004) [34].

4.4. W/S Ratio

Parameter W/S was determined using a spin label MSL, which covalently binds proteins and analyzed by EPR spectroscopy (Brucker 300 Spectrometer, Ettlingen, Germany). The exact procedure for performing the experiment is described in the article by Maćczak et al. (2017) [21].

4.5. Internal Viscosity

The TEMPAMINE spin label was used to determine the intracellular environment of RBCs [35]. The analysis was conducted using Brucker 300 Spectrometer (Ettlingen, Germany). The changes in the parameter studied were calculated and expressed as percentages of control. The exact procedure for performing the experiment is described in the article by Maćczak et al. (2017) [21].

4.6. Thiol Groups Level

The number of thiol groups in the erythrocyte membranes was determined using the method of Ellman et al. (1959) [36]. 5.5′-Dithiobis (2-nitrobenzoic) acid reacts with protein thiol groups. This reaction releases the 5-thio-2-nitrobenzoic anion having an intense yellow color, which is determined spectrophotometrically at 412 nm wavelength. The procedure of determination of the thiol group level was previously described by Maćczak et al. (2017) [21]. Results are expressed as -SH nmol/mg proteins and presented as percentages of control.

4.7. ATP Level

Intracellular ATP level in RBCs is determined by oxidative decarboxylation of luciferin by firefly luciferase in the presence of ATP and magnesium ions with bioluminescence emission. The emission is linearly related to the intracellular ATP concentration [37]. The measurements were made at the wavelength of 590 nm using fluorimeter (Fluoroskan Ascent FL, Thermo Fisher Scientific, Vantaa, Finland). The exact procedure for performing the experiment is described in an article by Maćczak et al. (2017) [21].

4.8. Lipid Peroxidation

Lipid peroxidation in erythrocyte membranes is determined according to the method of Stocks and Dormandy (1971) [38]. Lipid peroxidation is analyzed by measuring of formation of thiobarbituric acid reactive substances (TBARS). The absorbance is determined colorimetrically using BioTek ELx808 reader (Winooski VT, USA) at the wavelength of 532 nm. Lipid peroxidation is expressed in absorbance units of TBARS products and is shown as a percentage of control.

4.9. Osmotic Fragility

The osmotic resistance (fragility) was determined by the method of Dacia and Lewis (1975) [39]. A small number of erythrocytes are placed in a NaCl solution at a concentration of 0.2 to 0.9%. Osmotic resistance is determined by measuring the hemoglobin released from erythrocytes by the colorimetric method using the BioTek ELx808 reader (Winooski VT, USA) at λ = 540 nm. Osmotic resistance is assessed on the basis of the hemolysis curve shift, in the graph of percentage of hemolysis vs. NaCl concentration. Before performing the assay, RBCs were incubated with test compounds for 3 h. The exact procedure for performing the experiment is described in the publication by Maćczak et al. (2017) [21].

4.10. Morphological Changes of Erythrocytes (FSC and SSC Parameter)

The flow cytometry technique was used to assess the size and shape of the erythrocytes (LSR II Becton Dickinson). Data were recorded for a total of 10,000 events per sample. Results are presented as percentages of control. This method was described by Bukowska et al. (2011) [28].

4.11. Microscopic Analysis

Microscopic analysis was completed using the phase contrast microscope (Olympus, Japan) at the magnification of 600×. Images were taken following a 48 h incubation of RBCs with analyzed compounds. After incubation, RBCs were suspended in Ringer's buffer at the final concentration of 0.02%, placed on a Petri dish and pictures were taken.

4.12. Statistical Analysis

Results are presented as means ± standard deviations of 4–6 experiments (blood donors); each experiment performed was the mean of 2–3 replicates. The statistical analysis was described in the previous article by Jarosiewicz et al. (2020) [26].

5. Conclusions

Both TBBPA and TBBPS were found to cause changes in the erythrocyte cell membrane. Both compounds increase the fluidity of the hydrophilic region of the RBC membrane. TBBPA strongly damages proteins (changes in the S, W/S ratio, level of thiol groups, and levels of tryptophan oxidation in membranes and in human albumin, as shown in previous studies) [26]. In our opinion it is the main target of this retardant. It was also shown that TBBPS contributed to lipid peroxidation only at its highest concentration of 250 μg/mL, which may indicate that the peroxidation process will be a secondary process to the induction of ROS and protein oxidation by these compounds [22]. Both compounds also caused changes in the shape and size of erythrocytes, which are associated with damage to the cell membrane, hemolysis and incorporation of these compounds into the structure of the membrane. In addition, the induced decrease in the level of ATP would contribute to a decrease in cell survival. It is worth noting that changes in the structure and function of the cell membrane were observed for significantly lower concentrations in the case of RBC incubation with TBBPA than with TBBPS, occurring only at occupational and not epidemiological exposure. The obtained data indicate a low toxicity of TBBPS only at very high concentrations (in contrast to TBBPA), and therefore, a low toxicological risk posed by this retardant to human erythrocytes.

Author Contributions: Conceptualization, M.J. and B.B.; methodology, B.B. and P.D.; software, M.J. and P.D.; validation, M.J., P.D. and B.B.; formal analysis, M.J., P.D. and B.B.; investigation M.J., P.D. and B.B.; resources, M.J., P.D., B.B. and B.H.; data curation, M.J.; writing—original draft preparation, M.J.; writing—review and editing, M.J., P.D. and B.B.; visualization, M.J., P.D. and P.J.; supervision, B.B.; project administration, M.J.; funding acquisition, M.J. All authors have read and agreed to the published version of the manuscript.

Funding: The research was conducted within the framework of National Science Centre, Poland, Project No. 2017/27/N/NZ7/00548.

Institutional Review Board Statement: The study was conducted according to the guidelines of the Declaration of Helsinki, and approved by the Bioethics Committee of the University of Lodz No. 7/KBBN-UŁ/II/2015.

Informed Consent Statement: The leukocyte-platelets buffy coat also containing erythrocytes was purchased from Blood Bank in Lodz, Poland. All procedures related to blood donation were executed at the Regional Centre of Blood Donation and Blood Treatment in Lodz, Poland. The blood donor recruitment was at the Centre, according to national legal procedures and European Union regulations (incl. the regulation (EU) 2016/679 OF THE EUROPEAN PARLIAMENT AND OF THE COUNCIL of 27 April 2016 on the protection of natural persons regarding the processing of personal data and on the free movement of such data).

Conflicts of Interest: The authors declare no conflict of interest.

Abbreviations

BFRs	brominated flame retardants
TBBPA	tetrabromobisphenol A
TBBPS	tetrabromobisphenol S
ATP	adenosine triphosphate
RBCs	erythrocytes
FSC	forward scatter channel
SSC	side scatter channel
EPR	electron paramagnetic resonance
ROS	reactive oxygen species

References

1. Zhou, X.; Guo, J.; Zhang, W.; Zhou, P.; Deng, J.; Lin, K. Tetrabromobisphenol A contamination and emission in printed circuit board production and implications for human exposure. *J. Hazard. Mater.* **2014**, *273*, 27–35. [CrossRef] [PubMed]
2. Covaci, A.; Voorspoels, S.; Abdallah, M.A.E.; Geens, T.; Harrad, S.; Law, R.J. Analytical and environmental aspects of the flame retardant tetrabromobisphenol-A and its derivatives. *J. Chromatogr. A.* **2009**, *1216*, 346–363. [CrossRef]
3. Lai, D.Y.; Kacew, S.; Dekant, W. Tetrabromobisphenol A (TBBPA): Possible modes of action of toxicity and carcinogenicity in rodents. *Food Chem. Toxicol.* **2015**, *80*, 206–214. [CrossRef] [PubMed]
4. Tang, J.; Feng, J.; Li, X.; Li, G. Levels of flame retardants HBCD, TBBPA and TBC in surface soils from an industrialized region of East China. *Environ. Sci. Proc. Impacts* **2014**, *16*, 1015–1021. [CrossRef]
5. Gorga, M.; Martínez, E.; Ginebreda, A.; Eljarrat, E.; Barceló, D. Determination of PBDEs, HBB, PBEB, DBDPE, HBCD, TBBPA and related compounds in sewage sludge from Catalonia (Spain). *Sci. Total Environ.* **2013**, *444*, 51–59. [CrossRef]
6. Kowalski, B.; Mazur, M. The simultaneous determination of six flame retardants in water samples using SPE pre-concentration and UHPLC-UV method. *Water Air Soil Pollut.* **2014**, *225*, 1–9. [CrossRef] [PubMed]
7. Xie, Z.; Ebinghaus, R.; Lohmann, R.; Heemken, O.; Caba, A.; Püttmann, W. Trace determination of the flame retardant tetrabromobisphenol A in the atmosphere by gas chromatography–mass spectrometry. *Anal. Chim. Acta* **2007**, *584*, 333–342. [CrossRef]
8. Abdallah, M.A.E.; Harrad, S.; Covaci, A. Hexabromocyclododecanes and tetrabromobisphenol—A in indoor air and dust in Birmingham, UK: Implications for human exposure. *Environ. Sci. Technol.* **2008**, *42*, 6855–6861. [CrossRef]
9. Wang, W.; Abualnaja, K.O.; Asimakopoulos, A.G.; Covaci, A.; Gevao, B.; Johnson-Restrepo, B.; Kumosani, T.A.; Malarvannan, G.; Minh, T.B.; Moon, H.-B.; et al. A comparative assessment of human exposure to tetrabromobisphenol A and eight bisphenols including bisphenol A via indoor dust ingestion in twelve countries. *Environ. Int.* **2015**, *83*, 183–191. [CrossRef] [PubMed]
10. Nakao, T.; Akiyama, E.; Kakutani, H.; Mizuno, A.; Aozasa, O.; Akai, Y.; Ohta, S. Levels of tetrabromobisphenol A, tribromobisphenol A, dibromobisphenol A, monobromobisphenol A, and bisphenol A in Japanese breast milk. *Chem. Res. Toxicol.* **2015**, *28*, 722–728. [CrossRef]

11. Kim, U.J.; Oh, J.E. Tetrabromobisphenol A and hexabromocyclododecane flame retardants in infant–mother paired serum samples, and their relationships with thyroid hormones and environmental factors. *Environ. Pollut.* **2014**, *184*, 193–200. [CrossRef]
12. Dunnick, J.K.; Sanders, J.M.; Kissling, G.E.; Johnson, C.L.; Boyle, M.H.; Elmore, S.A. Environmental chemical exposure may contribute to uterine cancer development: Studies with tetrabromobisphenol A. *Toxicol. Pathol.* **2015**, *43*, 464–473. [CrossRef]
13. McCollum, C.W.; Riu, A. Obesity: An Effect of Environmental Pollutants? In Proceedings of the Endocrine Society's 94th Annual Meeting and Expo, Houston, TX, USA, 23–26 June 2012.
14. Barret, J. Warm Reception? *Halogenated BPA Flame Retardants and PPARγ Activation. Environ. Health Perspect.* **2011**, *119*, 398.
15. Qu, G.; Liu, A.; Hu, L.; Liu, S.; Shi, J.; Jiang, G. Recent advances in the analysis of TBBPA/TBBPS, TBBPA/TBBPS derivatives and their transformation products. *Trends Anal. Chem.* **2016**, *83*, 14–24. [CrossRef]
16. Xu, H.; Li, Y.; Lu, J.; Lu, J.; Zhou, L.; Chovelon, J.M.; Ji, Y. Aqueous photodecomposition of the emerging brominated flame retardant tetrabromobisphenol S (TBBPS). *Environ. Pollut.* **2021**, *271*, 116406. [CrossRef] [PubMed]
17. Liu, A.; Shi, J.; Shen, Z.; Lin, Y.; Qu, G.; Zhao, Z.; Jiang, G. Identification of unknown brominated bisphenol s congeners in contaminated soils as the transformation products of tetrabromobisphenol S derivatives. *Environ. Sci. Technol.* **2018**, *52*, 10480–10489. [CrossRef] [PubMed]
18. Li, A.; Zhuang, T.; Shi, W.; Liang, Y.; Liao, C.; Song, M.; Jiang, G. Serum concentration of bisphenol analogues in pregnant women in China. *Sci. Total Environ.* **2020**, *707*, 136100. [CrossRef]
19. Farag, M.R.; Alagawany, M. Erythrocytes as a biological model for screening of xenobiotics toxicity. *Chem.-Biol. Interact.* **2018**, *279*, 73–83. [CrossRef] [PubMed]
20. Duchnowicz, P.; Pilarski, R.; Michałowicz, J.; Bukowska, B. Changes in Human Erythrocyte Membrane Exposed to Aqueous and Ethanolic Extracts from Uncaria tomentosa. *Molecules* **2021**, *26*, 3189. [CrossRef]
21. Maćczak, A.; Duchnowicz, P.; Sicińska, P.; Koter-Michalak, M.; Bukowska, B.; Michałowicz, J. The in vitro comparative study of the effect of BPA, BPS, BPF and BPAF on human erythrocyte membrane; perturbations in membrane fluidity, alterations in conformational state and damage to proteins, changes in ATP level and Na^+/K^+ ATPase and AChE activities. *Food Chem. Toxicol.* **2017**, *110*, 351–359. [CrossRef]
22. Arosiewicz, M.; Michałowicz, J.; Bukowska, B. In vitro assessment of eryptotic potential of tetrabromobisphenol A and other bromophenolic flame retardants. *Chemosphere* **2019**, *215*, 404–412. [CrossRef]
23. Liguori, I.; Russo, G.; Curcio, F.; Bulli, G.; Aran, L.; Della-Morte, D.; Gargiulo, G.; Testa, G.; Cacciatore, F.; Bonaduce, D.; et al. Oxidative stress, aging, and diseases. *Clin. Interv. Aging* **2018**, *13*, 757. [CrossRef] [PubMed]
24. Krisko, A.; Radman, M. Protein damage, ageing and age-related diseases. *Open Biol.* **2019**, *9*, 180249. [CrossRef]
25. Pocernich, C.B.; Cardin, A.L.; Racine, C.L.; Lauderback, C.M.; Butterfield, D.A. Glutathione elevation and its protective role in acrolein-induced protein damage in synaptosomal membranes: Relevance to brain lipid peroxidation in neurodegenerative disease. *Neurochem. Int.* **2001**, *39*, 141–149. [CrossRef]
26. Jarosiewicz, M.; Miłowska, K.; Krokosz, A.; Bukowska, B. Evaluation of the Effect of Selected Brominated Flame Retardants on Human Serum Albumin and Human Erythrocyte Membrane Proteins. *Int. J. Mol. Sci.* **2020**, *21*, 3926. [CrossRef] [PubMed]
27. Bukowska, B. Toxicity of 2, 4-Dichlorophenoxyacetic Acid-Molecular Mechanisms. *Pol. J. Environ. Stud.* **2006**, *15*, 365–374.
28. Bukowska, B.; Michałowicz, J.; Wojtaszek, A.; Marczak, A. Comparison of the effect of phenoxyherbicides on human erythrocyte membrane (in vitro). *Biologia* **2011**, *66*, 379–385. [CrossRef]
29. Stasiuk, M.; Kijanka, G.M.; Kozubek, A. Transformations of erythrocytes shape and its regulation. *Postepy Biochem.* **2009**, *55*, 425–433. [PubMed]
30. Michałowicz, J. Pentachlorophenol and its derivatives induce oxidative damage and morphological changes in human lymphocytes (in vitro). *Arch. Toxicol.* **2010**, *84*, 379–387. [CrossRef]
31. Sheetz, M.P.; Singer, S.J. Biological membranes as bilayer couples. A molecular mechanism of drug-erythrocyte interactions. *Proc. Natl. Acad. Sci. USA* **1974**, *71*, 4457–4461. [CrossRef]
32. Jarosiewicz, M.; Duchnowicz, P.; Włuka, A.; Bukowska, B. Evaluation of the effect of brominated flame retardants on hemoglobin oxidation and hemolysis in human erythrocytes. *Food Chem. Toxicol.* **2017**, *109*, 264–271. [CrossRef]
33. Dodge, J.T.; Mitchell, C.; Hanahan, D.J. The preparation and chemical characteristics of hemoglobin-free ghosts of human erythrocytes. *Arch. Biochem. Biophys.* **1963**, *100*, 119–130. [CrossRef]
34. Koter, M.; Franiak, I.; Strychalska, K.; Broncel, M.; Chojnowska-Jezierska, J. Damage to the structure of erythrocyte plasma membranes in patients with type-2 hypercholesterolemia. *Int. J. Biochem. Cell Biol.* **2004**, *36*, 205–215. [CrossRef]
35. Morse, P.D. Use of the spin label tempamine for measuring the internal viscosity of red blood cells. *Biochem. Biophys. Res. Commun.* **1977**, *77*, 1486–1491. [CrossRef]
36. Ellman, G. Tissue sulfhydryl groups. *Arch. Biochem. Biophys.* **1959**, *82*, 70–77. [CrossRef]
37. Stanly, P.; Williams, S. Use of the liquid scintillation spectrometer for determining adenosine triphosphate by the luciferase enzyme. *Anal. Biochem.* **1969**, *29*, 381–392. [CrossRef]
38. Stocks, J.; Dormandy, T.L. The autoxidation of human red cell lipids induced by hydrogen peroxide. *Br. J. Haematol.* **1971**, *20*, 95–111. [CrossRef] [PubMed]
39. Dacia, J.V.; Lewis, S.M. Practical Hematology, 5th ed.Churchill Livingston: Edingurgth, UK; London, UK; New York, NY, USA, 1975; pp. 32–34.

International Journal of
Molecular Sciences

MDPI

Article

Natural Mutations Affect Structure and Function of gC1q Domain of Otolin-1

Rafał Hołubowicz *, **Andrzej Ożyhar** and **Piotr Dobryszycki** *

Department of Biochemistry, Molecular Biology and Biotechnology, Faculty of Chemistry, Wrocław University of Science and Technology, Wybrzeże Wyspiańskiego 27, 50-370 Wrocław, Poland; andrzej.ozyhar@pwr.edu.pl

* Correspondence: rafal.holubowicz@pwr.edu.pl (R.H.); piotr.dobryszycki@pwr.edu.pl (P.D.); Tel.: +48-71-320-63-34 (R.H.); +48-71-320-63-32 (P.D.)

Abstract: Otolin-1 is a scaffold protein of otoliths and otoconia, calcium carbonate biominerals from the inner ear. It contains a gC1q domain responsible for trimerization and binding of Ca^{2+}. Knowledge of a structure–function relationship of gC1q domain of otolin-1 is crucial for understanding the biology of balance sensing. Here, we show how natural variants alter the structure of gC1q otolin-1 and how Ca^{2+} are able to revert some effects of the mutations. We discovered that natural substitutions: R339S, R342W and R402P negatively affect the stability of apo-gC1q otolin-1, and that Q426R has a stabilizing effect. In the presence of Ca^{2+}, R342W and Q426R were stabilized at higher Ca^{2+} concentrations than the wild-type form, and R402P was completely insensitive to Ca^{2+}. The mutations affected the self-association of gC1q otolin-1 by inducing detrimental aggregation (R342W) or disabling the trimerization (R402P) of the protein. Our results indicate that the natural variants of gC1q otolin-1 may have a potential to cause pathological changes in otoconia and otoconial membrane, which could affect sensing of balance and increase the probability of occurrence of benign paroxysmal positional vertigo (BPPV).

Keywords: analytical ultracentrifugation; C1q; calcium binding proteins; circular dichroism; genetic variation; otoconia; otolin-1; OTOL1; site-directed mutagenesis; thermal shift assay

Citation: Hołubowicz, R.; Ożyhar, A.; Dobryszycki, P. Natural Mutations Affect Structure and Function of gC1q Domain of Otolin-1. *Int. J. Mol. Sci.* **2021**, 22, 9085. https://doi.org/10.3390/ijms22169085

Academic Editors: Istvan Simon and Csaba Magyar

Received: 27 July 2021
Accepted: 19 August 2021
Published: 23 August 2021

1. Introduction

C1q superfamily encompasses short chain collagen-like proteins engaged in a wide variety of biological processes: immune recognition (C1q) [1], metabolic control (adiponectin) [2], endochondral ossification (collagen X) [3], formation of subendothelial and subcorneal matrices (collagen VIII) [4], cell adhesion in the retinal pigment epithelium (RPE) (Complement C1q tumor necrosis factor-related protein 5—C1QTNF5) [5] and more. Over the years, many disease causing mutations of the proteins from the C1q superfamily were detected. Many of them involve the globular C-terminal domain (gC1q), which is responsible for trimerization, which is usually Ca^{2+}-dependent, and for interactions with the macromolecular ligands. Here, we focus on the missense mutations, which result in a substitution of a single amino acid. Clinically important mutations also involve frameshifts, which have more pronounced effects, insertion–deletion polymorphisms (indels), which result in excision or insertion of a DNA fragment, and mutations involving non-coding sequences, for example introns [6,7]. Typically, the pathogenic missense mutations of the gC1q domain interrupt trimerization, which results in the inability to form biologically active multimers and results in the lack of protein secretion or, in milder cases, secretion of defective, incorrectly folded protein. C1q protein, which initiates a classical complement pathway upon recognition of immune ligands, is a hexameric assembly of heterotrimers composed of chains A, B and C. G244R variant of chain B is associated with C1q deficiency, a rare genetic disease associated with systemic lupus erythematosus and increased susceptibility to bacterial infections [8]. In the case of adiponectin, R112C and I164T are examples of variants, which impair trimerization of adiponectin and secretion of the protein into circulation, which leads to reduced adiponectin levels and ultimately to a diabetic phenotype [9]. In the case of collagen X, various mutations in the gC1q domain (conventionally called NC1 for collagens) are associated with Schmid metaphyseal chondrodysplasia (spondylometaphyseal dysplasia), a rare genetic disease characterized by short stature, long bone growth abnormalities and waddling gait [10–13]. S163R variant of C1QTNF5 is involved in pathogenesis of late-onset retinal macular degeneration due to the weakening of the intracellular connections in RPE mediated by C1QTNF5. Moreover, mutated C1QTNF5 has decreased stability leading to its aggregation, which contributes to local tissue damage [5].

Otolin-1 is a protein from the C1q superfamily, which is a crucial component of the otoconial membrane and organic matrix of otoconia. Otoconia are small, numerous calcium carbonate biominerals, which appear as "ear dust" embedded in a gelatinous membrane. They are formed before birth. The otoconial membranes are connected to the hair cells of the sensory epithelia in the utricle and saccule, which are part of the vestibule in the inner ear. Aggregated otoconia move in response to the movements of the body, contributing together with semicircular canals to the sense of balance [14]. Interestingly, fish have analogous biominerals, otoliths, which in contrast are large, grow continuously during life and are involved in hearing [15,16]. Otolin-1 was first indirectly found through comparative analysis of amino acid content of organic matrices of otoliths from many species of fish, which showed exceptional conservation of amino acid composition and high content of hydroxyproline [17]. The *OTOL1* gene was cloned in 2002 for chum salmon *Oncorhynchus keta* [18] and in 2010 for mouse [19] and since then, sequences of otolin-1 from other organisms were inferred from homology.

Although the protein was cloned nearly 20 years ago, still only limited information is known regarding its structure and function in the inner ear. Ablation of otolin-1 in zebrafish resulted in formation of detached, often fused otoliths [20]. There are no reports showing the effects of knockdown of otolin-1 in mammals such as mice. Otolin-1 interacts with otoconin-90 (Oc90), another abundant otoconial matrix protein, through the globular gC1q domain and collagen-like domain [19,21]. Together with Oc90, it influenced the formation of calcite in vitro, which led to formation of barrel-like shape crystals resembling natural otoconia instead of rhombohedral, which appear in the absence of proteins with biomineralization activity. Otolin-1 and Oc90 had distinct effects on formation of calcite. Oc90 seems to increase the nucleation rate of calcium carbonate and inhibit growth of the crystals, whereas otolin-1 increased the rate of growth of the crystals. Nevertheless, such artificial otoconia were much larger than the natural biominerals, therefore the mechanisms of their synthesis in vivo depend on additional factors. In the same study, it was also shown that otolin-1 can form a hexagonal, fibrillary matrix, which predisposes it to form an organic scaffold of otoliths and otoconia [22]. It is important to note that in nature, not only calcite, the most stable polymorph of calcium carbonate, is produced, but aragonite, vaterite and amorphous calcium carbonate are found in the biominerals [23]. Otoliths of teleost fish are a good example, as they may contain aragonite or vaterite, depending on the species and growth conditions of the fish [16,24]. For rainbow trout (*Oncorhynchus mykiss*), it was shown that the high molecular weight aggregate extracted from the otolith matrix, which contained otolin-1, is necessary for formation of aragonite—a native polymorph of calcium carbonate. However, otolin-1 alone was not enough to drive the formation of aragonite [25]. Biomineralization of otoconia and otoliths is therefore a complex process, which depends on otolin-1, other proteins such as Oc90, and multiple other factors.

In our previous studies on the gC1q domain of otolin-1, we showed that it can form trimers; however, Ca^{2+} are required to form stable oligomers. We discovered that gC1q domain of human otolin-1 (hOtolC1q) forms stable oligomers at lower Ca^{2+} concentrations than the zebrafish analog, dOtolC1q, which relates to the differences in composition of endolymph in mammals and fish [26]. The mechanism of trimerization of hOtolC1q involves neutralization of repulsive charge at the axis of a trimer, which normally occurs due to binding of Ca^{2+} [27]. In this work, we analyzed the influence of identified natural variants of hOtolC1q on ability to form stable trimers and respond to increasing concentration of Ca^{2+}, which is crucial for function of otolin-1 as an otoconial matrix protein and a constituent of the otoconial membrane. We were able to classify the variants according to the extent of their influence on the structure of hOtolC1q, and our results will enable to interpret clinical symptoms, which could be associated with the occurrence of mutations in *OTOL1* gene. We hypothesize that the mutations can disrupt the delicate homeostasis of otoconia and contribute to earlier occurrence of pathologies such as benign paroxysmal positional vertigo (BPPV).

2. Results and Discussion

During the database search, we found many single nucleotide variants (SNVs), including two single nucleotide polymorphisms (SNPs—SNVs with prevalence in the population of 1% or more [7]) in human *OTOL1* gene fragment encoding gC1q domain of otolin-1 (hOtolC1q). Then, we checked the position of affected residues in the small angle X-ray scattering (SAXS)-derived model of hOtolC1q trimer [26] and drew suppositions, how the mutations could affect structure and function of hOtolC1q. E470A (rs3921595) SNP was present in nearly 50% of sequencing reads. Since we suspected that due to its acidic properties E470 could contribute to a Ca^{2+} binding site, it was a subject of our previous analysis [27]. R339S (rs540167726) is a rarer SNP with maximal frequency of 2.5%. In the primary sequence, R339 is near the beginning of the gC1q domain and is placed at the base of a trimer (Figure 1a). It is modeled adjacent to E471 (Figure 1c), therefore it can form stabilizing ionic and hydrogen interactions; however, the importance of these interactions may be minor, as R339 is often replaced by other residues even in mammals (mouse as an example in Figure 1g, more examples in the Supplementary File S1). Although R339 is poorly conserved between the classes, we were interested how the substitution would affect hOtolC1q. Overall, this SNP was predicted to be neutral (Table 1). Out of the rarer variants of hOtolC1q, which were identified in multiple sequencing reads, R342W (rs200878802), R402P (rs760999493) and Q426R (rs1243409251) seemed to have a potentially significant impact. Side chain of R342 is exposed to the solvent near the boundary between the gC1q protomers (Figure 1d). Wider comparison of the mammalian sequences of otolin-1 showed that this residue is often substituted with glutamine (murine example in Figure 1g), even in apes (Supplementary File S1). However, substitution with tryptophan would have much more pronounced effect compared to glutamine, as it would introduce a hydrophobic aromatic moiety in place of a hydrogen bond donor/acceptor. This could affect the formation of trimers and modify the surface properties of the protein. R402 is located in the middle of a β-strand adjacent to a strand containing E417, which together with D425 forms a known Ca^{2+} binding site (Figure 1e). The side chain of R402 is predicted to be at the trimerization surface. Thus, substitution of this residue with proline could have a very strong detrimental effect on folding of the gC1q domain and binding of Ca^{2+}. The malformation of the β-strand could propagate further, affecting the whole 10 β-barrel assembly typical for the C1q superfamily of proteins, especially near the Ca^{2+} binding site. Moreover, the substitution could affect the interactions between the protomers. Q426 follows D425 in the sequence, and its side chain is predicted to be at the trimerization interface (Figure 1f). Thus, substitution to arginine could affect the binding of Ca^{2+} and trimerization of the gC1q domain, although the effect should be weaker than for R402P.

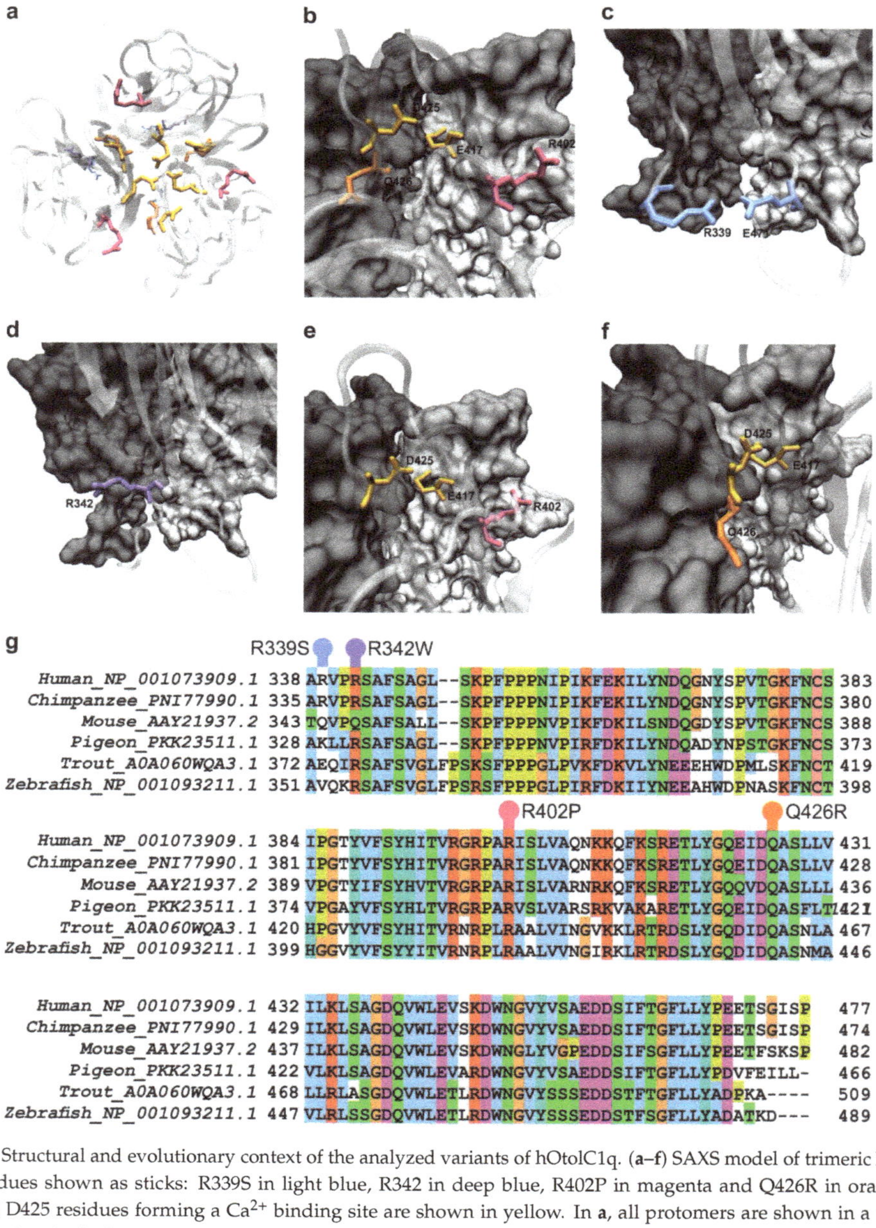

Figure 1. Structural and evolutionary context of the analyzed variants of hOtolC1q. (**a**–**f**) SAXS model of trimeric hOtolC1q with residues shown as sticks: R339S in light blue, R342 in deep blue, R402P in magenta and Q426R in orange. The E417 and D425 residues forming a Ca^{2+} binding site are shown in yellow. In **a**, all protomers are shown in a "cartoon" representation, in (**b**–**f**), one protomer is shown as translucent protein backbone, two others as light and dark gray protein surface projections. (**a**)—overall view of the gC1q trimer, (**b**)—Ca^{2+} binding site at the top of the trimer, (**c**)—R339, near the base of the trimer, with adjacent E471, (**d**)—R342 near the contact surface of the protomers, (**e**)—R402, and (**f**)—Q426, both near the Ca^{2+} binding site and the contact surface of the protomers. The visualizations were made using VMD [28]. (**g**)—Multiple sequence alignment showing the conservation of the gC1q domain of otolin-1 among the classes of the vertebrates. Investigated residues are highlighted above the alignment. The alignment was done using ClustalX [29] and visualized using Jalview [30].

Table 1. Selected known single nucleotide variants of hOtolC1q, their prevalence and predictions of deleteriousness. The data were retrieved from the Ensembl database, and for SNPMuSiC independently calculated based on the SAXS-derived model of hOtolC1q trimer (https://soft.dezyme.com/, accessed 16 August 2021) [31].

Variant dbSNP ID	Highest Population MAF	SIFT	PolyPhen	CADD	REVEL	MetaLR	Mutation Assessor	SNP MuSiC
R339S	0.025	0.5 (Tolerated)	0.097(Benign)	3 (Likely Benign)	0.129 (Likely Benign)	0.532 (Damaging)	0.268 (Low Impact)	−0.53 (Neutral)
R342W	1.159×10^{-4}	0 (Deleterious)	0.880 (Possibly Damaging)	22 (Likely Benign)	0.326 (Likely Benign)	0.762 (Damaging)	0.904 (Medium Impact)	0.17 (Deleterious)
R402P	1.394×10^{-4}	0 (Deleterious)	0.797 (Possibly Damaging)	22 (Likely Benign)	0.518 (Likely Disease Causing)	0.674 (Damaging)	0.792 (Medium Impact)	0.42 (Deleterious)
Q426R	4.643×10^{-4}	0.02 (Deleterious)	0.969 (Probably Damaging)	23 (Likely Benign)	0.587 (Likely Disease Causing)	0.777 (Damaging)	0.758 (Medium Impact)	0.16 (Deleterious)

MAF—mean allele frequency, prevalence of the variant in the population. SIFT score has a scale from 0 to 1. Variants with scores below 0.05 are predicted to be deleterious. PolyPhen score has a scale from 0 to 1. Variants with scores up to 0.446 are predicted to be benign, from 0.447 to 0.908 to be possibly damaging, and with scores higher than 0.908 to be probably damaging. CADD provides a ranking with higher scores more likely to be deleterious, the customary boundary is set at 30. REVEL score ranges from 0 to 1 and variants with higher scores are predicted to be more likely to be pathogenic. MetaLR classifies the variants as 'tolerated' or 'damaging'; a score between 0 and 1 is also provided and variants with higher scores are more likely to be deleterious. Mutation assessor gives a prediction, which is one of 'neutral', 'low', 'medium' and 'high', and the rank score, which is between 0 and 1 where variants with higher scores are more likely to be deleterious. For SNP MuSiC, positive score predicts the variant to be deleterious, negative to be neutral. SNP MuSiC also predicts solvent accessibility and effect on thermodynamic and thermal stabilities (results not shown).

The computational predictions accompanying the entries in the Ensembl database and independently conducted by us using SNPMuSiC suite (Table 1) suggested that all the rarer variants could be deleterious. The differences between the predictions obtained using different algorithms are too large to propose a relative degree of severity of the variants. CADD and MetaLR predictors gave results inconsistent with the other algorithms, as they did not differentiate the variants to benign or deleterious. However, as for REVEL and Mutation Assessor, CADD and MetaLR scores for R339S were lower than for the other variants, therefore it provides a rationale to differentiate this mutation as milder than the others. Predictions of varying severity of the mutations provided a motivation to produce the mutated hOtolC1q variants and subject them to analyses, which would reveal how the mutations affect the solution structure and Ca^{2+}-dependent trimerization of hOtolC1q.

The typical feature of the proteins from the C1q superfamily is trimerization, usually Ca^{2+}-dependent [1,3,32–34]. We used sedimentation velocity analytical ultracentrifugation (SV AUC) to see how the mutations could affect the assembly of gC1q trimers of hOtolC1q. We conducted the experiment for protein concentrations in the range of 0.1 to 0.5 mg/mL to properly characterize weak self-interactions already observed for wild type hOtolC1q [26]. There, sedimentation coefficient distributions ($c(s)$) calculated for varying concentrations of hOtolC1q centrifuged in the absence of Ca^{2+} were wide, with peaks between 2.0 and 2.5 S, and shifted continuously with increasing concentration from lower to higher sedimentation coefficients. The effect was even more pronounced for the zebrafish analogue, dOtolC1q. This phenomenon is characteristic for loosely bound complexes, which associate and dissociate rapidly during the SV AUC experiment [35]. The oligomerization of hOtolC1q seems to occur sequentially and follow a formula:

$$A_n + A \rightleftharpoons A_{n+1} \quad (1)$$

where A is a protein monomer and the superscript indicates the stoichiometry of the oligomer. Fast kinetics of association and dissociation result in observation of intermediate species with sedimentation coefficients and apparent molecular weights of hOtolC1q between dimer and trimer, and even between monomer and dimer for dOtolC1q. A tendency of Ca^{2+}-free hOtolC1q to form heavy aggregates was also noted. When 10 mM Ca^{2+} were added, a conformational change occurred which led to stabilization of the trimers at all tested protein concentrations. The trimers appeared in the $c(s)$ distributions as a sharp peak at 2.55 S. Ca^{2+} ions also diminished the tendency of hOtolC1q to form heavy aggregates. Here, the experiment for hOtolC1q was replicated to serve as a control and the $c(s)$ distributions are shown in the background of plots in Figure 2. In the case of R339S, the equilibrium of oligomerization in the absence of Ca^{2+} was slightly shifted to lighter forms compared to hOtolC1q (Figure 2a, Table S1), which we interpret as destabilization of the gC1q trimer. Conversely, Q426R variant apparently stabilized the gC1q trimer in the absence of Ca^{2+}, as the equilibrium was shifted towards heavier forms (Figure 2d, Table S1). Moreover, hOtolC1q Q426R did not form heavy aggregates in the absence of Ca^{2+}. The apparent beneficial effect of this variant is especially interesting if we consider that the most algorithms predicted it to be the most damaging (Table 1). Additionally, 10 mM $CaCl_2$ diminished the effects of R339S and Q426R variants, as the $c(s)$ distributions were identical as for hOtolC1q, showing the presence of homogenous trimers with no heavier aggregates (Figure 2e,h). R342W mutation was in contrast damaging, as it predisposed hOtolC1q to form heavy aggregates both in the absence and in the presence of Ca^{2+} (Figure 2b,f) in a proportion higher than for wild type hOtolC1q. In the absence of Ca^{2+}, we observed discrete populations of dimers and tetramers, with increased proportion of tetramers at higher protein concentrations. Apparently, for this mutant, when Ca^{2+} is absent, the oligomerization mechanism switches from sequential association of monomers to association of dimers:

$$2A \rightleftharpoons A_2 \rightleftharpoons A_4 \ldots A_n \quad (2)$$

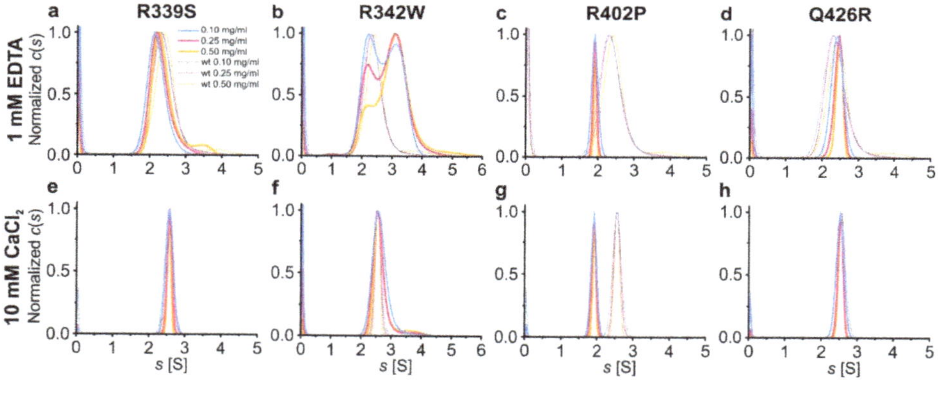

Figure 2. Influence of the mutations on oligomerization of hOtolC1q. Variants of hOtolC1q: (**a**,**e**) R339S, (**b**,**f**) R342W, (**c**,**g**) R402P, (**d**,**h**) Q426R were subjected to sedimentation velocity analytical ultracentrifugation at concentrations of 0.1–0.5 mg/mL in the presence of (**a**–**d**) 1 mM EDTA or (**e**–**h**) 10 mM Ca^{2+}. The $c(s)$ distributions are shown as solid lines. The dashed lines in the background show the $c(s)$ distributions calculated for wild type hOtolC1q.

Interestingly, the Ca^{2+} apparently rescued the correct oligomerization mechanism, as described by the Equation (1), because trimers were found for hOtolC1q R342W with 10 mM Ca^{2+}. However, Ca^{2+} did not completely protect hOtolC1q R342W from aggregation, as the aggregate trace was still detected. R402P mutation had the most striking effect on the oligomerization of hOtolC1q—this variant was dimeric both in the absence and in the presence of Ca^{2+} (Figure 2c,g). Dimerization of hOtolC1q R342W and R402P did not involve the unique cysteine residue present in hOtolC1q (Figure 1g), as $c(s)$ distributions calculated for samples centrifuged with 1 mM DTT were identical to those obtained in the absence of the reducing agent (Figure S1).

SV AUC showed that the natural variants, especially R402P, had a major influence on the assembly of gC1q trimers of hOtolC1q. To gain more detailed insight into the structural change induced by the mutations, we applied circular dichroism spectroscopy (CD) (Figure 3) with the secondary structure estimation using CDPro (Figure S2). As for SV AUC, the experiment for hOtolC1q was replicated as a control (Figure 3a). The CD spectrum of hOtolC1q in the absence of Ca^{2+} indicates that the polypeptide chain is folded into β-sheets, as a negative band is present near 215 nm. The protein also contains a substantial amount of disordered regions, because the ellipticity decreases below 210 nm. There is also a notable signal attributed to aromatic side chains with a positive ellipticity maximum at 233 nm. In the presence of at least 1 mM Ca^{2+}, structural change attributed to increase in β-strand content caused by binding of Ca^{2+} can be observed: position of the minimum shifts from 215 to 218 nm, and ellipticity is sharply increasing below 215 nm. Ellipticity near 233 nm also increased in response to added Ca^{2+} possibly due to structural rearrangements around the indole moieties of tryptophan side chains [26]. Similar features can be observed in the spectra of R339S mutant (Figure 3b). The band at 215 nm present in the absence of Ca^{2+} is slightly deeper than for the wild type hOtolC1q, but in the presence of 10 mM Ca^{2+}, the spectra of R339S and the native form are identical (Figure 3b). This result is consistent with SV AUC, where small differences were also observed in the absence of Ca^{2+} and none in the presence of Ca^{2+}. Similar changes appeared for the Q426R mutant (Figure 3e); however, the Ca^{2+}-induced conformational change became apparent at 10 mM Ca^{2+} instead of 1 mM. This indicates that the substitution near the Ca^{2+}-binding site weakened the affinity of hOtolC1q towards Ca^{2+}. The spectra of hOtolC1q R342W do not have the positive band at 233 nm and have a deeper negative band at 215–218 nm (Figure 3c). Interestingly, 10 mM Ca^{2+} instead of 1 mM was required to induce the structural change here as well. This shows that the mutation at the base of a trimer, at the opposite side from the Ca^{2+} binding site, can have a pronounced effect on binding of Ca^{2+}. As in the case of SV AUC, the most striking effect was noted for the R402P mutant, which was completely insensitive to the presence of Ca^{2+} (Figure 3d). The spectrum also shows no signal around 233 nm and a sharp decrease in ellipticity below 210 nm, which suggests that the degree of disorder compared to the wild type hOtolC1q was increased, probably due to the disruption of the β-strand containing R402, and possibly due to further alterations. Taken together, the CD spectra for hOtolC1q saturated with 10 mM Ca^{2+} (Figure 3f), which is biologically relevant since otolin-1 is present in the matrix of calcium carbonate otoconia, show that considering the secondary structure, R339S and Q426R mutations are benign, R342W is deleterious, and R402P may severely disrupt the function of hOtolC1q.

An even more detailed view of the changes caused by the natural variants can be obtained using thermal shift assay (TSA). We previously used this technique to discover the striking stabilization of hOtolC1q with Ca^{2+}, evidenced by transition temperature (T_m) change from 40 to over 95 °C. The results were consistent with temperature-dependent changes in the CD spectra [26]. Using TSA, we also found striking effects of alanine mutations in the Ca^{2+} binding site of otolin-1, which did not always lead to the destabilization of the protein [27]. The experiment replicated here confirmed that native hOtolC1q was slightly stabilized with 0.1 mM Ca^{2+} and strongly stabilized at higher concentrations—the T_m increased to 66 °C in 0.1 mM Ca^{2+} and to more than 95 °C in 100 mM Ca^{2+} (Figure 4a). R339S was slightly, but consistently less stable than hOtolC1q—the T_m difference was near 2 °C under all tested conditions (Figure 4b). The slight decrease of T_m can be associated with ablation of interactions between R339 and E471 (Figure 1c). hOtolC1q was substantially destabilized by the R342W substitution, as the T_m was decreased to 37 °C in the absence of Ca^{2+} (Figure 4c). Moreover, this mutant was stabilized at 10 mM Ca^{2+}, compared to 1 mM Ca^{2+} for wild-type hOtolC1q, which is consistent with the occurrence of the secondary structure change at 10 mM Ca^{2+}. Ultimately, R342W had a stability similar to hOtolC1q at 10–100 mM concentrations of Ca^{2+}, showing that Ca^{2+} mitigated the detrimental effect of the mutation. It was also noticeable that the fluorescent signal of SYPRO Orange bound to R342W was much weaker than for other variants, and the transitions were not clear. Modification of the surface properties of hOtolC1q by the mutation apparently interfered with binding of the SYPRO Orange probe. Q426R and, to a lower extent, R402P, were more stable than hOtolC1q in the absence of Ca^{2+} (T_m of 57.2 °C and 46.8 °C, respectively, Figure 4d,e). However, the Ca^{2+} stabilized Q426R more weakly than hOtolC1q (T_m 72.2 °C compared to 86.9 °C at 10 mM Ca^{2+}), and did not stabilize R402P at all. Together with the results of SV AUC, this shows that Q426R mutation is stabilizing at low concentrations of Ca^{2+}, but detrimental at higher concentrations, albeit not damaging enough to prevent trimerization of hOtolC1q. The results of TSA are also fully compatible with the results of CD, which showed that R342W and Q426R mutants are less sensitive to Ca^{2+} than the native hOtolC1q. The summary of the T_m for all the tested variants is provided in Figure 4f and Table S2.

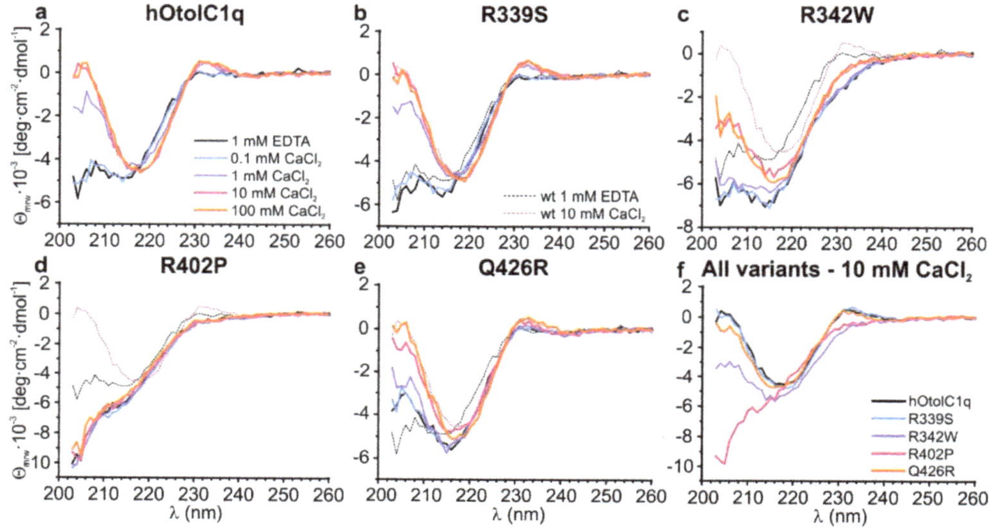

Figure 3. Changes in the secondary structure of (**a**) hOtolC1q introduced by the mutations: (**b**) R339S, (**c**) R342W, (**d**) R402P and (**e**) Q426R. Circular dichroism spectra were collected for 0.20 mg/mL proteins in the absence and in the presence of 0.1–100 mM Ca^{2+}. The panel (**f**) contains a comparison of the spectra for all tested variants at 10 mM Ca^{2+}.

TSA also provides interesting insight into the extent of exposition of hydrophobic regions on the surface of a protein, which was exhibited for gC1q domains of C1q and collagen X [1,3]. Affinity of hOtolC1q to hydrophobic compounds in the native state is evident as SYPRO Orange emits a strong fluorescence before the protein becomes unfolded. The further increase of fluorescence attributed to denaturation appears upon heating, when hydrophobic regions from the core of the protein become exposed and accessible for SYPRO Orange (Figure 4a). This is equivalent to binding of 8-anilino-1-naphthalenesulfonic acid (ANS), a fluorescent probe used specifically to probe the affinity of proteins to hydrophobic compounds [36,37]. ANS was in fact used in a prototypical TSA experiment [38] before SYPRO Orange was adopted due to its superior compatibility with existing qPCR devices [39]. In the case of hOtolC1q, the exposition of hydrophobic side chains decreases upon binding of Ca^{2+}, as 10–100 mM Ca^{2+} decrease the baseline fluorescence at 20 °C (Figure 4a). While R339S mutant shows similar behavior to hOtolC1q, the increase of fluorescence of SYPRO Orange during denaturation of the R342W is poor, decreasing the robustness of the analysis for R342W mutant (Figure 4c). Interestingly, R402P and Q426R mutations decreased the baseline fluorescence (in the case of R402P—to a background level), which shows that structural alterations caused by these mutations resulted in inaccessibility of the hydrophobic surface groups of hOtolC1q in the native state (Figure 4d,e). This could hamper the interactions with other macromolecules in the otoconial matrix or otoconial membrane and impair proper formation and anchoring of otoconia during the embryonic development [40]. Overall, beside the primary evidence of change of thermal stability, TSA contributes to the observations that mutations and binding of Ca^{2+} cause pronounced structural changes in hOtolC1q affecting the whole globular trimer.

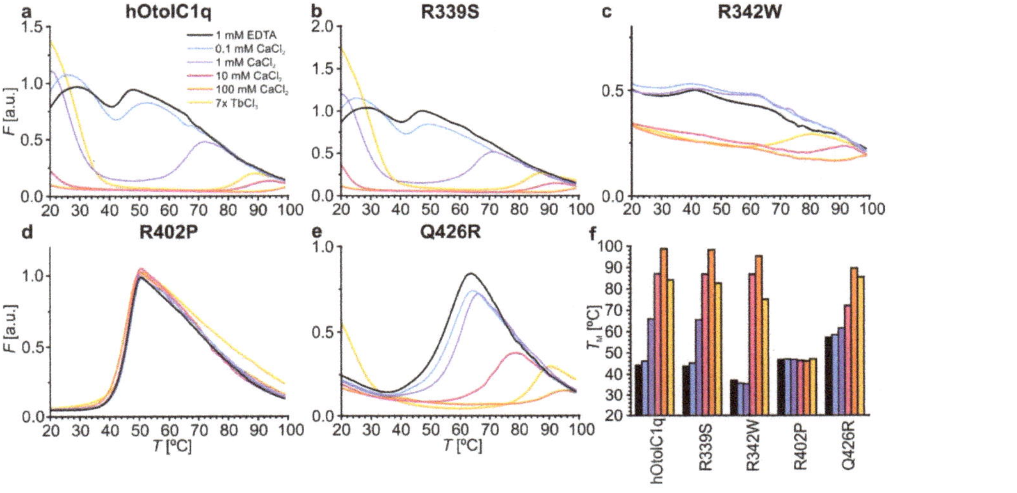

Figure 4. Changes in the thermal stability of hOtolC1q introduced by the mutations, analyzed by thermal shift assay (TSA). (**a**) Native hOtolC1q (control), (**b**) R339S, (**c**) R342W, (**d**) R402P and (**e**) Q426R. The T_m values are aggregated in the bar graph (**f**) and in Table S2.

The SV AUC, CD and TSA analyses showed that the mutations affected not only the solution structure of hOtolC1q, but also decreased (R342W, Q426R) or diminished (R402P) the ability of hOtolC1q to bind Ca^{2+}. We sought to gain a more detailed insight into these effects by conducting a Tb^{3+} binding assay, which allows to estimate the relative affinity of the protein to Ca^{2+} [27]. In the case of hOtolC1q, the direct measurement of binding of Ca^{2+} was not possible due to the irreversible precipitation of the protein during decalcification procedure involving either incubation with EDTA/EGTA and exhaustive dialysis against a decalcified solution, or direct incubation with buffered metal-binding Chelex resin. As Tb^{3+} tend to strongly bind to the Ca^{2+}-binding proteins [41,42], these ions could displace the trace Ca^{2+} from buffers and host cells and allow to conduct a comparative analysis of affinity of the proteins to Ca^{2+}. We observed that all the mutations decreased the affinity of hOtolC1q to Ca^{2+}: the dissociation constant (K_d) was increasing in the order of hOtolC1q < R339S < R342W < Q426R < R402P (Figure 5a). The binding of Tb^{3+} was equivalent to Ca^{2+}. hOtolC1q, R339S and Q426R responded to Tb^{3+} similarly as to Ca^{2+} by forming homogenous trimers (Figure 5b). Their CD spectra in the presence of Tb^{3+} were also identical as in the presence of Ca^{2+} (compare Figures 5c and 3f). As expected, the structural changes occurred at low concentration of Tb^{3+}, 35–82 μM, depending on the experiment. Interestingly, Q426R seemed to bind Tb^{3+} more preferentially than the other variants, as 35 μM Tb^{3+} stabilized it more strongly than 10 mM Ca^{2+}. In the case of R342W, Tb^{3+} had an additional effect of intensifying the aggregation (compare Figure 5b with Figure 2f). Interestingly, despite the lack of responsiveness to Ca^{2+}, R402P mutant seemed do bind Tb^{3+}. Apparent affinity to Tb^{3+} was actually greater than for dOtolC1q, which responded to Ca^{2+} at higher concentrations than hOtolC1q [26,27]. Apparently, despite losing the ability to bind Ca^{2+}, the R402P mutant retained some affinity to Tb^{3+}. The binding seems to be non-specific though, as Tb^{3+} did not alter the secondary structure, induce trimerization or increase the thermal stability of the R402P mutant (Figures 4d and 5b,c). Non-specific binding of lanthanide ions, including Tb^{3+}, was identified for many proteins by X-ray crystallography [43]. Although the Tb^{3+} binding assay alone is not sufficient to determine the absolute affinity of a protein to Ca^{2+}, it is useful for comparative analyses of the variants of the same protein from the same organism, when direct measurement of affinity to Ca^{2+} is not available. Here, CD and TSA were useful supplementary techniques. Moreover, SV AUC provided a mechanistic insight into the effects of mutations on the Ca^{2+}-dependent assembly of hOtolC1q, and supported the observation that R402P mutant is effectively unable to bind Ca^{2+}.

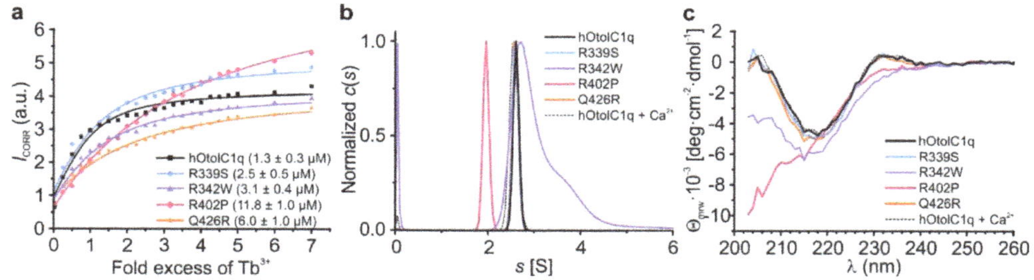

Figure 5. Binding of Tb^{3+} by hOtolC1q and its variants, analyzed by fluorometric titration, sedimentation velocity analytical ultracentrifugation and circular dichroism. (**a**) The results of a titration experiment, in which 3.7 μM proteins were treated with appropriate excess of $TbCl_3$. The fluorescence intensity data were measured 15 min after addition of each portion of Tb^{3+}, corrected for background fluorescence and fitted to a single binding site per monomer model. Apparent K_d values are given in the parentheses in the legend. (**b**) The *c*(*s*) distributions calculated from sedimentation velocity data obtained for 0.25 mg/mL (14.7 μM) hOtolC1q variants in the presence of 7-fold molar excess of $TbCl_3$. The distributions for R339S and Q426R variants overlap with that calculated for hOtolC1q. Distribution of hOtolC1q with 10 mM Ca^{2+} is shown as a dashed line in the background for comparison. (**c**) The circular dichroism spectra recorded for 0.2 mg/mL (11.8 μM) hOtolC1q variants in the presence of 7-fold molar excess of $TbCl_3$. Spectrum obtained for hOtolC1q with 10 mM Ca^{2+} is shown in the background as a dashed line.

Although the availability of the phenotype data, which can be associated with specific genetic variants, is continuously increasing, the algorithms which depend solely on the protein sequences do not allow to reliably predict the effects of the mutations. From the variants of hOtolC1q clearly predicted to be deleterious: R342W, R402P and Q426R, Q426R seemed like a candidate for the most detrimental of the three. However, the R402P variant was clearly the most damaging for the protein. It is worth noting that this variant was correctly predicted as the most damaging by a model-dependent algorithm SNP MuSiC. This highlights the importance of structural studies of proteins involving not only atomic-level structure determination with nuclear magnetic resonance, X-ray crystallography or cryoelectron microscopy, but also molecular shape determination with less precise techniques such as SAXS or its sister method, small angle neutron scattering (SANS). Together with advanced computational 3D structure prediction methods, SAXS and SANS can give enough structural information to correctly predict the effects of the mutations on the structure and function of the proteins. This is especially important in the analysis of intrinsically disordered proteins, which lack a defined structure to a varying extent, and thus determination of their 3D structure at atomic resolution may be impossible [44]. In our case, CD and TSA gave detailed information regarding the effects of the mutations of the gC1q domain of otolin-1. SV AUC gave more general, but very important insight on a larger scale, as it showed how the trimer assembly and aggregation propensity of hOtolC1q were affected. This is an example of the advantage of SV AUC as a preferred method for analysis of mutated oligomeric proteins. Overall, the molecular arrangement of gC1q trimer seems to be resilient against relatively benign mutations, but severely affected by the extensive disruption of the secondary structure within the protomer (R402P) or by major modifications of solvent-exposed moieties (R342W).

R339S polymorphism of hOtolC1q is potentially benign, as it may affect structure and function of otolin-1 to a small extent and only at low concentrations of Ca^{2+}. We noted similar effect for a prevalent polymorphism of otolin-1, E470A, which similarly to R339S slightly decreased the stability of hOtolC1q trimer in the absence of Ca^{2+} and slightly increased its tendency to form heavy aggregates [27]. Although these changes do not seem to be significant, they may negatively affect a decades long function of otolin-1 in the inner ear. According to the accumulated knowledge, otoconia and otoconial membrane do not regenerate, and the susceptibility of the otoconia to detach and incidentally accumulate in the semicircular canals leading to BPPV steadily increases during life [45–47]. Although it is normal that small amounts of otolin-1 leak from the labyrinth, patients suffering from BPPV have increased levels of otolin-1 in the serum [48,49]. It is important to note that beside the otoconial matrix, which is embedded in the solid calcium carbonate otoconium, otolin-1 is found in a fibrillary network interconnecting the otoconia [19,21,50], which makes it exposed to eventual pathological decreased level of Ca^{2+} in the endolymph. Destabilization of the otoconia and otoconial membrane, and resulting increased rate of release of otolin-1, may thus be a driving force of BPPV. Even the minor additional weakening of otolin-1 network caused by R339S and E470A mutations could accelerate the degradation of otolith organ enough to be a contributing factor to the earlier onset of BPPV, because they would make otolin-1 more sensitive to transient decreases in concentration of Ca^{2+} in the endolymph.

The rarer R342W and Q426R variants have strongly decreased the responsiveness of hOtolC1q to Ca^{2+} as they seemed to stabilize at approximately 10 mM of Ca^{2+} instead of 0.1–1 mM. Therefore, even in the healthy state with normal Ca^{2+} concentration in the endolymph (92–133 μM in guinea pig endolymph, possibly similar in humans) [51] these variants could weaken the network formed by otolin-1, induce early degradation of otolith organ and cause frequent BPPV at younger age. To remain stable, R342W and Q426R would require at least 1 mM Ca^{2+} in the endolymph, a concentration that is observed in hydropic ears serving as models for Ménière's disease, which is characterized by the endolymphatic hydrops, attacks of vertigo and progressive hearing loss [51,52]. R342W and Q426R also modify surface properties of hOtolC1q, possibly interrupting protein–protein interactions in the otoconial matrix and otoconial membrane. R402P variant has a severe destabilizing effect on hOtolC1q, even preventing hOtolC1q from forming the trimers. As the network formed by otolin-1 seems to be interconnected by the globular heads of otolin-1 [22], such disruption would distort the protein matrix and cause a dysfunction of the otolith organ. However, lack of the clinical data related to the investigated variants of otolin-1 rule out the formulation of definitive conclusions. The results of our experiments should, therefore, bring attention to genetic variation of otolin-1 in patients with inner ear disorders, especially suffering from BPPV and other manifestations of imbalance in younger age. Protein–protein interactions in the otoconial membrane and in the otoconial matrix are another challenging area of research, which remains to be studied. Definite identification of the proteins involved, characterization of these interactions and effects of mutations would improve our understanding of the biomineralization mechanisms of otoconia and otoliths.

3. Materials and Methods

3.1. Accession Numbers

Human OTOL1 gene Ensembl accession ID: ENSG00000182447
Human otolin-1 Uniprot accession ID: A6NHN0
Human otolin-1 R339S SNP variant ID: rs540167726 (A > C)
Human otolin-1 R342W SNP variant ID: rs200878802 (C > T)
Human otolin-1 R402P SNP variant ID: rs760999493 (G > C)
Human otolin-1 Q426R SNP variant ID: rs1243409251 (A > G)

3.2. Key Resources

Synthetic cDNA encoding full-length human otolin-1 was codon optimized for *Escherichia coli* and provided by GeneArt (currently Thermo Fisher Scientific, Warsaw, Poland). Nucleotide primers were provided by Genomed (Warsaw, Poland). pQE-80L plasmid expression vector was from Qiagen (Hilden, Germany). *Escherichia coli* Top10 cells, *Dpn*I enzyme, DNA ladders, protein markers, and LB broth were from Thermo Fisher Scientific. One-fusion DNA Polymerase was from GeneOn (Ludwigshafen am Rhein, Germany; distributed by ABO, Gdańsk, Poland). BlueStain sensitive and SimplySafe stains were from EurX (Gdańsk, Poland). Agar, agarose, tris(hydroxymethyl)aminomethane (Tris), ethylenediaminetetraacetic acid (EDTA), carbenicillin, isopropyl β-D-1-thiogalactopyranoside (IPTG), NaCl, glycerol, 2-mercaptoethanol, imidazole, glycine, sodium dodecyl sulfate (SDS) and $CaCl_2$ were from Carl Roth (Karlsruhe, Germany). *Escherichia coli* BL21(DE3) cells, TB broth, 4-(2-hydroxyethyl)-1-piperazineethanesulfonic acid (HEPES), phenylmethylsulfonyl fluoride (PMSF), DNase I, RNase A, terbium(III) chloride hexahydrate, xylenol orange disodium salt, dithiothreitol (DTT) and SYPRO Orange were from Sigma (currently Merck, Warsaw, Poland). Empty Tricorn and Superdex 200 Increase 10/300 GL columns were from (GE Healthcare Life Sciences, currently Cytiva, Warsaw, Poland). TALON® Metal Affinity resin was from Takara Bio (Mountain View, CA, USA; distributed by Biokom, Janki, Poland).

3.3. Single Nucleotide Polymorphisms and Variants

Ensembl genome browser (https://www.ensembl.org/index.html, accessed 16 August 2021) was queried for known SNPs in human otolin-1 gene (OTOL1, ENSG00000182447). Boundaries of the gC1q domain were retrieved from Uniprot database (A6NHN0) as 338–477. The entries were accompanied by mutation severity predictions made using SIFT (https://sift.bii.a-star.edu.sg/, accessed 16 August 2021) [53], PolyPhen2 (http://genetics.bwh.harvard.edu/pph2/, accessed 16 August 2021) [54], CADD (https://cadd.gs.washington.edu/, accessed 16 August 2021) [55], REVEL (https://sites.google.com/site/revelgenomics/, accessed 16 August 2021) [56], MetaLR (https://sites.google.com/site/jpopgen/dbNSFP, accessed 16 August 2021) [57] and Mutation assessor (http://mutationassessor.org/r3/, accessed 16 August 2021) [58] tools. Additionally, for all investigated mutations, SNP MuSiC (https://soft.dezyme.com/, accessed 16 August 2021) tool was used to predict effects on protein stability [31]. Model of gC1q trimer, which was used as a template, was based on already published ensemble optimization method (EOM) analysis conducted on the basis of SAXS data [26]. Default parameters were used in all predictions. The structure model was visualized using VMD software (University of Illinois, https://www.ks.uiuc.edu/Research/vmd/, version 1.9.3, accessed 20 August 2021) [28].

3.4. Preparation of Mutated gC1q Genes

Synthetic cDNA of hOtolC1q, which was previously subcloned into pQE-80L plasmid expression vector [26], was used as a template in modified QuickChange®, which was conducted as described [59]. One-fusion DNA polymerase was used in the mutagenic polymerase chain reaction (PCR). For the calculation of the annealing temperatures of the primers, the concentration of KCl in the reaction mixture was assumed to be 0.1 M, as in the assay buffer of the polymerase. Plasmids were propagated in *Escherichia coli* TOP10 cells. Progress of the cloning was followed by agarose electrophoresis with SimplySafe stain. All mutated genes were analyzed by DNA sequencing (Genomed).

3.5. Protein Expression and Purification

Escherichia coli BL21(DE3) cells were chemically transformed by heat shock and grown on plates containing LB broth with 1.5% agar and 100 μg/mL carbenicillin at 37 °C overnight. Single colonies were picked and used to inoculate starter cultures containing 100 mL of TB broth with carbenicillin, which were incubated overnight at 37 °C, 200 rpm. Portions of 500 mL TB with carbenicillin were inoculated with 2% volume of starter culture and incubated at 29 °C, 200 rpm. After reaching the optical density at 600 nm of at least 0.5, cultures were cooled to 15 °C and the expression of the protein of interest was induced by 0.5 mM IPTG. The culture was continued overnight (16–18 h) at 15 °C, 200 rpm. Cells were collected by centrifugation at 5000× *g* at 4 °C for 15 min and resuspended in H10Na500G5 buffer (HEPES 10 mM, pH 7.0 (20 °C), NaCl 500 mM, glycerol 5% (*v*/*v*)) with freshly added 1 mM 2-mercaptoethanol. The cells were kept frozen at −80 °C.

Cell lysis was initiated by thawing in a room temperature water bath. After thawing, 0.2 mg/mL PMSF, 20 μg/mL DNase I and 20 μg/mL RNase A were added. The lysis was achieved by applying 10 sonication cycles for 30 s with 1 min breaks in a Cole-Parmer CPX 500 ultrasonic processor with a microtip and amplitude set at 35% (Cole-Parmer, Vernon Hills, IL, USA). The cell suspension was cooled in ice to maintain the temperature below 10 °C. Lysates were clarified by centrifugation at 18,500× *g* for 30 min at 4 °C and incubated with 1 mL TALON® Metal Affinity resin for 1 h in a cold room (4–6 °C) in an orbital mixer set at 5 rpm. The resin was separated by centrifugation at 700× *g* for 5 min at 4 °C, washed with 20 bed volumes of H10Na500G5 (without the 2-mercaptoethanol), centrifuged again and packed in a glass Tricorn column. The column was connected to ÄKTA Avant chromatography system (GE Healthcare Life Sciences) with flow set at 1 mL/min. Contaminants were washed away with 20 bed volumes of H10Na500G5 and subsequently with 20 bed volumes of the buffer with 30 mM imidazole. Mutated hOtolC1q was eluted with the buffer containing 200 mM imidazole. The eluate was concentrated in Amicon Ultra centrifuge filters with 10 kDa cutoff (Merck) and subjected to gel filtration using Superdex 200 Increase 10/300 GL column operated at 0.75 mL/min with H10Na500G5 as a mobile phase. Pure fractions were identified by SDS-PAGE with acrylamide percentage of 4% in a stacking gel and 12% in a resolving gel in a Laemmli buffer system (Tris-glycine-SDS) (Figure S3) [60]. Pure protein samples were stored at –80 °C. For subsequent experiments, protein concentration was determined by measuring absorbance at 280 nm with elution buffer as a reference. The protein extinction coefficients and molecular weights were estimated using ProtParam tool (https://web.expasy.org/protparam/, accessed 16 August 2021) [61].

3.6. Tb^{3+} Binding Fluorescence

Binding of Tb^{3+} ions to hOtolC1q and its mutants was assessed using steady-state fluorescence. Terbium (III) chloride was dissolved in MilliQ water to a final concentration of approximately 0.5 M. Exact concentration of $TbCl_3$ was determined by titration of diluted stock solution with EDTA in the presence of xylenol orange. Aliquots of diluted $TbCl_3$ were added to 2 mL 3.7 μM protein solution in a 10 × 10 mm quartz SUPRASIL® cuvette (Hellma Analytics, Müllheim, Germany) and incubated for 15 min at room temperature. Subsequently, fluorescence emission at 520-580 nm was recorded using an excitation wavelength of 280 nm using a Fluorolog-SPEX fluorimeter (HORIBA Scientific, Jobin-Yvon, Kyoto, Japan) equipped with a Peltier heating accessory set at 20 °C. The bandwidth was set at 5 nm for both excitation and emission monochromators. A cut-off filter absorbing below 350 nm was installed in the emission path. Obtained fluorescence intensities were processed and fitted to a model based on work by Gonzalez et al. [62,63]. Data analysis was conducted as described [27].

3.7. Circular Dichroism

Circular dichroism of 0.2 mg/mL proteins in H10Na500G5 with 1 mM EDTA, 0.1 mM $CaCl_2$, 1 mM $CaCl_2$, 10 mM $CaCl_2$, 100 mM $CaCl_2$ or 7-fold excess of $TbCl_3$ was measured in 1 mm quartz SUPRASIL® cuvettes (Hellma Analytics, Müllheim, Germany) using Jasco J-815 spectropolarimeter (Jasco, Easton, MD, USA) with a Peltier temperature control accessory set at 20 °C. The proteins were incubated with the additives at room temperature for at least 1 h before the measurements. The spectra were collected between 200 and 260 nm every 1 nm at scanning speed of 50 nm/min with five accumulations. Data, for which photomultiplier voltage was below 600 V, were analyzed. CD spectra of the proteins were corrected for buffer background signal and normalized for protein composition and concentration using an equation [64]:

$$\theta_{mrw} = \frac{\theta \cdot MRW}{10 \cdot c \cdot l} \left[\frac{\text{deg} \cdot \text{cm}^2}{\text{dmol}} \right] \quad (3)$$

where θ_{mrw} is a mean residue ellipticity, θ—ellipticity [degrees], *MRW*—mean residual weight of a protein [g/mol], *c*—protein concentration [g/L] and *l*—optical pathlength of a cuvette [cm]. The secondary structure content was estimated using CDPro [65].

3.8. Analytical Ultracentrifugtion

Sedimentation velocity analytical ultracentrifugation (SV AUC) was conducted in a Beckman Coulter ProteomeLab XLI analytical ultracentrifuge (Beckman Coulter, Brea, CA, USA) with an An60Ti rotor and assembled cells with two-channel 12 mm charcoal filled Epon® centerpieces and quartz windows, or sapphire windows for samples containing DTT. The proteins were analyzed at concentrations of 0.1, 0.25 and 0.5 mg/mL in H10Na500G5 with 1 mM EDTA or 10 mM $CaCl_2$. Additional measurements were made for 0.25 mg/mL protein with EDTA and $CaCl_2$ supplemented with 1 mM DTT. Effect of Tb^{3+} was analyzed by centrifuging 0.25 mg/mL protein with 7-fold molar excess of $TbCl_3$. Assembled cells with the samples were preincubated in the ultracentrifuge for 3 h at 20 °C and then centrifuged at 50,000 rpm (approximately 200,000× *g* at the bottom of the cell) overnight. The absorbance scans at 280 nm were collected continuously with 0.003 cm resolution. The scans were time-corrected [66] and analyzed in SEDFIT (version, 16.1c, October 2018, available at https://sedfitsedphat.nibib.nih.gov/, accessed 16 August 2021) using a continuous *c*(*s*) distribution model [67] with at least 20 points per 1 S. Partial specific volumes of the proteins, densities and dynamic viscosities of the solvents were calculated using SEDNTERP (version 3.0.3, 14 March 2021, available at http://www.jphilo.mailway.com/download.htm, accessed 16 August 2021). Maximum entropy regularization with $p = 0.95$ was used. Simplex and Marquardt–Levenberg algorithms were alternately used until the RMSD converged. Among the results of the calculations were sedimentation coefficients (*s*), sedimentation coefficients corrected for water at 20 °C ($s_{20,w}$), weight-averaged sedimentation coefficients ($\overline{s_{20,w}}$), apparent molecular weights (MW_{app}) and frictional ratios (f/f_0). *c*(*s*) distributions were visualized using GUSSI (version 1.4.2, 24 July 2018, available at https://www.utsouthwestern.edu/labs/mbr/software/, accessed 16 August 2021) [68] and Origin Pro 9.0 software.

3.9. Thermal Shift Assay

Thermal shift assay (TSA) was conducted as described [27]. Five μM solutions of the proteins in H10Na500G5 were supplemented with SYPRO Orange at concentration of 5× (hOtolC1q, R339S, R402P, Q426R) or 10× (R342W). The measurements were done in the presence of 1 mM EDTA, 0.1 mM $CaCl_2$, 1 mM $CaCl_2$, 10 mM $CaCl_2$, 100 mM $CaCl_2$, and 7-fold molar excess of $TbCl_3$. Final sample volume was 20 μL. The samples and the non-protein controls (Figure S4) were aliquoted into a 96-well plate in triplicate, covered with optically clear foil and incubated at room temperature for at least 1 h before the measurements. Fluorescence of SYPRO Orange was measured using Applied Biosystems ImageQuant5 qPCR thermal cycler (Thermo Fisher Scientific) with optical filters set as x1-m3 (excitation at 470 ± 15 nm, emission at 587 ± 10 nm) between 20 and 99 °C during heating at 0.033 °C/s. The data were analyzed using Protein Thermal Shift software (Thermo Fisher Scientific). Transition temperatures (T_m) were determined from the derivative of fluorescence with increasing temperature (dF/dT).

Supplementary Materials: The following are available online at https://www.mdpi.com/article/10.3390/ijms22169085/s1, Supplementary File S1.pdf—multiple sequence alignment of mammalian sequences of gC1q domain of otolin-1 found during NCBI BLAST search. Supplementary File S2.pdf contains Figures S1–S4, Tables S1 and S2. Figure S1. Dithiothreitol (DTT) does not affect the oligomerization of hOtolC1q R342W and R402P. Figure S2. Estimation of the secondary structure content of hOtolC1q and its mutants. Figure S3. Purification of hOtolC1q and its mutants. Figure S4. Background fluorescence in the thermal shift assay. Table S1. Parameters derived from the sedimentation velocity analytical ultracentrifugation. Table S2. Transition temperature (T_m) values (in °C) determined using the thermal shift assay.

Author Contributions: Conceptualization, R.H., A.O. and P.D.; methodology, R.H., A.O. and P.D.; validation—R.H., A.O.; investigation—R.H., resources—R.H., A.O. and P.D., writing—original draft, R.H.; writing—review and editing—A.O. and P.D., visualization—R.H., supervision—A.O. and P.D., project administration—A.O., funding acquisition—R.H., A.O. and P.D. All authors have read and agreed to the published version of the manuscript.

Funding: This work was supported by the National Science Centre, Poland (grant number 2017/27/N/NZ1/01319) and by a statutory activity subsidy from the Polish Ministry of Science and High Education for the Faculty of Chemistry of Wrocław University of Science and Technology.

Institutional Review Board Statement: Not applicable.

Informed Consent Statement: Not applicable.

Data Availability Statement: The data and materials underlying this article will be shared on request to one of the corresponding authors.

Acknowledgments: Sylwia Groborz and Martyna Uścița are kindly acknowledged for preparing mutated constructs of hOtolC1q.

Conflicts of Interest: The authors declare no competing interest.

Abbreviations

ANS	8-anilino-1-naphthalenesulfonic acid
BPPV	benign paroxysmal positional vertigo
C1QTNF5	complement C1q tumor necrosis factor-related protein 5
CD	circular dichroism spectroscopy
f/f_0	frictional ratio
dOtolC1q	gC1q domain of zebrafish otolin-1
DTT	dithiothreitol
EDTA	ethylenediaminetetraacetic acid
EOM	ensemble optimization method
gC1q	globular C-terminal domain
HEPES	4-(2-hydroxyethyl)-1-piperazineethanesulfonic acid
hOtolC1q	gC1q domain of human otolin-1
IPTG	isopropyl β-D-1-thiogalactopyranoside
K_d	dissociation constant
MW_{app}	apparent molecular weight
Oc90	otoconin-90
PCR	polymerase chain reaction
PMSF	phenylmethylsulfonyl fluoride
RPE	retinal pigment epithelium
$s_{20,w}$	sedimentation coefficient corrected for water at 20°C
SANS	small angle neutron scattering
SAXS	small angle X-ray scattering
SDS	sodium dodecyl sulfate
SNP	single nucleotide polymorphism
SV AUC	sedimentation velocity analytical ultracentrifugation
T_m	transition temperature
Tris	tris(hydroxymethyl)aminomethane
TSA	thermal shift assay

References

1. Thielens, N.M.; Tedesco, F.; Bohlson, S.S.; Gaboriaud, C.; Tenner, A.J. C1q: A fresh look upon an old molecule. *Mol. Immunol.* **2017**, *89*, 73–83. [CrossRef]
2. Xu, X.; Huang, X.; Zhang, L.; Huang, X.; Qin, Z.; Hua, F. Adiponectin protects obesity-related glomerulopathy by inhibiting ROS/NF-κB/NLRP3 inflammation pathway. *BMC Nephrol.* **2021**, *22*, 218. [CrossRef]
3. Bogin, O.; Kvansakul, M.; Rom, E.; Singer, J.; Yayon, A.; Hohenester, E. Insight into Schmid Metaphyseal Chondrodysplasia from the Crystal Structure of the Collagen X NC1 Domain Trimer. *Structure* **2002**, *10*, 165–173. [CrossRef]
4. Zhou, J.; Song, Y.; Gan, W.; Liu, L.; Chen, G.; Chen, Z.; Luo, G.; Zhang, L.; Zhang, G.; Wang, P.; et al. Upregulation of COL8A1 indicates poor prognosis across human cancer types and promotes the proliferation of gastric cancer cells. *Oncol. Lett.* **2020**, *20*, 34. [CrossRef]
5. Tu, X.; Palczewski, K. The macular degeneration-linked C1QTNF5 (S163) mutation causes higher-order structural rearrangements. *J. Struct. Biol.* **2014**, *186*, 86–94. [CrossRef] [PubMed]
6. Pandey, A.K.; Williams, R.W. Genetics of Gene Expression in CNS. *Int. Rev. Neurobiol.* **2014**, *116*, 195–231.
7. Petrosino, M.; Novak, L.; Pasquo, A.; Chiaraluce, R.; Turina, P.; Capriotti, E.; Consalvi, V. Analysis and Interpretation of the Impact of Missense Variants in Cancer. *Int. J. Mol. Sci.* **2021**, *22*, 5416. [CrossRef] [PubMed]
8. Schejbel, L.; Skattum, L.; Hagelberg, S.; Åhlin, A.; Schiller, B.; Berg, S.; Genel, F.; Truedsson, L.; Garred, P. Molecular basis of hereditary C1q deficiency—Revisited: Identification of several novel disease-causing mutations. *Genes Immun.* **2011**, *12*, 626–634. [CrossRef]
9. Waki, H.; Yamauchi, T.; Kamon, J.; Ito, Y.; Uchida, S.; Kita, S.; Hara, K.; Hada, Y.; Vasseur, F.; Froguel, P.; et al. Impaired multimerization of human adiponectin mutants associated with diabetes. Molecular structure and multimer formation of adiponectin. *J. Biol. Chem.* **2003**, *278*, 40352–40363. [CrossRef] [PubMed]
10. Wu, H.; Wang, S.; Li, G.; Yao, Y.; Wang, N.; Sun, X.; Fang, L.; Jiang, X.; Zhao, J.; Wang, Y.; et al. Characterization of a novel COL10A1 variant associated with Schmid-type metaphyseal chondrodysplasia and a literature review. *Mol. Genet. Genom. Med.* **2021**, *9*, e1668.
11. Zhang, C.; Liu, J.; Iqbal, F.; Lu, Y.; Mustafa, S.; Bukhari, F.; Lou, H.; Fu, R.; Wu, Z.; Yang, X.; et al. A missense point mutation in COL10A1 identified with whole-genome deep sequencing in a 7-generation Pakistan dwarf family. *Heredity* **2018**, *120*, 83–89. [CrossRef]
12. Skworc, A.; Osiadło, G.; Sławska, H.; Jezela-Stanek, A.; Marciniak, S. Wpływ leczenia usprawniającego na rozwój ruchowy pacjentki z chondrodysplazją przynasadową typu Schmida—Opis przypadku. *Pediatr. Pol.* **2017**, *92*, 214–217. [CrossRef]
13. Ikegawa, S.; Nishimura, G.; Nagai, T.; Hasegawa, T.; Ohashi, H.; Nakamura, Y. Mutation of the Type X Collagen Gene (COL10A1) Causes Spondylometaphyseal Dysplasia. *Am. J. Hum. Genet.* **1998**, *63*, 1659–1662. [CrossRef] [PubMed]
14. Athanasiadou, D.; Jiang, W.; Reznikov, N.; Rodríguez-Navarro, A.B.; Kröger, R.; Bilton, M.; González-Segura, A.; Hu, Y.; Nelea, V.; McKee, M.D. Nanostructure of mouse otoconia. *J. Struct. Biol.* **2020**, *210*, 107489. [CrossRef]
15. Schulz-Mirbach, T.; Ladich, F.; Plath, M.; Heß, M. Enigmatic ear stones: What we know about the functional role and evolution of fish otoliths. *Biol. Rev.* **2019**, *94*, 457–482. [CrossRef]
16. Thomas, O.R.B.; Swearer, S.E. Otolith Biochemistry—A Review. *Rev. Fish. Sci. Aquac.* **2019**, *27*, 458–489. [CrossRef]
17. Degens, E.T.; Deuser, W.G.; Haedrich, R.L. Molecular structure and composition of fish otoliths. *Mar. Biol.* **1969**, *2*, 105–113. [CrossRef]
18. Murayama, E.; Ohira, T.; Davis, J.G.; Takagi, Y.; Nagasawa, H.; Greene, M.I. Fish otolith contains a unique structural protein, otolin-1. *Eur. J. Biochem.* **2002**, *269*, 688–696. [CrossRef]
19. Deans, M.R.; Peterson, J.M.; Wong, G.W. Mammalian Otolin: A Multimeric Glycoprotein Specific to the Inner Ear that Interacts with Otoconial Matrix Protein Otoconin-90 and Cerebellin-1. *PLoS ONE* **2010**, *5*, e12765. [CrossRef]
20. Murayama, E.; Herbomel, P.; Kawakami, A.; Takeda, H.; Nagasawa, H. Otolith matrix proteins OMP-1 and Otolin-1 are necessary for normal otolith growth and their correct anchoring onto the sensory maculae. *Mech. Dev.* **2005**, *122*, 791–803. [CrossRef]
21. Yang, H.; Zhao, X.; Xu, Y.; Wang, L.; He, Q.; Lundberg, Y.W. Matrix Recruitment and Calcium Sequestration for Spatial Specific Otoconia Development. *PLoS ONE* **2011**, *6*, e20498. [CrossRef] [PubMed]
22. Moreland, K.T.; Hong, M.; Lu, W.; Rowley, C.W.; Ornitz, D.M.; De Yoreo, J.J.; Thalmann, R. In Vitro Calcite Crystal Morphology Is Modulated by Otoconial Proteins Otolin-1 and Otoconin-90. *PLoS ONE* **2014**, *9*, e95333. [CrossRef] [PubMed]
23. Chen, M.; McNeill, A.S.; Hu, Y.; Dixon, D.A. Elucidation of Bottom-Up Growth of $CaCO_3$ Involving Prenucleation Clusters from Structure Predictions and Decomposition of Globally Optimized $(CaCO_3)$ n Nanoclusters. *ACS Nano* **2020**, *14*, 4153–4165. [CrossRef] [PubMed]
24. Austad, B.; Vøllestad, L.A.; Foldvik, A. Frequency of vateritic otoliths and potential consequences for marine survival in hatchery-reared Atlantic salmon. *J. Fish Biol.* **2021**, *98*, 1401–1409. [CrossRef]
25. Tohse, H.; Saruwatari, K.; Kogure, T.; Nagasawa, H.; Takagi, Y. Control of Polymorphism and Morphology of Calcium Carbonate Crystals by a Matrix Protein Aggregate in Fish Otoliths. *Cryst. Growth Des.* **2009**, *9*, 4897–4901. [CrossRef]
26. Hołubowicz, R.; Wojtas, M.; Taube, M.; Kozak, M.; Ożyhar, A.; Dobryszycki, P. Effect of calcium ions on structure and stability of the C1q-like domain of otolin-1 from human and zebrafish. *FEBS J.* **2017**, *284*, 4278–4297. [CrossRef]
27. Hołubowicz, R.; Ożyhar, A.; Dobryszycki, P. Molecular mechanism of calcium induced trimerization of C1q-like domain of otolin-1 from human and zebrafish. *Sci. Rep.* **2021**, *11*, 12778. [CrossRef] [PubMed]

28. Humphrey, W.; Dalke, A.; Schulten, K. VMD: Visual molecular dynamics. *J. Mol. Graph.* **1996**, *14*, 33–38. [CrossRef]
29. Larkin, M.A.; Blackshields, G.; Brown, N.P.; Chenna, R.; McGettigan, P.A.; McWilliam, H.; Valentin, F.; Wallace, I.M.; Wilm, A.; Lopez, R.; et al. Clustal W and Clustal X version 2.0. *Bioinformatics* **2007**, *23*, 2947–2948. [CrossRef]
30. Waterhouse, A.M.; Procter, J.B.; Martin, D.M.A.; Clamp, M.; Barton, G.J. Jalview Version 2—A multiple sequence alignment editor and analysis workbench. *Bioinformatics* **2009**, *25*, 1189–1191. [CrossRef]
31. Ancien, F.; Pucci, F.; Godfroid, M.; Rooman, M. Prediction and interpretation of deleterious coding variants in terms of protein structural stability. *Sci. Rep.* **2018**, *8*, 4480. [CrossRef] [PubMed]
32. Kvansakul, M.; Bogin, O.; Hohenester, E.; Yayon, A. Crystal structure of the collagen α1(VIII) NC1 trimer. *Matrix Biol.* **2003**, *22*, 145–152. [CrossRef]
33. Tu, X.; Palczewski, K. Crystal structure of the globular domain of C1QTNF5: Implications for late-onset retinal macular degeneration. *J. Struct. Biol.* **2012**, *180*, 439–446. [CrossRef] [PubMed]
34. Pascolutti, R.; Erlandson, S.C.; Burri, D.J.; Zheng, S.; Kruse, A.C. Mapping and engineering the interaction between adiponectin and T-cadherin. *J. Biol. Chem.* **2020**, *295*, 2749–2759. [CrossRef]
35. Ebel, C.; Birck, C. Sedimentation Velocity Methods for the Characterization of Protein Heterogeneity and Protein Affinity Interactions. In *Multiprotein Complexes*; Springer: New York, NY, USA, 2021; pp. 155–171.
36. Semisotnov, G.V.; Rodionova, N.A.; Razgulyaev, O.I.; Uversky, V.N.; Gripas', A.F.; Gilmanshin, R.I. Study of the "molten globule" intermediate state in protein folding by a hydrophobic fluorescent probe. *Biopolymers* **1991**, *31*, 119–128. [CrossRef]
37. Stryer, L. The interaction of a naphthalene dye with apomyoglobin and apohemoglobin. *J. Mol. Biol.* **1965**, *13*, 482–495. [CrossRef]
38. Pantoliano, M.W.; Petrella, E.C.; Kwasnoski, J.D.; Lobanov, V.S.; Myslik, J.; Graf, E.; Carver, T.; Asel, E.; Springer, B.A.; Lane, P.; et al. High-density miniaturized thermal shift assays as a general strategy for drug discovery. *J. Biomol. Screen.* **2001**, *6*, 429–440. [CrossRef]
39. Lo, M.; Aulabaugh, A.; Jin, G.; Cowling, R.; Bard, J.; Malamas, M.; Ellestad, G. Evaluation of fluorescence-based thermal shift assays for hit identification in drug discovery. *Anal. Biochem.* **2004**, *332*, 153–159. [CrossRef]
40. Lundberg, Y.W.; Xu, Y.; Thiessen, K.D.; Kramer, K.L. Mechanisms of otoconia and otolith development. *Dev. Dyn.* **2015**, *244*, 239–253. [CrossRef]
41. Ye, Y.; Lee, H.-W.; Yang, W.; Yang, J.J. Calcium and lanthanide affinity of the EF-loops from the C-terminal domain of calmodulin. *J. Inorg. Biochem.* **2005**, *99*, 1376–1383. [CrossRef]
42. Wang, C.L.; Aquaron, R.R.; Leavis, P.C.; Gergely, J. Metal-binding properties of calmodulin. *Eur. J. Biochem.* **1982**, *124*, 7–12. [CrossRef]
43. Pidcock, E.; Moore, G.R. Structural characteristics of protein binding sites for calcium and lanthanide ions. *JBIC J. Biol. Inorg. Chem.* **2001**, *6*, 479–489. [CrossRef]
44. Necci, M.; Piovesan, D.; Tosatto, S.C.E. Critical assessment of protein intrinsic disorder prediction. *Nat. Methods* **2021**, *18*, 472–481. [CrossRef] [PubMed]
45. Dror, A.A.; Taiber, S.; Sela, E.; Handzel, O.; Avraham, K.B. A mouse model for benign paroxysmal positional vertigo with genetic predisposition for displaced otoconia. *Genes Brain Behav.* **2020**, *19*, e12635. [CrossRef] [PubMed]
46. Walther, L.E.; Blödow, A.; Buder, J.; Kniep, R. Principles of calcite dissolution in human and artificial otoconia. *PLoS ONE* **2014**, *9*, e102516. [CrossRef] [PubMed]
47. Han, D.-G.; Kim, D.-J. The evolutionary hypothesis of benign paroxysmal positional vertigo. *Med. Hypotheses* **2020**, *134*, 109445. [CrossRef]
48. Irugu, D.V.K.; Singh, A.; Yadav, H.; Verma, H.; Kumar, R.; Abraham, R.A.; Ramakrishnan, L. Serum otolin-1 as a biomarker for benign paroxysmal positional vertigo: A case-control study. *J. Laryngol. Otol.* **2021**, *135*, 589–592. [CrossRef]
49. Wu, Y.; Han, W.; Yan, W.; Lu, X.; Zhou, M.; Li, L.; Guan, Q.; Fan, Z. Increased Otolin-1 in Serum as a Potential Biomarker for Idiopathic Benign Paroxysmal Positional Vertigo Episodes. *Front. Neurol.* **2020**, *11*, 367. [CrossRef] [PubMed]
50. Zhao, X.; Yang, H.; Yamoah, E.N.; Lundberg, Y.W. Gene targeting reveals the role of Oc90 as the essential organizer of the otoconial organic matrix. *Dev. Biol.* **2007**, *304*, 508–524. [CrossRef]
51. Salt, A.N.; Inamura, N.; Thalmann, R.; Vora, A. Calcium gradients in inner ear endolymph. *Am. J. Otolaryngol.* **1989**, *10*, 371–375. [CrossRef]
52. Nakashima, T.; Pyykkö, I.; Arroll, M.A.; Casselbrant, M.L.; Foster, C.A.; Manzoor, N.F.; Megerian, C.A.; Naganawa, S.; Young, Y.-H. Meniere's disease. *Nat. Rev. Dis. Prim.* **2016**, *2*, 16028. [CrossRef] [PubMed]
53. Sim, N.-L.; Kumar, P.; Hu, J.; Henikoff, S.; Schneider, G.; Ng, P.C. SIFT web server: Predicting effects of amino acid substitutions on proteins. *Nucleic Acids Res.* **2012**, *40*, W452–W457. [CrossRef] [PubMed]
54. Adzhubei, I.; Jordan, D.M.; Sunyaev, S.R. Predicting Functional Effect of Human Missense Mutations Using PolyPhen-2. *Curr. Protoc. Hum. Genet.* **2013**, *76*, 7.20.1–7.20.41. [CrossRef]
55. Rentzsch, P.; Witten, D.; Cooper, G.M.; Shendure, J.; Kircher, M. CADD: Predicting the deleteriousness of variants throughout the human genome. *Nucleic Acids Res.* **2019**, *47*, D886–D894. [CrossRef] [PubMed]
56. Ioannidis, N.M.; Rothstein, J.H.; Pejaver, V.; Middha, S.; McDonnell, S.K.; Baheti, S.; Musolf, A.; Li, Q.; Holzinger, E.; Karyadi, D.; et al. REVEL: An Ensemble Method for Predicting the Pathogenicity of Rare Missense Variants. *Am. J. Hum. Genet.* **2016**, *99*, 877–885. [CrossRef]

57. Dong, C.; Wei, P.; Jian, X.; Gibbs, R.; Boerwinkle, E.; Wang, K.; Liu, X. Comparison and integration of deleteriousness prediction methods for nonsynonymous SNVs in whole exome sequencing studies. *Hum. Mol. Genet.* **2015**, *24*, 2125–2137. [CrossRef]
58. Reva, B.; Antipin, Y.; Sander, C. Predicting the functional impact of protein mutations: Application to cancer genomics. *Nucleic Acids Res.* **2011**, *39*, e118. [CrossRef]
59. Liu, H.; Naismith, J.H. An efficient one-step site-directed deletion, insertion, single and multiple-site plasmid mutagenesis protocol. *BMC Biotechnol.* **2008**, *8*, 91. [CrossRef]
60. Laemmli, U.K. Cleavage of Structural Proteins during the Assembly of the Head of Bacteriophage T4. *Nature* **1970**, *227*, 680–685. [CrossRef]
61. Gasteiger, E.; Hoogland, C.; Gattiker, A.; Duvaud, S.; Wilkins, M.R.; Appel, R.D.; Bairoch, A. Protein Identification and Analysis Tools on the ExPASy Server. In *The Proteomics Protocols Handbook*; Humana Press: Totowa, NJ, USA, 2005; pp. 571–607.
62. Gonzalez, W.G.; Ramos, V.; Diaz, M.; Garabedian, A.; Molano-Arevalo, J.C.; Fernandez-Lima, F.; Miksovska, J. Characterization of the Photophysical, Thermodynamic, and Structural Properties of the Terbium(III)–DREAM Complex. *Biochemistry* **2016**, *55*, 1873–1886. [CrossRef]
63. Gonzalez, W.G.; Pham, K.; Miksovska, J. Modulation of the Voltage-gated Potassium Channel (Kv4.3) and the Auxiliary Protein (KChIP3) Interactions by the Current Activator NS5806. *J. Biol. Chem.* **2014**, *289*, 32201–32213. [CrossRef] [PubMed]
64. Martin, S.R.; Schilstra, M.J. Circular Dichroism and Its Application to the Study of Biomolecules. *Methods Cell Biol.* **2008**, *84*, 263–293. [PubMed]
65. Sreerama, N.; Woody, R.W. Estimation of Protein Secondary Structure from Circular Dichroism Spectra: Comparison of CONTIN, SELCON, and CDSSTR Methods with an Expanded Reference Set. *Anal. Biochem.* **2000**, *287*, 252–260. [CrossRef]
66. Zhao, H.; Ghirlando, R.; Piszczek, G.; Curth, U.; Brautigam, C.A.; Schuck, P. Recorded scan times can limit the accuracy of sedimentation coefficients in analytical ultracentrifugation. *Anal. Biochem.* **2013**, *437*, 104–108. [CrossRef]
67. Schuck, P. Size-Distribution Analysis of Macromolecules by Sedimentation Velocity Ultracentrifugation and Lamm Equation Modeling. *Biophys. J.* **2000**, *78*, 1606–1619. [CrossRef]
68. Brautigam, C.A. Calculations and Publication-Quality Illustrations for Analytical Ultracentrifugation Data. *Methods Enzymol.* **2015**, *562*, 109–133. [PubMed]

International Journal of
Molecular Sciences

Article

Structural Contour Map of the Iota Carbonic Anhydrase from the Diatom *Thalassiosira pseudonana* Using a Multiprong Approach

Erik L. Jensen [1,†], Véronique Receveur-Brechot [1], Mohand Hachemane [1], Laura Wils [1], Pascale Barbier [2], Goetz Parsiegla [1], Brigitte Gontero [1,*] and Hélène Launay [1,*]

[1] Aix Marseille Univ, CNRS, BIP, UMR 7281, IMM, FR 3479, 31 Chemin J. Aiguier, CEDEX 20, 13 402 Marseille, France; jensen@ibpc.fr (E.L.J.); veronique.brechot@imm.cnrs.fr (V.R.-B.); mohand-said.HACHEMANE@etu.univ-amu.fr (M.H.); laura.wils@hotmail.fr (L.W.); goetz.parsiegla@imm.cnrs.fr (G.P.)

[2] Aix Marseille Univ, CNRS, INP, Inst Neurophysiopathol, 13 402 Marseille, France; pascale.barbier@univ-amu.fr

* Correspondence: bmeunier@imm.cnrs.fr (B.G.); helene.LAUNAY@univ-amu.fr (H.L.)

† Current address: UMR 7141 Centre National de la Recherche Scientifique (CNRS), Laboratory of Chloroplast Biology and Light Sensing in Microalgae, Institut de Biologie Physico-Chimique, Sorbonne Université, 75005 Paris, France.

Citation: Jensen, E.L.; Receveur-Brechot, V.; Hachemane, M.; Wils, L.; Barbier, P.; Parsiegla, G.; Gontero, B.; Launay, H. Structural Contour Map of the Iota Carbonic Anhydrase from the Diatom *Thalassiosira pseudonana* Using a Multiprong Approach. *Int. J. Mol. Sci.* **2021**, 22, 8723. https://doi.org/10.3390/ijms22168723

Academic Editors: Istvan Simon and Csaba Magyar

Received: 21 July 2021
Accepted: 11 August 2021
Published: 13 August 2021

Publisher's Note: MDPI stays neutral with regard to jurisdictional claims in published maps and institutional affiliations.

Abstract: Carbonic anhydrases (CAs) are a family of ubiquitous enzymes that catalyze the interconversion of CO_2 and HCO_3^-. The "iota" class (ι-CA) was first found in the marine diatom *Thalassiosira pseudonana* (tpι-CA) and is widespread among photosynthetic microalgae and prokaryotes. The ι-CA has a domain COG4875 (or COG4337) that can be repeated from one to several times and resembles a calcium–calmodulin protein kinase II association domain (CaMKII-AD). The crystal structure of this domain in the ι-CA from a cyanobacterium and a chlorarachniophyte has been recently determined. However, the three-dimensional organization of the four domain-containing tpι-CA is unknown. Using biophysical techniques and 3-D modeling, we show that the homotetrameric tpι-CA in solution has a flat "drone-like" shape with a core formed by the association of the first two domains of each monomer, and four protruding arms formed by domains 3 and 4. We also observe that the short linker between domains 3 and 4 in each monomer confers high flexibility, allowing for different conformations to be adopted. We propose the possible 3-D structure of a truncated tpι-CA containing fewer domain repeats using experimental data and discuss the implications of this atypical shape on the activity and metal coordination of the ι-CA.

Keywords: analytical ultracentrifugation; CO_2 concentrating mechanism; diffusion-ordered NMR spectroscopy; electrospray ionization mass spectrometry; homotetramer; manganese; metalloprotein; photosynthesis; small-angle X-ray scattering

1. Introduction

Carbonic anhydrases (CAs; EC 4.2.1.1) are widespread enzymes found in all domains of life [1,2]. They all catalyze the same reversible reaction of CO_2 hydration to form HCO_3^-. Several classes of CAs have been described so far, named using the Greek letters α-, β-, γ-, δ-, ζ-, η-, θ- and ι- [1,3–6]. CAs participate in numerous cellular processes, such as pH regulation, ion transport and cell metabolism [7], and in CO_2 concentrating mechanisms in photosynthetic organisms [2,8]. Living organisms may possess one to several different CA classes encoded in their genomes, reflecting the diversity of CAs [2,8,9].

Until recently, nearly all CAs have been described as metalloenzymes, which commonly use Zn^{2+} as a metal cofactor; however, some CAs from the γ-, δ- and ζ- classes are cambialistic and are able to replace Zn^{2+} by Fe^{2+}, Co^{2+} and Cd^{2+}, respectively [4,10]. Although the different CA classes catalyze the same reaction, they share little or no apparent evolutionary relationship in terms of amino acid sequence or structure [1], including the amino acids involved in the coordination of their metal ion cofactor and catalytic site as well as their oligomeric state [7,11].

The most recently described CAs from the new ι- class was first found in the marine diatom *Thalassiosira pseudonana* [6,12]. This ι-CA is an important component of the diatom CO_2-concentrating mechanisms (CCMs) that are essential for carbon fixation in many photosynthetic organisms [13]. In addition, ι-CAs are widespread among marine photosynthetic microalgae and non-photosynthetic prokaryotes, which suggest an important role of this CA in global carbon biogeochemical cycling [6]. The ι-CAs have metal cofactors that differ from one organism to the other. While ι-CA from *T. pseudonana* was shown to use Mn^{2+} instead of Zn^{2+}, a recently reported ι-CA homolog from the bacterium *Burkholderia territorii* is highly specific for Zn^{+2} [14]. Moreover, a novel type of ι-CA from a cyanobacterium (*Anabaena* sp. PCC7120) and a chlorarachniophyte alga (*Bigelowiella natans*) that is able to act without any metal cofactor was recently described [15]. These differences in metal cofactor preference of ι-CA homologs might be related to specific features of their structure. The amino acid sequence of the ι-CA from diatoms is characterized by a domain (COG4875) from the nuclear transport factor 2 (NTF2) family, which is found in proteins with a broad range of biological functions [16]. This domain is also highly similar to the calcium-calmodulin protein kinase II association domain (CaMKII-AD) and can be repeated one to several times along the protein sequence. The amino acid sequence of the ι-CA from *T. pseudonana* has four domain repetitions, but in other algal species, it can contain two or three (or more) repetitions [6,15]. Interestingly, most sequences from prokaryotes contain only one domain—in contrast to sequences from eukaryotes—which may reflect an evolutionary trait in species from different domains in the Tree of Life. The CaMKII-AD has a known role in protein oligomerization [17] as well as in NTF2 proteins, which are known to form dimers [18], and in both cases, the domain structure is characterized by a cone-shaped cavity formed by an angled arrangement of a β-sheet and α-helices [16,19]. A similar structure has been predicted from the dimeric ι-CA from *B. territorii*, which contains one domain [14]. Similarly, the X-ray crystal structure of the COG4337 domains that compose the ι-CA from *Anabaena* and *B. natans* confirmed a high structural homology with the CaMKII-AD from *Xanthomonas campestris* (PDB: 3H51) [15]. However, there is still not a resolved structure from a full-length multi-domain-containing ι-CA from diatoms based on experimental data. This information could help to determine the 3D arrangement of multiple CaMKII-AD-containing proteins as well as their multimeric organization.

In this work, we describe some structural features of the four-domain-repeats containing ι-CA from *T. pseudonana* (tpι-CA) using different biophysical techniques. We constructed a model of the homotetrameric form of tpι-CA using predicted 3D models integrating small-angle X-ray scattering (SAXS) and nuclear magnetic resonance (NMR) approaches. In addition, based on these data, we also proposed a model for other ι-CAs that contain three or fewer domain repetitions.

2. Results

2.1. Oligomerization State of the tpι-CA

Two oligomeric forms of the recombinant tpι-CA in solution have previously been observed [6] (Figure 1b), which were named the "high molecular mass" (HMM) and "low molecular mass" (LMM) forms as their real molecular masses were not determined. The elution volumes of both forms on size exclusion chromatography (SEC) were much smaller than expected for tpι-CA monomers, indicating a higher oligomerization state for both forms. Congo red spectral shift assay experiments were used to exclude the possibility that the HMM form resulted from a denatured and amyloid-like aggregated form of tpι-CA. Our results showed that Congo red does not bind to tpι-CA and, thus, suggest that the protein does not form fibrils in solution in our conditions (Figure 1a). Besides an absorption at 280 nm, the HMM form unexpectedly also showed an absorption at 260 nm, indicating the presence of nucleic acid in this sample. Agarose gel electrophoresis and ethidium bromide staining confirmed the presence of DNA or RNA in this form (not shown). Using SEC, we showed that, when the protein sample was treated with Benzonase, a nuclease that attacks and degrades all forms of nucleic acids, the amount of HMM form decreased while the amount of LMM increased (Figure 1b). We speculate that this DNA/RNA binding is very likely to be unspecific and not physiological relevant because tpι-CA is located in the vicinity of chloroplast membranes [6]. Consequently, further structural characterization was performed instead on the LMM form using dynamic light scattering (DLS) and analytical ultracentrifugation (AUC). We observed that this LMM form was principally monodisperse. The LMM form has an hydrodynamic radius of 8.81 ± 0.4 nm, determined by DLS, and a sedimentation coefficient standard $S^0_{20,W}$ (20 °C in water and extrapolated to protein concentration equal to zero) of 8.5 S, determined by AUC (Figure 1c,d).

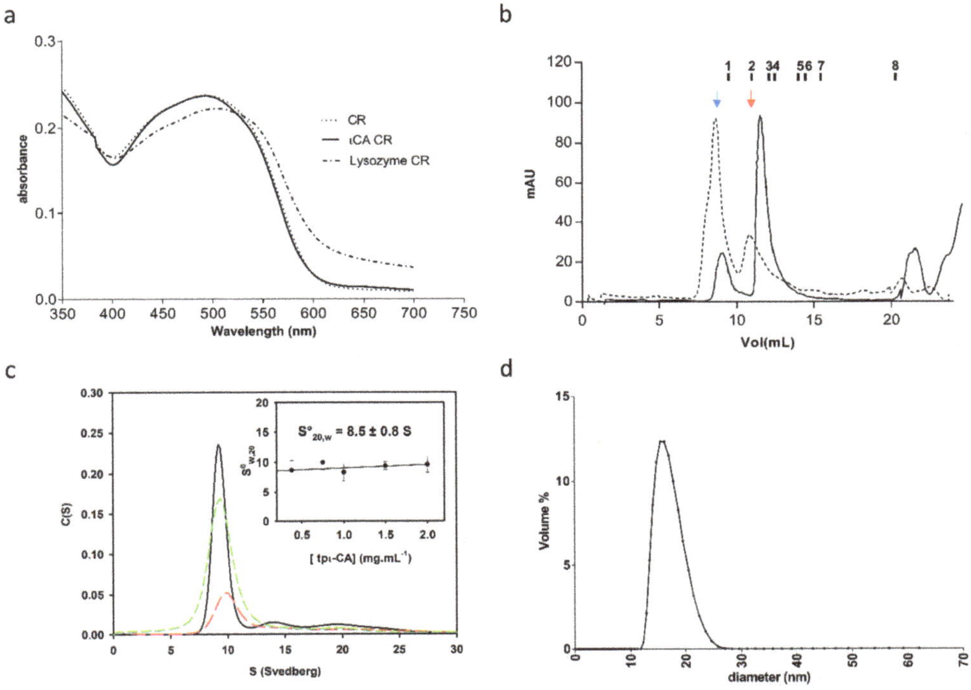

Figure 1. Oligomerization state of the tpι-CA. (**a**) Congo red (CR) spectral shift assay from purified tpι-CA. The spectrum of CR alone and mixed with a tpι-CA sample is shown. The presence of fibrils is shown by a shift in the spectrum of a CR-lysozyme control previously heated at 55 °C for 5 min prior to assay. (**b**) Size exclusion chromatography of recombinant tpι-CA. The HMW form (dotted line) produces the LMM form (plain line) upon treatment with Benzonase. The elution volume of the LMM form is 10.71 mL which corresponds to an apparent MW of 280 kDa. The elution volumes of standard proteins are indicated above the profile: 1—blue dextran, 2—ferritin (440 kDa), 3—catalase (240 kDa), 4—aldolase (158 kDa), 5—Bovine Serum Albumin (BSA) dimer (136 kDa), 6—BSA monomer (68 kDa), 7—ovalbumin (43 kDa) and 8—Cytochrome C (12.5 kDa). Blue and red arrows show peaks corresponding to the HMM and LMM, respectively. (**c**) Sedimentation velocity experiment in an analytical ultracentrifugation (AUC) performed on the purified LMM form at different concentrations: 2.0 (black), 1.5 (green dashed) and 0.7 (red dashed) mg mL^{-1}. Standard sedimentation coefficient $S^0_{20,W}$ determination is obtained by extrapolating the $S_{20,W}$ value at protein concentration equal to zero, as shown in the inset. (**d**) Dynamic light scattering (DLS) curve of the LMM form.

Electrospray ionization mass spectrometry (ESI-MS) was used under non-denaturing conditions to probe the oligomerization state of the tpι-CA in solution. We observed a distribution of multiple charged ions of tpι-CA from 28 to 32 that corresponded to a calculated averaged neutral molecular mass of 260 kDa, indicating that the oligomerization state of tpι-CA is a homotetramer (Table 1). In addition, after glutaraldehyde-induced protein cross-linking, SDS-PAGE and Western blot analysis also showed the presence of a homotetrameric form of ~240 kDa, together with possible trimeric and dimeric intermediates (~120 and ~180 kDa, respectively; Figure 2). The homotetrameric state of tpι-CA is also in agreement with the observed apparent molecular mass of 280 kDa (Figure 1b) observed by SEC.

Table 1. Molecular mass and oligomeric state of the full-length tpι-CA determined by ESI-MS.

Charge State	***m/z***		**Delta *m/z***
	Theoretical	**Experimental** *	
28	9285.00	9288.00	3.00
29	8964.86	8970.22	5.36
30	8666.07	8670.32	4.25
31	8386.55	8390.12	3.57
32	8124.50	8127.45	2.95
Deduced multimeric mass (Da)	**Oligomer state**	**Deduced monomer mass (theoretical) (Da)**	**Error (ppm)**
260,066.20	4	65,016.55 (64,988)	439

* As an example, an m/z of 9288 with a charge state of 28 gives a ((9288 × 28) − 28) or 260,036 Da molecular mass.

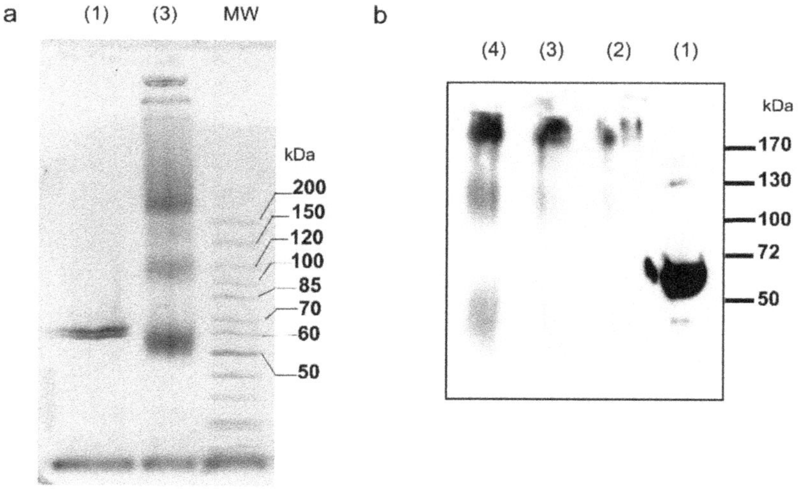

Figure 2. Glutaraldehyde-induced protein cross-linking. (**a**) SDS-PAGE and (**b**) Western blot of the following samples: (1) untreated purified tpι-CA, 5 μg; (2–4) cross-linked purified tpι-CA, 2, 5 and 10 μg, respectively. MW: molecular weight markers.

2.2. Characterization of the Secondary Structure of tpι-CA and Its Domain Variants

As previously described, the ι-CA is widely distributed among living organisms, and the number of COG4875 domain repetitions contained within its sequence may vary among different species [6]. The ι-CA protein from *T. pseudonana* contains four repetitions of the COG4875 domain along its full amino acid sequence, excluding a chloroplast-targeting signal peptide on its N-terminus. These four COG4875 domains shared more than 60% of amino acid identity (Figure 3a) and more than 40% identity with the putative calcium/calmodulin protein kinase II association domain from *X. campestris* (CaMKII-AD: PDB 3H51). In contrast, alignment with a NTF2 protein family domain (from *Rattus norvegicus*: PDB 1OUN), to which the COG4875 is also predicted as a family member, showed less than 20% identity (Figure 3a). These results indicate that the amino acid sequences of the four domains are highly similar.

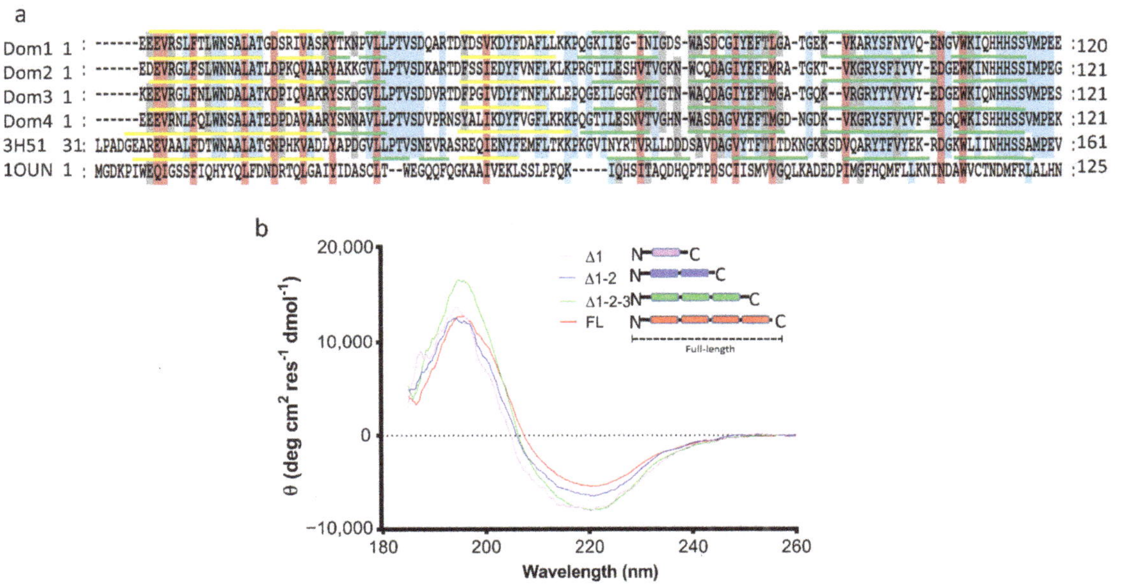

Figure 3. Analysis of the secondary structure. (**a**) Alignment of the amino acid sequence of each domain contained in the full-length tpι-CA; 3H51 and 1OUN correspond to the PDB sequences of the CaMKII-AD from *Xanthomonas campestris* and the NTF2 from *Rattus norvegicus*, respectively. Alignments were performed using MEGA4 software and analyzed using GeneDoc (University of Pittsburg. Available online: http://www.psc.edu/biomed/genedoc (accessed on 15th June 2021)). Shading levels correspond to the conservation of amino acids: Red, above 80% of identity; blue, 70% identity; and light grey, 60% identity. The yellow and green lines above each sequence represent the presences of α-helices and β-strands, respectively. The secondary structures of the four tpι-CA domains were predicted using PSIPred software; for 3H51 and 1OUN, the secondary structures were obtained from their crystal structures. (**b**) Circular dichroism spectra of full-length tpι-CA and of the variants containing one, two or three domain repetitions.

The secondary structure of each domain contained in the tpι-CA was predicted using PSIPred webserver [20]. These predictions showed a similar proportion of α-helices (19%), β-strands (32%) and coils (49%) to that from experimentally determined crystal structures from the CaMKII-AD (3H51) and NTF2 (1OUN) protein domains (Table 2). We experimentally confirmed the secondary structural content of tpι-CA using circular dichroism (CD; Figure 3b) and showed that the protein in solution contains 39% of β-strands; 54% coils, which are highly similar to the predicted secondary structure population; and a slightly lower content of α-helices (7%) compared with the predictions.

Table 2. Proportion of secondary structural elements in tpι-CA, derived either from the experimental CD data or from prediction.

Tp-ιCA Domain-Repeat Variant	Experimental CD (DichroWeb Analysis)			Predicted Secondary Structure (PSIPred)		
	β-Strand	α-Helix	Coil	β-Strand	α-Helix	Coil
FL	0.39	0.07	0.54	0.32	0.19	0.49
Δ1-2-3	0.35	0.18	0.48	0.33	0.17	0.50
Δ1-2	0.37	0.08	0.55	0.34	0.17	0.49
Δ1	0.31	0.19	0.50	0.32	0.16	0.52

In order to determine whether the secondary structure of independent COG4875 domains varies within the full-length tpι-CA, we analyzed the secondary structure of several truncated forms of the tpι-CA containing different numbers of domain repetitions by deleting the increasing number of domains from the C-terminus, hereafter referred to as the Δ1-2-3 (composed of domains 1, 2 and 3), Δ1-2 (composed of domains 1 and 2) and Δ1 (composed of domain 1 only) variants. The CD spectrum of each variant (Figure 3b) was analyzed using Dichroweb server [21,22]. The analyses showed that all variants have similar contents of β-strands (31–39%) and unstructured coils (48–55%) in agreement to predictions from PSIPred (Table 2) and that they possess a variable and low content of α-helices (8–19%). This result suggests that the overall secondary structure of the tpι-CA might remain invariable regardless of the number of individual COG4875 domains.

2.3. Domain Organization in Tetrameric tpι-CA

The domain repetition in tpι-CA raises the question of their respective organization. We used small-angle X-ray scattering (SAXS) to determine the global structure of the tpι-CA in solution. Size-exclusion chromatography coupled with SAXS on tpι-CA gave rise to a single elution peak, as expected from the abovementioned SEC data (Figure 1b). A Guinier analysis of the X-ray scattering data indicated a radius of gyration of this LMM form of 66.5 ± 0.6 Å and the distance distribution computed from the scattering curve indicated a maximum dimension (D_{max}) of 250 Å, suggesting that the protein has a very anisotropic shape. The molecular mass inferred from the data was 292 ± 30 kDa, corresponding to a tetrameric tpι-CA, as observed using MS-ESI (Table 1), cross-linking and SEC (Figures 1b and 2). The global envelope computed from the scattering data is an atypical flat shape with four protruding arms and a ring-like structure in the center. Sixteen copies of the homology models of the COG4875 domains can be accommodated in this global envelope, with eight domains in the doughnut-shaped center of the SAXS envelope and two domains per arm (Figure 4). This global envelope indicates that the four domains are not equivalent and that one moiety constitutes the oligomer interface while the other is exposed to the solvent.

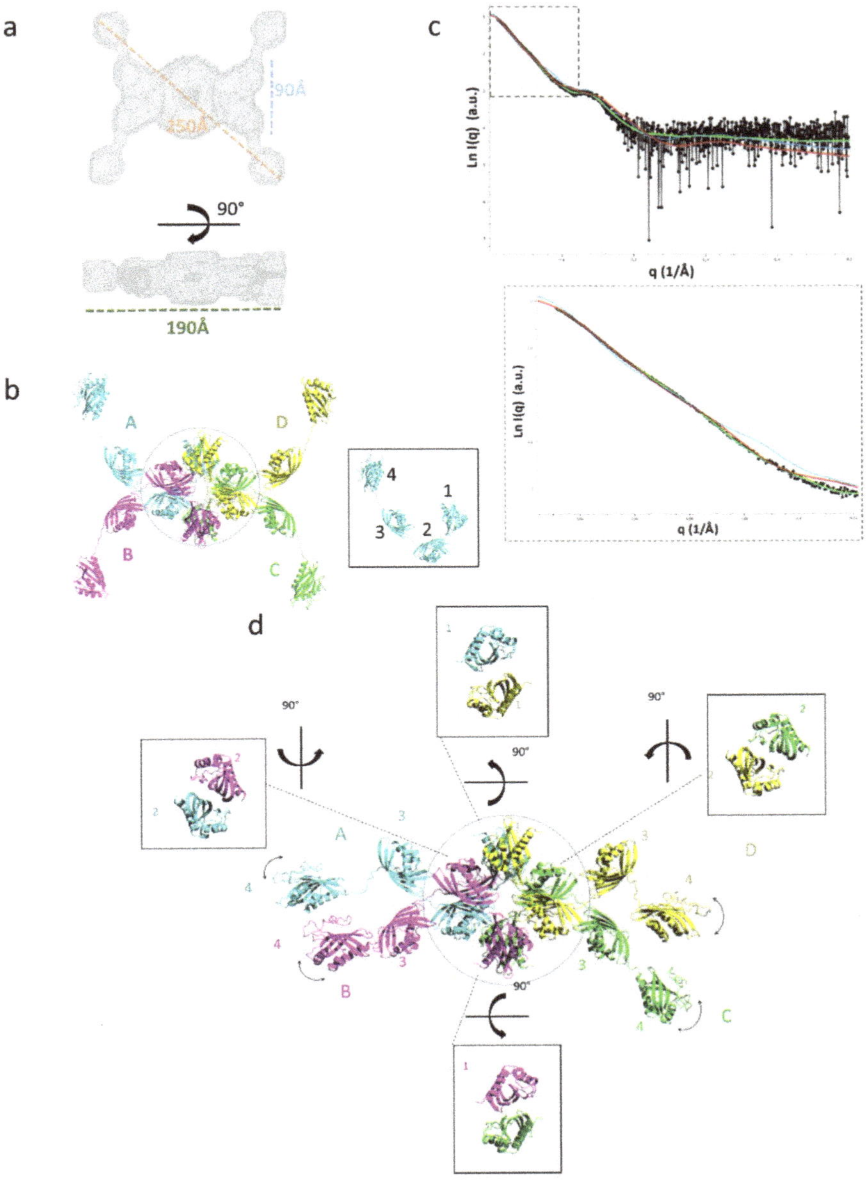

Figure 4. Three-dimensional modeling of the tetrameric full-length tpι-CA. (**a**) Ab initio SAXS-based model envelope inferred from the experimental data. *Top*: front view; *bottom*: lateral view. (**b**) Proposed model of the tetramer based on a SAXS envelope. Colored letters are used to designate each monomer. The box on the right shows the orientation of each domain within one monomer. (**c**) *Top*: The experimental scattering curve (black circles) is compared with the calculated scattering curves of the shape determined by DAMMIF (green curve, χ^2 of 2.1), of the initial atomic model (blue curve, χ^2 of 6.2) and of a model generated by CORAL (red curve, χ^2 of 2.1); *bottom*: A zoom on the data at low q (dashed box) is shown. (**d**) The CORAL model is represented in ribbons, with one color per monomeric chain, as in (**b**). The interface of each of the domain pairs contained in the central core (circled) is shown.

In order to ascertain which moiety was involved in the oligomerization, we measured the hydrodynamic properties of the truncated domain variants presented above. The size exclusion profiles and hydrodynamic radii determined from Diffusion Ordered Spectroscopy-NMR (DOSY-NMR) indicated that these truncated forms remained oligomeric (Figure 5). We thus placed N-terminal domain 1 at the oligomeric interface at the center of the SAXS global envelope and the C-terminal domain 4 in the protruding arms. We then built an atomic model by sequence-based homology modeling using the crystal structure of the *X. campestris* CaMKII-AD (3H51).

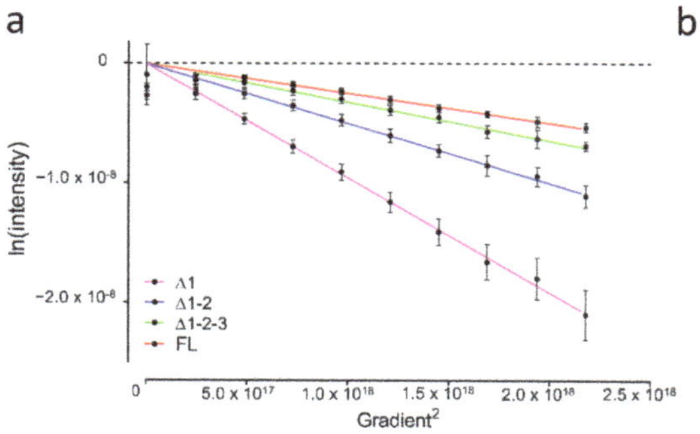

Tp-ιCA Domain-repeat variant	Experimental D_t (x 10^{-11} m^2 s^{-1})	Experimental R_h (nm)	HYDROPRO D_t (x 10^{-11} m^2 s^{-1})	HYDROPRO R_h (nm)
FL	2.2 ± 0.1	8.5 ± 0.3	2.52	8.52
Δ1-2-3	3.2 ± 0.2	6.6 ± 0.4	4.00	5.37
Δ1-2	5.0 ± 0.3	4.3 ± 0.3	5.52	3.88
Δ1	9.6 ± 0.5	2.2 ± 0.2	8.70	2.47

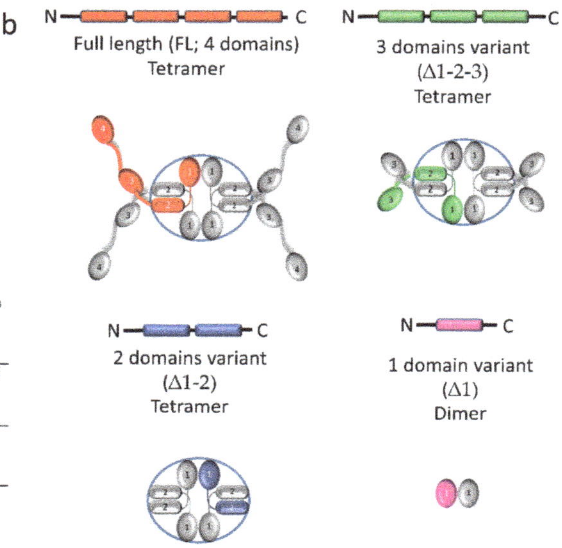

Figure 5. Domain organization of tpι-CA. (**a**) Translational diffusion coefficient (D_t) of the different domain variants produced from the full-length (four domain-containing) tpι-CA obtained from DOSY-NMR. *Top*: Logarithm of the NMR signal intensity as a function of the square of the gradient strength. *Bottom*: Table showing experimental D_t for the inferred R_h using the Stokes–Einstein relation as well as the computed D_t and R_h from the homology model using HYDROPRO software. The calculation for the full-length construct and the domain variant constructs Δ1-2-3 and Δ1-2 are performed on the tetrameric forms. The calculation of the domain variant Δ1 is performed on the dimeric form. (**b**) Schematic models of the different tpι-CA domain variants from which the HYDROPRO calculation was performed. Only one monomer was colored in each model.

The doughnut-shaped center accommodates four copies of the domains 1 and 2. In the X-ray structures of ι-CA domains from *Anabaena* (7C5W and 7C5V) and from *B. natans* (7C5Y and 7C5X), and of the *X. campestris* CaMKII-AD (3H51), the domain-domain interface is composed of two antiparallel β-sheets, and this interface was conserved in our homology model of tpι-CA. This β-sandwich domain-domain interface was observed twice between two domain 1s and twice between two domain 2s. This interface between two domain 1s associates the monomers A and D, and the monomers B and C of the tetrameric tpι-CA (Figure 4). Conversely, the domain 2 pairs do not connect the same monomers: the domain 2 interfaces are between monomers A and B, and monomers C and D. This means that each monomer faces two different chains in its domains 1 and 2. This "turning" or entangled scaffold allows for the tetramerization of tpι-CA (Figure 4b). The short linker between domains 1 and 2 is embedded in the doughnut-shaped center. The SAXS envelope of the arm accommodates domains 3 and 4, together with the linkers between domains 2 and 3 and between domains 3 and 4.

2.4. Flexibility in the Protruding Arms Brought by the Linkers

This "drone-like" homology model is coherent with all of the biophysical data and indicates that, despite their high homology in an amino-acid and secondary structure composition, the four domains are not equivalent within the tpι-CA structure. Domains 1 and 2 are involved in dimeric interfaces, while domains 3 and 4 are more exposed to the solvent. The linkers between the domains might be the determining factor controlling this peculiar domain organization. Indeed, the homology between the three linkers (i.e., between domains 1 and 2, domains 2 and 3, and domains 3 and 4) is lower than the homology between domains (Figure 6a), as expected for disordered regions. These linkers are predicted to be flexible linkers using the disorder predictor IU-pred2A [23] and other predictors (Figure 6b) and are expected to bring a high level of flexibility in the tpι-CA arms. The predicted flexibility of the linker 3–4 is higher than that of the other two linkers, as expected from the presence of two proline and three charged residues (Figure 6a). We also calculated the theoretical scattering curve of our atomic model and compared it with the experimental scattering curve using CRYSOL. The fit to the data was fair, with a χ^2 of 6.2, revealing that the model is good but that there may be some flexibility in the overall architecture of the tetramer, accounting for the slight discrepancies between the two curves in the low-q-region (Figure 4c).

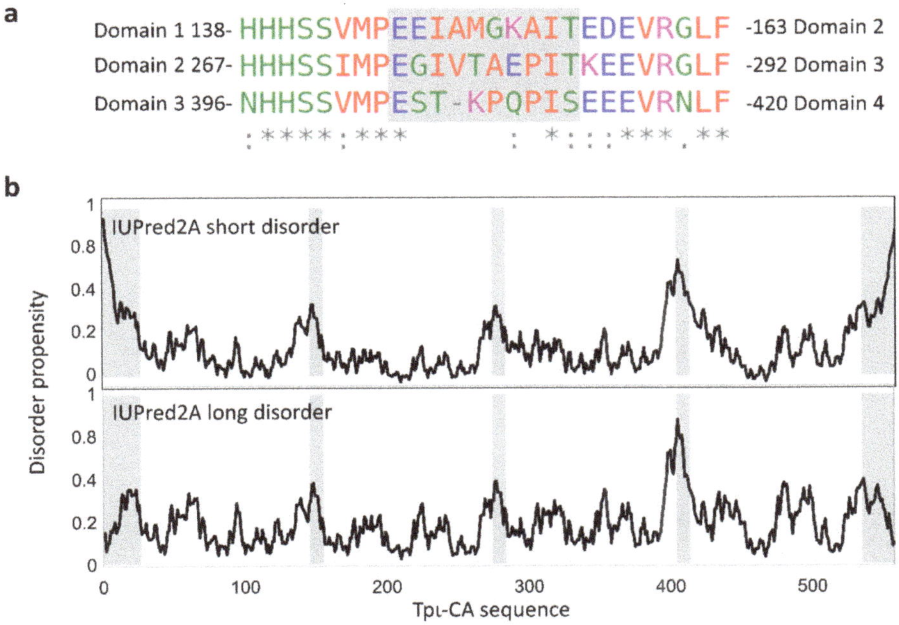

Figure 6. Flexibility of the interdomain linkers. (**a**) Clustalω alignments of the linkers between domains 1 and 2, domains 2 and 3, and domains 3 and 4. The linker residues are shaded in grey. (**b**) IUPred2A disorder prediction for short disordered regions (top) or long disordered regions (bottom). Only the linker between domains 3 and 4 is predicted to be a long-disordered region.

Since the global (and average) SAXS envelope did not account for this putative flexibility, we introduced a flexibility between domains 3 and 4 in our atomic homology model using the program **CO**mplexes with **RA**ndom **L**oops (CORAL) and compared the back-calculated theoretical scattering curves with the experimental SAXS data [24]. Better fits to the data were obtained when flexibility was allowed for this linker compared with rigid models (χ^2 of 2.1 vs. 6.2, respectively; Figure 4c,d). In the generated structures, the domains 4 were localized in a range of positions that confirm the flexibility of the linkers between domains 3 and 4 (Figure 4b,d). The experimental SAXS data arise from the ensemble of possible conformers, and these average data were best reproduced when the localization of domain 4 was not constrained. This confirms the dynamic nature of the protruding arms of the "drone-like" structure. The presence of a highly flexible linker between domains 3 and 4 might act as a string that competes with a possible domain 4–omain 4 dimerization.

2.5. Experimental Validation of the "Drone-Like" Structural Model

In order to validate this domain organization and the atypical "drone-like" shape, we first computed the hydrodynamic radius of the model using HYDROPRO [25], and compared the calculated hydrodynamic properties with the experimentally measured translational diffusion coefficient obtained from DOSY-NMR; they were identical within uncertainty (Figure 5). We further confirmed the position of the domains by analyzing the truncated domain variants mentioned above (Section 2.2). The experimental hydrodynamic properties of the domain-4-deleted construct (Δ1-2-3) were identical to that computed from the model in which the domain 4 was deleted from the full-length sequence. The experimental hydrodynamic radius of the two-domain construct (Δ1-2) is typical of a spherical tetramer, as expected from the central doughnut shape. The experimental hydrodynamic radius of the domain 1 (Δ1) alone is close to that computed from a dimer, as expected from the dimer interface in our model and in other ι-CA domains.

3. Discussion

CAs are often cited as a good example of convergent evolution, in which unrelated enzymes evolve to catalyze the same ubiquitous reaction. The different classes forming the CA family are surprisingly diverse in primary, secondary, tertiary and even quaternary structures [11,26]. Here, we studied the features of the overall three-dimensional shape of the recently discovered ι-CA, based on the amino acid sequence of the four repeated domain-containing proteins from *T. pseudonana*. Using a battery of biophysical approaches, we proposed a model of the folding of a homotetrameric tpι-CA in solution, which was also used to infer the structure of the same protein containing fewer domain repetitions (three, two and one).

Based on our results, we confirmed that the previously described LMM form [6] corresponds to a stable tetrameric form in solution. Due to the characteristic subcellular localization of the ι-CA towards the periphery of the plastid of photosynthetic eukaryotes [6,15], it is unlikely that the HMM-nucleic acid form occurs in vivo and, thus, could be an unspecific artefactual association of multiple ι-CA monomers together with nucleic acids. However, the possibility that the ι-CA could interact with other cellular components (e.g., other proteins and lipids) cannot be discarded, in particular because both LMM and HMM are active and catalyze CO_2 protonation [6]. The nucleic acid-bound HMM might mimic other forms induced by interaction with other negatively charged surfaces, such as galactolipids [27] that are abundant in plastid membranes. Such high molecular weight forms of CA with undetermined mass were also observed for other CAs such as a stromal β-CA, PtCA1, from the diatom *Phaeodactylum tricornutum*, which can form aggregated structures when purified [28]. Its association within large clumped macromolecular complexes was confirmed in vivo, as was also shown for the homologous PtCA2 [29]. This complex formation was possible through a C-terminal amphipathic α-helix exposed to the solvent that does not participate in the dimerization of the PtCA [29] and is also present in other β-CAs from other diatom species [30]. Interactants of the PtCA1 and PtCA2 have not yet been found, but an interaction with lipids (e.g., galactolipids) or carbohydrates has been hypothesized [29]. In this same context, further studies are necessary to show whether the tpι-CA is also able to form a complex with other cell components and which structural feature would allow this.

The ESI-MS, SEC and SAXS data indicated that the LMM form of tpι-CA is a tetramer. This is in contrast with the multi-domain ι-CA from the chlorarachniophyte alga *B. natans* that forms a homodimer [15]. The SAXS and DOSY-NMR data indicated that the global envelope has an atypical "drone-like" shape with a central core (domains 1 and 2) and four protruding arms (domains 3 and 4), and this also differs from the X-ray proposed structure of *B natans* ι-CA, which is an elongated dimer [15] (Figure 7).

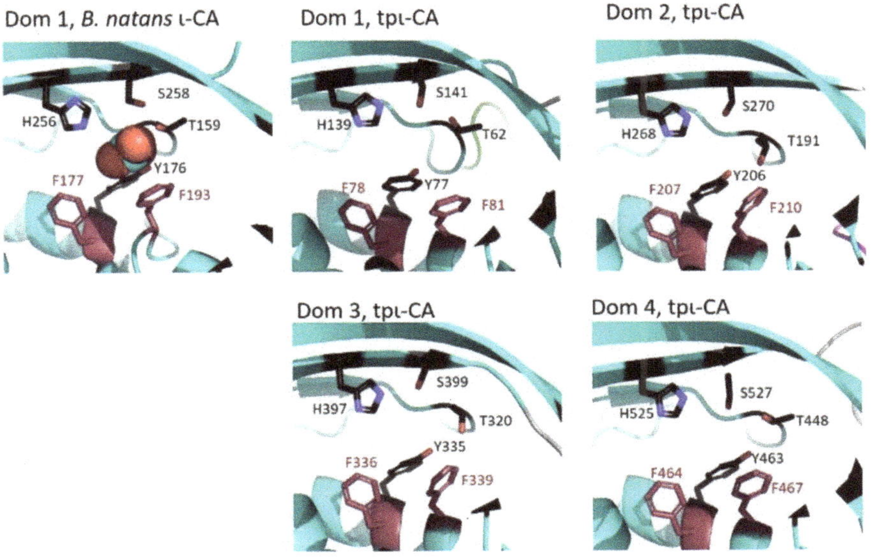

Figure 7. Structure of the ι-CA active site in domain 1 of *B. natans* ι-CA (pdb 7C5X and of each of the four domains of tpι-CA (homology model).

The CD data indicated that the secondary structure composition of each of the four tpι-CA domains is similar to other COG4875 domains, as expected from the high sequence conservation within this class. The secondary structure composition is also very close to the COG4337 domains, which are constituents of a metal-free ι-CA from *B. natans* and the cyanobacterium *Anabaena sp.* PCC7120. The high sequence identity with COG4875 domains for which X-ray structures are available enabled us to build a homology atomic model that we constrained within the SAXS average envelope. In the three COG4337 domains of ι-CA from *B. natans*, the active site for the CO_2 protonation was composed of the residues Thr159/322/486, Tyr176/339/503, His256/420/584 and Ser258/422/586, with the last two being part of the specific HHHSS sequence and which are all oriented to the core of the domains [15]. In all tpι-CA domains, all these residues are present (Thr62/191/320/448, Tyr77/206/335/463, His139/268/397/525 and Ser141/270/399/527, Figure 7) and oriented towards the core of the domain. Notably, both *B. natans* ι-CA and tpι-CA are inhibited by Zn^{2+}, and this peculiar property might be specific to these catalytic residues. Moreover, as in the ι-CA from *B. natans*, in tpι-CA, CO_2 can be protonated even in the absence of metal ion. In CAs, CO_2 is proposed to be positioned near phenylalanine residues. Phe177/340/504 and Phe193/357/521 in *B. natans* ι-CA are proximal to the active site and are conserved in all four tpι-CA domains (Phe78/207/336/464 and Phe81/210/339/467, Figure 7). The CA activity of the four-domain tpι-CA as well as the variant constructs Δ1-2-3 and Δ1-2, has been previously confirmed [6]; however, whether all domains contribute to the overall CA activity or to metal binding in a particular tetrameric conformation is still unknown and must be further investigated.

We validated the localization of the domains in our model by comparing the predicted hydrodynamic properties of the FL and domain truncated variants with experimentally measured diffusion coefficients by DOSY-NMR. Back calculation of the SAXS curve from the model fitted the experimental data better when the C-terminal domain 4 localization was unconstrained, indicating that the protruding arms are flexible, and this flexibility might be conferred by the linker between domains 3 and 4 that was predicted to be a long-disordered linker by IUPred2A. This particular linker is also predicted to be disordered in the ι-CA sequences from other diatom species with four-domains, including *Cyclotella cryptica*, *Fistulifera solaris* and *Thalassiosira oceanica*. However, it is absent in the C-terminal linkers from homologous sequences having less domain repeats (data not shown). This suggests that the protruding and flexible arms observed in the SAXS envelope from the tetrameric tpι-CA is a particular feature of the four-domain ι-CA and, in agreement with our proposed models (Figure 5b), does not exist in other homologous sequences with fewer domains.

Tpι-CA domains 1 and 2 are associated in a dimer with their β-sheet surface, which includes the specific HHHSS sequence [14], embedded in a β-sandwich interface. This interface also contains conserved Arg and Cys residues (Arg122/251/380/508 and Cys105/231). Interestingly, only domains 1 and 2 possess the cysteine residues that might stabilize the dimer interface. The distance between the His269 residues in the domain 2–domain 2 β-sandwich is less than 4.5 Å and similar to what was found between the His257 residues in *B. natans* ι-CA (Figure 8). On the contrary, in the domain 1–domain 1 β-sandwich, the distance between the His140 residues is more than 7 Å (Figure 8). This larger interface also includes highly conserved Asp and Glu residues from the neighboring domains 2 (Asp259 and Glu260, Figure 8) and the His139 and His140 residues that are the homologues of the residues that were predicted to coordinate metal [6,14]. In the present dataset of Mn-bound proteins, metal coordination involved mainly His, Glu and Asp residues [31,32]. We hypothesize that the pair of His140 (domain 1), pair of Asp259 and pair of Glu260 (domain 2) residues that are all localized within 10 Å in our homology model might be involved in Mn^{2+} coordination, and this can be further tested by site-directed mutagenesis.

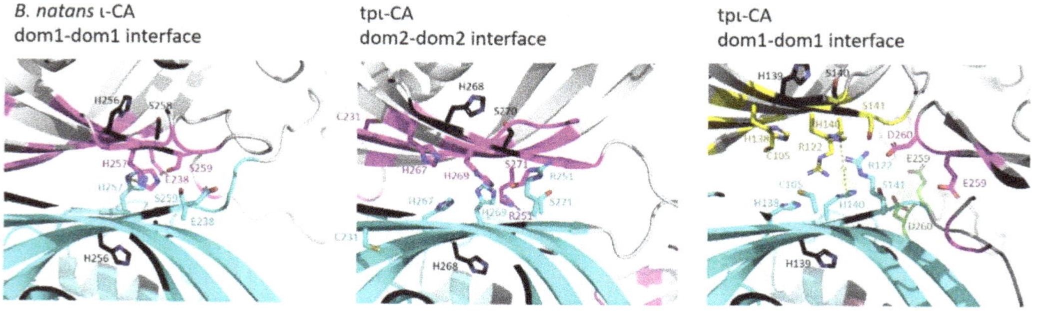

Figure 8. Structure of the ι-CA β-sandwich interface between the domain 1s of *B. natans* ι-CA (pdb 7C5X), between domains 2 and 1 of tpι-CA. The monomer A is colored blue, and the residues within 5 Å of the monomer A are colored magenta for those of monomer B, green for those of monomer C and yellow for those of monomer D. The distance between His257 residues in *B. natans* ι-CA is 4.5 Å, between His269 residues in the domains 2 of tpι-CA is 3.5 Å and between His 140 in the domains 1 of tpι-CA is 7.1 Å.

In our model, the domain 4s do not interact and the β-sheet surface that composes the β-sandwich domain-dimerization interface in the other COG4875 domains is protected by the C-terminal extension (Figure 9). This C-terminal extension has a peculiar amino-acid composition with a high number of hydrophobic residues (oriented toward the β-sheet surface) surrounded by glutamic acid residues exposed to the solvent. The C-terminal extension might act as a "gate" that prevents domain 4-domain 4 dimerization.

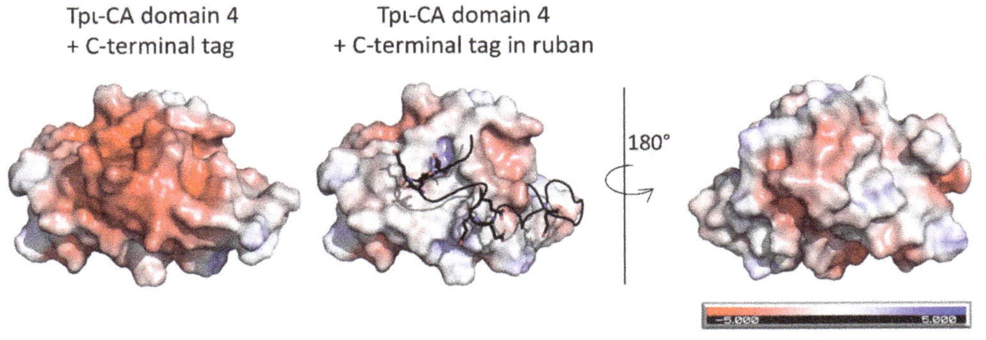

Figure 9. Electrostatic surface of the tpι-CA domain 4 with the C-terminal extension (**left**) or without (**right**). The C-terminal extension is represented in cartoon with the Glu residues in stick. The color of the surface corresponds to the electrostatic charge, which ranges from −5 (red) to +5 (blue).

4. Conclusion

A comparison of the tpι-CA structure with other existing CA structures revealed that, while there are elements of the fold that resemble previously known structures, the overall fold is novel as was anticipated from the absence of sequence conservation. The tpι-CA monomer has an unusual four domain repeat with a non-compact appearance. The tetramer has a drone-shape, comprising a doughnut core flanked by four protruding extensions mediated by the flexibility of the last linker. Its atypical shape might be linked to its localization in the appressed intermembrane space of the chloroplast endoplasmic reticulum (CER) [2]. It is intriguing to speculate that the drone structure of tpι-CA may be linked to its CO_2/HCO_3^- delivery function as is the case for delivery by human-made drones.

5. Materials and Methods

5.1. Protein Expression and Purification

The DNA sequence of the four-domain-containing ι-CA from *T. pseudonana* was produced synthetically (GeneCust, Ellange, Luxembourg) based on the cDNA sequence without the nucleotides coding for the signal peptide and inserted between the NdeI and XhoI restriction sites of a pET-28a+vector so that the protein was fused to a His-tag on its N-terminus. The same procedure was performed for the ι-CA variants containing three, two and one domains, always removing domains from the C-terminal extremity. The resulting vectors containing either the ι-CA gene or its domain variants were cloned in the *Escherichia coli* strain BL21-C41(DE3). The expression of recombinant proteins in *E. coli* was induced by 1 mM Isopropyl β-D-1-thiogalactopyranoside (IPTG) at 37 °C for 5 h. Cell pellets were resuspended in a buffer containing 50 mM sodium phosphate, 10 mM imidazole and 50 mM NaCl buffer (pH 8), plus lysozyme and protease inhibitor cocktail. Cells were broken by sonication. Lysates were centrifuged for 30 min at 16,000 g and 4 °C, and the supernatant was loaded onto a Ni-NTA column (height 6 cm and diameter 1.5 cm). The column was washed with a buffer containing 0.15 M imidazole. Elution of the ι-CA was performed with a buffer containing 0.35 M imidazole.

5.2. Size Exclusion Chromatography (SEC)

SEC chromatography was performed on a Superdex Increase S200 10 × 300 (mm × mm) at 4 °C. Five hundred microliters of the sample were loaded at varying concentrations ranging from 50 μM to 200 μM. The elution volumes were monitored by absorbance at 215 nm and 280 nm. When mentioned, 25 U mL^{-1} of benzonase (Sigma) with 1 mM $MgCl_2$ was added to the sample, and nucleic acid digestion was performed for 1 h at 37 °C followed by an overnight incubation at 4 °C. When mentioned, 3 U μL^{-1} of thrombin was added to the sample, and proteolysis of the His-tag was performed by incubation overnight at 4 °C.

5.3. *Congo Red Assay*

A solution of 1 mM Congo red (CR) was prepared in a buffer of 20 mM Tris and 50 mM NaCl (pH 8). A mix of 70 μM of Congo red and 10 μM of the protein sample was prepared, and the absorption spectrum between 300 and 700 nm wavelengths was recorded using a Perkin Elmer Lambda 25 UV/Vis spectrophotometer. The spectrum of the protein solution without CR was subtracted to the spectrum of the protein–CR mix. The spectrum of the CR solution alone was also recorded for comparison. A control of fibril formation was performed using lysozymes heated at 55 °C for 5 min, as described previously [33].

5.4. *Protein Cross-Linking*

One hundred micrograms (100 μg) of purified protein extract in 20 mM Na_2HPO_4 buffer (pH 8.0) was mixed with 0.01% glutaraldehyde. The mixture was then incubated for 10 to 15 min at room temperature, and then, the reaction was stopped by the addition of 80 mM Tris-HCl (pH 8.0). The reaction was immediately mixed with a Laemmli buffer for further SDS-PAGE and Western blot analysis.

5.5. *Protein Analysis*

Protein samples from the purified protein were mixed with the sample buffer (62.5 mM Tris, 2.5% SDS, 0.002% bromophenol blue, 10% glycerol, 20 mM DTT and pH 6.8) and denatured by heating at 85 °C for 5–10 min. Samples of 20 to 30 μg protein extracts or 2 to 5 μg of purified proteins were loaded onto 12% acrylamide/bis-acrylamide gels and run at 120 volts until the migration front reached the bottom of the gel. Electrophoresis was performed in a Bio-Rad Mini Protean III system (Bio-Rad, Hercules, California, United States) using a buffer of 50 mM Tris, 380 mM glycine and 10% SDS.

After electrophoresis, the gel was either stained with Coomassie blue or used for Western blot analysis. For Western blot, the proteins were transferred to a nitrocellulose membrane (0.2 μm; Carl Roth Gmbh, Karlsruhe, Germany) with active transfer at 80 volts for 1 h. The transfer buffer contained 25 mM Tris, 121 mM glycine and 20% ethanol. After transfer, the membrane was blocked with a solution of 5% low fat milk in TBS-T (50 mM Tris, 150 mM NaCl, 0.05% Tween-20 and pH 7.6). The membrane was then incubated for 1 h at room temperature or overnight at 7 °C with the primary antibody (1:1000 dilution). The secondary antibody (anti-rabbit IgG horseradish peroxidase, 1:10,000 dilution) was incubated for 1 h at room temperature. The membrane was revealed with the Enhanced Chemiluminescence technique and then visualized with a digital imaging system (ImageQuant LAS 4000 mini, GE, Chicago, Illinois, United States).

5.6. *Analytical Ultracentrifugation (AUC)*

Sedimentation velocity experiments were carried out at 40,000 rpm and 20 °C in a Beckman Optima-XL-A analytical ultracentrifuge using 1.2 cm or 0.3 cm double sector centerpieces in an AN50Ti rotor. Scans were acquired in the continuous mode at 280 nm in the range of 0.1 to 1 absorption. All ι-CA samples were in a 20 mM Tris and 50 mM NaCl (pH 8) buffer. At 20 °C, the partial specific volume of ι-CA, the solvent density and the viscosity calculated with SEDNTERP (jphilo. Available online: http://www.jphilo.mailway.com/index.htm (accessed on 10th December 2018)) [34] were 0.734501 mL g^{-1}, 1.0009 g cm^{-3} and 0.01002 poise, respectively. The data recorded from moving boundaries were analyzed in terms of continuous size distribution functions of sedimentation coefficient, C(S), using the program SEDFIT [35]. The sedimentation coefficient was measured at different protein concentrations, and the standard sedimentation coefficient was obtained by extrapolation to a concentration of the protein equal to zero. All S values of ι-CA were corrected to standard conditions—i.e., 20 °C in water—by SEDFIT.

5.7. Electrospray Mass Spectrometry (ESI-MS)

Purified recombinant tpι-CA in 20 mM Tris-HCl, 50 mM NaCl, and pH 8 was buffer-exchanged using a micro Bio-Spin column Bio-Gel P6 (Bio-Rad) against a 500 mM aqueous ammonium acetate pH 8 solution for native mass spectrometry (MS) at a final protein concentration of 8 μM measured by a Thermo Scientific nanodrop 2000 C (at lambda 280 nm, epsilon 110.356 μM^{-1} cm^{-1}). The MS parameters used in the electrospray Q-ToF mass spectrometer (Synapt G1, Waters) were set as source temperature 20 °C, capillary voltage 1.5 kV, sampling cone 140, extraction cone 4, trap collision energy 40, transfer collision energy 30 and m/z window 6,000 to 10,000 to detect the oligomers of tpι-CA. The neutral molecular mass was manually calculated from spectra by averaging adjacent m/z from five consecutive charge-state assigned peaks. The deduced molecular mass was compared with the theoretical mass of ι-CA, which has been deduced from the sequence including the His-tag (GSSHHHHHHSSGLV . . . : MW = 64,988 Da), as confirmed by N-ter sequencing (data not shown).

5.8. Circular Dichroism (CD)

The purified recombinant protein was prepared at 2 μM in filtered 20 mM Na_2HPO_4 buffer (pH 8). The circular dichroism (CD) spectra were recorded from 260 to 180 nm in a Jasco J-815 CD Spectrometer (JASCO. Easton, Maryland, United States) at 25 °C in a 2 mm path quartz cuvette. The raw values of ellipticity (mdeg) were converted into mean residue molar ellipticity (θ) using the following formula:

$$\theta\left(\text{deg}cm^2dmol^{-1}res^{-1}\right) = \frac{MRW \times E(\text{deg})}{d \times c \times 10}$$

where "*MRW*" corresponds to the mean residue weight, "*E*" is the raw ellipticity, "*d*" is the path length of the cuvette (cm) and "*c*" is the sample concentration (g cm^{-3}). The contents of α-sheets and β-helices were estimated using DichroWeb software [22].

5.9. Dynamic Light Scattering (DLS)

Purified ι-CA was analyzed by DLS using a Zetasizer Nano ZS (Malvern Instruments, Malvern, United Kingdom). The proteins samples prepared at 3 to 5 mg mL^{-1} in a buffer of 20 mM Tris and 50 mM NaCl (pH 8) were centrifuged for 15 min at 14,000 rpm at 4 °C prior to DLS measurement. Three measurements consisting of 10 runs, each 10 s, were performed on ι-CA sample with a scatter angle of 173 degrees.

5.10. Size Exclusion Chromatography Coupled to Small-Angle X-Ray Scattering (SEC-SAXS)

The SEC-SAXS experiments were performed at the SWING beamline of SOLEIL Synchrotron (Gif-Sur-Yvette, France). A HPLC column Agilent Bio SEC-5, 500 Å (length: 300 mm, particle size: 5 μm) was used prior to SAXS measurements. All experiments were performed at 15 °C. The sample-to-detector (Eiger 4M) distance was adjusted to 2 m, giving access to scattering vector $q = 4\pi/\lambda \cdot \sin\theta$ (where 2θ is the scattering angle and λ is the wavelength, equal to 1.033 Å) ranging from 0.011 to 0.50 $Å^{-1}$. Two-hundred frames of 990 ms with 10 ms dead-time were recorded during the first minutes of the elution for the background signal. The signal of the protein was recorded during all protein elutions.

The raw data were processed using the dedicated in-house software Foxtrot. The buffer signal was subtracted, and careful inspection of the protein scattering data allowed us to average the identical scattering curves recorded during protein elution. Data analysis was performed using the ATSAS suite of [36]. The Rg was obtained via PRIMUS using the Guinier approximation, and the distance distribution function P(r) was obtained via GNOM. The molecular mass was assessed using SAXS-MoW [37]. Ab initio 3D models corresponding to the scattering envelopes were calculated using DAMMIF [24] with P4 symmetry, and using GASBOR with the number of amino acids of the protein monomer as input and with P4 symmetry [24]. The atomic models were assessed and refined using CRYSOL [36] and CORAL [24]. CORAL started with the atomic model as input and was allowed to move the domains 4 independently and to build new CA traces between the domains 3 and 4. The SAXS data have been deposited at SAS BDB (draft ID 3391).

5.11. Diffusion-Ordered Spectroscopy Nuclear Magnetic Resonance (DOSY-NMR)

DOSY-NMR was performed on the purified ι-CA and its domain variants. A 500 μL sample was prepared at 30 to 80 μM in a buffer with 27 mM Tris, 45 mM NaCl, 10% D_2O and 1 μL DSS (4,4-dimethyl-4-silapentane-1-sulfonic acid) at pH 8. The measurements were repeated with 10 increasing strengths for a duration of 2.8 ms. A time delay of 200 ms was used to allow the molecules to diffuse before gradient decoding and diffusion. The experiments were performed at 298 K on a 600 MHz Bruker advance II spectrometer equipped with a cryo-probe. The spectra were transformed by NMRPipe [38], and the diffusion coefficients were calculated using Octave (GNU Octave. Available online: https://www.gnu.org/software/octave/ (accessed on 18th December 2018)). The diffusion coefficient (D_t) obtained was used to calculate the hydrodynamic radius (R_h) of each sample using the Einstein–Stokes equation:

$$D_t = \frac{k_B T}{6\pi\eta R_h}$$

where k_B is the Boltzmann constant (1.380×10^{-23} kg m^2 s^{-2} K^{-1}), T is the temperature (in Kelvin) and η is the viscosity of the medium.

The translational diffusion coefficient of the structural models was computed using HYDROPRO version 10 [25]. Standard parameters were used for the HYDROPRO calculation; the radius of primary elements was set to 2.9 Å, six beads sizes were used, which ranged from 10 to 20 Å for the full-length construct, from 7 to 14 Å for the Δ123 construct, from 5 to 10 Å for the Δ12 construct and from 2 to 4 Å for the Δ1 construct; the temperature was set to 293 K to match the experimental diffusion measurements; and the viscosity of the solvent was set to 0.01 poises.

5.12. D-Modeling

Homodimer models of domains 1, 2 and 3 were built using the homology modelling server SWISS-MODEL with the homodimer structure of the putative Calcium/Calmoduline-dependent kinase II association domain from *Xanthomonas campestri* (3H51.pdb) as a template [39]. The non-dimeric C-terminal domain 4, where a mostly hydrophobic C-terminal extension covers the monomer interface, which is responsible for the dimerization of the other domains, was built using the I-Tasser server [40]. The homodimer domains 1, 2 and 3 and the monomeric C-terminal domain were assembled manually in the SAXS envelope using the PyMol program. In this step, individual N- and C-terminals of each domain were carefully oriented to allow for their connection to the full-length monomer, constraining their position in the overall structure. Linker regions between domains were added and minimized, and some colliding loops were moved using the Wincoot structural modeling program [41]. The overall geometry of a connected 4 domain monomer was refined using the ModRefiner program [42] and the tetrameric assembly reconstructed with this refined model. Finally, the whole tetrameric assembly was refined using the SWISS-MODEL server with the tetrameric model from the previous step as the starting template.

Author Contributions: Conceptualization, E.L.J., B.G., H.L. and V.R.-B.; experimental design, H.L. (NMR), V.R.-B. (SAXS), P.B. (UCA) and G.P. (model design); experiments, E.L.J., M.H. and L.W.; supervision, B.G., H.L., V.R.-B. (SAXS) and P.B. (UCA); writing—original draft preparation, E.L.J., B.G. and H.L.; writing—SAXS, (V.R.-B.); writing—UCA (P.B.). All authors have discussed the results, read and agreed to the published version of the manuscript.

Funding: This research was funded by CNRS and Aix-Marseille Université, but received no external funding.

Institutional Review Board Statement: Not applicable.

Informed Consent Statement: Not applicable.

Data Availability Statement: Not applicable.

Acknowledgments: The proteomic platform from the Institut de Microbiologie de la Méditerranée, Marseille-Protéomique, IBiSA and Aix-Marseille Université is gratefully acknowledged for its contribution to the mass spectrometry and N-ter sequencing experiments. Fanny Alamelle is thanked for the IT services. We acknowledge SOLEIL for the provision of the synchrotron radiation facilities, and we thank Aurélien Thureau for the assistance in using beamline SWING. E.L.J. acknowledges the support of the BECAS CHILE program from the "Agencia Nacional de Investigación y Desarrollo (ANID, former CONICYT)". The authors are grateful to Pr Stephen C Maberly for checking the English.

Conflicts of Interest: The authors declare no conflict of interest.

References

1. Lionetto, M.G.; Caricato, R.; Giordano, M.E.; Schettino, T. The complex relationship between metals and carbonic anhydrase: New insights and perspectives. *Int. J. Mol. Sci.* **2016**, *17*, 127. [CrossRef] [PubMed]
2. Jensen, E.L.; Maberly, S.C.; Gontero, B. Insights on the functions and ecophysiological relevance of the diverse carbonic anhydrases in microalgae. *Int. J. Mol. Sci.* **2020**, *21*, 2922. [CrossRef] [PubMed]
3. Del Prete, S.; Vullo, D.; Fisher, G.M.; Andrews, K.T.; Poulsen, S.A.; Capasso, C.; Supuran, C.T. Discovery of a new family of carbonic anhydrases in the malaria pathogen *Plasmodium falciparum*—The η-carbonic anhydrases. *Bioorgan. Med. Chem. Lett.* **2014**, *24*, 4389–4396. [CrossRef] [PubMed]
4. Xu, Y.; Feng, L.; Jeffrey, P.D.; Shi, Y.; Morel, F.M.M. Structure and metal exchange in the cadmium carbonic anhydrase of marine diatoms. *Nature* **2008**, *452*, 56–61. [CrossRef]
5. Kikutani, S.; Nakajima, K.; Nagasato, C.; Tsuji, Y.; Miyatake, A.; Matsuda, Y. Thylakoid luminal θ-carbonic anhydrase critical for growth and photosynthesis in the marine diatom *Phaeodactylum tricornutum*. *Proc. Natl. Acad. Sci. USA* **2016**, *113*, 9828–9833. [CrossRef]
6. Jensen, E.L.; Clement, R.; Kosta, A.; Maberly, S.C.; Gontero, B. A new widespread subclass of carbonic anhydrase in marine phytoplankton. *ISME J.* **2019**, *13*, 2094–2106. [CrossRef] [PubMed]
7. Supuran, C.T. Structure and function of carbonic anhydrases. *Biochem. J.* **2016**, *473*, 2023–2032. [CrossRef]
8. DiMario, R.J.; Machingura, M.C.; Waldrop, G.L.; Moroney, J.V. The many types of carbonic anhydrases in photosynthetic organisms. *Plant Sci.* **2018**, *268*, 11–17. [CrossRef]
9. Tsuji, Y.; Nakajima, K.; Matsuda, Y. Molecular aspects of the biophysical CO_2-concentrating mechanism and its regulation in marine diatoms. *J. Exp. Bot.* **2017**, *68*, 3763–3772. [CrossRef] [PubMed]
10. MacAuley, S.R.; Zimmerman, S.A.; Apolinario, E.E.; Evilia, C.; Hou, Y.M.; Ferry, J.G.; Sowers, K.R. The archetype γ-class carbonic anhydrase (cam) contains iron when synthesized in vivo. *Biochemistry* **2009**, *48*, 817–819. [CrossRef]
11. DiMario, R.J.; Clayton, H.; Mukherjee, A.; Ludwig, M.; Moroney, J.V. Plant Carbonic Anhydrases: Structures, Locations, Evolution, and Physiological Roles. *Mol. Plant* **2017**, *10*, 30–46. [CrossRef] [PubMed]
12. Clement, R.; Lignon, S.; Mansuelle, P.; Jensen, E.; Pophillat, M.; Lebrun, R.; Denis, Y.; Puppo, C.; Maberly, S.C.; Gontero, B. Responses of the marine diatom *Thalassiosira pseudonana* to changes in CO_2 concentration: A proteomic approach. *Sci. Rep.* **2017**, *7*, 42333. [CrossRef] [PubMed]
13. Giordano, M.; Beardall, J.; Raven, J.A. CO_2 concentrating mechanisms in algae: Mechanisms, environmental modulation, and evolution. *Annu. Rev. Plant Biol.* **2005**, *56*, 99–131. [CrossRef]
14. Del Prete, S.; Nocentini, A.; Supuran, C.T.; Capasso, C. Bacterial ι-carbonic anhydrase: A new active class of carbonic anhydrase identified in the genome of the Gram-negative bacterium *Burkholderia territorii*. *J. Enzym. Inhib. Med. Chem.* **2020**, *35*, 1060–1068. [CrossRef]
15. Hirakawa, Y.; Senda, M.; Fukuda, K.; Yu, H.Y.; Ishida, M.; Taira, M.; Kinbara, K.; Senda, T. Characterization of a novel type of carbonic anhydrase that acts without metal cofactors. *BMC Biol.* **2021**, *19*, 1–15. [CrossRef]
16. Eberhardt, R.Y.; Chang, Y.; Bateman, A.; Murzin, A.G.; Axelrod, H.L.; Hwang, W.C.; Aravind, L. Filling out the structural map of the NTF2-like superfamily. *BMC Bioinform.* **2013**, *14*, 1–11. [CrossRef]
17. Rosenberg, O.S.; Deindl, S.; Comolli, L.R.; Hoelz, A.; Downing, K.H.; Nairn, A.C.; Kuriyan, J. Oligomerization states of the association domain and the holoenyzme of Ca^{2+}/CaM kinase II. *FEBS J.* **2006**, *273*, 682–694. [CrossRef] [PubMed]
18. Stewart, M.; Kent, H.M.; Mccoy, A.J. Structural basis for molecular recognition between nuclear transport factor 2 (NTF2) and the GDP-bound form of the Ras-family GTPase Ran. *J. Mol. Biol.* **1998**, *277*, 635–646. [CrossRef]
19. Vognsen, T.; Kristensen, O. Crystal structure of the Rasputin NTF2-like domain from *Drosophila melanogaster*. *Biochem. Biophys. Res. Commun.* **2012**, *420*, 188–192. [CrossRef]
20. Jones, D.T. Protein secondary structure prediction based on position-specific scoring matrices. *J. Mol. Biol.* **1999**, *292*, 195–202. [CrossRef]

21. Whitmore, L.; Wallace, B.A. Protein secondary structure analyses from circular dichroism spectroscopy: Methods and reference databases. *Biopolymers* **2008**, *89*, 392–400. [CrossRef]
22. Whitmore, L.; Wallace, B.A. DICHROWEB, an online server for protein secondary structure analyses from circular dichroism spectroscopic data. *Nucleic Acids Res.* **2004**, *32*, 668–673. [CrossRef]
23. Mészáros, B.; Erdös, G.; Dosztányi, Z. IUPred2A: Context-dependent prediction of protein disorder as a function of redox state and protein binding. *Nucleic Acids Res.* **2018**, *46*, W329–W337. [CrossRef] [PubMed]
24. Petoukhov, M.V.; Franke, D.; Shkumatov, A.V.; Tria, G.; Kikhney, A.G.; Gajda, M.; Gorba, C.; Mertens, H.D.T.; Konarev, P.V.; Svergun, D.I. New developments in the ATSAS program package for small-angle scattering data analysis. *J. Appl. Crystallogr.* **2012**, *45*, 342–350. [CrossRef]
25. Ortega, A.; Amorós, D.; De La Torre, J.G. Prediction of hydrodynamic and other solution properties of rigid proteins from atomic- and residue-level models. *Biophys. J.* **2011**, *101*, 892–898. [CrossRef] [PubMed]
26. Liljas, A.; Laurberg, M. A wheel invented three times. *EMBO Rep.* **2000**, *1*, 16–17. [CrossRef] [PubMed]
27. Sahaka, M.; Amara, S.; Wattanakul, J.; Gedi, M.A.; Aldai, N.; Parsiegla, G.; Lecomte, J.; Christeller, J.T.; Gray, D.; Gontero, B.; et al. The digestion of galactolipids and its ubiquitous function in Nature for the uptake of the essential a-linolenic acid. *Food Funct.* **2020**, *11*, 6710–6744. [CrossRef]
28. Satoh, D.; Hiraoka, Y.; Colman, B.; Matsuda, Y. Physiological and molecular biological characterization of intracellular carbonic anhydrase from the marine diatom *Phaeodactylum tricornutum*. *Plant. Physiol.* **2001**, *126*, 1459–1470. [CrossRef] [PubMed]
29. Kitao, Y.; Matsuda, Y. Formation of macromolecular complexes of carbonic anhydrases in the chloroplast of a marine diatom by the action of the C-terminal helix. *Biochem. J.* **2009**, *688*, 681–688. [CrossRef] [PubMed]
30. Maberly, S.C.; Gontero, B.; Puppo, C.; Villain, A.; Severi, I.; Giordano, M. Inorganic carbon uptake in a freshwater diatom, *Asterionella formosa* (Bacillariophyceae): From ecology to genomics. *Phycologia* **2021**, 1–12. [CrossRef]
31. Udayalaxmi, S.; Gangula, M.R. Investigation of manganese metal coordination in proteins: A comprehensive PDB analysis and quantum mechanical study. *Struct. Chem.* **2020**, *31*, 1057–1064.
32. Zheng, H.; Chruszcz, M.; Lasota, P.; Lebioda, L.; Minor, W. Data mining of metal ion environments present in protein structures. *J. Inorg. Biochem.* **2008**, *102*, 1765–1776. [CrossRef] [PubMed]
33. Antimonova, O.I.; Grudinina, N.A.; Egorov, V.V.; Polyakov, D.S. Interaction of the Dye Congo Red with Fibrils of Lysozyme, Beta2-Microglobulin, and Transthyretin. *Cell Tissue Biol.* **2016**, *10*, 468–475. [CrossRef]
34. Laue, T.M.; Shah, B.D.; Ridgeway, T.M.; Pelletier, S.L. Computer-aided interpretation of analytical sedimentation data for proteins. In *Analytical Ultracentrifugation in Biochemistry and Polymer Science*; Harding, S.E., Rowe, A.J., Horton, J.C., Eds.; Royal Society of Chemistry: Cambridge, UK, 1992; pp. 90–125. ISBN 0851863450.
35. Schuck, P.; Rossmanith, P. Determination of the sedimentation coefficient distribution by least-squares boundary modeling. *Biopolymers* **2000**, *54*, 328–341. [CrossRef]
36. Franke, D.; Petoukhov, M.V.; Konarev, P.V.; Panjkovich, A.; Tuukkanen, A.; Mertens, H.D.T.; Kikhney, A.G.; Hajizadeh, N.R.; Franklin, J.M.; Jeffries, C.M.; et al. ATSAS 2.8: A comprehensive data analysis suite for small-angle scattering from macromolecular solutions. *J. Appl. Crystallogr.* **2017**, *50*, 1212–1225. [CrossRef]
37. Piiadov, V.; de Araújo, E.A.; Neto, M.O.; Craievich, A.F.; Polikarpov, I. SAXSMoW 2.0: Online calculator of the molecular weight of proteins in dilute solution from experimental SAXS data measured on a relative scale. *Protein Sci.* **2019**, *28*, 454–463. [CrossRef]
38. Delaglio, F.; Grzesiek, S.; Vuister, G.W.; Zhu, G.; Pfeifer, J.; Bax, A. NMRPipe: A multidimensional spectral processing system based on UNIX pipes. *J. Biomol. NMR* **1995**, *6*, 277–293. [CrossRef]
39. Waterhouse, A.; Bertoni, M.; Bienert, S.; Studer, G.; Tauriello, G.; Gumienny, R.; Heer, F.T.; de Beer, T.A.P.; Rempfer, C.; Bordoli, L.; et al. SWISS-MODEL: Homology modelling of protein structures and complexes. *Nucleic Acids Res.* **2018**, *46*, W296–W303. [CrossRef]
40. Yang, J.; Zhang, Y. I-TASSER server: New development for protein structure and function predictions. *Nucleic Acids Res.* **2015**, *43*, W174–W181. [CrossRef]
41. Emsley, P.; Lohkamp, B.; Scott, W.G.; Cowtan, K. Features and development of Coot. *Acta Crystallogr. Sect. D Biol. Crystallogr.* **2010**, *66*, 486–501. [CrossRef]
42. Xu, D.; Zhang, Y. Improving the physical realism and structural accuracy of protein models by a two-step atomic-level energy minimization. *Biophys. J.* **2011**, *101*, 2525–2534. [CrossRef] [PubMed]

International Journal of
Molecular Sciences

MDPI

Article

Architecture Insight of Bifidobacterial α-L-Fucosidases

José Antonio Curiel *, Ángela Peirotén, José María Landete, Ana Ruiz de la Bastida, Susana Langa and Juan Luis Arqués

Departamento de Tecnología de Alimentos, Instituto Nacional de Investigación y Tecnología Agraria y Alimentaria (INIA-CSIC), Carretera de La Coruña Km 7.5, 28040 Madrid, Spain; angela.peiroten@inia.es (Á.P.); landete.josem@inia.es (J.M.L.); ana.ruiz@inia.es (A.R.d.l.B.); langa.susana@inia.es (S.L.); arques@inia.es (J.L.A.)

* Correspondence: joseantonio.curiel@inia.es; Tel.: +34-91-347-68-86

Abstract: Fucosylated carbohydrates and glycoproteins from human breast milk are essential for the development of the gut microbiota in early life because they are selectively metabolized by bifidobacteria. In this regard, α-L-fucosidases play a key role in this successful bifidobacterial colonization allowing the utilization of these substrates. Although a considerable number of α-L-fucosidases from bifidobacteria have been identified by computational analysis, only a few of them have been characterized. Hitherto, α-L-fucosidases are classified into three families: GH29, GH95, and GH151, based on their catalytic structure. However, bifidobacterial α-L-fucosidases belonging to a particular family show significant differences in their sequence. Because this fact could underlie distinct phylogenetic evolution, here extensive similarity searches and comparative analyses of the bifidobacterial α-L-fucosidases identified were carried out with the assistance of previous physicochemical studies available. This work reveals four and two paralogue bifidobacterial fucosidase groups within GH29 and GH95 families, respectively. Moreover, *Bifidobacterium longum* subsp. *infantis* species exhibited the greatest number of phylogenetic lineages in their fucosidases clustered in every family: GH29, GH95, and GH151. Since α-L-fucosidases phylogenetically descended from other glycosyl hydrolase families, we hypothesized that they could exhibit additional glycosidase activities other than fucosidase, raising the possibility of their application to transfucosylate substrates other than lactose in order to synthesis novel prebiotics.

Keywords: bifidobacteria; fucosidases; glycosyl hydrolases; conserved domains; human milk

Citation: Curiel, J.A.; Peirotén, Á.; Landete, J.M.; Ruiz de la Bastida, A.; Langa, S.; Arqués, J.L. Architecture Insight of Bifidobacterial α-L-Fucosidases. *Int. J. Mol. Sci.* **2021**, 22, 8462. https://doi.org/10.3390/ijms22168462

Academic Editors: Istvan Simon and Csaba Magyar

Received: 29 June 2021
Accepted: 4 August 2021
Published: 6 August 2021

1. Introduction

The impact of human milk glycobiome on the gut microbiota of infants is well established [1]. While a great part of the components of breast milk provide nutrients to the infant, human milk oligosaccharides (HMOs) and human milk glycoproteins (HMGs) selectively favor the colonization and growth of bifidobacteria in the infant intestine, contributing to the development of the gut microbiota [1,2]. In this regard, *Bifidobacterium* species are considered key actors in the multifaceted process of gut development and maturation of the immune system [3]. In fact, during the first months of birth, the loss of bifidobacteria or the gain of other bacteria can significantly alter the progression of the healthy microbial community with negative consequences for the infant, including a predisposition to autoimmune and/or metabolic diseases such as allergies and childhood obesity [4,5]. Concerning to that, fucosylated HMOs (FHMOs) and fucosylated HMGs (FHMGs) constitute a great part of the glycobiome of the breast milk [6] (Figure 1) and have been proposed to be essential in the development of the microbiota [7].

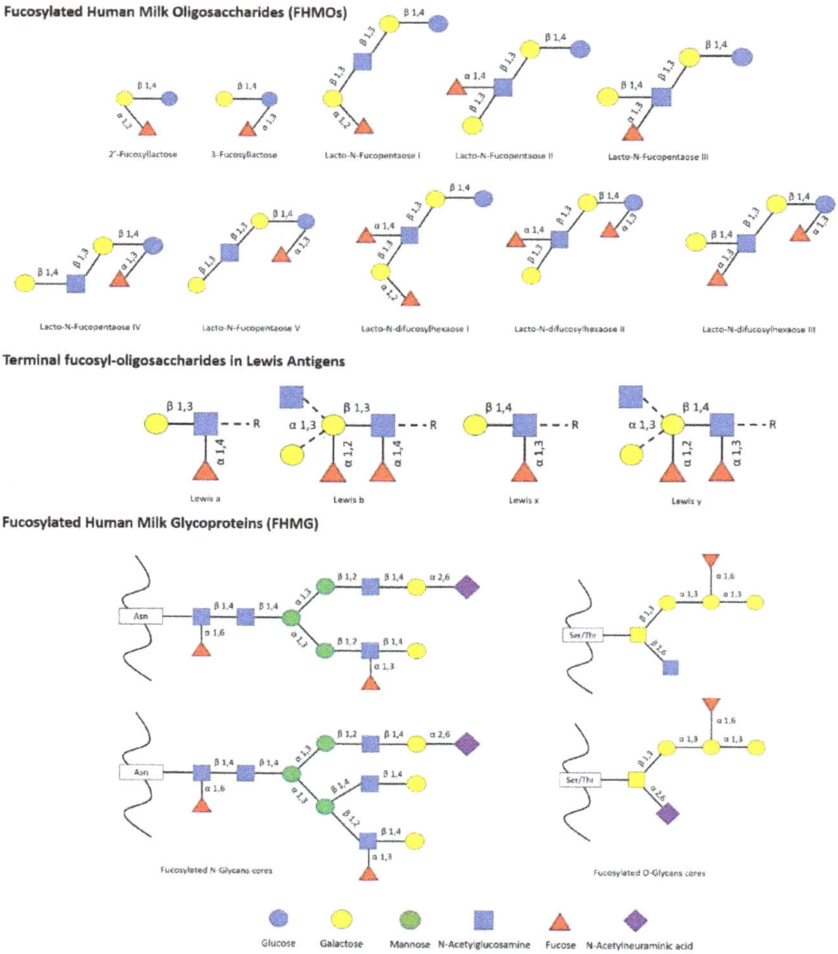

Figure 1. List of main fucosylated human milk oligosaccharides (FHMOs) and fucosylated human milk glycoproteins (FHMG) reported [1,6,7].

FHMOs constitute the largest fraction of human milk oligosaccharides, and although they show a small number of different conformations, they can make up to 70% of the total in an individual mother's milk [6]. The fucosylated trisaccharide 2′-fucosyllactose is the most abundant FHMO, representing from 12 to 45% of the total HMO content in breastmilk, while 3-fucosyllactose is less abundant, from 0.5% to 3% [8]. On the other hand, there are several FMHGs investigated, and contrary to FHMOs, they appear at lower concentration but show a higher number of different forms, including lactoferrin (17%), immunoglobulins IgG (<1%), IgM (<1%), and secretory IgA (11%) [9–11]. Both FHMOs and FHMGs stand out for their ability to stimulate the growth of bifidobacteria [7,12], whose metabolism transforms fucosylated oligosaccharides into short-chain fatty acids (SCFAs) such as acetate, formate, lactate, and pyruvate [13], which in turn stimulate the immune system by inducing the differentiation of T-regulatory cells via inhibition of histone deacetylase [14].

The great influence of fucosylated compounds present in breast milk on bifidobacteria is due to their ability to metabolize them, being α-L-fucosidases (henceforth, fucosidases) indispensable tools that allow shaping the gut microbiome in the first months of life.

According to CAZy database, hitherto, more than 10,000 sequences have been identified in silico as α-L-fucosidases, belonging to a wide variety of organisms from archaea to fungi and plants. However, the vast majority of fucosidase sequences have been described in bacteria and belong to more than 2000 bacteria species (www.cazy.org). This database classifies fucosidases into three families (GH29, GH95, and GH151) according to their catalytic structures. GH29 fucosidases act through a retaining mechanism and have a broader substrate specificity, including hydrolysis of Fuc-α1,3/4/6 linkages [15]. Moreover, family GH29 fucosidases have been subclassified into two subfamilies. The subfamily A contains α-fucosidases with relatively relaxed substrate specificities, able to hydrolyze *p*-nitrophenyl-α-L-fucopyranoside (pNP-fucose), while the members of subfamily B are specific to α1,3/4-glycosidic linkages and are practically unable to hydrolyze pNP-fucose [16]. Although GH29 fucosidases also could exhibit hydrolysis of Fuc-α1,2 linkages, that activity is mainly attributed to GH95 family, which catalyzes the hydrolysis of fucose linkages through an inverting mechanism, resulting in the inversion of the anomeric configuration [17,18]. Finally, GH151 family has poor activity on fucosylated substrates; this is the reason why it is currently questioned as to whether they are genuine fucosidases [19–21].

Even though species of the *Bifidobacterium* genus dominate the infant gut microbiota in early life, and given the importance of their metabolism of fucosylated conjugates, there are only a few bifidobacterial species studied extensively at both cellular and genomics level for their ability to utilize fucosylated carbohydrates, including *B. bifidum* and *Bifidobacterium longum* subsp. *infantis* [22,23]. However, different strain-dependent metabolic abilities have been unraveled for the use of fucosylated conjugates and are likely determined by their fucosidases' diversity [24]. Indeed, agreeing with the evolution and phylogenetics of fucosidases previously studied in metazoan fucosidases [25], bifidobacterial fucosidase sequences listed in CAZy reveal substantial in silico differences regarding to their conserved domains, even those ones clustered in the same GH, revealing different adaptation/specialization ranges as well as their origin. Therefore, this work addresses the diverse conserved architectures of bifidobacterial fucosidases and cluster them by activity and phylogenetic evolution in order to propose a novel classification within the GH groups already listed in CAZy.

2. Results

2.1. *Bifidobacterial GH29 Fucosidases*

GH29 fucosidases from bifidobacteria listed in CAZy are shown in Table S1. Based on in silico studies concerning conserved domains released by NCBI Conserved Domains Database (CDD), bifidobacterial GH29 fucosidases could be classified into four different phylogenetic groups (Table S1). That differentiation was also confirmed through sequence homology PCA and cluster analyses (Figure 2).

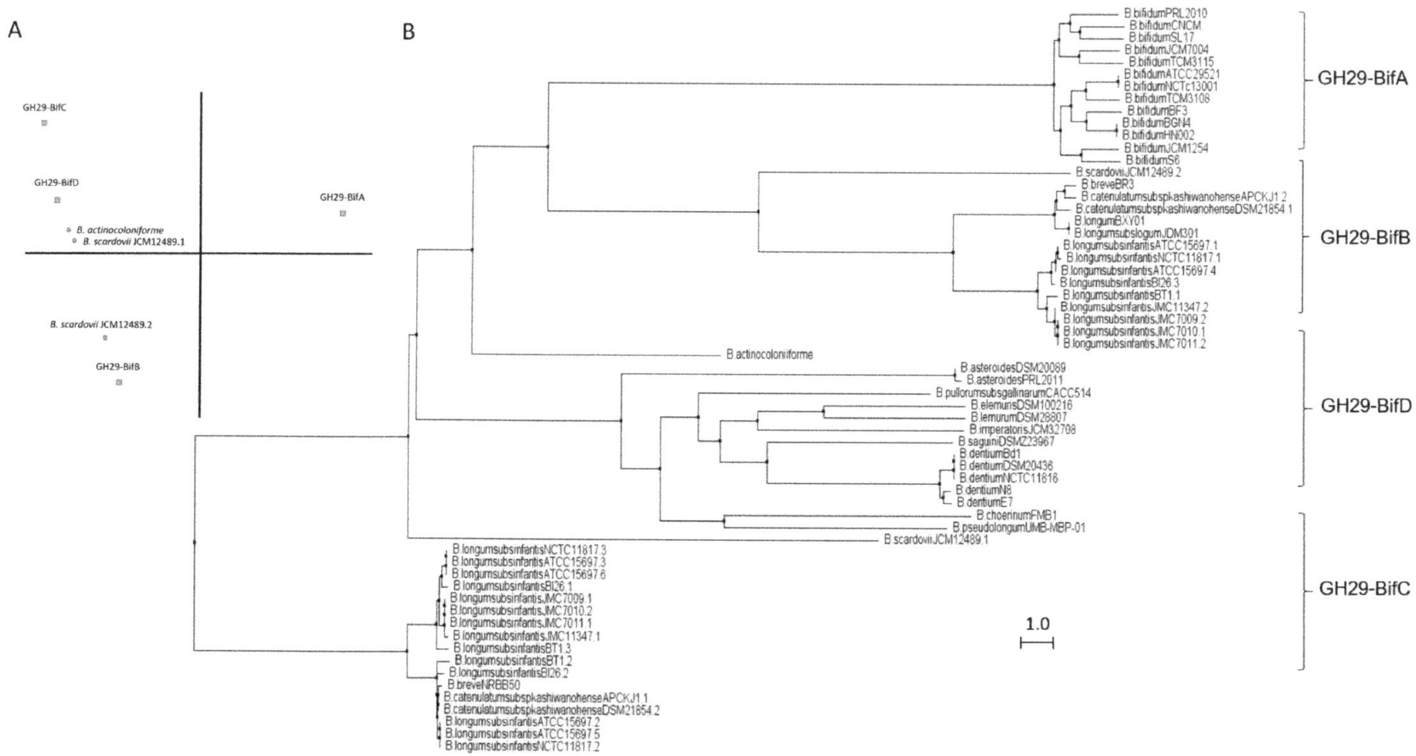

Figure 2. Phylogenetic analysis of bifidobacterial GH29 fucosidases. PCA (**A**) and cladogram tree (**B**) distributions of bifidobacterial GH29 fucosidase sequences listed in CAZy, released from Jalview 2.11.1.4 software using the neighbor-joining method.

The enzymes included in the proposed GH29-BifA, only found in *B. bifidum* strains, are characterized as large membrane-bound fucosidases (AfuC super family domain; NCBI CDD accession number cl34656) and exhibit an accessory F5/F8 type C domain family (NCBI CDD accession number cl23730), probably involved in recognizing galactose or N-acetyllactosamine [26]. Interestingly, while InterPro database (EMBL-EBI) recognized the F5/F8 type C domain (IPR000421), it interpreted the AfuC domain as Glyco_Hydro_29 domain (IPR000933), probably due to the degree of updating of both databases (Table S1). In addition, Ashida et al., 2009 identified a second putative sugar-binding domain in GH29 fucosidase AfcB from *B. bifidum* JCM1254, domain that is frequently found in membrane-bound or cell-wall-associated proteins and denominated FIVAR [27]. Those results were here confirmed by SOSUI and HMMTOP databases, which allowed the identification of two putative transmembrane helices in GH29-BifA fucosidases (Table S1). Therefore, it has been suggested that both accessory F5/F8 type C and FIVAR domains allow the extracellular character of GH29 fucosidases in *B. bifidum* and could enhance affinity toward fucosyl conjugates [27]. Moreover, in all the N-terminal regions of GH29-BifA fucosidases, hydrophobic sequences predicted by SignalP-5.0 to be putative signal peptide with potential cleavage sites were observed (Table S1).

Concerning the AfuC/Glyco_hydro_29 domain, the only representative GH29 fucosidase of GH29-BifA purified and characterized, which is AfcB from *B. bifidum* ATCC 1254, is able to hydrolyze 3-fucosyllactose, Lewis blood group substances (a, b, x, and y types), and lacto-N-fucopentaose II and III. However, the enzyme did not act on glycoconjugates containing α1,2-fucosyl residue or on synthetic pNP-fucose [27].

Supporting the in silico characterization of GH29-BifA fucosidases, several studies confirm the ability of *B. bifidum* to extracellularly hydrolyze FHMOs [28]. However, *B. bifidum* appears to prefer the utilization of lactose when growing on FHMO, probably releasing fucose to the environment [28]. This incapacity to consume fucose may be due to the lack of specific transporters. Nevertheless, the extracellular fucosidase activity of *B. bifidum* could facilitate the establishment of the bifidobacteria community, allowing them to consume the released fucose residues [29].

In contrast to GH29-BifA, the rest of the GH29 fucosidases from bifidobacteria do not have either putative signal peptides or transmembrane helices and consequently their mode of action can be considered intracellular. Indeed, GH29-BifB fucosidases are characterized by exhibiting an AfuC super family/Glyco_Hydro_29 domain (NCBI CDD accession number cl34656/IPR000933) such as GH29-BifA fucosidases but lacking F5/8 type C and FIVAR domains. Due to the presence of the same fucosidase domain in both groups of fucosidases (GH29-BifA and GH29-BifB), similar metabolic capacities could be affirmed. In fact, the only characterized bifidobacterial GH29-BifB fucosidase (Blon_2336 from *Bifidobacterium longum* subsp. *infantis* ATCC 15697) revealed similar activity to AfcB from *B. bifidum* ATCC 1254 (GH29-BifA) against Fuc-α1,3 glucosidic, Fuc-α1,3GlcNAc, and Fuc-α1,4GlcNAc linkages [21]. These GH29-BifB fucosidases appear to be distributed along strains of different species, contrary to GH29-BifA fucosidases, and frequently, strains that exhibit GH29-BifB fucosidases also show GH29-BifC fucosidases, which are duplicated in some of the sequenced strains (Table S1). Actually, the duplication of GH29 fucosidases has been reported previously and plays an important role in fucosidases evolution [30].

GH29-BifC fucosidases are characterized by showing conserved α-Amylase catalytic domain family (NCBI CDD accession number cl38930). It must be taken into account that this superfamily is present in a large number of GHs able to hydrolyze α1,4/6 glycosidic bonds, although in turn they have specific domains unlike the GH29-BifC fucosidases of bifidobacteria [31]. However, since GH29-BifC fucosidases can catalyze the transformation of fucosidic α1-2Gal/3GlcNAc linkages in LNFP I and III, respectively, and mainly Fuc-α1,6 GlcNAc linkages [32], activity non described in the above fucosidase groups, it is difficult to ensure that its catalytic family proposed is α-Amylase catalytic (NCBI CDD) or Glyco_Hydro_29 (InterPro) (Table S1). In this sense, InterPro database (EMBL-EBI) indicated the presence in GH29-BifC fucosidases of a second catalytic family denominated FUC_metazoa_typ (IPR016286) that is close to eukaryotic fucosidases (Table S1). Probably the presence of this domain is key for these fucosidases to be considered as the most unspecific and versatile fucosidases of bifidobacteria since a wide range of substrates has been reported for two different GH29-BifC fucosidases from *B. longum* subsp. *infantis* ATCC 15697 [21,27].

Both GH29-BifB/C fucosidases described in *B. longum* subsp. *infantis* strains are likely found in the cytosol. Therefore, efficient transport of oligosaccharides is needed, unlike *B. bifidum* [13,21]. In this context, genomic studies carried out on *B. longum* subsp. *infantis* ATCC 15697 have unraveled several putative fucose permeases that may facilitate environmental scavenging when soluble fucose is encountered.

In order to elucidate the roles and fitness of the bifidobacterial community to shape the gut microbiome and taking into account the relevance of fucosidases in this regard, their features mentioned above should be updated and expanded to avoid ambiguities in the catalytic domains and relate them to their metabolic properties. Certainly, the rest of the enzymes from different bifidobacterial species need to be characterized in order to reliably distinguish the properties of each group of fucosidases for determining the interaction and mode of actions of bifidobacteria during gut colonization. In this sense, the role of GH29-BifD of fucosidases remains unknown despite having been sequenced and identified in certain *Bifidobacterium* species (Table S1). Unlike to GH29-BifC, GH29-BifD fucosidases exhibit specific α-L-fucosidase main domain (NCBI CDD accession number cl38930). Surprisingly, their accession number is matching with superfamily AmyAc family of group II, suggesting a better accurate and updated in silico annotation. However, InterPro database (EMBL_EBI) indicates both catalytic domain Glyco_Hydro_29 and FUC_metazoa_typ (InterPro IPR000933 and IPR016286, respectively). Nevertheless, physicochemical properties, substrate specificity confirmation, and their correlation with catalytic domains are still pending to be characterized.

2.2. Bifidobacterial GH95 Fucosidases

Similar to GH29 bifidobacterial fucosidases and according to architecture domains, bifidobacterial GH95 fucosidases collected on CAZy could also be subclassified into two main groups (Table S2; Figure 3).

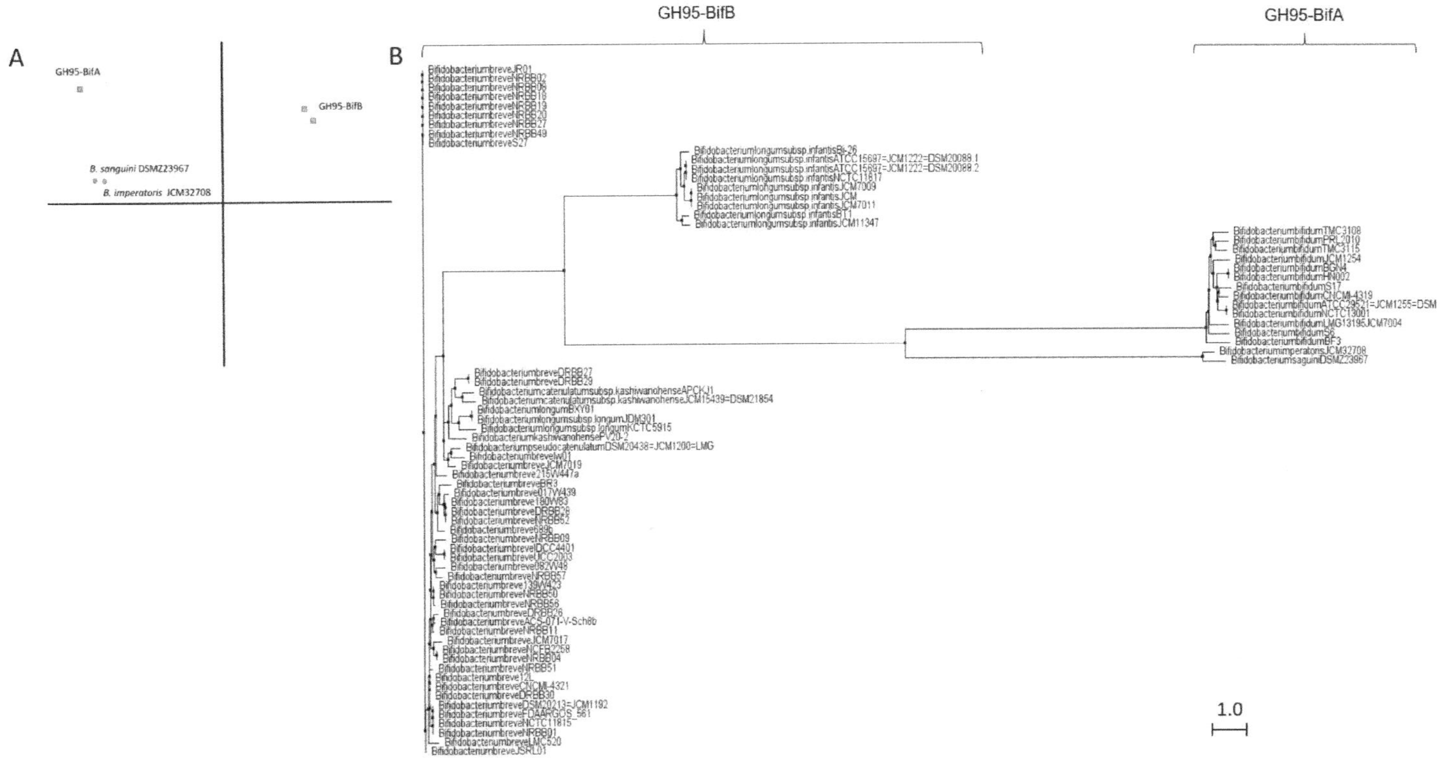

Figure 3. Phylogenetic analysis of bifidobacterial GH95 fucosidases. PCA (**A**) and cladogram tree (**B**) distributions of bifidobacterial GH95 fucosidase sequences listed in CAZy, released from Jalview 2.11.1.4 software using the neighbor-joining method.

The extracellular character observed in GH29-BifA fucosidases from *B. bifidum* strains is also reflected in their GH95 fucosidases, which are characterized by a putative signal peptide and two predicted transmembrane helices. Among GH95 fucosidases, those features are only found in the proposed GH95-BifA fucosidases from *B. bifidum* with the exception of *Bifidobacterium saguini* DSMZ 23967 fucosidase (Genbank QTB91571.1), which exhibited two putative transmembrane helices (Table S2).

The proposed GH95-BifA was characterized according to the NCBI CDD database by exhibiting Glycosyl hydrolase 65 N-terminal (accession number cl22392) as main catalytic domain, while InterPro database analysis (EMBL-EBI) revealed a Glycosyl hydrolase 95 N-terminal (IPR027414) (Table S2). The observed ambiguous prediction on the catalytic architecture could be due to the lack of updating and mismatch annotations. Nevertheless, a common evolutionary origin for GH65 and GH95 families, among others, with conservation of their putative catalytic amino acid residues, was noticed and likely influenced the in silico results [18]. Nevertheless, and contrary to GH65 family, the only GH95-BifA representative fucosidase recombinantly produced and characterized (AfcA from *B. bifidum* JCM1254) showed great activity against Fuc-α1,2 Gal linkages, mainly hydrolyzing 2′-Fucosyllactose and lacto-N-fucopentaose I [17,33].

On the other hand, while NCBI CDD database detected two YjdB overlapping domains (accession number cl35007), whose functions are still uncharacterized but in turn contain Ig-like domain, InterPro database noticed Ig-like_Bact and Bacterial Ig-like group 2 (BIG2) domains instead (accession number IPR022038 and IPR003343, respectively) (Table S2). Despite this coincidence, only the position of one domain practically matches in both databases (YjdB and BIG2) (Table S2). In addition, InterPro identifies Ig-like_Bact near to N-terminal unlike NCBI CDD, and probably GH95-BifA sequences could exhibit up to three accessory domains.

It should be noted that, although the function of BIG2 domain has not been unraveled, it has been hypothesized to participate in facilitating the protrusion of the AfcA catalytic GH95 domain from the cell surface to allow its extracellular activity and degrade the fucosyl residues present on glycoconjugates of enterocytes [17]. This fact could lead one to define AfcA as a bifidobacterial tool for protecting the host's health through modifying α1,2 fucosylated Lewis antigen receptors b and y, recognized by gut pathogens such as *Helicobacter pylori* [34], and norovirus [35]. Taking into account the conserved domains, GH95 fucosidases from *B. imperatoris* and *B. saguini* could be close to being clustered within the GH95-BifA (Table S2). The extracellular character of *B. imperatoris* and *B. saguini* fucosidases could even be affirmed since signal peptides and transmembrane helices are found, although they have not yet been characterized. Indeed, cladogram phylogenetic analysis revealed that both fucosidases actually exhibit more similarities with GH95-BifA (Figure 3).

Beyond GH95-BifA, there are a large number of intracellular GH95 fucosidases from *Bifidobacterium breve* and *B. longum* subsp. *infantis* strains in silico categorized by showing a glycosyl hydrolase 65 N-terminal domain (cl22392; NCBI CDD). They share the catalytic domain with GH95-BifA without exhibiting accessory BIG2 (Table S2). Nevertheless, InterPro database managed to identify a catalytic domain of greater length than in the GH95-BifA sequences, denominated Alpha_L_Fuco family (IPR016518). The presence of this domain could be the key for *B. breve* and *B. longum* subsp. *infantis* GH95 fucosidases to show phylogenetic differences with GH95-BifA as shown by the PCA and cladogram analyses (Figure 3), and therefore are clustered in GH95-BifB.

Unfortunately, no *B. breve* GH95-BifB fucosidases have yet been characterized, although the described hydrolytic activity of *B. breve* on Fuc-α1,2 Gal linkages supports the presence of a functional GH95 fucosidase [36]. Blon_2335 from *B. longum* subsp. *infantis* is the only representative of GH95-BifB that has been characterized [21]. In that study, Blon_2335 showed a strong preference for Fuc-α1,2 linkages (2′-FL, LNFP-I), although it partially cleaved Fuc-α1,3 linkages (3-FL), unlike AfcA from *B. bifidum* [21]. Because AfcA structural exploration revealed its catalytic reaction as a α1,2 fucosidase [18], and since both AfcA and Blon_2335 fucosidases show catalytic architecture differences, further studies concerning crystallization of Blon_2335 are needed in order to elucidate its ability for hydrolyzing both Fuc-α1,2 and Fuc-α1,3 linkages. Structure elucidation could also explain the substantial differences between the GH95-BifB fucosidases from *B. breve* and *B. longum* subsp. *infantis*, also observed in PCA and cladogram (Figure 3), despite presenting the same conserved architecture (Table S2).

2.3. *Bifidobacterial GH151 Fucosidases*

GH151 enzymes form the smallest group of fucosidases (Table S3) and although there are still doubts about their fucosidase activity, *B. longum* subsp. *infantis* ATCC 15697 counts, with a GH151 enzyme (Blon_0346) that exhibits probed Fuc-α1,2 Gal activity [21]. Interestingly, bifidobacterial GH151 fucosidases are quite divergent from the fucosidases classified in other GH families [21] and all of them belong to *B. longum* subsp. *infantis* species although they show little differences in their sequences (Figure 4). While no signal peptide or transmembrane helices were observed, CDD architecture analyses revealed AmyAc_family superfamily and A4_beta-galactosidase_middle_domain, although some sequences are also identified as containing GanA superfamily domain as well (Table S3).

GH151 enzymes probably have domains closest to GH29-BifC fucosidases, identified by containing conserved AmyAc superfamily domain and likely the ability to hydrolyze α glycosidic linkages [31]. However, because GH151 accessory domains shown (Table S3), they could be considered as potential non-specific beta galactosidase enzymes with the capacity to hydrolyze Fuc-α1,2 Gal linkages as occurs with Blon_0346. Nevertheless, further studies in order to elucidate their subjacent activity, substrate specificity, and conformational structure are needed to understand their role in the hydrolysis of fucosylated carbohydrates.

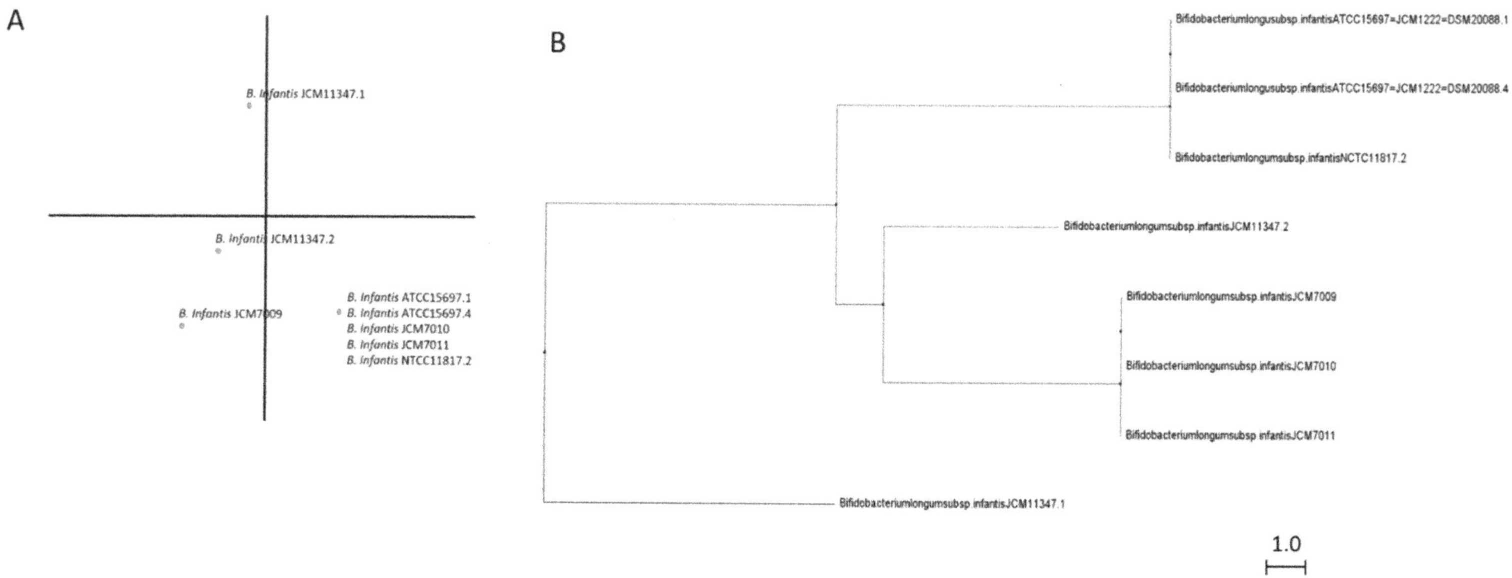

Figure 4. Phylogenetic analysis of bifidobacterial GH151 fucosidases. PCA (**A**) and cladogram tree (**B**) distributions of bifidobacterial GH151 complete fucosidase sequences listed in CAZy, released from Jalview 2.11.1.4 software using the neighbor-joining method.

3. Discussion

Breast milk, beyond its nutritional function, provides the necessary pillars for the initial establishment of the gut microbiota in newborns. In this regard, FHMOs and FHMGs stand out for their ability to stimulate the growth of bifidobacteria [8,12], which in turn produce SCFAs such as acetate, formate, lactate, and pyruvate [13], stimulating the immune system [14], and serving as an energy source for colonocytes [37].

Although only a few bifidobacterial species have been studied extensively at both cellular and genomics level for their ability to utilize fucosylated carbohydrates such as *B. bifidum* and *B. longum* subsp. *infantis* [22,23], their success in colonizing the gut is due to the different strain-dependent metabolic abilities developed for the use of both FHMOs and FHMGs [24]. Therefore, fucosidases play a key role in the bifidobacterial gut establishment. Concerning to that, *B. bifidum* strains show two extracellular fucosidases belonging to GH29 and GH95 families. Both fucosidases cover the hydrolysis of Fuc-α1,3Glu; Fuc-α1,3/4GlcNAc; and Fuc-α1,2 Gal linkages [17,18,27,33]. Since *B. bifidum* prefers the utilization of lactose [28], 2′-fucosyllactose could be its target substrate for its extracellular fucosidases, releasing to the environment lactose and fucose, the last could be also liberated from blood Lewis a, b, x, and y antigens [27]. For all the above, *B. bifidum* fucosidases could be considered altruistic and essential for microbial gut establishment through promoting bifidobacterial mutualism and carbohydrate syntrophy in the infant gut [38]. Given that bifidobacteria are able to metabolize lactose, and species such as *B. longum* subsp. *infantis* or *B. breve* can metabolize fucose, their growth is improved under the presence of fucosidases from *B. bifidum*. Thus, Gotoh et al. (2018) suggested that extracellular fucosidases from *B. bifidum* could be crucial during the development of a bifidobacteria-rich microbiota in the breastfed infant gut, by providing fucosylated conjugate degradants [33]. On the other hand, *B. bifidum* fucosidases contribute to the protection of the host through the modification of Lewis antigens [27].

Regarding the catalytic domains of the *B. bifidum* fucosidases, it should be noted that GH29-BifA present orthologous fucosidases in other bifidobacterial species clustered in GH29-BifB/D, and they probably all have a common phylogenetic lineage (Figure 2). However, this statement has only been functionally corroborated through the characterization of the enzymes AfcB (GH29-BifA) and Blon_2336 (GH29-BifB), due to lack of results of GH29-BifD fucosidases.

Conversely, GH95-BifA fucosidases as well as those grouped in GH95-BifB, and according to CDD database observations (Table S2), could phylogenetically descend from either an evolutionary specialization or non-specification of glycosidases clustered in GH65. Indeed, this in silico observation agrees with the crystallization results obtained for the structure AfcA from *B. bifidum* [18]. According to that, both GH65 and GH95 enzymes share an α/α 6 barrel fold with inverting mechanism and glutamate^{566} as catalytic proton donor. Moreover, Nagae et al. (2007) compared the structures between families GH65 and GH95, revealing conservation of the general acid residues, except for catalytic acid/base aspartate^{766}, which is shifted in AfcA [18]. That shifting was also found in the rest of the bifidobacterial GH95 fucosidases (data not shown), and agreeing with the above mentioned authors, the reaction mechanisms of bifidobacterial GH95 fucosidases differ from those of the GH65 family [18].

The other species widely studied for its fucosidase activity is *B. longum* subsp. *infantis*. Actually, it is the only species of bifidobacteria that exhibits GH29, GH95, and GH151 fucosidases that have been recombinantly purified and characterized [21]. Those fucosidases allow *B. longum* subsp. *infantis* to use a wide range of substrates, hydrolyzing Fuc-α1,3Glu; Fu-cα1,2/3Gal; and Fuc-α1,3/4/6GlcNAc linkages [21,32]. As previously commented, *B. longum* subsp. *infantis* GH29-BifB fucosidases are orthologous with those classified in GH29-BifA. However, this species also shows GH29-duplicated fucosidases, clustered in the GH29-BifC, with different architecture and paralogs from those of GH29-BifB (Figure 3). Taking into account the fucosidase duplication and in agreement with You et al. (2019), *B. longum* subsp. *infantis* GH29-BifC fucosidases could have evolved from a different glycosyl hydrolase [30]. According to CDD database observations (Table S1) and because their predicted structure is composed by a β/α 6 barrel fold with retaining mechanism and glutamate as catalytic proton donor, GH29-BifC fucosidases from *B. longum* subsp. *infantis* could descend from GH13 glycosidases (α-amylases).

GH29-BifC fucosidases, similar to GH95-BifB, which is probably phylogenetically originated from GH65 family as described above, need to have their structural crystallization further explored in order to elucidate their origins and evolution pathway. In addition, GH29-BifC fucosidases show similarities with metazoan fucosidases according to the InterPro database (Table S1), including aspartate^{224} and glutamate^{270} residues (data not shown), which play the role of the catalytic nucleophile and catalytic acid/base, respectively, in metazoan fucosidases [25].

Finally, GH151 fucosidases are exclusively present in *B. longum* subsp. *infantis*. This fact could suggest a fourth pathway of fucosidases phylogenetic evolution in that species closely related to GH29-BifC fucosidases, since they present a N-terminal α amylase catalytic domain. In addition, Blon_0346 was originally classified as a member of GH29 family due to their fucosidase activity despite low similarity [21]. However, GH151 enzymes may be the result of a branch in the evolution of GH29-BifC fucosidases, since they show a GH42 beta galactosidase trimerization architecture instead of conserved features of metazoan fucosidases.

4. Materials and Methods

4.1. Identification and Selection of Fucosidase Sequences

Complete bifidobacterial fucosidase protein sequences belonging to GH29, GH95, and GH151 families were retrieved from CAZy database [19]. Fucosidase sequences were used as probes in PSI-BLAST searches [39] against the NCBI [40], Swiss-Prot [41], and Ensembl [42] protein databases.

4.2. Protein Sequence, Alignment, and Phylogenetic Analysis of α-L-Fucosidases

Fucosidase sequences were analyzed using SignalP-5.0 [43], with default options to predict signal peptide sequences: SOSUI [44] and HMMTOP [45] with default parameters for the prediction of transmembrane helices. NCBI Conserved Domains Database (CDD) [46] and InterPro databases (EMBL_EBI) [47] were used to predict the domain architecture. Inferred fucosidase amino acid sequences were aligned using Clustal Omega web version [48]. All sequences belonging to the same GH families were considered in phylogenetic analyses. Neighbor-joining method cladogram and PCA analyses were performed using the program Jalview 2.11.1.4 [49].

5. Conclusions

This is the first study that explores phylogenetically the three families of the bifidobacterial fucosidases: GH29, GH95, and GH151, through their conserved architecture, showing that *B. bifidum* and *B. longum* subsp. *infantis* reveal two and four different phylogenetic lineages, respectively, belonging to different fucosidase families. On the other hand, given the differences in the catalytic architecture observed in this work, the bifidobacterial fucosidases belonging to the GH29 and GH95 families could be subclassified into four and two groups, respectively.

Taking into account that the observations described in this work were obtained in silico and supported by current characterization results from some *B. bifidum* and *B. longum* subsp. *infantis* fucosidases, further studies regarding structural characterization and physicochemical properties of more fucosidases identified by computational analysis are needed in order to validate the novel classification of bifidobacterial fucosidases here proposed.

Concerning to *B. longum* subsp. *infantis* fucosidases, which evolved from different GH families such as GH29-BifC, GH95-BifB, and GH151, and given that their conserved architecture presents vestiges of ancestral glycosidases GH13, GH65, and GH42, respectively, as well as *B. Bifidum* GH95-BifA fucosidases phylogenetically descended from GH65, deepening substrate spectrum analyses could determine their underlying roles in those species. In this context, and since some fucosidases have been used to transfucosylate carbohydrates or glycoconjugates, the application of these evolved and hypothetically non-specific *B. longum* subsp. *infantis* fucosidases mentioned above can open a new perspective towards the synthesis of novel fucosylated conjugates by using different substrates beyond lactose for synthetizing 2′-fucosyllactose. This vision is oriented towards the supply those novel fucosylated conjugates to adults in combination with fucosidase producer bifidobacteria in order to maintain a healthy microbiota or to reestablish it from dysbiosis states as described previously [50,51]. In this regard, it would be important to elucidate phylogenetically, as well as structurally and physicochemically, the fucosidases of many other gut microorganism genera, as for instance *Lactobacillus*, *Bacteroides*, and *Akkermansia*, with the aim to reveal the whole gut fucosidase interaction.

Supplementary Materials: The following are available online at https://www.mdpi.com/article/10.3390/ijms22168462/s1.

Author Contributions: Conceptualization, J.A.C.; methodology, J.A.C. and J.M.L.; validation and formal analysis, J.A.C. and Á.P.; investigation, J.A.C., Á.P., J.M.L., S.L. and A.R.d.l.B.; writing—original draft preparation, J.A.C. and J.M.L.; writing—review and editing, J.A.C. and Á.P.; visualization and supervision, J.A.C. and J.L.A.; project administration, J.A.C.; funding acquisition, J.A.C. All authors have read and agreed to the published version of the manuscript.

Funding: This research was funded by Spanish Ministry of Science and Innovation—Ramón y Cajal program, grant number RYC2019-026368-I.

Institutional Review Board Statement: Not applicable.

Informed Consent Statement: Not applicable.

Data Availability Statement: Not applicable.

Acknowledgments: J.A.C. has a postdoctoral contract with the research program "Ramón y Cajal" (RYC2019-026368-I) and A.R.d.l.B. is recipient of a Pre-doctoral contract (PRE2018-086293), both from Spanish Ministry of Science and Innovation.

Conflicts of Interest: The authors declare no conflict of interest.

Abbreviations

FHMG	Fucosylated human milk glycoprotein
FHMO	Fucosylated human milk oligosaccharide
Fuc	Fucose
Gal	Galactose
GH	Glycosyl hydrolase
GlcNAc	N-acetylglucosamine
Glu	Glucose
HMG	human milk glycoprotein
HMO	human milk oligosaccharide
LNFP	Lacto-N-Fucopentaose
pNP-fucose	*p*-nitrophenyl-α-L-fucopyranoside
SCFAs	Short-chain fatty acids.

References

1. Zivkovic, A.M.; German, J.B.; Lebrilla, C.B.; Mills, D.A. Human milk glycobiome and its impact on the infant gastrointestinal microbiota. *Proc. Natl. Acad. Sci. USA* **2011**, *108*, 4653–4658. [CrossRef]
2. Milani, C.; Mancabelli, L.; Lugli, G.A.; Duranti, S.; Turroni, F.; Ferrario, C.; Mangifesta, M.; Viappiani, A.; Ferreti, P.; Corfer, V.; et al. Exploring vertical transmission of bifidobacteria from mother to child. *Appl. Environ. Microbiol.* **2015**, *81*, 7078–7087. [CrossRef]
3. Turroni, F.; Peano, C.; Pass, D.A.; Foroni, E.; Severgnini, M.; Claesson, M.J.; Kerr, C.; Hourihane, J.; Murray, D.; Fuligni, F.; et al. Diversity of bifidobacteria within the infant gut microbiota. *PLoS ONE* **2012**, *7*, e36957. [CrossRef] [PubMed]

4. O'Neill, I.; Schofield, Z.; Hall, L.J. Exploring the role of the microbiota member *Bifidobacterium* in modulating immune-linked diseases. *Emerg. Top. Life Sci.* **2017**, *1*, 333–349.
5. Wampach, L.; Heintz-Buschart, A.; Fritz, J.V.; Ramiro-Garcia, J.; Habier, J.; Herold, M.; Nayaranasamy, S.; Kaysen, A.; Hogan, A.H.; Bindl, L.; et al. Birth mode determines earliest strainconferred gut microbiome functions and immunostimulatory potential. *Nat. Commun.* **2018**, *9*, 1–14. [CrossRef]
6. Wu, S.; Tao, N.; German, J.B.; Grimm, R.; Lebrilla, C.B. Development of an annotated library of neutral human milk oligosaccharides. *J. Proteome Res.* **2010**, *9*, 4138–4151. [CrossRef] [PubMed]
7. Bai, Y.; Tao, J.; Zhou, J.; Fan, Q.; Liu, M.; Hu, Y.; Xu, Y.; Zhang, L.; Yuan, J.; Li, W.; et al. Fucosylated human milk oligosaccharides and N-glycans in the milk of Chinese mothers regulate the gut microbiome of their breast-fed infants during different lactations stages. *MSystems* **2018**, *3*, e00206-18. [CrossRef]
8. McGuire, M.K.; Meehan, C.L.; McGuire, M.A.; Williams, J.E.; Foster, J.; Sellen, D.W.; Prentice, A.M. What's normal? Oligosaccharide concentrations and profiles in milk produced by healthy women vary geographically. *Am. J. Clin. Nutr.* **2017**, *105*, 1086–1100. [CrossRef] [PubMed]
9. Huang, J.; Guerrero, A.; Parker, E.; Strum, J.S.; Smilowitz, J.T.; German, J.B.; Lebrilla, C.B. Site-specific glycosylation of secretory immunoglobulin A from human colostrum. *J. Proteome Res.* **2015**, *14*, 1335–1349. [CrossRef] [PubMed]
10. Li, W.; Yu, R.; Ma, B.; Yang, Y.; Jiao, X.; Liu, Y.; Cao, H.; Dong, W.; Liu, L.; Ma, K.; et al. Core fucosylation of IgG B cell receptor is required for antigen recognition and antibody production. *J. Immunol.* **2015**, *194*, 2596–2606. [CrossRef]
11. Peterson, R.; Cheah, W.Y.; Grinyer, J.; Packer, N. Glycoconjugates in human milk: Protecting infants from disease. *Glycobiology* **2013**, *23*, 1425–1438. [CrossRef]
12. Korpela, K.; Salonen, A.; Hickman, B.; Kunz, C.; Sprenger, N.; Kukkonen, K.; Savilahti, E.; Kuitunen, M.; de Vos, W.M. Fucosylated oligosaccharides in mother's milk alleviate the effects of caesarean birth on infant gut microbiota. *Sci. Rep.* **2018**, *8*, 1–7. [CrossRef] [PubMed]
13. Zabel, B.E.; Gerdes, S.; Evans, K.C.; Nedveck, D.; Singles, S.K.; Volk, B.; Budinoff, C. Strain-specific strategies of 2′-fucosyllactose, 3-fucosyllactose, and difucosyllactose assimilation by *Bifidobacterium longum* subsp. *infantis* Bi-26 and ATCC 15697. *Sci. Rep.* **2020**, *10*, 1–18. [CrossRef] [PubMed]
14. Park, J.; Kim, M.; Kang, S.G.; Jannasch, A.H.; Cooper, B.; Patterson, J.; Kim, C.H. Short-chain fatty acids induce both effector and regulatory T cells by suppression of histone deacetylases and regulation of the mTOR–S6K pathway. *Mucosal Immunol.* **2015**, *8*, 80–93. [CrossRef] [PubMed]
15. Grootaert, H.; Van Landuyt, L.; Hulpiau, P.; Callewaert, N. Functional exploration of the GH29 fucosidase family. *Glycobiology* **2020**, *30*, 735–745. [CrossRef] [PubMed]
16. Shaikh, F.A.; Van Bueren, A.L.; Davies, G.J.; Withers, S.G. Identifying the catalytic acid/base in GH29 α-L-Fucosidase subfamilies. *Biochemistry* **2013**, *52*, 5857–5864. [CrossRef]
17. Katayama, T.; Sakuma, A.; Kimura, T.; Makimura, Y.; Hiratake, J.; Sakata, K.; Yamanoi, T.; Kumagai, H.; Yamamoto, K. Molecular cloning and characterization of *Bifidobacterium bifidum* 1, 2-α-L-fucosidase (AfcA), a novel inverting glycosidase (glycoside hydrolase family 95). *J. Bacteriol.* **2004**, *186*, 4885–4893. [CrossRef]
18. Nagae, M.; Tsuchiya, A.; Katayama, T.; Yamamoto, K.; Wakatsuki, S.; Kato, R. Structural basis of the catalytic reaction mechanism of novel 1, 2-α-L-fucosidase from *Bifidobacterium bifidum*. *J. Biol. Chem.* **2007**, *282*, 18497–18509. [CrossRef]
19. Lombard, V.; Ramulu, H.G.; Drula, E.; Coutinho, P.M.; Henrissat, B. The carbohydrate-active enzymes database (CAZy) in 2013. *Nucleic Acids Res.* **2014**, *42*, 490–495. [CrossRef]
20. Lezyk, M.; Jers, C.; Kjaerulff, L.; Gotfredsen, C.H.; Mikkelsen, M.D.; Mikkelsen, J.D. Novel α-L-fucosidases from a soil metagenome for production of fucosylated human milk oligosaccharides. *PLoS ONE* **2016**, *11*, e0147438. [CrossRef] [PubMed]
21. Sela, D.A.; Garrido, D.; Lerno, L.; Wu, S.; Tan, K.; Eom, H.J.; Joachimiak, A.; Lebrilla, C.B.; Mills, D.A. *Bifidobacterium longum* subsp. *infantis* ATCC 15697 α-fucosidases are active on fucosylated human milk oligosaccharides. *Appl. Environ. Microbiol.* **2012**, *78*, 795–803. [CrossRef]
22. Angeloni, S.; Ridet, J.L.; Kusy, N.; Gao, H.; Crevoisier, F.; Guinchard, S.; Kochhar, S.; Sigrist, H.; Sprenger, N. Glycoprofiling with micro-arrays of glycoconjugates and lectins. *Glycobiology* **2005**, *15*, 31–41. [CrossRef]
23. Bergström, A.; Skov, T.H.; Bahl, M.I.; Roager, H.M.; Christensen, L.B.; Ejlerskov, K.T.; Molgaard, C.; Michaelsen, K.F.; Licht, T.R. Establishment of intestinal microbiota during early life: A longitudinal, explorative study of a large cohort of Danish infants. *Appl. Environ. Microbiol.* **2014**, *80*, 2889–2900. [CrossRef] [PubMed]
24. Turroni, F.; Bottacini, F.; Foroni, E.; Mulder, I.; Kim, J.H.; Zomer, A.; Sánchez, B.; Bidossi, A.; Ferrarini, A.; Giubellini, V.; et al. Genome analysis of *Bifidobacterium bifidum* PRL2010 reveals metabolic pathways for host-derived glycan foraging. *Proc. Natl. Acad. Sci. USA* **2010**, *107*, 19514–19519. [CrossRef]
25. Intra, J.; Perotti, M.E.; Pavesi, G.; Horner, D. Comparative and phylogenetic analysis of α-l-fucosidase genes. *Gene* **2007**, *392*, 34–46. [CrossRef] [PubMed]
26. Abbott, D.W.; Eirin-Lopez, J.M.; Boraston, A.B. Insight into ligand diversity and novel biological roles for family 32 carbohydrate-binding modules. *Mol. Biol. Evol.* **2008**, *25*, 155–167. [CrossRef]
27. Ashida, H.; Miyake, A.; Kiyohara, M.; Wada, J.; Yoshida, E.; Kumagai, H.; Katayama, T.; Yamamoto, K. Two distinct α-L-fucosidases from *Bifidobacterium bifidum* are essential for the utilization of fucosylated milk oligosaccharides and glycoconjugates. *Glycobiology* **2009**, *19*, 1010–1017. [CrossRef] [PubMed]

28. Garrido, D.; Ruiz-Moyano, S.; Lemay, D.G.; Sela, D.A.; German, J.B.; Mills, D.A. Comparative transcriptomics reveals key differences in the response to milk oligosaccharides of infant gut-associated bifidobacteria. *Sci. Rep.* **2015**, *5*, 1–18.
29. Asakuma, S.; Hatakeyama, E.; Urashima, T.; Yoshida, E.; Katayama, T.; Yamamoto, K.; Kumagai, H.; Ashida, H.; Hirose, J.; Kitaoka, M. Physiology of consumption of human milk oligosaccharides by infant gut-associated bifidobacteria. *J. Biol. Chem.* **2011**, *286*, 34583–34592. [CrossRef]
30. You, J.; Lin, S.; Jiang, T. Origins and evolution of the α-L-fucosidases: From bacteria to metazoans. *Front. Microbiol.* **2019**, *10*, 1756. [CrossRef]
31. Janeček, Š.; Svensson, B.; MacGregor, E.A. α-Amylase: An enzyme specificity found in various families of glycoside hydrolases. *Cell. Mol. Life Sci.* **2014**, *71*, 1149–1170. [CrossRef]
32. Ashida, H.; Fujimoto, T.; Kurihara, S.; Nakamura, M.; Komeno, M.; Huang, Y.; Katayama, T.; Kinoshita, T.; Takegawa, K. 1,6-α-L-Fucosidases from *Bifidobacterium longum* subsp. *infantis* ATCC 15697 involved in the degradation of core-fucosylated N-glycan. *J. Appl. Glycosci.* **2020**, *67*, 23–29. [CrossRef]
33. Gotoh, A.; Katoh, T.; Sakanaka, M.; Ling, Y.; Yamada, C.; Asakuma, S.; Urashima, T.; Tomabechi, Y.; Katayama-Ikegami, A.; Kurihara, S.; et al. Sharing of human milk oligosaccharides degradants within Bifidobacterial communities in faecal cultures supplemented with *Bifidobacterium bifidum*. *Sci. Rep.* **2018**, *8*, 1–14. [CrossRef] [PubMed]
34. Moran, A.P. Relevance of fucosylation and Lewis antigen expression in the bacterial gastroduodenal pathogen *Helicobacter pylori*. *Carbohydr. Res.* **2008**, *343*, 1952–1965. [CrossRef]
35. Huang, P.; Farkas, T.; Marionneau, S.; Zhong, W.; Ruvoën-Clouet, N.; Morrow, A.L.; Altaye, M.; Pickering, L.K.; Newburg, D.S.; LePendu, J.; et al. Noroviruses bind to human ABO, Lewis, and secretor histo-blood group antigens: Identification of 4 distinct strain-specific patterns. *J. Infect. Dis.* **2003**, *188*, 19–31. [CrossRef]
36. Sakanaka, M.; Gotoh, A.; Yoshida, K.; Odamaki, T.; Koguchi, H.; Xiao, J.Z.; Kitaoka, M.; Katayama, T. Varied pathways of infant gut-associated *Bifidobacterium* to assimilate human milk oligosaccharides: Prevalence of the gene set and its correlation with bifidobacteria-rich microbiota formation. *Nutrients* **2020**, *12*, 71. [CrossRef]
37. Corrêa-Oliveira, R.; Fachi, J.L.; Vieira, A.; Sato, F.T.; Vinolo, M.A.R. Regulation of immune cell function by short-chain fatty acids. *Clin. Transl. Immunol.* **2016**, *5*, e73. [CrossRef]
38. Motherway, M.O.C.; O'Brien, F.; O'Driscoll, T.; Casey, P.G.; Shanahan, F.; van Sinderen, D. Carbohydrate syntrophy enhances the establishment of *Bifidobacterium breve* UCC2003 in the neonatal gut. *Sci. Rep.* **2018**, *8*, 1–10.
39. Altschul, S.F.; Madden, T.L.; Schäffer, A.A.; Zhang, J.; Zhang, Z.; Miller, W.; Lipman, D.J. Gapped BLAST and PSI-BLAST: A new generation of protein database search programs. *Nucleic Acids Res.* **1997**, *25*, 3389–3402. [CrossRef] [PubMed]
40. Wheeler, D.L.; Barrett, T.; Benson, D.A.; Bryant, S.H.; Canese, K.; Chetvernin, V.; Church, D.M.; DiCuccio, M.; Edgar, R.; Federhen, S.; et al. Database resources of the national center for biotechnology information. *Nucleic Acids Res.* **2007**, *36*, 13–21. [CrossRef]
41. Gasteiger, E.; Gattiker, A.; Hoogland, C.; Ivanyi, I.; Appel, R.D.; Bairoch, A. ExPASy: The proteomics server for in-depth protein knowledge and analysis. *Nucleic Acids Res.* **2003**, *31*, 3784–3788. [CrossRef] [PubMed]
42. Yates, A.D.; Achuthan, P.; Akanni, W.; Allen, J.; Allen, J.; Alvarez-Jarreta, J.; Amode, M.R.; Armean, I.M.; Azov, A.G.; Bennett, R.; et al. Ensembl 2020. *Nucleic Acids Res.* **2020**, *48*, 682–688. [CrossRef] [PubMed]
43. Almagro Armenteros, J.J.; Tsirigos, K.D.; Sønderby, C.K.; Petersen, T.N.; Winther, O.; Brunak, S.; von Heijne, G.; Nielsen, H. SignalP 5.0 improves signal peptide predictions using deep neural networks. *Nat. Biotechnol.* **2019**, *37*, 420–423. [CrossRef]
44. Hirokawa, T.; Boon-Chieng, S.; Mitaku, S. SOSUI: Classification and secondary structure prediction system for membrane proteins. *Bioinformatics* **1998**, *14*, 378–379. [CrossRef]
45. Tusnady, G.E.; Simon, I. Topology of membrane proteins. *J. Chem. Inf. Comput. Sci.* **2001**, *41*, 364–368. [CrossRef] [PubMed]
46. Lu, S.; Wang, J.; Chitsaz, F.; Derbyshire, M.K.; Geer, R.C.; Gonzales, N.R.; Gwadz, M.; Hurwitz, D.I.; Marchler, G.H.; Song, J.S.; et al. CDD/SPARCLE: The conserved domain database in 2020. *Nucleic Acids Res.* **2020**, *48*, 265–268. [CrossRef]
47. Blum, M.; Chang, H.Y.; Chuguransky, S.; Grego, T.; Kandasaamy, S.; Mitchell, A.; Nuka, G.; Paysan-Lafosse, T.; Qureshi, M.; Raj, S.; et al. The InterPro protein families and domains database: 20 years on. *Nucleic Acids Res.* **2021**, *49*, 344–354. [CrossRef] [PubMed]
48. McWilliam, H.; Li, W.; Uludag, M.; Squizzato, S.; Park, Y.M.; Buso, N.; Cowley, A.P.; Lopez, R. Analysis tool web services from the EMBL-EBI. *Nucleic Acids Res.* **2013**, *41*, 597–600. [CrossRef] [PubMed]
49. Waterhouse, A.M.; Procter, J.B.; Martin, D.M.A.; Clamp, M.; Barton, G.J. Jalview version 2-A Multiple sequence alignment editor and analysis workbench. *Bioinformatics* **2009**, *25*, 1189–1191. [CrossRef]
50. Ryan, J.J.; Monteagudo-Mera, A.; Contractor, N.; Gibson, G.R. Impact of 2′-Fucosyllactose on Gut Microbiota Composition in Adults with Chronic Gastrointestinal Conditions: Batch Culture Fermentation Model and Pilot Clinical Trial Findings. *Nutrients* **2021**, *13*, 938. [CrossRef]
51. Elison, E.; Vigsnaes, L.K.; Krogsgaard, L.R.; Rasmussen, J.; Sørensen, N.; McConnell, B.; Hennet, T.; Sommer, M.O.A.; Bytzer, P. Oral supplementation of healthy adults with 2′-O-fucosyllactose and lacto-N-neotetraose is well tolerated and shifts the intestinal microbiota. *Brit. J. Nutr.* **2016**, *116*, 1356–1368. [CrossRef] [PubMed]

International Journal of *Molecular Sciences*

MDPI

Article

Nrf2, the Major Regulator of the Cellular Oxidative Stress Response, is Partially Disordered

Nadun C. Karunatilleke [1], Courtney S. Fast [2], Vy Ngo [3], Anne Brickenden [1], Martin L. Duennwald [3], Lars Konermann [1,2,*] and Wing-Yiu Choy [1,2,*]

1 Department of Biochemistry, The University of Western Ontario, London, ON N6A 5C1, Canada; nkarunat@uwo.ca (N.C.K.); abricken@uwo.ca (A.B.)

2 Department of Chemistry, The University of Western Ontario, London, ON N6A 5B7, Canada; courtfast@gmail.com

3 Department of Pathology and Laboratory Medicine, The University of Western Ontario, London, ON N6A 5C1, Canada; vngo23@uwo.ca (V.N.); martin.duennwald@schulich.uwo.ca (M.L.D.)

* Correspondence: konerman@uwo.ca (L.K.); jchoy4@uwo.ca (W.-Y.C.)

Abstract: Nuclear factor erythroid 2-related factor 2 (Nrf2) is a transcription regulator that plays a pivotal role in coordinating the cellular response to oxidative stress. Through interactions with other proteins, such as Kelch-like ECH-associated protein 1 (Keap1), CREB-binding protein (CBP), and retinoid X receptor alpha (RXRα), Nrf2 mediates the transcription of cytoprotective genes critical for removing toxicants and preventing DNA damage, thereby playing a significant role in chemoprevention. Dysregulation of Nrf2 is linked to tumorigenesis and chemoresistance, making Nrf2 a promising target for anticancer therapeutics. However, despite the physiological importance of Nrf2, the molecular details of this protein and its interactions with most of its targets remain unknown, hindering the rational design of Nrf2-targeted therapeutics. With this in mind, we used a combined bioinformatics and experimental approach to characterize the structure of full-length Nrf2 and its interaction with Keap1. Our results show that Nrf2 is partially disordered, with transiently structured elements in its Neh2, Neh7, and Neh1 domains. Moreover, interaction with the Kelch domain of Keap1 leads to protection of the binding motifs in the Neh2 domain of Nrf2, while the rest of the protein remains highly dynamic. This work represents the first detailed structural characterization of full-length Nrf2 and provides valuable insights into the molecular basis of Nrf2 activity modulation in oxidative stress response.

Keywords: oxidative stress; Nrf2; Keap1; nuclear magnetic resonance spectroscopy; hydrogen/deuterium exchange; mass spectrometry; circular dichroism; intrinsically disordered

Citation: Karunatilleke, N.C.; Fast, C.S.; Ngo, V.; Brickenden, A.; Duennwald, M.L.; Konermann, L.; Choy, W.-Y. Nrf2, the Major Regulator of the Cellular Oxidative Stress Response, is Partially Disordered. *Int. J. Mol. Sci.* **2021**, *22*, 7434. https://doi.org/10.3390/ijms22147434

Academic Editors: Istvan Simon and Csaba Magyar

Received: 18 June 2021
Accepted: 8 July 2021
Published: 11 July 2021

1. Introduction

Reactive oxygen species (ROS) from the environment or generated by the cellular metabolism can cause oxidative damage to proteins, DNA, and lipids, leading to diseases such as cancer, dementia, and cardiovascular disease, to name a few [1,2]. Nuclear factor erythroid 2-related factor 2 (Nrf2) is an essential transcription factor for protecting cells from these harmful effects [3–6]. Through binding to the antioxidant-responsive element (ARE) in their promoter regions, Nrf2 induces the expression of numerous cytoprotective genes and safeguards cells from tumorigenesis [4,7]. On the other hand, aberrant activation of Nrf2 is associated with poor prognosis and chemoresistance of many cancer types [8–14]. Genomic characterization of squamous cell lung cancers revealed that the Nrf2 antioxidant pathway is one of the most severely altered pathways [15]. Many mutations of Nrf2 are expected to affect its target recognition [16,17], resulting in its dysregulation [18,19]. Thus, pharmacological modulation of Nrf2 activity represents an attractive strategy for cancer treatment [20].

Nrf2 activity is regulated through the interactions with a suite of different proteins [4,21–27], including Kelch-like ECH-associated protein 1 (Keap1) [22,28]. The 70 kDa Keap1, which acts as the primary negative regulator of Nrf2, consists of three major functional domains: the N-terminal BTB domain, the IVR region, and the C-terminal Kelch domain. Under homeostatic conditions, dimeric Keap1 binds Nrf2 via the Kelch domains and recruits it to the Cullin-3 based E3 ligase complex for ubiquitination, leading to the proteasomal degradation of Nrf2 [22,29,30]. In the presence of oxidative stress, several redox-sensitive cysteines in Keap1 (e.g., C151, C273, and C288) are modified by ROS, resulting in protein conformational changes and disruption of Nrf2 binding [31,32]. This leads to the accumulation of Nrf2 in the nucleus and subsequent activation of ARE-dependent gene transcription.

Nrf2 is a member of the Cap 'n' Collar (CNC) family of basic leucine-zipper transcription factors. The 68 kDa human Nrf2 comprises seven Nrf2-ECH homology (Neh) functional domains, known as Neh1–7 [25,33]. Note that for historical reasons, the domain numbers are not ordered according to the sequence [34] (Figure S2A). The N-terminal Neh2 domain binds the Kelch domains of dimeric Keap1 via a high-affinity ETGE motif and a low-affinity DLG motif [22,35]. Following the Neh2 domain in the protein sequence, Neh4 and Neh5 are the transactivation domains that recognize the transcription co-activator CBP [23,36], whereas the Neh7 domain binds to the negative regulator RXRα [25,37]. The Neh6 domain interacts with the β-transducin repeat-containing protein 1, which leads to Keap1-independent ubiquitination and degradation of Nrf2 [38,39]. The Neh1 domain mediates heterodimerization with sMaf proteins for DNA-binding [4,40], whereas the C-terminal Neh3 region is another transactivation domain associated with chromo ATPase/helicase DNA-binding protein 6 [41].

Despite its critical role in the antioxidant response, little is known about the molecular structure of Nrf2. To date, only the structures of the isolated N-terminal Neh2 domain (residues 1–98) and part of the Neh1 DNA-binding domain (residues 445–523; PDB: 2LZ1) have been experimentally characterized [22,28,42,43]. Nuclear magnetic resonance (NMR) studies revealed that the Neh2 domain is intrinsically disordered (i.e., it does not adopt a stable folded conformation) yet possesses transient local structural elements. In particular, the region of residues 39–71 that links the two Keap1-binding motifs (i.e., the DLG and ETGE elements) displays significant helical propensity [22].

The Neh1 DNA-binding domain comprises three regions: the CNC homology region, the basic DNA recognition motif, and the leucine-zipper region [33]. The solution structure of a 79-residue fragment (PDB: 2LZ1; residues 445–523 of human Nrf2) representing only part of the Neh1 domain (lacking the C-terminal leucine-zipper dimerization region) was solved using NMR spectroscopy [43]. The structure contains 4 α-helices (H1–H4; residues 455–465, 469–475, 478–489, and 491–505), whereas the remaining 34 residues (~43% of the structure) are disordered.

Gaining a mechanistic understanding of how Nrf2 is regulated through interactions with different binding partners requires a multi-pronged approach. In this work, we combined bioinformatics tools with various biophysical techniques, including hydrogen/deuterium exchange mass spectrometry (HDX-MS), circular dichroism (CD) spectropolarimetry, and NMR spectroscopy to investigate the structural properties of full-length Nrf2 (from now on referred to as FL-Nrf2). This is in contrast to earlier studies that were limited to truncated constructs [22,42,43]. Further, HDX-MS was used to characterize the interactions of Nrf2 with the Kelch domain of Keap1 (from now on referred to as "Kelch"). Intriguingly, our results reveal that FL-Nrf2 is partially disordered yet possesses several transiently structured elements. Upon binding with Kelch, the DLG and ETGE binding motifs in the Neh2 domain of Nrf2 became protected, yet the rest of the protein remained highly dynamic. These unique structural properties may be involved in regulating the interactions of Nrf2 with other proteins and thus determine its function in response to oxidative stress.

2. Results

2.1. Full-Length Nrf2 (FL-Nrf2) is Predicted to Be Partially Disordered

FL-Nrf2 is an acidic protein with pI ~4.8. At physiological pH, it is highly charged (~−40 at pH 7). Using the optimized expression and purification protocols outlined under Materials and Methods, we were able to produce ~0.3 mg of recombinant full-length protein with high purity from 1 L of M9 culture. Intriguingly, although the molecular weight of FL-Nrf2 is only 70.4 kDa (including the His-tag), it runs with an apparent MW of ~110 kDa on SDS-PAGE gels as already reported previously [44] (Figure S1). The high net charge and aberrant SDS-PAGE migration behavior suggest that FL-Nrf2 may be partially disordered [45,46].

We used bioinformatics tools to further investigate the potential disorder of FL-Nrf2 (Figure S2). PONDR-FIT [47], a meta-protein disorder predictor that combines the results of six different methods, indicates that many regions of FL-Nrf2 are disordered (i.e., disorder disposition > 0.5, Figure S2A). In particular, the N-terminal Neh2 domain is predicted to be largely unstructured, in agreement with published data [22]. Meanwhile, local structured elements are expected to be present in the Neh4 (112–134) and Neh5 domains (183–201), the two transactivation domains that bind CBP [23,36]. Neh6 (338–388) and Neh3 (562–605) are predicted to be mainly unstructured, whereas Neh7 (209–316) and Neh1 (435–562) appear to be partially disordered.

We also applied s2D, another sequence-based disorder predictor, to further examine the conformational propensities of FL-Nrf2. s2D predicts not only disordered regions but also estimates the secondary structure at the residue level [48]. The method predicted that ~90% of the residues in FL-Nrf2 predominantly adopt a random coil conformation (Figure S2B). While no residues were identified to sample the β-strand conformations, 57 residues (20–23, 479–486, 492–502, and 533–566) were predicted to have a helical propensity of >46%. Notably, the longest stretch (residues 533–566) that showed a preferentially helical conformation is in the Neh1 domain. DisEMBL [49] and IUPread2A [50], two other sequence-based structural predictors, were also used to characterize FL-Nrf2, all of which agree with the PONDR-FIT and s2D data (Figure S2C).

2.2. CD and NMR Experiments Confirm that FL-Nrf2 is Intrinsically Disordered

We next used CD and NMR techniques to validate our bioinformatics findings. CD spectropolarimetry was applied for assessing the secondary structure of FL-Nrf2 at 5, 10, 25, and 35 °C. The negative bands at 208 and 222 nm indicate the presence of α-helical structural elements (Figure 1A). Spectral deconvolution indicates that at 25 °C, there are around 27% α-helical, 21% β-strand and 52% disordered/turn structures, implying that Nrf2 is indeed partially disordered (Figure 1B; Table S1). While lowering the temperature to 5 or 10 °C did not considerably alter the secondary structure content, the percentage of α-helical structure dropped to ~18% when the temperature was increased to 35 °C.

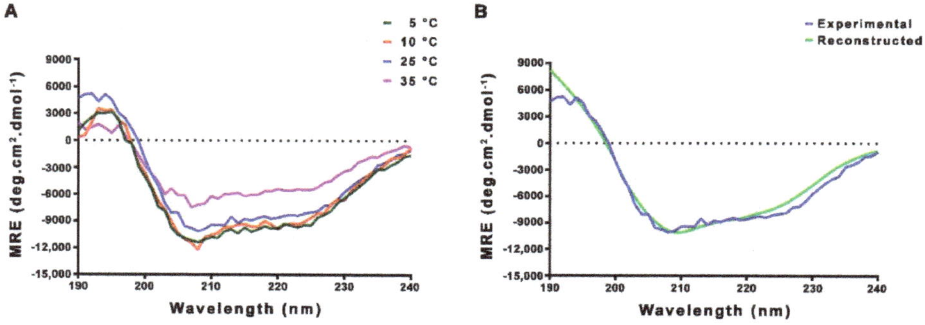

Figure 1. (**A**) CD spectra of FL-Nrf2 recorded at 5, 10, 25, and 35 °C. (**B**) Deconvolution of FL-Nrf2 CD spectrum at 25 °C using the CONTINLL program in DichroWeb, with protein reference data set 4 (optimized for 190–240 nm). The NRMSD between the experimental and reconstructed CD data is 0.13.

We used NMR spectroscopy to further verify the disordered nature of FL-Nrf2. Figure 2 shows the ^1H-^{15}N HSQC spectra of FL-Nrf2 acquired at 5, 10, 25, and 35 °C. At all four temperatures, despite the presence of some well-dispersed peaks with relatively weak intensities, most of the observed backbone amide signals are crowded in a narrow region between 7.8 and 8.7 ppm in the ^1H dimension. This lack of ^1H resonance dispersion indicates that many parts of FL-Nrf2 do not adopt stable structures and undergo rapid conformational interconversion [51–54]. In addition, the disordered nature of FL-Nrf2 does not change significantly at lower temperatures. For comparison, we also acquired the HSQC spectrum of FL-Nrf2 at 35 °C in the presence of 6 M urea, under which conditions the protein was expected to be largely unfolded (Figure S3). The similarity of all these spectra therefore suggest that FL-Nrf2 is already extensively unfolded even in the absence of urea. Although the signal overlap hampers further site-specific structural analyses of FL-Nrf2 by NMR spectroscopy, our CD and NMR results nevertheless demonstrate that FL-Nrf2 is significantly disordered.

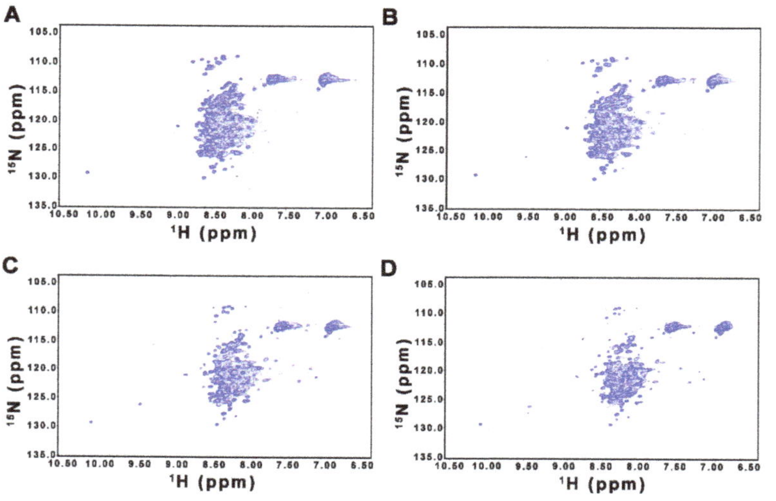

Figure 2. ^1H-^{15}N HSQC spectra of FL-Nrf2 (20 μM) in 50 mM ammonium acetate buffer, 0.5 mM TCEP (pH 6.5) recorded at (**A**) 5 °C, (**B**) 10 °C, (**C**) 25 °C, and (**D**) 35 °C. Most of the observed backbone amide signals are crowded in a narrow region between 7.8 and 8.7 ppm in the ^1H dimension at all four temperatures.

The solution NMR structure of a 79-residue fragment (residues 445–523 of human Nrf2; denoted as Neh1–2LZ1; PDB accession: 2LZ1) representing part of the Neh1 domain shows that this segment is partially folded in isolation. Therefore, some well-dispersed peaks in the ^{1}H-^{15}N HSQC spectrum of FL-Nrf2 could originate from the Neh1–2LZ1 region. To test this possibility, we performed NMR and CD analyses on isolated Neh1–2LZ1. Figure 3 shows the ^{1}H-^{15}N HSQC NMR spectrum of Neh1–2LZ1 in 50 mM ammonium acetate (pH 6.5) with 50 mM arginine, the same condition used for solving the Neh1–2LZ1 structure. It has been shown that arginine can increase the stability and solubility of some proteins [55,56]. For completeness, we also acquired HSQC data of Neh1–2LZ1 in the absence of arginine (Figure 3). Both spectra contain well-dispersed peaks and are very similar, suggesting that isolated Neh1–2LZ1 is, to some extent, structured both in the presence and absence of arginine.

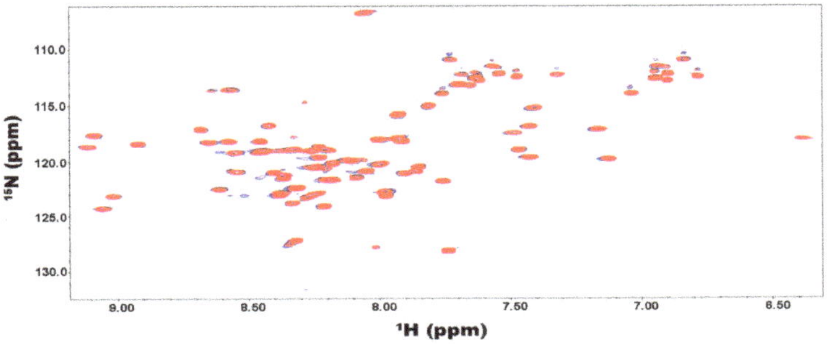

Figure 3. ^{1}H-^{15}N HSQC NMR spectra of Neh1–2LZ1 (100 μM) in the presence (red) and absence (blue) of 50 mM arginine in 50 mM ammonium acetate buffer, 1 mM DTT, pH 6.5. The spectra were recorded at 25 °C.

The CD spectrum of Neh1–2LZ1 depicts two negative minima at 208 and 222 nm, as well as a positive band at 195 nm (Figure S4), illustrating the existence of helical content. Deconvolution analysis estimated the population of α-helical, β-strand, and disordered/turn structures to be 39%, 14%, and 47%, respectively, consistent with the partially disordered nature of Neh1–2LZ1.

2.3. HDX-MS Reveals that Many Regions of FL-Nrf2 are Highly Dynamic

HDX-MS experiments were performed to further probe the conformational dynamics of FL-Nrf2 in a spatially resolved manner. Backbone amide hydrogen/deuterium exchange (HDX) coupled with ESI-MS is a powerful method for studying protein behavior in solution [57,58]. Regions that are involved in hydrogen-bonding networks or occluded from the solvent will undergo slow exchange. In contrast, regions that are disordered and solvent-accessible will undergo fast exchange upon protein exposure to D_2O. The exchange process is due to dynamic fluctuations that disrupt backbone amide hydrogen bonds. By measuring deuterium incorporation, it is possible to determine the relative flexibility of backbone segments of a protein [59].

Pepsin digestion of FL-Nrf2 yielded 101 peptides resulting in 86.3% sequence coverage (Figure S5B). Fast HDX kinetics were observed throughout the entire protein, where the majority of the peptides exhibit ~100% deuteration after 12 s. This lack of protection indicates that FL-Nrf2 is significantly disordered (Figure S6). Notably, several peptides in distinct parts of the protein showed somewhat slower deuteration. For instance, the peptide covering residues 54–74 in the Neh2 domain only displayed complete deuteration after >24 s. This observation is consistent with earlier findings that even though the Neh2 domain is disordered, helical propensity exists between residues 39 and 71 [22]. Other regions that showed somewhat slower exchange are 235–249, 417–434, and 512–537. They are located in the Neh7 domain, the linker between the Neh6 and Neh1 domains, and the Neh1 domain, respectively.

Interestingly, the regions corresponding to three of the four helices (H1-H3; residues 455–465, 469–475, and 478–489) in the Neh1–2LZ1 structure did not show higher protection, suggesting that these helical regions undergo extensive conformational fluctuations, which would facilitate deuterium uptake. Another possible explanation is that the Neh1–2LZ1 region is less stable in FL-Nrf2.

2.4. Binding of the Kelch Domain of Keap1 ("Kelch") Triggers Conformational Changes Localized in the Neh2 Domain of Nrf2

We also investigated how FL-Nrf2 interacts with Kelch, which is part of the negative regulator Keap1. Previous studies showed that the isolated Neh2 domain of FL-Nrf2 binds Kelch at two sites: the high- affinity ETGE motif (around residues 76–84; K_d ~5 nM) and the low-affinity DLG motif (around residues 17–46; K_d ~1 μM) [22]. Here we used HDX-MS to further dissect the effects of Kelch-binding on the conformational dynamics of FL-Nrf2. In these experiments, the FL-Nrf2 concentration was held constant at 2 μM, while the Kelch concentration varied from 2 to 4 and 6 μM, which corresponded to FL-Nrf2:Kelch ratios of 1:1, 1:2, and 1:3, respectively (Figure 4). By testing these different concentration ratios, it should be possible to selectively saturate either only the high-affinity motif or both binding sites due to their distinct K_d values. Significant changes in deuterium uptake were observed in the Neh2 domain upon the addition of Kelch. In the presence of 1:1 Kelch, a considerable reduction in FL-Nrf2 deuterium uptake was displayed around the ETGE binding motif and a small reduction in deuterium uptake around the DLG motif, consistent with the binding affinities of the two sites. For a 1:2 ratio, a substantial reduction in deuterium uptake was observed around the DLG binding site and an even more significant reduction in HDX close to the ETGE motif. Further reduction in deuterium uptake around both sites was noted at a 1:3 ratio. Notably, the addition of Kelch did not slow down the deuterium uptake in any other domains of FL-Nrf2.

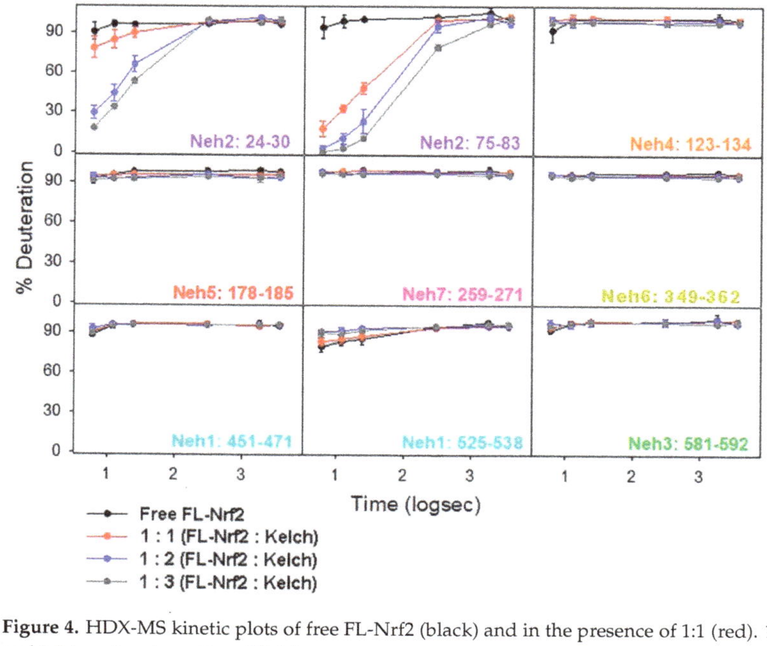

Figure 4. HDX-MS kinetic plots of free FL-Nrf2 (black) and in the presence of 1:1 (red). 1:2 (blue), and 1:3 (gray) molar ratios of Kelch.

Figure 5 shows % HDX differences at t = 6 s of FL-Nrf2 after addition of Kelch in 1:1, 1:2, and 1:3 ratios. In these maps, blue regions indicate less HDX than for free FL-Nrf2, whereas red regions indicate more HDX (more dynamic) than for free FL-Nrf2. These plots clearly display the enhanced protection of the Neh2 domain that was caused by increasing the Kelch concentration. Surprisingly, all the other Nrf2 domains experience slightly higher flexibility at the earliest time point upon binding the Kelch domain (red hues). These data suggest that when FL-Nrf2 binds to the Kelch domain, the rest of the protein becomes somewhat more dynamic. Our results, therefore, reveal that Kelch binds the Neh2 domain of FL-Nrf2 in a highly selective manner without triggering folding transitions in the rest of the protein.

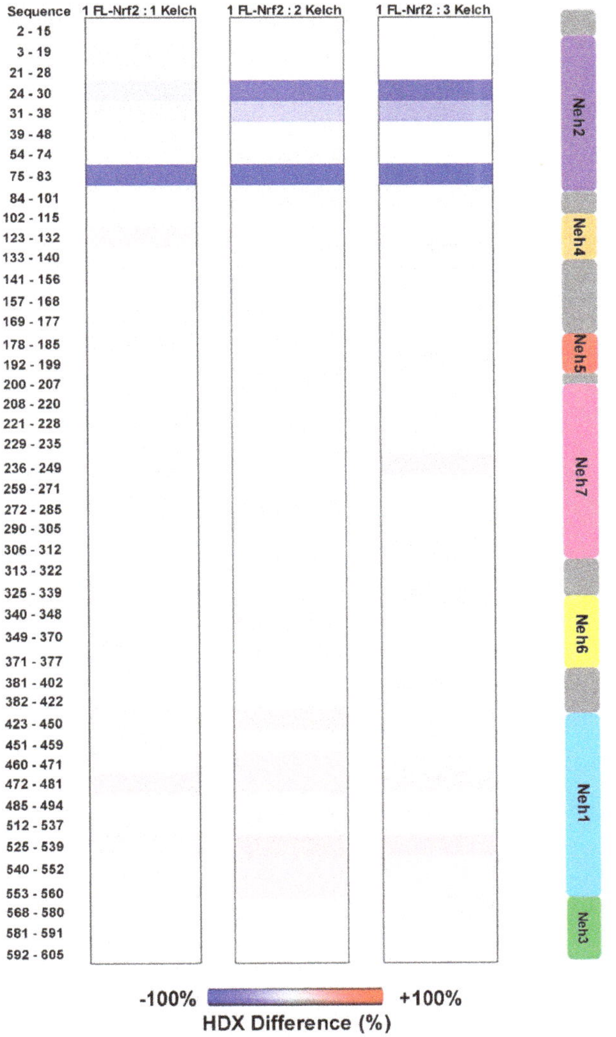

Figure 5. HDX-MS% difference plots of FL-Nrf2 upon addition of Kelch in 1:1, 1:2, and 1:3 ratios relative to free FL-Nrf2 at t = 6 s. Blue represents reduced deuteration, and red represents enhanced deuteration after the addition of Kelch. The side panel indicates the domain organization of FL-Nrf2.

2.5. Effects of Nrf2-Binding on the Deuterium Uptake of the Kelch Domain

The deuteration kinetics of free Kelch (Figure 6; black) display slow uptake throughout its entire sequence, consistent with its folded β-propeller structure [28,60]. For 1:1 and 2:1 (Kelch: FL-Nrf2) conditions, the overall HDX kinetics remained the same throughout the entire Kelch sequence. Peptides 335–341, 375–393, and 571–581 are the only three regions that showed a slight reduction in deuterium uptake when FL-Nrf2 was added. Peptide 375–393 contains residue N382 that is known to form a hydrogen bond with the ETGE motif of the Neh2 domain of Nrf2 [28,60]. Our HDX data are consistent with this behavior, as peptide 375–393 displayed a reduction in deuterium uptake indicative of protection of this site.

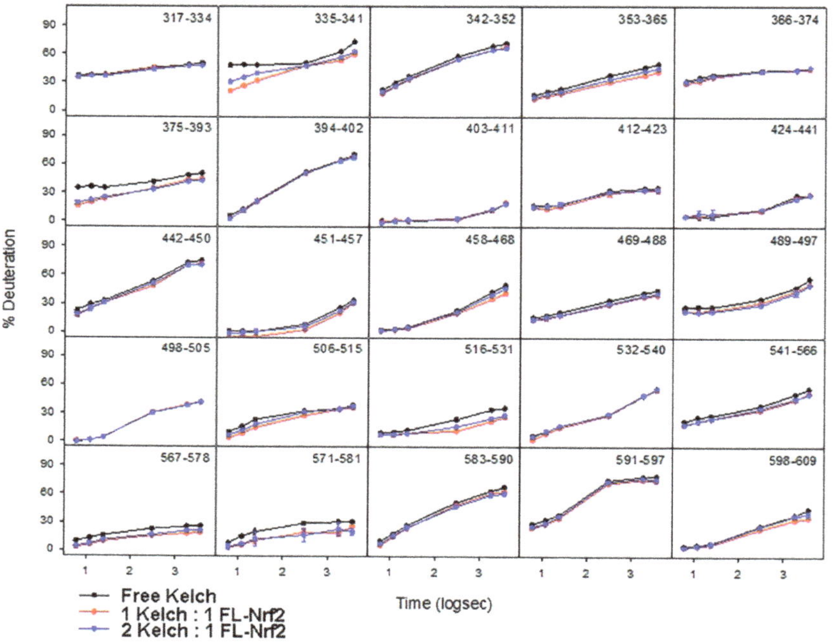

Figure 6. HDX-MS kinetic plots for free Kelch (black), 1:1 (red), and 2:1 (blue) Kelch:FL-Nrf2.

To understand the effect of Nrf2-binding on the global dynamics of Kelch, the HDX difference of the 1:1 state was plotted at time points 0.1, 0.4, and 30 min compared to free Kelch (Figure 7). At 0.1 and 0.4 min, a large reduction in deuterium uptake was observed in peptides 375–393, consistent with hydrogen bond formation of this site when the Kelch domain interacts with FL-Nrf2. Overall, the data show that there is a general stabilization of the entire Kelch domain upon binding to FL-Nrf2.

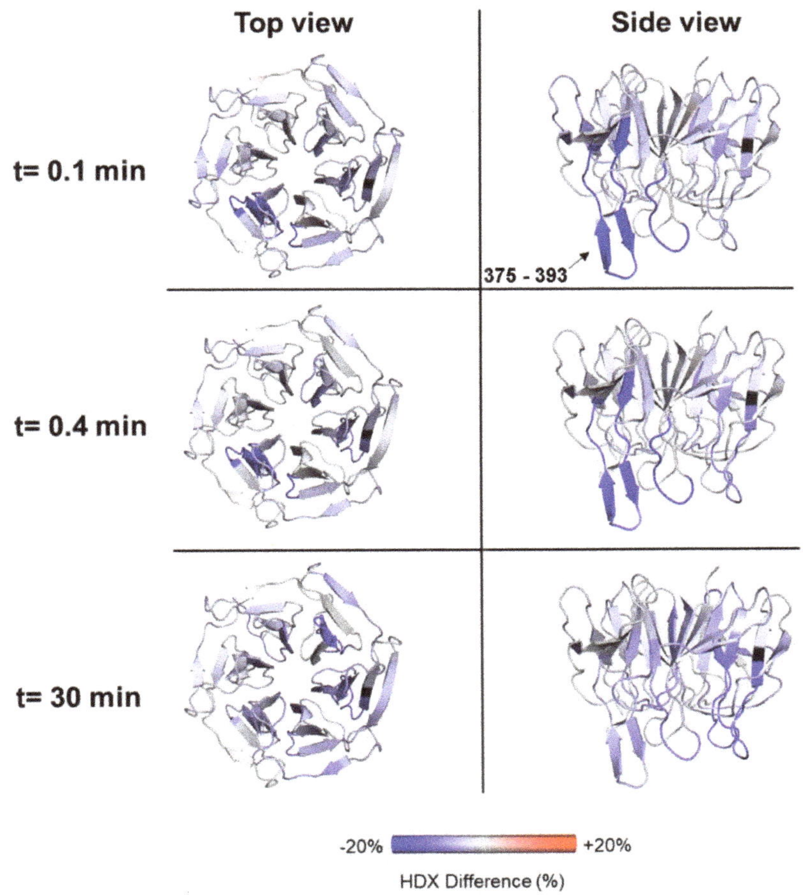

Figure 7. HDX % difference plots for Kelch upon addition of FL-Nrf2 at 0.1, 0.4, and 30 min compared to free Kelch. Blue represents reduced deuteration after the addition of FL-Nrf2.

3. Discussion

Nrf2 is a key transcription factor that orchestrates cellular responses to oxidative stress [3,6,61]. Aberrant activation of Nrf2 has been shown to play a pivotal role in pathogenesis and chemoresistance for many types of cancer [18,62]. Recent studies revealed that dysregulation of Nrf2 is also associated with neurodegenerative disorders and cardiovascular disease [63–66]. The involvement of Nrf2 in these human diseases makes the pharmacological modulation of Nrf2 activity a promising therapeutic strategy. Indeed, since the structures of Kelch in complex with the ETGE and DLG peptides derived from Nrf2 became available [28,42,60], tremendous efforts have been devoted to the design of small molecules and peptides that can inhibit the Nrf2-Kelch interaction with high specificity [26,67]. Unraveling the mechanism of Nrf2 binding to other regulators, such as CBP and RXRα, will no doubt open up additional opportunities for developing effective therapeutic strategies. Our work represents the first structural characterization of Nrf2 in the full-length context, which is a critical step toward this goal.

We have used various bioinformatics tools to predict the structural characteristics of FL-Nrf2. The results suggest that FL-Nrf2 is significantly disordered, although local structural elements exist in specific regions of the protein. Our bioinformatics findings are supported by extensive biophysical characterization using CD, NMR, and HDX-MS techniques. The CD results confirm that FL-Nrf2 is partially disordered but displays a significant helical propensity. The results are consistent with the lack of peak dispersion observed in the ^{1}H-^{15}N HSQC NMR spectrum of FL-Nrf2.

Our HDX-MS results provide a clearer picture of the FL-Nrf2 dynamics. With an ~86% overall sequence coverage, we were able to probe the conformational dynamics of different FL-Nrf2 domains. Even though the majority of the peptides produced by pepsin digestion showed fast deuterium uptake, a few regions were found to be moderately protected. Unexpectedly, peptides in the Neh1–2LZ1 region did not display high protection. It is possible that even though isolated Neh1–2LZ1 is partially structured, conformational fluctuations still allow for relatively fast deuterium uptake.

The data presented in this work highlight the importance of using full-length Nrf2 to uncover the structural and dynamic characteristics of this protein, as opposed to earlier studies that used fragments or protein truncations. By using the FL-Nrf2 construct, we uncovered important conformational properties that provide insights into the biological role of Nrf2. Our results show that free FL-Nrf2 is highly dynamic throughout its entire sequence. When FL-Nrf2 interacts with Kelch, a large reduction in deuterium uptake is observed in the DLG and ETGE binding motifs of FL-Nrf2. Other parts of FL-Nrf2 became slightly more dynamic, as indicated by an increase in HDX. We also found that the interaction of FL-Nrf2 with Kelch resulted in a stabilization of the entire Kelch domain.

The disordered nature of FL-Nrf2 has substantial implications for its biological function. Intrinsically disordered proteins are highly abundant in all organisms [68]. Like FL-Nrf2, many of these proteins are involved in gene transcription and signal transduction [69,70]. Distinct from globular proteins, disordered proteins do not adopt a well-defined structure under physiological conditions. Instead, they exist as a large population of conformations in dynamic equilibria that can shift upon changes in the environment [51,71]. The structural plasticity of FL-Nrf2 can confer functional advantages. For instance, the lack of a stable tertiary fold allows FL-Nrf2 to bind multiple targets using a number of linear motifs located in different protein regions, either simultaneously or sequentially, without steric restrictions [72,73]. This aligns with our HDX-MS results showing that the effects of Kelch-binding on the structure of FL-Nrf2 are localized to the Neh2 domain in a highly specific manner. Further, the conformational dynamics of FL-Nrf2 can also have substantial consequences for its target recognition. Upon complex formation, the unfavorable entropy loss due to the folding into more stable bound-state conformations of FL-Nrf2 must be offset by strong enthalpic interactions with the binding partner. This enthalpy-entropy compensation confers Nrf2 the ability to bind distinct targets with high specificity and low affinity, which is essential for its regulation through various protein-protein interactions [74–76].

4. Materials and Methods

4.1. Protein Expression and Purification of Full-Length Human Nrf2

The construct of full-length human Nrf2 (FL-Nrf2; purchased from Invitrogen®) cloned into the Gateway Destination Vector pDEST17 was transformed into *E. coli* (Rosetta 2(DE3) pLysS) cells for protein expression. The cell culture was incubated in M9 minimal media (47.8 mM of Na_2HPO_4, 22.0 mM of KH_2PO_4, 8.6 mM of NaCl, 0.1 mM of $CaCl_2$, 2.0 mM of $MgSO_4$, 10 mg of biotin, 10 mg of thiamin, 4.0 g of glucose, and 1.0 g of NH_4Cl; pH 7.4) at 37 °C until the OD600 reached ~0.8. Protein over-expression was induced with 1 mM isopropyl-β-D-thiogalactopyranoside (IPTG). To avoid purifying Nrf2 from inclusion bodies through refolding procedures, we have tested four different expression temperatures (15, 22, 30, and 37 °C) and two induction times (5 and 18 h) to identify conditions that maximize the amount of FL-Nrf2 in the soluble fraction. Our data showed that induction at higher temperatures (i.e., 30 and 37 °C) for 5 or 18 h resulted in the majority of FL-Nrf2 in the insoluble fraction. In contrast, most of the protein was found in the soluble fraction when cells were induced at 15 °C for 18 h.

The cell pellets were resuspended using solubilization buffer (20 mM Tris-HCl, 150 mM NaCl, 1 mM EDTA, 5 mM 2-mercaptoethanol, pH 8.1). Lysozyme was added to the solubilized cell suspension, and the mixture was incubated for 30 min at 37 °C. The incubated sample was homogenized using an Avestin EmulsiFlex-C5 homogenizer. A SigmaFast Protease Inhibitor Cocktail tablet (EDTA-free) and 1 mM PMSF (100 μL per 10 mL of the lysed sample) were added to the sample. Final concentrations of imidazole (10 mM) and NaCl (500 mM) were adjusted, and the sample was centrifuged at 40,000× *g* for 30 min at 4 °C. The supernatant was collected, and the pH was adjusted to 7.4–7.8. The sample was loaded onto equilibrated Ni-Sepharose 6 fast flow beads (GE Healthcare) and incubated for 2 h at room temperature. The sample was then washed with 400 mL of primary wash buffer (20 mM Tris-HCl, 500 mM NaCl, 80 mM imidazole, 5 mM 2-mercaptoethanol, pH 7.8), followed by 10 mL (5 mL × 2) of secondary wash buffer (20 mM Tris-HCl, 500 mM NaCl, 150 mM imidazole, 5 mM 2-mercaptoethanol, pH 7.8). The protein was then eluted using 5-mL fractions of elution buffer (20 mM Tris-HCl, 500 mM NaCl, 1.5 M imidazole, 5 mM 2-mercaptoethanol, pH 7.8), and the eluate was monitored using Bradford assay (Bio-Rad, Hercules, CA, USA). Fractions containing FL-Nrf2 were pooled and dialyzed overnight into the dialysis buffer (50 mM ammonium acetate, 500 μM TCEP, pH 6.5). The final protein concentration was determined by the Lowry assay. By using this new protocol, we were able to obtain about 0.3 mg of purified FL-Nrf2 from a 1 L M9 minimal media culture. Notably, purified FL-Nrf2 does not have very high solubility. The solubility of FL-Nrf2 was analyzed by the sedimentation assay (the procedure of the assay is outlined in Supplemental Materials) to partition the soluble and aggregated protein molecules into supernatant and pellet for analysis. The results show that FL-Nrf2 in 50 mM ammonium acetate and 0.5 mM TCEP (pH 6.5) was only present in the supernatant but not in the pellet (Figure S7), confirming that the protein does not aggregate at concentrations <20 μM. Therefore, samples with a concentration of <20 μM were used in our studies.

4.2. Expression and Purification DNA-Binding Neh1 Domain of Nrf2 (Neh1–2LZ1, Residues 445–523)

The Neh1–2LZ1 construct was purchased from the Northeast Structural Genomics Consortium. It was cloned into the Gateway Destination Vector pDEST17 with a tobacco etch virus (TEV) protease cleaving site and transformed into BL21(DE3) for expression. The cell culture was incubated in M9 minimal media at 37 °C until the OD600 reached ~0.8. Protein over-expression was induced with 1 mM IPTG. Cells were then grown overnight at 17 °C before harvest by centrifugation.

The cell pellets were resuspended in denaturing Ni^{2+} binding buffer (25 mM Tris-HCl, 250 mM NaCl, 8 M urea, 5 mM 2-mercaptoethanol, pH 8.0). The cell suspension was homogenized by Dounce homogenization and sonication. The mixture was then centrifuged at 50,000× *g* at room temperature for 40 min. Ni-Sepharose 6 fast flow beads (GE Healthcare) pre-equilibrated with binding buffer were added to the supernatant, and the mixture was incubated for 2 h at room temperature. The mixture was loaded onto a column and washed with 200 mL of primary wash buffer (25 mM Tris-HCl, 250 mM NaCl, 10 mM imidazole, 8 M urea, 5 mM 2-mercaptoethanol, pH 8.0), followed by a wash with 200 mL of secondary wash buffer (25 mM Tris-HCl, 250 mM NaCl, 10 mM imidazole, 5 mM 2-mercaptoethanol, pH 8.0). The protein was eluted using 5 mL fractions of elution buffer (25 mM Tris-HCl, 500 mM NaCl, 750 mM imidazole, 5 mM 2-mercaptoethanol, pH 7.8), and eluted fractions were monitored using Bradford assay. Fractions containing Neh1–2LZ1 were pooled and dialyzed overnight into the HEPES buffer (20 mM HEPES, 5 mM 2-mercaptoethanol, pH 8.0) at 4 °C. The next day, the buffer was refreshed, and the sample was dialyzed for another 4 h before the protein concentration was determined using Bradford assay. TEV protease was then added accordingly (1 mg of TEV protease/25 mg of protein). Following overnight incubation at room temperature, the sample was diluted into the HEPES buffer and was loaded onto a pre-equilibrated SP-Sepharose (GE Healthcare) column and incubated at room temperature for one hour. After incubation, the sample was washed with 200 mL of the third wash buffer (20 mM HEPES, 50 mM NaCl,

5 mM 2-mercaptoethanol, pH 8.0). Finally, protein was eluted using the elution buffer (20 mM HEPES, 500 mM NaCl, 5 mM 2-mercaptoethanol, pH 8.0) in 5 mL fractions. Eluted fractions were pooled and dialyzed into the NMR buffer (50 mM ammonium acetate, 1 mM DTT, pH 6.5) in the presence or absence of 50 mM of arginine.

4.3. Expression and Purification Kelch Domain of Human Keap1

The pET15b plasmid of human Keap1-Kelch cDNA, a kind gift from Dr. Mark Hannink at the University of Missouri-Columbia, was transformed into *E. coli* BL21 (DE3) cells. The expression and purification were carried out using the procedure described in Khan et al. [77].

4.4. CD Spectropolarimetry

CD experiments were performed using a Jasco J-810 spectropolarimeter. Spectra of FL-Nrf2 (~0.3 mg/mL) and Neh1–2LZ1 construct (~0.1 mg/mL) were recorded in 50 mM ammonium acetate buffer, pH 6.5. FL-Nrf2 spectra were recorded at 5, 10, 25, and 35 °C. For each spectrum, 20 accumulated scans were obtained at 20 nm/min rate. For Neh1–2LZ1, the data were recorded at 20 °C with 20 accumulated scans (20 nm/min). The CD data were deconvoluted using the CONTINLL program in DichroWeb, with protein reference data set 4 (optimized for 190–240 nm) [78].

4.5. NMR Spectroscopy

^{1}H-^{15}N HSQC NMR spectra of FL-Nrf2 were acquired on Varian Inova 600-MHz spectrometers (UWO Biomolecular NMR Facility) at 5, 10, 25, and 35 °C in 50 mM ammonium acetate buffer (pH 6.5) using BioPack. Each data set was recorded with 160 scans and a relaxation delay of 1.0 s. For Neh1–2LZ1 (in ammonium acetate buffer, pH 6.5), spectra were recorded in the presence and absence of 50 mM arginine. Each spectrum was recorded with 32 scans, and a relaxation delay of 1.0 s. A total of 1 mM 4,4-dimethyl-4-silapentane-1-sulfonic acid (DSS) was added to the samples for chemical shift referencing. Data were processed and analyzed using NMRPipe and NMRViewJ, respectively [79,80].

4.6. HDX-MS

HDX-MS experiments were performed in 50 mM sodium phosphate buffer (90% D_2O, pH 7.0) with 100 mM NaCl and at a final protein concentration of 2 μM for experiments on isolated FL-Nrf2 and Kelch. In Nrf2-Kelch-binding experiments, the FL-Nrf2 concentration was kept at 2 μM, and the Kelch concentration was varied from 2, 4, and 6 μM for 1:1, 1:2, and 1:3 (FL-Nrf2:Kelch) binding experiments. Aliquots were removed after 0.1, 0.2, 0.4, 5, 30, and 60 min and were quenched by lowering the pH to 2.5 using 20% (v/v) formic acid, followed by flash freezing in liquid nitrogen. The samples were thawed and injected into a nanoACQUITY UPLC equipped with HDX technology (Waters, Milford, MA, USA). Online digestion was carried out using a POROS pepsin column (2.1 × 30 mm, Life Technologies/Applied Biosystems) held at 15 °C. Peptic peptides were trapped on a C18 BEH130 VanGuard column (5 × 1 mm, 1.7 μm) for three minutes at 80 μL/min and separated on a C8 column (50 × 2.1 mm, 1.7 μm) at 100 μL/min using a water/acetonitrile gradient with 0.1% formic acid. The LC outflow was directed to a Waters Synapt Q-TOF G2 mass spectrometer. The ion source was operated at+2.8 kV and a cone voltage of 20 V. The desolvation and source temperatures were 250 and 80 °C, respectively. Mass spectra were acquired in resolution mode. Ion mobility was employed to aid in separating overlapping isobaric peaks.

Peptide identification was performed using three separate label-free MSE acquisitions with analysis using Protein Lynx Global Server 2.5.3. DynamX 3.0 was used for HDX data analysis. Deuterium uptake levels were corrected for artificial in-exchange and back-exchange using controls that represent minimum exchange under quench conditions (m_0)

and fully deuterated samples (m_{100}), respectively. Percentage deuteration values for each peptide at time t was calculated according to

$$\%D(t) = \frac{m_t - m_0}{m_{100} - m_0} \times 100\% \quad (1)$$

Pepsin digestion yielded 101 peptides for Nrf2, corresponding to 86.3% coverage. Digestion of Kelch yielded 109 peptides (99.7% coverage, see Figure S5B,D).

Supplementary Materials: The following are available online at https://www.mdpi.com/article/10.3390/ijms22147434/s1, Figure S1: SDS-PAGE analysis of the FL-Nrf2 at different purification steps, Figure S2: Bioinformatics predictions of the structural properties of FL-Nrf2, Figure S3: ^{1}H-^{15}N HSQC spectrum of FL-Nrf2 recorded at 35 °C in the presence of 6 M urea, Figure S4: The CD spectrum of Neh1–2LZ1 recorded at 20 °C, Figure S5: Peptide coverage of FL-Nrf2 and Kelch domain of Keap1, Figure S6: HDX data suggest the majority of FL-Nrf2 is not protected, Figure S7: The (A) SSD-AGE and (B) SDS-PAGE results of the sedimentation assay show that the FL-Nrf2 does not form aggregates at low concentrations, Table S1: Estimated secondary structural contents presence in FL-Nrf2 at different temperatures, The procedure of the sedimentation assay.

Author Contributions: N.C.K., C.S.F., L.K. and W.-Y.C. designed the study. N.C.K. and A.B. performed the protein expression and purification for all the experiments. N.C.K. performed the NMR and CD experiments. C.S.F. performed the HDX-MS experiments. V.N. performed the sedimentation assays. N.C.K., C.S.F., V.N., M.L.D., L.K. and W.-Y.C. analyzed the data. N.C.K., C.S.F., L.K. and W.-Y.C. wrote the manuscript. All authors reviewed the manuscript. All authors have read and agreed to the published version of the manuscript.

Funding: This work was supported by Discovery Grants (RGPIN-2019-06711 awarded to W.-Y.C. and RGPIN-2018-04243 awarded to L.K.) from the Natural Sciences and Engineering Research Council (NSERC) of Canada.

Institutional Review Board Statement: Not applicable.

Informed Consent Statement: Not applicable.

Data Availability Statement: The data presented in this study are available within the article text, figures, and Supplementary Materials.

Acknowledgments: We thank the Biomolecular NMR Facility and Biomolecular Interactions and Conformations Facility (University of Western Ontario) for their assistance and use of equipment.

Conflicts of Interest: The authors declare no conflict of interest.

References

1. Van Dam, L.; Dansen, T.B. Cross-Talk between Redox Signalling and Protein Aggregation. *Biochem. Soc. Trans.* **2020**, *48*, 379–397. [CrossRef] [PubMed]
2. Juan, C.A.; de la Lastra, J.M.P.; Plou, F.J.; Pérez-Lebeña, E. The Chemistry of Reactive Oxygen Species (ROS) Revisited: Outlining Their Role in Biological Macromolecules (DNA, Lipids and Proteins) and Induced Pathologies. *Int. J. Mol. Sci.* **2021**, *22*, 4642. [CrossRef] [PubMed]
3. Taguchi, K.; Motohashi, H.; Yamamoto, M. Molecular Mechanisms of the Keap1–Nrf2 Pathway in Stress Response and Cancer Evolution. *Genes Cells* **2011**, *16*, 123–140. [CrossRef] [PubMed]
4. Itoh, K.; Chiba, T.; Takahashi, S.; Ishii, T.; Igarashi, K.; Katoh, Y.; Oyake, T.; Hayashi, N.; Satoh, K.; Hatayama, I.; et al. An Nrf2/Small Maf Heterodimer Mediates the Induction of Phase II Detoxifying Enzyme Genes through Antioxidant Response Elements. *Biochem. Biophys. Res. Commun.* **1997**, *236*, 313–322. [CrossRef] [PubMed]
5. Ma, Q. Role of Nrf2 in Oxidative Stress and Toxicity. *Annu. Rev. Pharmacol. Toxicol.* **2013**, *53*, 401–426. [CrossRef]
6. Baird, L.; Dinkova-Kostova, A.T. The Cytoprotective Role of the Keap1-Nrf2 Pathway. *Arch. Toxicol.* **2011**, *85*, 241–272. [CrossRef] [PubMed]
7. Jaramillo, M.C.; Zhang, D.D. The Emerging Role of the Nrf2-Keap1 Signaling Pathway in Cancer. *Genes Dev.* **2013**, *27*, 2179–2191. [CrossRef]
8. Homma, S.; Ishii, Y.; Morishima, Y.; Yamadori, T.; Matsuno, Y.; Haraguchi, N.; Kikuchi, N.; Satoh, H.; Sakamoto, T.; Hizawa, N.; et al. Nrf2 Enhances Cell Proliferation and Resistance to Anticancer Drugs in Human Lung Cancer. *Clin. Cancer Res.* **2009**, *15*, 3423–3432. [CrossRef]

9. Zhang, P.; Singh, A.; Yegnasubramanian, S.; Esopi, D.; Kombairaju, P.; Bodas, M.; Wu, H.; Bova, S.G.; Biswal, S. Loss of Kelch-like ECH-Associated Protein 1 Function in Prostate Cancer Cells Causes Chemoresistance and Radioresistance and Promotes Tumor Growth. *Mol. Cancer Ther.* **2010**, *9*, 336–346. [CrossRef]
10. Furfaro, A.L.; Traverso, N.; Domenicotti, C.; Piras, S.; Moretta, L.; Marinari, U.M.; Pronzato, M.A.; Nitti, M. The Nrf2/HO-1 Axis in Cancer Cell Growth and Chemoresistance. *Oxidative Med. Cell. Longev.* **2016**, *2016*, 1958174. [CrossRef]
11. Takahashi, T.; Sonobe, M.; Menju, T.; Nakayama, E.; Mino, N.; Iwakiri, S.; Nagai, S.; Sato, K.; Miyahara, R.; Okubo, K.; et al. Mutations in Keap1 Are a Potential Prognostic Factor in Resected Non-Small Cell Lung Cancer. *J. Surg. Oncol.* **2010**, *101*, 500–506. [CrossRef]
12. Hayden, A.; Douglas, J.; Sommerlad, M.; Andrews, L.; Gould, K.; Hussain, S.; Thomas, G.J.; Packham, G.; Crabb, S.J. The Nrf2 Transcription Factor Contributes to Resistance to Cisplatin in Bladder Cancer. *Urol. Oncol.* **2014**, *32*, 806–814. [CrossRef] [PubMed]
13. Liu, M.; Yao, X.D.; Li, W.; Geng, J.; Yan, Y.; Che, J.P.; Xu, Y.F.; Zheng, J.H. Nrf2 Sensitizes Prostate Cancer Cells to Radiation via Decreasing Basal ROS Levels. *Biofactors* **2015**, *41*, 52–57. [CrossRef]
14. Hartikainen, J.M.; Tengström, M.; Kosma, V.M.; Kinnula, V.L.; Mannermaa, A.; Soini, Y. Genetic Polymorphisms and Protein Expression of NRF2 and Sulfiredoxin Predict Survival Outcomes in Breast Cancer. *Cancer Res.* **2012**, *72*, 5537–5546. [CrossRef]
15. The Cancer Genome Atlas Research Network. Comprehensive Genomic Characterization of Squamous Cell Lung Cancers. *Nature* **2012**, *489*, 519–525. [CrossRef] [PubMed]
16. Shibata, T.; Kokubu, A.; Saito, S.; Narisawa-Saito, M.; Sasaki, H.; Aoyagi, K.; Yoshimatsu, Y.; Tachimori, Y.; Kushima, R.; Kiyono, T.; et al. NRF2 Mutation Confers Malignant Potential and Resistance to Chemoradiation Therapy in Advanced Esophageal Squamous Cancer. *Neoplasia* **2011**, *13*, 864–873. [CrossRef] [PubMed]
17. Shibata, T.; Ohta, T.; Tong, K.I.; Kokubu, A.; Odogawa, R.; Tsuta, K.; Asamura, H.; Yamamoto, M.; Hirohashi, S. Cancer Related Mutations in NRF2 Impair Its Recognition by Keap1-Cul3 E3 Ligase and Promote Malignancy. *Proc. Natl. Acad. Sci. USA* **2008**, *105*, 13568–13573. [CrossRef] [PubMed]
18. Menegon, S.; Columbano, A.; Giordano, S. The Dual Roles of NRF2 in Cancer. *Trends Mol. Med.* **2016**, *22*, 578–593. [CrossRef]
19. Lau, A.; Villeneuve, N.F.; Sun, Z.; Wong, P.K.; Zhang, D.D. Dual Roles of Nrf2 in Cancer. *Pharmacol. Res.* **2008**, *58*, 262–270. [CrossRef]
20. Panieri, E.; Buha, A.; Telkoparan-Akillilar, P.; Cevik, D.; Kouretas, D.; Veskoukis, A.; Skaperda, Z.; Tsatsakis, A.; Wallace, D.; Suzen, S.; et al. Potential Applications of NRF2 Modulators in Cancer Therapy. *Antioxidants (Basel)* **2020**, *9*, 193. [CrossRef]
21. Zhang, D.D. Mechanistic Studies of the Nrf2-Keap1 Signaling Pathway. *Drug Metab. Rev.* **2006**, *38*, 769–789. [CrossRef]
22. Tong, K.I.; Katoh, Y.; Kusunoki, H.; Itoh, K.; Tanaka, T.; Yamamoto, M. Keap1 Recruits Neh2 through Binding to ETGE and DLG Motifs: Characterization of the Two-Site Molecular Recognition Model. *Mol. Cell. Biol.* **2006**, *26*, 2887–2900. [CrossRef]
23. Katoh, Y.; Itoh, K.; Yoshida, E.; Miyagishi, M.; Fukamizu, A.; Yamamoto, M. Two Domains of Nrf2 Cooperatively Bind CBP, a CREB Binding Protein, and Synergistically Activate Transcription. *Genes Cells* **2001**, *6*, 857–868. [CrossRef]
24. Sun, Z.; Chin, Y.E.; Zhang, D.D. Acetylation of Nrf2 by P300/CBP Augments Promoter-Specific DNA Binding of Nrf2 during the Antioxidant Response. *Mol. Cell. Biol.* **2009**, *29*, 2658–2672. [CrossRef]
25. Wang, H.; Liu, K.; Geng, M.; Gao, P.; Wu, X.; Hai, Y.; Li, Y.; Luo, L.; Hayes, J.D.; Wang, X.J.; et al. RXRα Inhibits the NRF2-ARE Signaling Pathway through a Direct Interaction with the Neh7 Domain of NRF2. *Cancer Res.* **2013**, *73*, 3097–3108. [CrossRef] [PubMed]
26. Chen, W.; Sun, Z.; Wang, X.J.; Jiang, T.; Huang, Z.; Fang, D.; Zhang, D.D. Direct Interaction between Nrf2 and P21(Cip1/WAF1) Upregulates the Nrf2-Mediated Antioxidant Response. *Mol. Cell* **2009**, *34*, 663–673. [CrossRef] [PubMed]
27. Kim, J.H.; Yu, S.; Chen, J.D.; Kong, A.N. The Nuclear Cofactor RAC3/AIB1/SRC-3 Enhances Nrf2 Signaling by Interacting with Transactivation Domains. *Oncogene* **2013**, *32*, 514–527. [CrossRef] [PubMed]
28. Lo, S.C.; Li, X.; Henzl, M.T.; Beamer, L.J.; Hannink, M. Structure of the Keap1:Nrf2 Interface Provides Mechanistic Insight into Nrf2 Signaling. *EMBO J.* **2006**, *25*, 3605–3617. [CrossRef] [PubMed]
29. McMahon, M.; Thomas, N.; Itoh, K.; Yamamoto, M.; Hayes, J.D. Dimerization of Substrate Adaptors Can Facilitate Cullin-Mediated Ubiquitylation of Proteins by a "Tethering" Mechanism: A Two-Site Interaction Model for the Nrf2-Keap1 Complex. *J. Biol. Chem.* **2006**, *281*, 24756–24768. [CrossRef] [PubMed]
30. Kobayashi, A.; Kang, M.-I.; Okawa, H.; Ohtsuji, M.; Zenke, Y.; Chiba, T.; Igarashi, K.; Yamamoto, M. Oxidative Stress Sensor Keap1 Functions as an Adaptor for Cul3-Based E3 Ligase to Regulate Proteasomal Degradation of Nrf2. *Mol. Cell. Biol.* **2004**, *24*, 7130–7139. [CrossRef] [PubMed]
31. Zhang, D.D.; Hannink, M. Distinct Cysteine Residues in Keap1 Are Required for Keap1-Dependent Ubiquitination of Nrf2 and for Stabilization of Nrf2 by Chemopreventive Agents and Oxidative Stress. *Mol. Cell. Biol.* **2003**, *23*, 8137–8151. [CrossRef] [PubMed]
32. Kansanen, E.; Kuosmanen, S.M.; Leinonen, H.; Levonen, A.L. The Keap1-Nrf2 Pathway: Mechanisms of Activation and Dysregulation in Cancer. *Redox Biol.* **2013**, *1*, 45–49. [CrossRef] [PubMed]
33. Moi, P.; Chan, K.; Asunis, I.; Cao, A.; Kan, Y.W. Isolation of NF-E2-Related Factor 2 (Nrf2), a NF-E2-like Basic Leucine Zipper Transcriptional Activator That Binds to the Tandem NF-E2/AP1 Repeat of the Beta-Globin Locus Control Region. *Proc. Natl. Acad. Sci. USA* **1994**, *91*, 9926–9930. [CrossRef]

34. Itoh, K.; Wakabayashi, N.; Katoh, Y.; Ishii, T.; Igarashi, K.; Engel, J.D.; Yamamoto, M. Keap1 Represses Nuclear Activation of Antioxidant Responsive Elements by Nrf2 through Binding to the Amino-Terminal Neh2 Domain. *Genes Dev.* **1999**, *13*, 76–86. [CrossRef] [PubMed]
35. Baird, L.; Llères, D.; Swift, S.; Dinkova-Kostova, A.T. Regulatory Flexibility in the Nrf2-Mediated Stress Response Is Conferred by Conformational Cycling of the Keap1-Nrf2 Protein Complex. *Proc. Natl. Acad. Sci. USA* **2013**, *110*, 15259–15264. [CrossRef] [PubMed]
36. Zhang, J.; Hosoya, T.; Maruyama, A.; Nishikawa, K.; Maher, J.M.; Ohta, T.; Motohashi, H.; Fukamizu, A.; Shibahara, S.; Itoh, K.; et al. Nrf2 Neh5 Domain Is Differentially Utilized in the Transactivation of Cytoprotective Genes. *Biochem. J.* **2007**, *404*, 459–466. [CrossRef]
37. Wu, J.; Wang, H.; Tang, X. Rexinoid Inhibits Nrf2-Mediated Transcription through Retinoid X Receptor Alpha. *Biochem. Biophys. Res. Commun.* **2014**, *452*, 554–559. [CrossRef]
38. McMahon, M.; Thomas, N.; Itoh, K.; Yamamoto, M.; Hayes, J.D. Redox-Regulated Turnover of Nrf2 Is Determined by at Least Two Separate Protein Domains, the Redox-Sensitive Neh2 Degron and the Redox-Insensitive Neh6 Degron. *J. Biol. Chem.* **2004**, *279*, 31556–31567. [CrossRef]
39. Chowdhry, S.; Zhang, Y.; McMahon, M.; Sutherland, C.; Cuadrado, A.; Hayes, J.D. Nrf2 Is Controlled by Two Distinct β-TrCP Recognition Motifs in Its Neh6 Domain, One of Which Can Be Modulated by GSK-3 Activity. *Oncogene* **2013**, *32*, 3765–3781. [CrossRef]
40. Katsuoka, F.; Motohashi, H.; Ishii, T.; Aburatani, H.; Engel, J.D.; Yamamoto, M. Genetic Evidence That Small Maf Proteins Are Essential for the Activation of Antioxidant Response Element-Dependent Genes. *Mol. Cell. Biol.* **2005**, *25*, 8044–8051. [CrossRef]
41. Nioi, P.; Nguyen, T.; Sherratt, P.J.; Pickett, C.B. The Carboxy-Terminal Neh3 Domain of Nrf2 Is Required for Transcriptional Activation. *Mol. Cell. Biol.* **2005**, *25*, 10895–10906. [CrossRef]
42. Fukutomi, T.; Takagi, K.; Mizushima, T.; Ohuchi, N.; Yamamoto, M. Kinetic, Thermodynamic, and Structural Characterizations of the Association between Nrf2-DLGex Degron and Keap1. *Mol. Cell. Biol.* **2014**, *34*, 832–846. [CrossRef] [PubMed]
43. Eletsky, A.; Pulavarti, S.V.S.R.K.; Lee, D.; Kohan, E.; Janjua, H.; Xiao, R.; Acton, T.B.; Everett, J.K.; Montelione, G.T.; Szyperski, T.; et al. Solution NMR Structure of the DNA-Binding Domain of Human NF-E2-Related Factor 2, Northeast Structural Genomics Consortium (NESG) Target HR3520O. *PDB* **2012**. [CrossRef]
44. Lau, A.; Tian, W.; Whitman, S.A.; Zhang, D.D. The Predicted Molecular Weight of Nrf2: It Is What It Is Not. *Antioxid. Redox Signal.* **2013**, *18*, 91–93. [CrossRef] [PubMed]
45. Tompa, P. Intrinsically Unstructured Proteins. *Trends Biochem. Sci.* **2002**, *27*, 527–533. [CrossRef]
46. Iakoucheva, L.M.; Kimzey, A.L.; Masselon, C.D.; Smith, R.D.; Dunker, A.K.; Ackerman, E.J. Aberrant Mobility Phenomena of the DNA Repair Protein XPA. *Protein Sci.* **2001**, *10*, 1353–1362. [CrossRef]
47. Xue, B.; Dunbrack, R.L.; Williams, R.W.; Dunker, A.K.; Uversky, V.N. PONDR-FIT: A Meta-Predictor of Intrinsically Disordered Amino Acids. *Biochim. Biophys. Acta* **2010**, *1804*, 996–1010. [CrossRef] [PubMed]
48. Sormanni, P.; Camilloni, C.; Fariselli, P.; Vendruscolo, M. The S2D Method: Simultaneous Sequence-Based Prediction of the Statistical Populations of Ordered and Disordered Regions in Proteins. *J. Mol. Biol.* **2015**, *427*, 982–996. [CrossRef]
49. Linding, R.; Jensen, L.J.; Diella, F.; Bork, P.; Gibson, T.J.; Russell, R.B. Protein Disorder Prediction: Implications for Structural Proteomics. *Structure* **2003**, *11*, 1453–1459. [CrossRef]
50. Mészáros, B.; Erdos, G.; Dosztányi, Z. IUPred2A: Context-Dependent Prediction of Protein Disorder as a Function of Redox State and Protein Binding. *Nucleic Acids Res.* **2018**, *46*, W329–W337. [CrossRef]
51. Mittag, T.; Forman-Kay, J.D. Atomic-Level Characterization of Disordered Protein Ensembles. *Curr. Opin. Struct. Biol.* **2007**, *17*, 3–14. [CrossRef] [PubMed]
52. Gall, C.; Xu, H.; Brickenden, A.; Ai, X.; Choy, W.Y. The Intrinsically Disordered TC-1 Interacts with Chibby via Regions with High Helical Propensity. *Protein Sci.* **2007**, *16*, 2510–2518. [CrossRef]
53. Mokhtarzada, S.; Yu, C.; Brickenden, A.; Choy, W.-Y. Structural Characterization of Partially Disordered Human Chibby: Insights into Its Function in the Wnt-Signaling Pathway. *Biochemistry* **2011**, *50*, 715–726. [CrossRef]
54. Yi, S.; Boys, B.L.; Brickenden, A.; Konermann, L.; Choy, W.-Y. Effects of Zinc Binding on the Structure and Dynamics of the Intrinsically Disordered Protein Prothymosin Alpha: Evidence for Metalation as an Entropic Switch. *Biochemistry* **2007**, *46*, 13120–13130. [CrossRef]
55. Tischer, A.; Lilie, H.; Rudolph, R.; Lange, C. L-Arginine Hydrochloride Increases the Solubility of Folded and Unfolded Recombinant Plasminogen Activator RPA. *Protein Sci.* **2010**, *19*, 1783–1795. [CrossRef] [PubMed]
56. Baynes, B.M.; Wang, D.I.; Trout, B.L. Role of Arginine in the Stabilization of Proteins against Aggregation. *Biochemistry* **2005**, *44*, 4919–4925. [CrossRef] [PubMed]
57. Iacob, R.E.; Engen, J.R. Hydrogen Exchange Mass Spectrometry: Are We out of the Quicksand? *J. Am. Soc. Mass Spectrom.* **2012**, *23*, 1003–1010. [CrossRef] [PubMed]
58. Percy, A.J.; Rey, M.; Burns, K.M.; Schriemer, D.C. Probing Protein Interactions with Hydrogen/Deuterium Exchange and Mass Spectrometry-a Review. *Anal. Chim. Acta* **2012**, *721*, 7–21. [CrossRef]
59. Konermann, L.; Pan, J.; Liu, Y.H. Hydrogen Exchange Mass Spectrometry for Studying Protein Structure and Dynamics. *Chem. Soc. Rev.* **2011**, *40*, 1224–1234. [CrossRef]

60. Padmanabhan, B.; Tong, K.I.; Ohta, T.; Nakamura, Y.; Scharlock, M.; Ohtsuji, M.; Kang, M.-I.; Kobayashi, A.; Yokoyama, S.; Yamamoto, M. Structural Basis for Defects of Keap1 Activity Provoked by Its Point Mutations in Lung Cancer. *Mol. Cell* **2006**, *21*, 689–700. [CrossRef]
61. Tonelli, C.; Chio, I.I.C.; Tuveson, D.A. Transcriptional Regulation by Nrf2. *Antioxid. Redox Signal.* **2018**, *29*, 1727–1745. [CrossRef]
62. Sporn, M.B.; Liby, K.T. NRF2 and Cancer: The Good, the Bad and the Importance of Context. *Nat. Rev. Cancer* **2012**, *12*, 564–571. [CrossRef] [PubMed]
63. Yamazaki, H.; Tanji, K.; Wakabayashi, K.; Matsuura, S.; Itoh, K. Role of the Keap1/Nrf2 Pathway in Neurodegenerative Diseases. *Pathol. Int.* **2015**, *65*, 210–219. [CrossRef] [PubMed]
64. Howden, R. Nrf2 and Cardiovascular Defense. *Oxidative Med. Cell. Longev.* **2013**, *2013*, 104308. [CrossRef]
65. Satta, S.; Mahmoud, A.M.; Wilkinson, F.L.; Yvonne Alexander, M.; White, S.J. The Role of Nrf2 in Cardiovascular Function and Disease. *Oxidative Med. Cell. Longev.* **2017**, *2017*, 9237263. [CrossRef] [PubMed]
66. Reuland, D.J.; McCord, J.M.; Hamilton, K.L. The Role of Nrf2 in the Attenuation of Cardiovascular Disease. *Exerc. Sport Sci. Rev.* **2013**, *41*, 162–168. [CrossRef]
67. Zhuang, C.; Wu, Z.; Xing, C.; Miao, Z. Small Molecules Inhibiting Keap1-Nrf2 Protein-Protein Interactions: A Novel Approach to Activate Nrf2 Function. *Medchemcomm* **2017**, *8*, 286–294. [CrossRef]
68. Ward, J.J.; Sodhi, J.S.; McGuffin, L.J.; Buxton, B.F.; Jones, D.T. Prediction and Functional Analysis of Native Disorder in Proteins from the Three Kingdoms of Life. *J. Mol. Biol.* **2004**, *337*, 635–645. [CrossRef] [PubMed]
69. Wright, P.E.; Dyson, H.J. Intrinsically Disordered Proteins in Cellular Signalling and Regulation. *Nat. Rev. Mol. Cell Biol.* **2015**, *16*, 18–29. [CrossRef]
70. Dunker, A.K.; Silman, I.; Uversky, V.N.; Sussman, J.L. Function and Structure of Inherently Disordered Proteins. *Curr. Opin. Struct. Biol.* **2008**, *18*, 756–764. [CrossRef] [PubMed]
71. Krzeminski, M.; Marsh, J.A.; Neale, C.; Choy, W.Y.; Forman-Kay, J.D. Characterization of Disordered Proteins with ENSEMBLE. *Bioinformatics* **2013**, *29*, 398–399. [CrossRef] [PubMed]
72. Berlow, R.B.; Dyson, H.J.; Wright, P.E. Functional Advantages of Dynamic Protein Disorder. *FEBS Lett.* **2015**, *589*, 2433–2440. [CrossRef] [PubMed]
73. Dunker, A.K.; Cortese, M.S.; Romero, P.; Iakoucheva, L.M.; Uversky, V.N. Flexible Nets. The Roles of Intrinsic Disorder in Protein Interaction Networks. *FEBS J.* **2005**, *272*, 5129–5148. [CrossRef] [PubMed]
74. Dyson, H.J.; Wright, P.E. Coupling of Folding and Binding for Unstructured Proteins. *Curr. Opin. Struct. Biol.* **2002**, *12*, 54–60. [CrossRef]
75. Wright, P.E.; Dyson, H.J. Intrinsically Unstructured Proteins: Re-Assessing the Protein Structure-Function Paradigm. *J. Mol. Biol.* **1999**, *293*, 321–331. [CrossRef]
76. Wright, P.E.; Dyson, H.J. Linking Folding and Binding. *Curr. Opin. Struct. Biol.* **2009**, *19*, 31–38. [CrossRef]
77. Khan, H.; Cino, E.A.; Brickenden, A.; Fan, J.; Yang, D.; Choy, W.Y. Fuzzy Complex Formation between the Intrinsically Disordered Prothymosin α and the Kelch Domain of Keap1 Involved in the Oxidative Stress Response. *J. Mol. Biol.* **2013**, *425*, 1011–1027. [CrossRef]
78. Whitmore, L.; Wallace, B.A. Protein Secondary Structure Analyses from Circular Dichroism Spectroscopy: Methods and Reference Databases. *Biopolymers* **2008**, *89*, 392–400. [CrossRef]
79. Delaglio, F.; Grzesiek, S.; Vuister, G.W.; Zhu, G.; Pfeifer, J.; Bax, A. NMRPipe: A Multidimensional Spectral Processing System Based on UNIX Pipes. *J. Biomol. NMR* **1995**, *6*, 277–293. [CrossRef]
80. Johnson, B.A. Using NMRView to Visualize and Analyze the NMR Spectra of Macromolecules. *Methods Mol. Biol.* **2004**, *278*, 313–352.

International Journal of
Molecular Sciences

MDPI

Article

A Concerted Action of UBA5 C-Terminal Unstructured Regions Is Important for Transfer of Activated UFM1 to UFC1

Nicole Wesch [1], Frank Löhr [1], Natalia Rogova [1], Volker Dötsch [1,2,*] and Vladimir V. Rogov [1,2,3,*]

1 Institute of Biophysical Chemistry, Center for Biomolecular Magnetic Resonance, Goethe-University Frankfurt, 60438 Frankfurt am Main, Germany; Wesch@bpc.uni-frankfurt.de (N.W.); Murph@bpc.uni-frankfurt.de (F.L.); Rogova@bpc.uni-frankfurt.de (N.R.)

2 Structural Genomics Consortium, Buchmann Institute for Life Sciences, Goethe-University Frankfurt, 60438 Frankfurt am Main, Germany

3 Institute of Pharmaceutical Chemistry, Goethe-University Frankfurt, 60438 Frankfurt am Main, Germany

* Correspondence: vdoetsch@em.uni-frankfurt.de (V.D.); rogov@pharmchem.uni-frankfurt.de (V.V.R.)

Abstract: Ubiquitin fold modifier 1 (UFM1) is a member of the ubiquitin-like protein family. UFM1 undergoes a cascade of enzymatic reactions including activation by UBA5 (E1), transfer to UFC1 (E2) and selective conjugation to a number of target proteins via UFL1 (E3) enzymes. Despite the importance of ufmylation in a variety of cellular processes and its role in the pathogenicity of many human diseases, the molecular mechanisms of the ufmylation cascade remains unclear. In this study we focused on the biophysical and biochemical characterization of the interaction between UBA5 and UFC1. We explored the hypothesis that the unstructured C-terminal region of UBA5 serves as a regulatory region, controlling cellular localization of the elements of the ufmylation cascade and effective interaction between them. We found that the last 20 residues in UBA5 are pivotal for binding to UFC1 and can accelerate the transfer of UFM1 to UFC1. We solved the structure of a complex of UFC1 and a peptide spanning the last 20 residues of UBA5 by NMR spectroscopy. This structure in combination with additional NMR titration and isothermal titration calorimetry experiments revealed the mechanism of interaction and confirmed the importance of the C-terminal unstructured region in UBA5 for the ufmylation cascade.

Keywords: UFM1; UBA5; UFC1; protein-protein interactions; NMR; complex structure

Citation: Wesch, N.; Löhr, F.; Rogova, N.; Dötsch, V.; Rogov, V.V. A Concerted Action of UBA5 C-Terminal Unstructured Regions Is Important for Transfer of Activated UFM1 to UFC1. *Int. J. Mol. Sci.* **2021**, 22, 7390. https://doi.org/10.3390/ijms22147390

Academic Editors: Istvan Simon and Csaba Magyar

Received: 17 June 2021
Accepted: 6 July 2021
Published: 9 July 2021

1. Introduction

UFM1 is a small ubiquitin-like (UBL) protein spanning 85 residues. Like other UBLs, it has a low sequence identity to ubiquitin, but shares its specific (β-grasp) fold [1,2]. Unlike other UBLs (except for SUMO), UFM1 has a single C-terminal glycine residue, by which UFM1 gets attached to target proteins using an E1-E2-E3 enzymatic cascade [1,3,4]. Initially, the UFM1 precursor protein gets processed by the two specific proteases UfSP1 and UfSP2 to expose the C-terminal glycine residue [5–7]. Processed UFM1 gets activated by UBA5 (E1), a member of the ubiquitin-activating protein family [8–10], from which activated UFM1 is transferred to the catalytic cysteine 116 of UFC1 (E2) [1,8,11]. The last step is the transfer of UFM1 to the target proteins mediated by the specific UFM1 ligase 1 (UFL1), showing no typical E3 ligases domain organization [1,12]. The mechanism of this step is largely unknown and other proteins could be required for UFL1 ligase activity as well [13–16].

The first identified target of UFM1 was Ufm1-binding protein 1 (UFBP1, also known as DDRGK1 or C20orf116) [12]. Since then, discovery of new targets for UFM1 and the characterization of functional consequences of their ufmylation has constantly increased. Recently, new ufmylation targets involved in cancer progression [16,17], DNA damage response [18,19], translation machinery [20] and ribosome functioning [13,14] have been identified. Taking in account the broad range of biological pathways affected by ufmylation, it is not surprising that impaired ufmylation can be connected to many human diseases [16,21–24] and seems to be essential for embryonic development [25–27].

The exact mechanism of ufmylation and the full range of physiological consequences are not well investigated yet. The key elements of the ufmylation cascade (UBA5, UFC1, UFL1) show significant evolutionary differences to the well characterized enzymatic UBL cascades (e.g., ubiquitin or NEDD8) resulting in a number of structural and functional deviations from the canonical E1-E2-E3 pathways [3,4,28]. In contrast to other E1 family members, UBA5 does not display the characteristic domain architecture [28]. This 404-residue protein possesses a single well-folded adenylation domain (residues 57–329), comprising the active site Cys250 and provides a platform for ATP binding and UFM1 activation [8,29]. Two UBA5 regions—the N-terminal (1–56) and the C-terminal (334–404) segments—appear to be important regulatory elements for the function of UBA5 and in the ufmylation cascade. The N-terminal segment 1–56 (absent in one of the two existing UBA5 splice isoforms) significantly enhances ATP binding and therefore increases efficiency and velocity of UFM1 activation. Additionally, the N-terminal extension accelerates UFM1 transfer to UFC1 from the UBA5~UFM1 conjugate in presence of ATP [30].

The UBA5 C-terminal part (Figure 1A) plays a complex regulatory role, consisting of a few conserved regions that mediate interaction of UBA5 with other key players in the ufmylation cascade [31]. The first sequence is a conserved region (R1, residues 334–348), interacting with UFM1 [10,29–32] and also with LC3/GABARAP proteins [31,33]. This region (called LIR/UFIM by its dual nature) is important for the initial binding of UFM1 to UBA5 [10,29,31,32] and for the following UFM1 activation in a *trans*-fashion [29]. *Trans*-activation means that UBA5 forms an active homodimer, like other non-canonical E1 enzymes, and UFM1 bound to the LIR/UFIM segment of one monomer exposes its C-terminal Gly83 residue to the catalytic Cys250 of the other monomer [29]. GABARAP (and to a lesser extend LC3) proteins interact with the same UBA5 region and inhibit UFM1 binding to UBA5, thus modulating the conjugation of UFM1 to UBA5 and to UFC1 in vitro [31]. No evidence for the activation of LC3/GABARAP proteins by UBA5 was found so far. However, we showed previously that interaction between GABARAP proteins and UBA5 facilitates membrane localization of the latter [33].

The second region (R2, residues 364–372) is significantly less conserved among different species than the first region, with only Gly367 being evolutionary invariant. The role of this region is not understood, and no interacting proteins could be identified so far. However, a A371T mutation in the human protein located in this region decreases the ability of UBA5 to activate UFM1, to transfer the activated UFM1 to UFC1 and to mediate UFBP1~UFM1 formation [25,34].

Another conserved region in UBA5 is located at it very C-terminus (R3, residues 393–404) and is predicted to have a helical conformation. Initially, it was postulated by analogy with canonical E1 enzymes that the UBA5 C-terminal part possesses an ubiquitin-fold domain, mediating UBA5 interaction with UFC1 [8,11]. Later it was shown that a short UBA5 peptide (residues 381–404) is solely responsible for this interaction [32]. UFC1, the only known E2 enzyme for UFM1, was characterized structurally [11,35] a few years after discovery of the UFM1 cascade [1]. The common architecture of E2 enzymes—four α-helices, four β-strands and one 3_{10}-helix (reviewed in [28])—is conserved for the UFC1 core (25–157). Lack of C-terminal α-helices and conserved motifs as well as the presence of an N-terminal α-helix, which stabilizes the UFC1 structure [11] result in structural differences, which classify UFC1 as a non-canonical E2 enzyme. Computational modeling (based on the existing crystal structure of the E1:E2 complex for the NEDD8 cascade) revealed that the second α-helix in UFC1 is the most probable site for interaction with UBA5. Indeed, the UFC1 K33A mutation significantly reduces both UBA5 binding and UFM1 transfer from UBA5 to UFC1 [11].

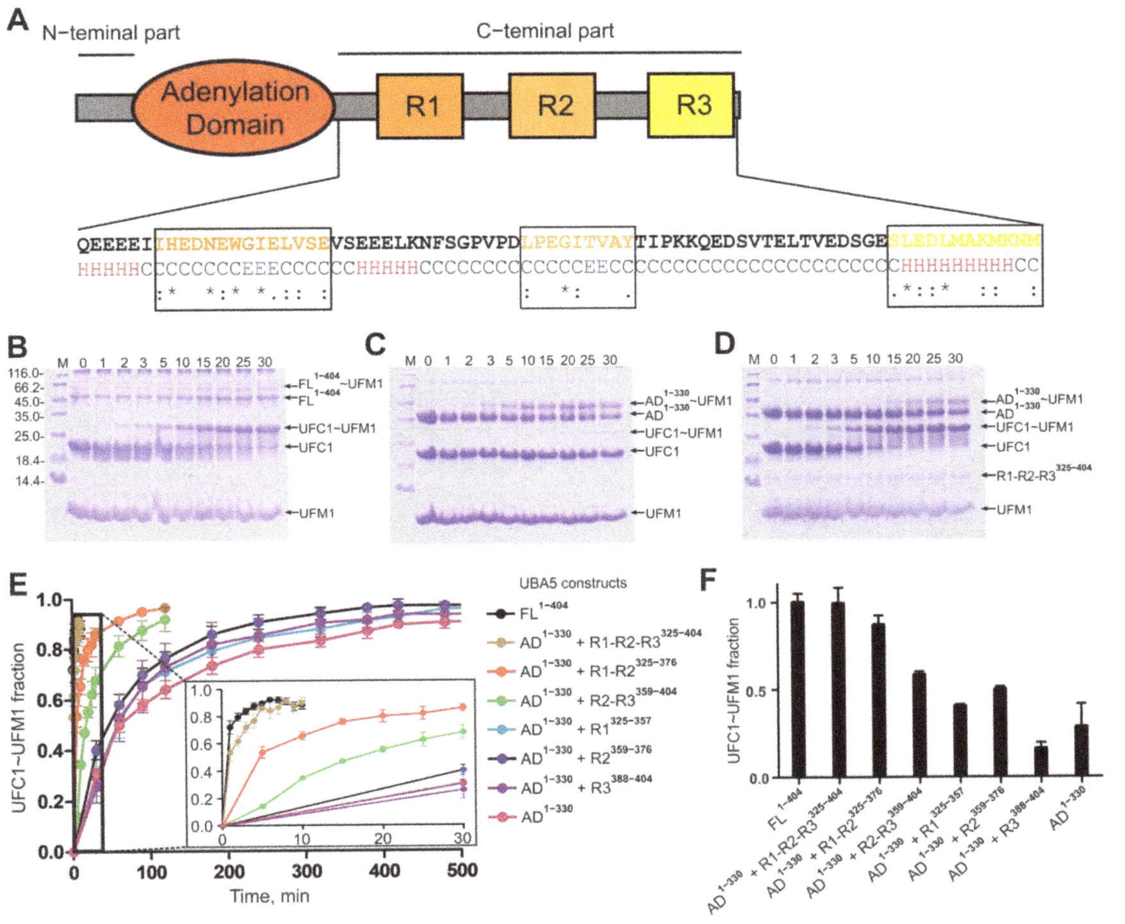

Figure 1. Role of C-terminal UBA5 regions on UFC1~UFM1 conjugation. (**A**) Overview of UBA5 conserved regions. Structure prediction (JPRED) and residue conservation are indicated below the C-terminal sequence (* indicates fully conserved residues; : indicates residues of high similarity; . indicates residues of low similarity). The different UBA5 C-terminal conserved regions are highlighted. (**B–D**) Gel electrophoresis of ufmylation assays including UBA5 FL$^{1-404}$ (**B**), AD$^{1-330}$ (**C**) and a mixture of UBA5 AD$^{1-330}$ and R1-R2-R3$^{325-404}$ (**D**) as E1 enzymes. Ufmylation was tracked over 30 min. Corresponding protein bands are labeled on the right side. (**E**) Ufmylation assays tracked over time with different UBA5 constructs indicated on the right side. The time points of 0–30 min are magnified. All assays were done as triplicates. Evaluation of UFC1~UFM1 conjugate was done via Western blotting. (**F**) Ufmylation assays quantified after 30 min reaction time. The fractions of the UFC1~UFM1 species are presented as bar diagram for each reaction mixture. For quantification of conjugated and unconjugated UFC1 coloc2 software implemented in ImageJ was used.

Despite these previous investigations, structural aspects and molecular mechanisms of the interaction between UBA5 and UFC1 are still largely unknown. Additionally, it is not clear, if other factors (e.g., UFM1 conjugated or bound to UBA5, or UFC1) could affect this interaction. In order to fill this gap, we systematically analyzed by isothermal titration calorimetry and NMR spectroscopy the interactions between different UBA5 fragments and UFC1, UFM1 and LC3/GABARAP proteins. Using this knowledge, we solved the solution structure of UFC1 in complex with an optimized C-terminal fragment of UBA5.

Finally, our biochemical experiments showed the importance of the UBA5:UFC1 interaction for effective ufmylation.

2. Results

2.1. The UBA5 C-Terminal Part Is a Regulatory Platform for the Ufmylation Cascade

In order to understand the importance of the whole UBA5 C-terminal part and the roles of its individual conserved regions, we cloned and expressed a set of constructs containing the whole C-terminus, individual conserved regions and their combinations (Table 1) and investigated their interaction with the key elements of the ufmylation cascade.

Table 1. A list of DNA constructs used in this study.

DNA Construct	Expressed Protein/Peptide	Short Description	References
pET39_Ub19_UBA5$^{1–404}$	FL$^{1–404}$	Full length UBA5, residues 1–404	[31]
pET39_Ub19_UBA5$^{325–404}$	R1-R2-R3$^{325–404}$	UBA5 C-terminal part, residues 325–404	[31]
pET39_Ub19_UFM1	UFM1	Full length UFM1, residues 2–83	[31]
pETm60_Ub3_LC3A	LC3A	LC3A, residues 4–120	[36]
pETm60_Ub3_LC3B	LC3B	LC3B, residues 5–120	[36]
pET39_Ub19_LC3C	LC3C	LC3C, residues 5–126	[36]
pET39_Ub19_GABARAP	GABARAP	GABARAP, residues 3–116	[36]
pETm60_Ub3_GABARAPL1	GABARAPL1	GABARAPL1, residues 2–116	[36]
pET39_Ub19_GABARAPL2	GABARAPL2	GABARAPL2, residues 3–116	[36]
pETm60_Ub	Ubiquitin	Ubiquitin, residues 1–76	[37]
pET39_Ub19_UFC1	UFC1	Full length UFC1, residues 1–167	This work
pET39_Ub19_UBA5$^{1–330}$	AD$^{1–330}$	UBA5 adenylation domain, residues 1–330	This work
pET39_Ub19_UBA5$^{325–376}$	R1-R2$^{325–376}$	UBA5 C-terminal regions R1 and R2, residues 325–376	This work
pET39_Ub19_UBA5$^{359–404}$	R2-R3$^{359–404}$	UBA5 C-terminal regions R2 and R3, residues 359–404	This work
pET39_Ub19_UBA5$^{325–357}$	R1$^{325–357}$	UBA5 C-terminal region R1, residues 325–357	This work
pET39_Ub19_UBA5$^{359–376}$	R2$^{359–376}$	UBA5 C-terminal region R2, residues 359–376	This work
pET39_Ub19_UBA5$^{388–404}$	R3$^{388–404}$	UBA5 C-terminal region R3, residues 388–404	This work
pET39_Ub19_UBA5$^{381–404W}$	R3$^{381–404W}$	Optimized R3, residues 381–404 with C-terminal W	This work
pET39_Ub19_UBA5$^{325–404}$ A371T	R1-R2-R3$^{325–404}$ A371T	UBA5 C-terminal part with A371T mutant (res. 325–404)	This work
pET39_Ub19_UBA5$^{325–404}$ A371E	R1-R2-R3$^{325–404}$ A371E	UBA5 C-terminal part with A371E mutant (res. 325–404)	This work
pET39_Ub19_UBA5$^{1–380}$	ΔR3$^{1–380}$	UBA5 with deleted R3 region, residues 1–380	This work
pNiC-CTH0_UBA5$^{1–404}$ C250K	FL$^{1–404}$ C250K	Full length UBA5 (res. 1–404) with C250K mutant	This work
pET39_Ub19_UBA5$^{1–330}$ C250K ^{0}	AD$^{1–330}$ C250K	UBA5 adenylation domain with C250K mutant	This work
pNiC-CTH0_UFC1	UFC1_His6	Full length UFC1 with C-terminal hexahistidine-tag	This work

First, we analyzed the effect of the UBA5 C-terminus on UFM1 transfer to UFC1 with an in vitro thioester formation assay (Figure 1B–E). Using UBA5 full length protein as E1 enzyme, we observed fast formation of a UFC1~UFM1 conjugate (~90% UFC1 was conjugated to UFM1 within 30 min, Figure 1B). When we used C-terminally truncated UBA5 (only the adenylation domain—AD, residues 1–330) as E1 enzyme, formation of a UFC1~UFM1 conjugate was significantly reduced (less than 5% UFC1~UFM1 conjugation was reached within 30 min; 7 h were needed to reach 80% UFC1~UFM1 conjugation, Figure 1C). However, transfer of UFM1 to UFC1 was rescued when we used an equimolar mixture of the UBA5 AD and the UBA5 C-terminal part as E1 enzyme. In this case, the ure 1D). These results indicate a crucial role of the UBA5 C-terminal part in the ufmylation cascade.

The most important regions in the UBA5 C-terminal parts—R1 (containing the LIR/UFIM sequence) and R3 (containing the UFC1 binding sequence)—seem to have a cumulative effect on the ability of UBA5 to transfer activated UFM1 on UFC1. Addition to the reaction mixture (UBA5 AD$^{1-330}$, UFC1, UFM1, ATP/Mg2^{+}) of UBA5 peptides lacking either the R1 or R3 sequences led to a reduced conjugation rate (Figure 1E and Supplementary Figure S1A). The results also indicate that the LIR/UFIM sequence is more important for the ufmylation cascade than the R3 site and that the conserved region R2 could also play an additive role in this process: the level of UFC1~UFM1 conjugates reached in reactions with AD$^{1-330}$/R1-R2$^{325-376}$ a higher level than when the R1$^{325-357}$ peptide was added alone. Similarly, the addition of the isolated R2$^{359-376}$ and R3$^{381-404W}$ peptides had virtually no effect on the ufmylation reaction (Figure 1E and Supplementary Figure S1B).

UBA5 mutations within the R2 sequence (A371T and it phosphomimicking variant A371E) did not affect significantly the formation of the UFC1~UFM1 conjugate (Supplementary Figure S1C), indicating that the mutation becomes important for downstream events in the ufmylation cascade—potentially during binding of UBA5 to the membrane-associated E3 ligase (UFL1), to targets (UFBP1 [12], ASC1 [16], p53 [17], etc.) or important for other regulatory events. However, in another assay, using a mixture of wild type and mutated full length UBA5 proteins, we observed a small but reproducible reduction of UFC1~UFM1 conjugation (Supplementary Figure S1D).

Taken together we were able to restore the UFM1 transfer to UFC1 with separated AD and C-terminal peptides. With the single AD and only one of the regions the reaction took 7 h. The reaction rate increased by addition of peptides containing two regions and was similar to the full length UBA5 containing the complete C-terminal part.

2.2. *Interactions between Different UBA5 C-Terminal Regions and UFC1, UFM1 and LC3/GABRAP Proteins*

To understand how the UBA5 C-terminus participates in the ufmylation cascade, we performed isothermal titration calorimetry (ITC) experiments, in which we titrated UBA5 C-terminal peptides (see Table 1) to the UBA5 AD, UFC1, UFM1 and representative LC3/GABARAP proteins (Figure 2A, Supplementary Figure S2A–D and Table 2). The ITC experiments revealed that the entire UBA5 C-terminus (R1-R2-R3$^{325-404}$) does not interact with the UBA5 AD, forming an independent UBA5 domain (Supplementary Figure S2A). The affinity of UFM1 to the R1-containing peptides (R1-R2-R3$^{325-404}$ and R1-R2$^{325-376}$, Supplementary Figure S2B and Table 2) does not change significantly compared to the affinity of the isolated R1$^{325-357}$ peptide [31], indicating that this interaction is completely located within the LIR/UFIM containing region.

In contrast, LC3/GABARAP proteins showed a 10-fold higher affinity to the R1-R2-R3$^{325-404}$ and R1-R2$^{325-376}$ peptides compared to the isolated LIR/UFIM motif (R1$^{337-348}$) characterized in [31,33]. The K_D values for interactions between R1-R2-R3$^{325-404}$ and GABARAPL2 (0.17 μM) or LC3B (3.7 μM) indicate the same subfamily-specific preferences that were reported previously (Supplementary Figure S2C,D).

The affinity of the interaction between UBA5 and UFC1 has not been characterized previously. In ITC experiments, the shortest UBA5 peptide spanning the R3 sequence (R3$^{388-404}$) bound to UFC1 with a K_D of >11 μM. The affinity increased 3-fold for R2-R3$^{359-404}$ and R1-R2-R3$^{325-404}$ peptides (K_D of 2.7 and 2.4 μM, respectively; Figure 2A and Table 2).

UFM1 and LC3/GABARAP proteins did not show interaction with the R2 region. However, R1-R2-R3$^{325-404}$ peptides containing A371T and A371E mutations showed some increase in affinity to LC3B and GABARAPL2 proteins but not to UFM1 and UFC1 (Supplementary Figure S2E,F, Table 2).

To understand the role of the UBA5 C-terminal region in coordination of the binding events reported above on the molecular level, we performed NMR titration experiments. In those experiments, we titrated non-labeled UFC1 and GABARAPL2 proteins to a ^{15}N-labeled R1-R2-R3$^{325-404}$ peptide. The NMR experiments revealed that the interaction between UFC1 and UBA5 is mediated mostly by the UBA5 residues 386–404. These residues (in contrast to the vast majority of the R1-R2-R3$^{325-404}$ resonances, which are not affected by addition of UFC1) showed a slow-to-intermediate exchange mode. The amide backbone resonances of these residues disappeared with small chemical shift perturbation

(CSP) at the earlier stages of titrations and did not appear again up to an 8-fold molar excess of UFC1 (Figure 2B, the full size spectra are presented in Supplementary Figure S3D). UBA5 residues 383–386, 400 and 403 appeared to be in intermediate exchange mode (their amide backbone resonances displayed CSP with intensity change, however, they became visible at the latest titration steps). It seems, that these UBA5 residues form additional interactions with UFC1. Interestingly, a subset of the residues within the R2 region (V370, A371, Y372 and T373) displayed moderate CSPs, however, below standard deviation level (Figure 2C), possibly indicating an influence of the UBA5 A371T mutation on the recognition of UFC1.

The GABARAPL2 titration to the R1-R2-R3$^{325–404}$ peptide revealed a complex behavior of interactions between these two polypeptides (Supplementary Figure S3A,B). At the earlier stages of titrations (until a molar ratio of 1:1) the R1-R2-R3$^{325–404}$ resonances showed significant CSPs (in slow-to-intermediate exchange mode), mostly within the LIR/UFIM region (residues D338-V349). Moderate CSPs (with magnitudes above one standard deviation level) can also be observed in sequences adjacent to the R1 peptide: I335 N-terminally, and E352-S358 C-terminally. However, increased concentrations of GABARAPL2 induce further CSPs over the entire R1-R2-R3$^{325–404}$ peptide sequence, including residues in R2 (A371-I374) and R3 (V382-G391, L394, D396, M398) regions. For the resonances within the R1 and adjacent sequences, the direction of the CSPs changed (Supplementary Figure S3A), while residues in R2/R3 regions approach the slow-exchange regime with increased CSP values. These observations indicate, that GABARAPL2 binds first to the LIR/UFIM region, and after saturation of this binding site, GABARAPL2 interacts with additional sites within the UBA5 C-terminus. According to this model, high concentration of GABARAPL2 could efficiently prohibit the UFC1~UFM1 conjugation, which was observed in ufmylation assays (Supplementary Figure S3C).

Table 2. Thermodynamic parameters of the interactions between UBA5 C-terminal regions and UBA5-interacting proteins.

Proteins	UBA5 Regions	ΔH (kcal mol^{-1})	ΔS (cal mol^{-1} K^{-1})	−T*ΔS (kcal mol^{-1})	ΔG (kcal mol^{-1})	K_A*10^{-6} (M^{-1})	K_D (μM)	N
UFM1	R1-R2-R3$^{325–404}$	−5.83 ± 0.20 *	4.41	−1.31	−7.15	0.17 ± 0.02	5.8	1.04 ± 0.03
	R1-R2$^{325–376}$	−5.61 ± 0.21	4.99	−1.49	−7.10	0.16 ± 0.01	6.2	0.96 ± 0.03
	R2$^{359–376}$	ND				ND	>100 **	ND
	R1-R2-R3$^{325–404}$ A371T	−10.99 ± 0.25	−13.3	3.96	−7.02	0.14 ± 0.01	7.1	1.12 ± 0.02
	R1-R2-R3$^{325–404}$ A371E	−11.42 ± 0.42	−15.3	4.56	−6.86	0.11 ± 0.01	9.2	1.01 ± 0.01
UFC1	R1-R2-R3$^{325–404}$	−7.04 ± 0.07	2.09	−0.62	−7.66	0.41 ± 0.02	2.4	1.03 ± 0.01
	R2-R3$^{359–404}$	−8.08 ± 0.10	1.59	−0.47	−7.60	0.37 ± 0.01	2.7	0.97 ± 0.01
	R3$^{388–404}$	−4.99 ± 0.22	6.91	−2.06	−7.05	0.03 ± 0.003	10	0.95 ± 0.03
	R2$^{359–376}$	ND				ND	>50 **	ND
	R1-R2-R3$^{325–404}$ A371T	−7.88 ± 0.08	−0.19	0.06	−7.82	0.54 ± 0.03	1.8	1.03 ± 0.008
	R1-R2-R3$^{325–404}$ A371E	−7.78 ± 0.05	0.36	−0.11	−7.88	0.60 ± 0.02	1.6	1.03 ± 0.005
	FL$^{1–404}$	−7.62 ± 0.01	1.26	−0.38	−8.00	0.72 ± 0.04	1.4	0.97 ± 0.009
	FL$^{1–404}$ C250K~Ufm1	−8.21 ± 0.01	−0.34	0.10	−8.10	0.87 ± 0.05	1.2	1.03 ± 0.009
GABARAPL2	R1-R2-R3$^{325–404}$	−8.64 ± 0.06	2.04	−0.61	−9.25	5.99 ± 0.49	0.17	0.97 ± 0.004
	R1-R2$^{325–376}$	−8.08 ± 0.05	4.44	−1.32	−9.41	7.87 ± 0.74	0.13	0.91 ± 0.003
	R2$^{359–376}$	ND				ND	>100 **	ND
	R1-R2-R3$^{325–404}$ A371T	−7.58 ± 0.07	7.76	−2.31	−9.89	17.90 ± 3.75	0.06	0.937 ± 0.005
	R1-R2-R3$^{325–404}$ A371E	−7.79 ± 0.05	7.23	−2.16	−9.95	19.60 ± 2.57	0.06	1.01 ± 0.003
GABARAP	R1-R2-R3$^{325–404}$	−0.93 ± 0.04	24.2	−7.22	−8.15	0.96 ± 0.16	1.1	0.99 ± 0.03
	R1-R2$^{325–376}$	−1.1 ± 0.02	23.1	−6.89	−7.99	0.72 ± 0.05	1.4	0.94 ± 0.01
LC3B	R1-R2-R3$^{325–404}$	−4.47 ± 0.10	8.86	−2.64	−7.33	0.24 ± 0.09	4.2	0.92 ± 0.02
	R1-R2$^{325–376}$	−4.23 ± 0.09	10.7	−3.19	−7.42	0.27 ± 0.14	3.7	0.98 ± 0.02
	R2$^{359–376}$	ND				ND	>100 *	ND
	R1-R2-R3$^{325–404}$ A371T	−3.76 ± 0.05	14.3	−4.26	−8.02	0.76 ± 0.04	1.3	0.937 ± 0.009
	R1-R2-R3$^{325–404}$ A371E	−3.93 ± 0.05	15.3	−4.56	−8.49	1.66 ± 0.12	0.6	0.944 ± 0.009
LC3A	R1-R2-R3$^{325–404}$	4.25 ± 10.5	10.5	−3.13	−7.38	0.26 ± 0.03	3.8	0.91 ± 0.04
	R1-R2$^{325–376}$	−3.81 ± 0.18	11.6	−3.46	−7.26	0.21 ± 0.02	4.7	0.94 ± 0.03
UBA5 AD$^{1–330}$	R1-R2-R3$^{325–404}$	ND				ND	-	ND
Ub	R1-R2-R3$^{325–404}$	ND				ND	-	ND

* Here and further the ± sign corresponds to a fitting error of the individual experiment. ** Estimated value.

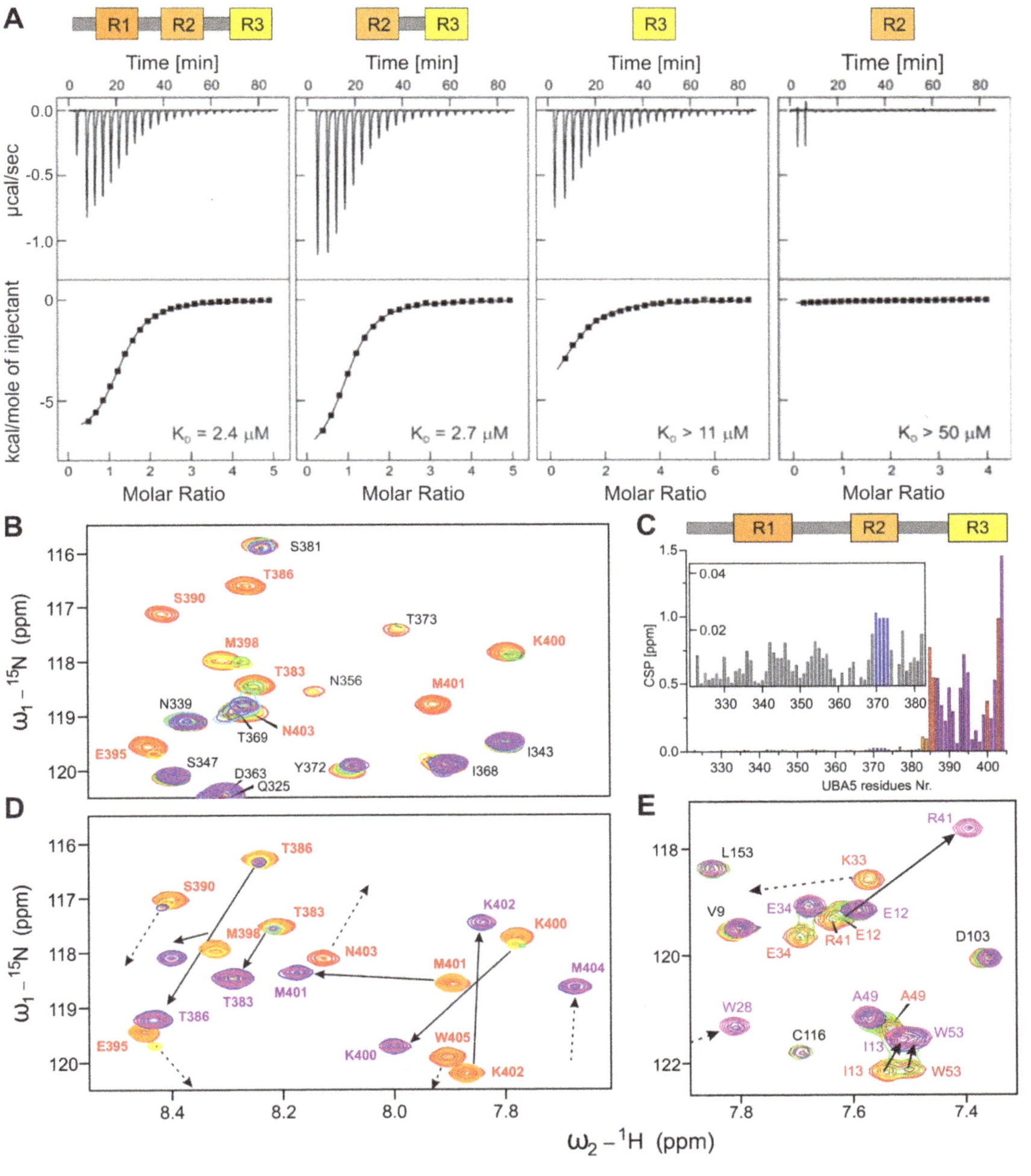

Figure 2. Interaction between UBA5 C-terminal fragments and UFC1 protein. (**A**) UFC1 binding to different UBA5 C-terminal peptides observed by ITC experiments. The upper graphs display the raw heat data; the lower graphs show the integrated heat per titration steps (black squares) with best-fit curve (line). The used peptides are graphically visualized above the corresponding titration profiles. K_D values are indicated. (**B**) NMR titration of ^{15}N-labeled R1-R2-R3$^{325–404}$ peptide with non-labeled UFC1. An overlay of representative areas of the [^{15}N,^{1}H] TROSY-HSQC spectra recorded at 500 MHz are presented. The increasing protein molar ratios are indicated with a rainbow color code from free R1-R2-R3$^{325–404}$ (red) to

8 molar excess of UFC1 (purple). (**C**) Mapping of CSPs induced by UFC1 on the R1-R2-R3$^{325–404}$ sequence. The CSP values (shown as bars) below standard deviation (SD), between 1xSD and 2xSD, and above 2xSD are labeled grey, yellow and red, respectively. The small box shows magnification of CSP diagram for UBA5 residues 325–382. The disappearing resonances within the core R3 sequence are also shown as purple bars; the CSP for the R2 residues around A371 are marked blue. (**D**) NMR titration of the ^{13}C,^{15}N-labeled R3$^{381–404W}$ peptide with non-labeled UFC1 protein performed at 800 MHz. The same spectral areas as in (**B**) are shown and the same color code is used. (**E**) NMR titration of ^{15}N,^{13}C-labeled UFC1 with non-labeled R3$^{381–404W}$ peptide recorded at 950 MHz. An overlay of representative areas of the [^{15}N,^{1}H] TROSY-HSQC spectra is presented. Titration steps are visualized in a rainbow color code. Most significant CSP are highlighted by arrows. Dashed arrows indicate that the initial or final peak position is outside of the presented area.

We could not observe any interactions between UFM1 and UFC1 proteins (using NMR titration of ^{15}N-labeled UFC1 with non-labeled UFM1 up to 1:2 molar ratio). Additionally, binding of UFC1 to the R3 region within the UBA5 C-terminus$^{325–404}$ did not initiate UFC1:UFM1 interactions as displayed by NMR experiments of ^{15}N-labeled UFC1 in complex with the R1-R2-R3$^{325–404}$ peptide titrated with non-labeled UFM1 until a 1:4 molar ratio. Furthermore, no interaction of ubiquitin to the UBA5 C-terminal region was observed, suggesting that the UBA5 C-terminus is specific for UFM1.

Taken together, we identified a UFC1-interacting region within the UBA5 C-terminus using ITC and NMR experiments. The region is slightly longer than the conserved R3 sequence which was detected previously and shows a micromolar affinity to UFC1. While UFM1 seems to bind only to the LIR/UFIM region of UBA5, LC3/GABARAP proteins interact with additional residues outside of the of the R1 sequence. LC3 and GABARAP subfamily proteins showed a 10-fold higher affinity to the complete UBA5 C-terminus compared to the isolated R1 peptide. Additionally, UFC1 showed interaction outside of the R3 region, binding residues within the R2 region. NMR titrations revealed that UFC1 and GABARAPL2 have a more complex binding mechanism to the UBA5 C-terminus, involving some residues in the R2 region. However; no direct interactions of all tested proteins to the isolated R2 peptide were observed.

2.3. Structure of UFC1 in Complex with the UBA5 R3 Peptide

To understand the interaction between UFC1 and UBA5 on a molecular level, we solved the NMR solution structure of UFC1 in complex with the UBA5 R3$^{381–404W}$ peptide. Based on the results of our ITC and NMR experiments, we optimized the R3 peptide sequence including residues 381–404 of UBA5 and an additional C-terminal tryptophan residue (at position 405), providing a possibility to calculate the peptide concentration by UV spectroscopy. The R3$^{381–404W}$ peptide displayed the expected ability to form a stable complex with UFC1. In contrast to the shorter R3$^{388–404}$ peptide or to the R1-R2-R3$^{325–404}$ peptide, the R3$^{381–404W}$ peptide showed re-appearance of all resonances at the latest titration steps with UFC1 (Figure 2D and Supplementary Figure S3E). Correspondingly, almost all backbone amide resonances of UFC1 became visible at the latest stages of titration with R3$^{381–404W}$ (Figure 2E and Supplementary Figure S3F), enabling us to solve the UFC1:R3$^{381–404W}$ complex structure. The structure is presented in Figure 3 and Supplementary Figure S4, structural statistics are given in Supplementary Table S1. The UFC1 structure in complex with the R3$^{381–404W}$ peptide is close to the previously published X-ray and NMR structures of free UFC1 (Supplementary Figure S4A, [11,35]). The most significant differences were observed in the orientation of the N-terminal α-helix α1 (residues 1–11), the conformation of the C-terminal UFC1 part (residues 156–167) and the flexible loop near the active-cite cysteine 116 (residues 91–124, Supplementary Figure S4B).

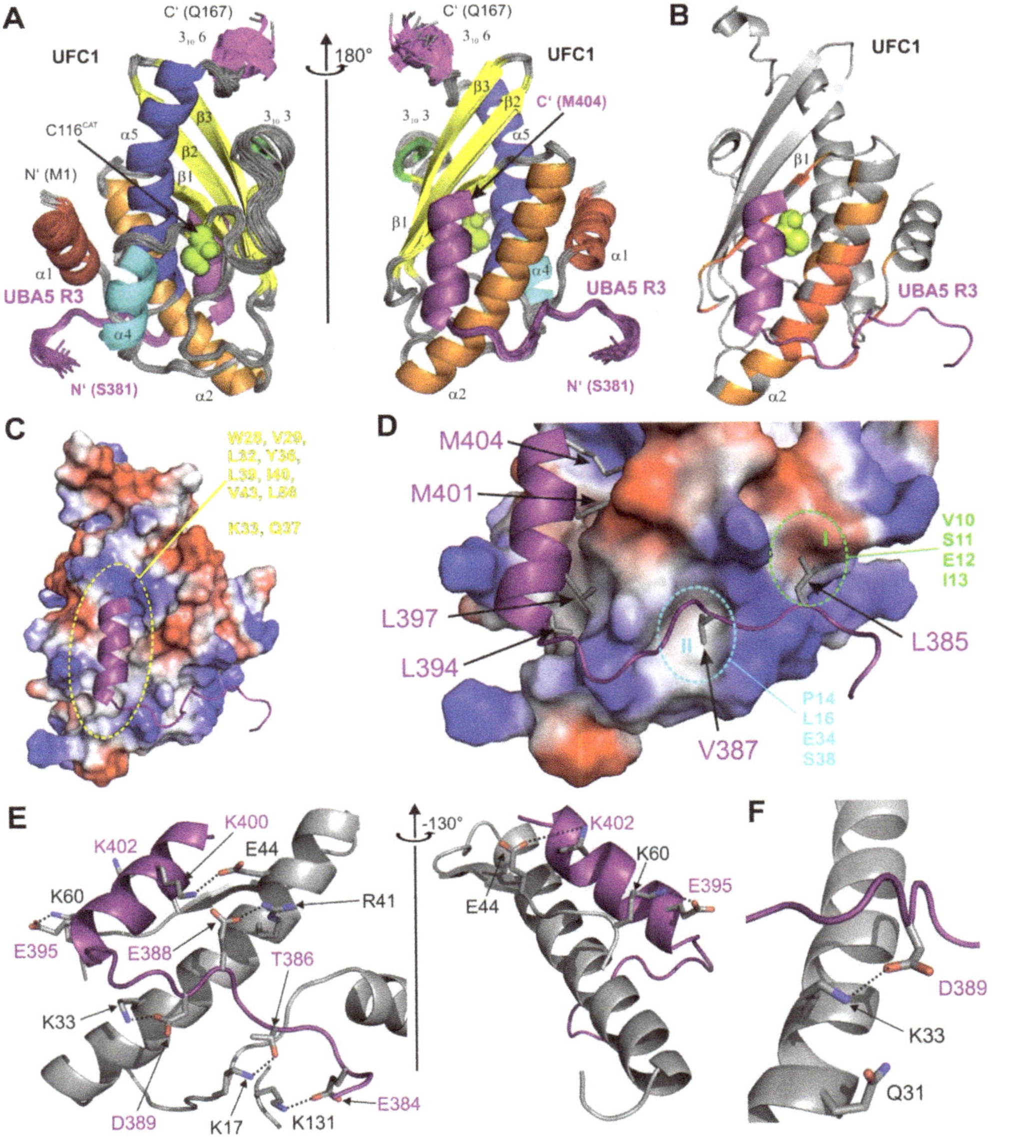

Figure 3. NMR structure of the complex between UFC1 and the UBA5 $R3^{381-404W}$ peptide. (**A**) NMR solution structure of the complex between UFC1 and $R3^{381-404W}$ peptide in two different orientations. All 20 conformers are superimposed over the structured UFC1 core (residues 3–162). All UFC1 secondary structure elements are marked by the following colors: α1—red; α2—orange; 3^{10} helix 3—green; α4—cyan, α5—blue; 3^{10} helix 6—magenta; all β-strands (β1, β2, β3) are yellow. $R3^{381-404W}$ chains are shown in purple. (**B**) Mapping of UFC1 CSPs upon titration with $R3^{381-404W}$ on a representative complex structure (conformer 6, the same orientation as in the A, right plot). The CSP values below standard deviation (SD), between 1xSD and 2xSD, and above 2xSD are labeled grey, yellow and red, respectively. Residues which were not assigned are presented in grey as well. (**C**) UFC1 molecule (conformer 6, the same orientation as in the A, right plot) is shown as a surface with calculated potentials, whereas the $R3^{381-404W}$ molecule is presented by ribbon diagram (purple). The large hydrophobic groove between UFC1 α-helix α2 and β-strand β1 is highlighted with a dashed yellow line. UFC1 residues

contributing to the groove formation are listed. (**D**) Hydrophobic patches on UFC1 surface mediating interactions with the UBA5 R3$^{381–404W}$ L385 and V387 side chains are shown as grey sticks. The UFC1 hydrophobic patches I and II are marked with dashed lines (green and magenta, respectively). UFC1 residues forming these patches are listed. (**E**) Polar interactions within the UFC1:R3$^{381–404W}$ complex. Intermolecular hydrogen bonds are shown as dashed lines. (**F**) Detailed view on the intermolecular hydrogen bond between UBA5 D389 and UFC1 K33. The UBA5 Q31 sidechain is also presented as sticks.

Residues 394–404 of the R3 region form the predicted [32] α-helix, residues 384–392 are in an extended conformation, well-defined and occupy a specific area on the UFC1 surface. Residues 381–383 seem disordered and do not interact specifically with any UFC1 residues. The amphiphilic R3 α-helix is aligned to the α2 α-helix of UFC1 (Figure 3A) on the side opposite to the catalytic cysteine (C116). The UFC1 resonances on the C116 side were not affected upon NMR titration experiments, leading to the suggestion that this side could interact with the adenylation domain during UFM1 transfer. Sidechains of the R3$^{381–404W}$ hydrophobic residues (L394, L397, M401 and M404) are placed into the large hydrophobic cleft formed by α-helix α2 and β-strand β1 of UFC1 (residues W28, V29, L32, Y36, L39, I40, V43, L56 and aliphatic moieties of K33 and Q37; Figure 3C). Two additional hydrophobic patches I and II (formed by residues within α-helices α1, α2 and the loop between them) accommodate UBA5 residues L385 and V386 (Figure 3D).

In addition to intermolecular hydrophobic interactions, the complex between UFC1 and the R3$^{381–404W}$ peptide is stabilized by a network of intermolecular hydrogen bonds and polar contacts (Figure 3E, all intermolecular contacts detected by the LigPlot software for the UFC1:R3$^{381–404W}$ complex are shown in Supplementary Figure S4C). The network covers almost all residues within the R3 region, which interact with the polar residues of UFC1 in the same area—α1, α2, loop between them and β-strand β1 (detailed information on the polar contacts is given in the Supplementary Figure S4C). The only additional UFC1 residue that forms intermolecular hydrogen bonds to the R3$^{381–404W}$ peptide outside of this UFC1 region, is K131, whose sidechain is in close proximity to the carboxyl group of UBA5 E384.

Previously, it was predicted that the UFC1:UBA5 interaction is mediated by the UFC1 α-helix α2 [11] and the point mutation K33A within this helix impaired UBA5 binding and UFM1 transfer to UFC1, whereas Q31A had no effect. In our structure we observed that the UFC1 K33 sidechain forms an intermolecular hydrogen bond with the UBA5 D389 sidechain (Figure 3F). In contrast, UFC1 Q31 is not in contact with any of the UBA5 R3 residue and could not affect the UBA5:UFC1 interaction.

In summary, the structure of UFC1 in complex with the R3$^{381–404W}$ peptide revealed that the C-terminal α-helical part of UBA5 is pivotal for the attraction of UFC1 to UBA5. In addition to the α-helical part, UBA5 residues L385 and V387 also play a role in the UBA5 interaction with UFC1. The UFC1 hydrophobic groove and hydrophobic patches I and II are the most important areas mediating the interaction. Intermolecular polar contacts and hydrogen bonds stabilize the observed complex. The sidechain of UFC1 K33 is involved in an intermolecular hydrogen bond formation (to UBA5 D389 as a counterpart), therefore, its substitution to alanine interferes with the UFC1 interaction to UBA5 [11].

2.4. Interactions within the Ufmylation Cascade

Our results so far describe the interaction of UFC1 with the UBA5 C-terminal region. However, the interaction between full length UBA5 and UFC1 could be more complex and could depend on UFM1 conjugation to UBA5 or UFC1. To answer the question if UBA5 can bind UFC1 via additional sites, we analyzed NMR spectra of UFC1 with a 2-fold excess of unlabeled UBA5 FL$^{1–404}$. We did not observe significant CSPs (shift or disappearance of the UFC1 resonances) in comparison to the spectra of the UFC1:R3$^{381–404W}$ complex (Supplementary Figure S5A).

Additionally, UBA5 lacking the R3 region (ΔR3$^{1–380}$) did not interact with UFC1 (as observed by NMR titration experiment, Supplementary Figure S5B) and significantly slowed down UFM1 transfer to UFC1 (Figure 4A, Supplementary Figure S5D). All these observations indicate that besides R3, UFC1 does not bind to any UBA5 regions efficiently.

However, even weak additional interactions could facilitate the UFC1~UFM1 conjugation as observed in this work for the UBA5 constructs lacking R3 (Figures 1C and 4A, Supplementary Figure S1).

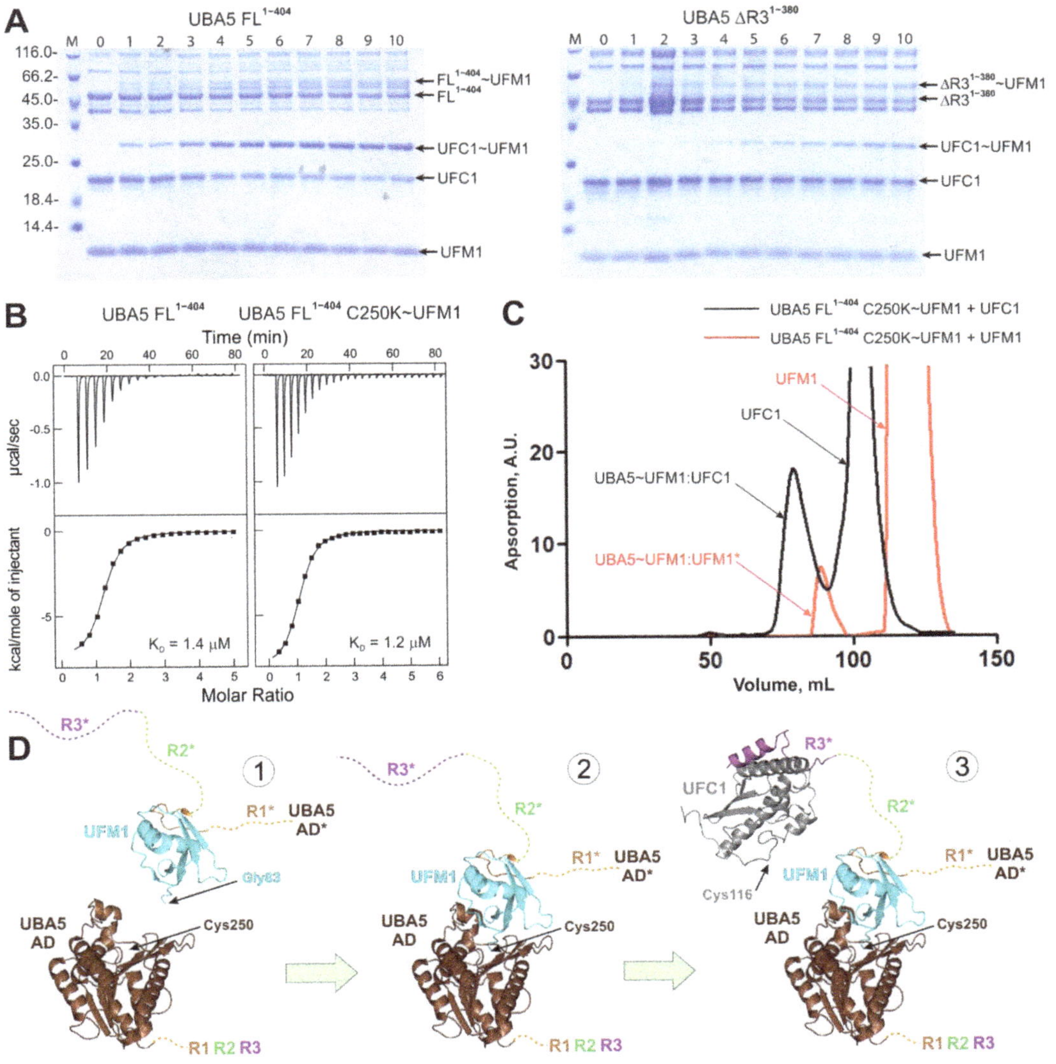

Figure 4. Interaction studies between full length UBA5 and UFC1 proteins. (**A**) Gel electrophoresis of ufmylation assays including UBA5 FL$^{1-404}$ (left plot) and UBA5 ΔR3$^{1-380}$. (**B**) UFC1 binding to full length UBA5 (left plot) and full length stable UBA5~UFM1 conjugate (right plot) observed by ITC experiments. The upper graph displays the raw heat data; the lower graph shows the integrated heat per titration steps (black squares) with best-fit curve (line). K_D values are indicated. (**C**) Gel-filtration profiles of the FL$^{1-404}$ C250K~UFM1 conjugates in presence of 4 times molar excess of UFM1 (red lines) and UFC1 (black lines). The peak subjected to electrophoretic analysis is indicated by an asterisk. (**D**) Scheme of reactions involving UBA5 in the ufmylation cascade. The structures of UBA5 AD (brown), UFM1 (cyan) and UFC1 (grey) are represented as ribbon diagrams; the UBA5 unstructured C-terminus containing regions R1 (orange), R2 (green) and R3 (violet) is shown as dashed lines. The structures were generated from PDB entry 5IAA [29]. * indicates regions of another UBA5 molecule involved in the *in trans* transfer of UFM1.

To investigate if conjugation of UFM1 to the UBA5 catalytic cysteine (C250) affects the UFC1:UBA5 interactions, we prepared full length UBA5 C250K mutant and stably conjugated UFM1 to it as reported before for a number of ubiquitin-specific E2 enzymes [38–40]. We compared the UFC1 spectra after addition of a twofold molar excess of FL^{1-404} and FL^{1-404} C250K~UFM1 constructs (Supplementary Figure S5C). Again, no significant enhancement of the UBA5:UFC1 interaction induced by the UBA5~UFM1 conjugation was observed. ITC experiments, in which we titrated UFC1 to FL^{1-404} and to FL^{1-404} C250K~UFM1 samples (Figure 4B, Table 2), showed small increases in their affinity to UFC1 in comparison to the R1-R2-$R3^{325-404}$ peptide (K_D values for R1-R2-$R3^{325-404}$, FL^{1-404}, FL^{1-404} C250K~UFM1 are 2.4, 1.4 and 1.2 μM, respectively).

UFM1 conjugation to UBA5 C250 did not prohibit UFM1 binding to the R1 region. The gel-filtration profile and following electrophoretic analysis of the fractions showed that the FL^{1-404}C250K~UFM1 but not the AD^{1-330} C250K~UFM1 peak contains non-conjugated UFM1 (Figure 4C and Supplementary Figure S5E).

3. Discussion

In this paper we analyzed the interactions between UBA5 and UFC1 enzymes within the ufmylation cascade and found that the unstructured UBA5 C-terminal part provides a platform for multiple protein–protein interactions affecting the efficiency of the activated UFM1 transfer from UBA5 to UFC1.

3.1. The UFC1:UBA5 Interaction

Our ITC and NMR titration experiments revealed that the interaction between UFC1 and UBA5 is mediated mostly by the relatively short and evolutionary conserved stretch of UBA5 residues (383–404). Using the optimized UBA5 construct ($R3^{381-404W}$ peptide), we solved the NMR structure of the UFC1:R3 complex. The complex structure in combination with the NMR and ITC titration experiments revealed that in addition to the core R3 region, residues in the region R2 contribute to the interaction. While the isolated R2 peptide does not interact with UFC1, the combination of R2 and R3 binds three times tighter than the R3 alone. This weak additional interaction also explains the results of the UFC1 ufmylation assay (Figure 1). Ability of the isolated UBA5 AD to transfer activated UFM1 on UFC1 gets rescued by addition of the R1-R2-R3 peptide. In this peptide the R1 sequence can bind to UFM1 conjugated to UBA5 and recruit via its exposed R3 peptide UFC1 to the complex (Figure 4D). In full length UBA5 this recruitment occurs similarly, resulting in very similar UFC1 ufmylation rates. Adding only the R2-R3 peptide to the UBA5 AD increases the reaction rate only slightly above the isolated individual R1, R2 or R3 peptides, because deletion of the R1 sequence prevents effective recruitment of UBA5 C-terminus in complex with UFC1 to the UFM1-charged AD. A stronger rescue effect is seen for the R1-R2 peptide, because the R2 peptide probably still can interact with UFC1 (Figure 2C) and thus increase the local concentration of UFC1 around the AD. In the full length UBA5 protein, this recruiting effect most likely occurs *in-trans* [29]. A dimer was found in the crystal structure of UBA5 in complex with UFM1 bound to the R1 region. The linker between the AD and the R1 sequence is too short for an *in-cis* transfer to the active site cysteine, but within the dimer UFM1 bound to R1 of one monomer can be adenylated by the other UBA5 molecule of the dimer. This mechanism was confirmed by clever mutational engineering showing that a forced monomer cannot activate UFM1. Similarly, a trans mechanism was proposed for the transfer to UFC1 as well (Figure 4D). In our NMR titration experiments the UFC1 catalytic cysteine C116 and neighboring residues were not affected upon titration with the R3 peptide and our complex structure revealed that the R3 peptide occupies the side of the UFC1 molecule opposite to C116, indicating that the UFC1 surface around C116 could be used by the UBA5 AD during UFM1 transfer. Note that our data alone did not exclude *in-cis* UFM1 transfer mode.

In general, we were able to observe relatively stable interactions between members of the ufmylation cascade only for the R1:UFM1 and R3:UFC1 interactions. All other interactions are so weak that they are hard to detect by NMR (additional R2 residues with UFC1) or cannot be characterized at all. This includes interaction of UFC1 with the UBA5 AD alone or charged with UFM1 as well as with isolated UFM1. These results suggest that transfer of UFM1 from the adenylation domain of UBA5 to UFC1 uses in addition to relatively strong interactions for recruitment of the necessary components very weak interactions for the transfer (hit-and-run model).

3.2. Interaction between GABARAPL2 and UBA5 C-Terminal Part

The GABARAP and LC3 subfamilies members were found to bind UBA5 via an atypical LIR (LIR/UFIM), an evolutionary conserved sequence within the UBA5 C-terminal part [31,33]. The ITC and NMR experiments revealed additional interactions next to the known binding site within the R1. UBA5 constructs including both R1 and R2 regions showed a 10fold higher binding affinity to all GABARAP and LC3 protein subfamily members. Binding preference towards the GABARAP subfamily proteins remains preserved [31,33]. NMR titration experiments disclosed a more complex binding mechanism of GABARAPL2 to the complete C-terminal UBA5 peptide. At earlier titration steps, UBA5 residues within R1 were strongly affected by GABARAPL2 binding. However, with increasing concentrations of GABARAPL2 conserved residues located mostly in R2 started to display significant CSPs as well. These additional interactions might become relevant when UBA5 gets recruited to a membrane and GABARAP proteins cluster in micro-domains. A high concentration of GABARAP proteins in combination with a reduction of the search space for interactions from three to two dimensions could allow simultaneous binding of several GABARAP proteins to the UBA5 C-terminus. Recruitment of UBA5 to the membrane of the endoplasmic reticulum (ER) has been observed [33], the exact role of this recruitment is subject for further investigations.

3.3. The Role of the A371T Mutation in the Ufmylation Cascade

Many diseases are associated with impaired ufmylation [16,21–24]. Ufmylation is essential for embryonic development [25–27]. The A371T mutation was described previously to be present in patients suffering from severe infantile-onset encephalopathy [25,34]. Further investigations showed slightly reduced UBA5 thioester conjugation with UFM1 and reduced enzymatic activity in trans-thioesterification of UFC1 in vivo for the UBA5 A371T mutant [25,34]. Our ITC experiments with C-terminal UBA5 peptides containing the A371T or its phosphomimicking A371E mutations (located in the R2 region) showed almost no influence on UFM1:UFC1 binding affinity. NMR titration of the wild type ^{15}N-labeled R1-R2-R3$^{325–404}$ peptide with UFC1 displayed some moderate CSPs for the A371 and residues around, indicating a minor role of the R2 sequence in UFC1 binding. In vitro ufmylation assays showed that R1-R2-R3$^{325–404}$ A371T and R1-R2-R3$^{325–404}$ A371E peptides have nearly the same trans-thioesterification efficiency compared to wild type R1-R2-R3$^{325–404}$ peptide in standard ufmylation assay conditions. However, reduction of ATP (to 25 μM) led to a reduction of the UFC1~UFM1 conjugate fraction for both mutated UBA5 peptides in comparison to wild type peptide, as reported previously [25,34].

Interestingly, we detected an increased affinity of R1-R2-R3$^{325–404}$ A371T and R1-R2-R3$^{325–404}$ A371E peptides to GABARAPL2 and LC3B proteins in ITC experiments. While GABARAPL2 showed a ~3-fold increased affinity to both mutated peptides in comparison to the wild type peptide, we detected a ~7-fold increased affinity for LC3B to the A371E mutant and a ~3-fold increased affinity to the A371T mutant. NMR titration experiments with wild type R1-R2-R3$^{325–404}$ peptide revealed that A371 and adjacent residues are involved in GABARAPL2 binding at high GABARAPL2 concentrations. Again, taking into account that GABARAP and LC3 protein family members are proposed to recruit UBA5 to the ER membrane and play a critical role in the regulation of the ufmylation pathway [33,41], these results lead to the assumption that the A371T mutation plays a minor role in the ufmylation reaction itself, but might affect UBA5 localization and thus influences target ufmylation.

4. Materials and Methods

4.1. DNA Constructs Used in This Study

Genes of proteins and UBA5 peptides were cloned into a pET39_Ub19 vector containing a modified ubiquitin tag [33] and a TEV cleavage site resulting in a N-terminal cloning artefact of the first three residues (GAM). UBA5 C250K and UFC1_His6 were cloned into pNiC-CTH0 vector with a C-terminal hexahistidine-tag cleavable by an introduced TEV cleavage site. For site-directed mutagenesis PfuUltra II fusion HS DNA polymerase (Agilent Technologies Germany, Frankfurt, Germany) was used according to the manufacturer's instructions. A comprehensive list of DNA constructs used in this study is given in Table 1.

4.2. Expression, Isolation and Purification of the Peptides and Proteins

All proteins and peptides were expressed in E.Coli T7 Express (New England Biolabs GmbH, Frankfurt, Germany) cells in LB or M9 (to obtain ^{15}N- and ^{13}C,^{15}N-labeled polypeptides) media according to the protocol described in [33,36]. For protein purification, bacterial cell pellets were resuspended in lysis buffer (50 mM Tris-HCl pH = 7.5, 100 mM NaCl, 5% glycerol, 5 mM PIC (protease inhibitor cocktail)) and lysed via sonication (TT13 Sonotrode, 40% amplitude, for 6 × 1 min with a 0.5/0.5-s pulse). Lysates were centrifuged for 45 min at 17,000× *g* at 4 °C. Supernatants were loaded onto a His Trap Fast Flow 5 mL column (GE Healthcare, München, Germany) equilibrated in loading buffer (50 mM Tris-HCl pH = 8.0, 250 mM NaCl, 1% glycerol, 20 mM imidazole). The column was washed with loading buffer for 5–10 CV and protein was eluted with elution buffer (50 mM Tris-HCl pH = 8, 250 mM NaCl, 1% glycerol, 400 mM imidazole). Simultaneous TEV cleavage (1 mg TEV protease was added to 100 mg peptides/proteins) and buffer exchange to loading buffer via dialysis was performed over night at 4 °C. After reverse IMAC, proteins were concentrated with conical concentrators (Millipore Merck, Darmstadt, Germany) and loaded on a Superdex 10/60 75 or 200 column (GE Healthcare, München, Germany) for further purification and equilibration with ITC/NMR buffer (25 mM HEPES pH = 7.5, 100 mM NaCl). For structural NMR spectroscopy, buffer containing 50 mM Tris-HCl pH = 7.5, 100 mM NaCl was used. Prior to NMR experiments, TCEP and protease inhibitors cocktail were added to the samples to final concentrations 1 and 5 mM, respectively. Purified peptides and protein were concentrated and stored at −80 °C. The protein and peptide concentrations were calculated from the UV absorption at 280 nm by Nanodrop spectrophotometer (Thermo Scientific, Langenselbold, Germany).

4.3. In Vitro Thioester Formation Assay

Ufmylation reaction assays were adopted from work of Xie [32]. Briefly, 70 μM UFM1, 20 μM UFC1 and 20 μM of different UBA5 constructs were mixed in reaction buffer (50 mM HEPES pH = 7.5, 100 mM NaCl, 5 mM $MgCl_2$). After starting the reaction with addition of 1 mM ATP, the reaction mix was incubated at 22 °C for the desired time. To quench the reaction and prepare electrophoretic samples, 1 μL of the reaction mix was added to 99 μL 1x non-reducing SDS loading buffer and frozen in liquid nitrogen. Sample content was visualized by sodium dodecyl sulfate polyacrylamide gel electrophoresis (SDS-PAGE). The transfer to polyvinylidene difluoride (PVDF) membrane was performed via a Trans-Blot® Turbo™ Transfer System (Bio-Rad, München, Germany). After transfer the membrane was blocked with TBST (Tris-buffered saline with Tween20 buffer, 20 mM Tris, 150 mM NaCl and 0.1% TWEEN 20) containing 5% *w*/*v* nonfat dry milk for 1 h, followed by α-UFC1 antibody incubation over night at 4 °C (ab189251 abcam, 1:10,000 in TBST containing 5% *w*/*v* nonfat dry milk). After washing with TBST the membrane was incubated with secondary antibody (anti-rabbit-HRP) for 1 h at RT and again washed with TBST. Detection was performed by addition of ECL solution. For quantification of UFC1 ufmylation coloc2 software implemented in ImageJ was used. To show the kinetic differences between $FL^{1–404}$ and $\Delta R3^{1–380}$ on UFC1 ufmylation, the reactions were started with 25 μM ATP.

For stable UBA5~UFM1 conjugation, 70 μM UFM1, 20 μM $FL^{1–404}$ C250K and 1 mM ATP were added to ufmylation reaction buffer (50 mM HEPES pH = 10.0, 100 mM NaCl,

5 mM $MgCl_2$). For NMR analysis, resulting complexes were concentrated and equilibrated with ITC/NMR buffer. To analyze complex formation by ufmylation assay 300 μL of sample were loaded onto a Superdex 200 10/300 column (GE Healthcare, München, Germany).

4.4. Isothermal Titration Calorimetry

All ITC experiments were performed at 25 °C using a VP-ITC microcalorimeter (Malvern Panalytical Ltd., Malvern, UK). Peptides in concentration of ~400 μM were titrated into 20–25 μM solutions of corresponding binding partner at a stirring speed of 307 rpm. The raw data were corrected on the dilution heat of peptides obtained in independent experiment (titration of the peptide in syringe into the ITC/NMR buffer in the measuring cell). Pre-titration delay was set to 180 s, interval between titration steps was experimentally adjusted to avoid kinetic contribution to the observed heat effects and set to 200 s. A single ITC profile was collected for each type of interaction. The ITC data were analyzed based on a "one-site" binding model with MicroCal ITC software implemented in Origin 7.0.

4.5. NMR Spectroscopy

All NMR experiments were performed at a sample temperature of 25 °C on Bruker 600, 700, 800, 900, and 950 MHz spectrometers equipped with cryogenic probes, and a 500 MHz spectrometer equipped with a room-temperature triple-resonance probe. All NMR spectra were analyzed with the Sparky 3.114 software (University of California, San Francisco, USA). For NMR titration experiments, the non-labeled UBA5 peptides were titrated to 100 μM ^{13}C,^{15}N-labeled UFC1 to a final molar ratio of 1:8 (UFC1:UBA5 peptide). Conversely, 100 μM ^{13}C,^{15}N-labeled UBA5 peptides were titrated with non-labeled UFC1 to a final molar ratio 1:4 (UBA5 peptide:UFC1). 2D ^{1}H-^{15}N correlation spectra ([^{15}N,^{1}H] TROSY-HSQC) were recorded at each titration point. The same types of spectra were recorded to estimate binding of ^{13}C,^{15}N-labeled UFC1 (75 μM) to non-labeled UBA5, UBA5~UFM1 and UFM1 at 1:2 molar ratios. CSP values, Δδ, were calculated for each individual amide group using the formula $\Delta\delta = [(\Delta\delta_N/5)^2 + \Delta\delta^2_{HN})]^{1/2}$.

For structural NMR spectroscopy, samples containing 1 mM ^{13}C,^{15}N-labeled UFC1 in the presence of 1 mM non-labeled $R3^{381-404W}$ and 0.3 mM ^{13}C,^{15}N-labeled $R3^{381-404W}$ in presence of 1.2 mM non-labeled UFC1 were used. As buffer condition 50 mM Tris pH = 7.5, 100 mM NaCl, 2 mM TCEP, 5 mM PIC, 5% D2O, 0.15 mM DSS was chosen. Backbone resonance assignment was performed using 3D BEST-TROSY versions [42,43] of HNCACB, HNCO, HN(CO)CACB and HN(CA)CO pulse sequences. Aliphatic ^{1}H and ^{13}C side-chain assignments resulted from (H)CC(CO)NH-TOCSY, and H(CCCO)NH-TOCSY experiments [44,45]. The assignment of aromatic side chain resonances was accomplished with amino-acid type specific versions of the (H)CB(CGCC-TOCSY)H^{ar} experiment [46] in conjunction with a [^{13}C,^{1}H]-ct-TROSY experiment [47,48] and an aromatic ^{13}C-resolved 3D NOESY-SOFAST-HMQC experiment was used for verification. To obtain distance restrains for structure calculations 3D ^{15}N- and ^{13}C- separated NOESY-HSQC spectra, recorded with a mixing time of 60 ms, were analyzed. To obtain intermolecular distance restrains, 3D F1-^{13}C/^{15}N-filtered NOESY-[$^{13}C_{ali}$,^{1}H]-HSQC, NOESY-[$^{13}C_{aro}$,^{1}H]-SOFAST-HMQC and NOESY-[^{15}N,^{1}H]-SOFAST-HMQC experiments (mixing time 150 ms) were performed [49]. The structure was calculated via CYANA [50] version 3.98 with automated peak assignment. Torsion angles were predicted based on chemical shift values by PREDITOR program [51]. Restrained energy refinement using OPALp [52] was performed for the 20 conformers with the lowest final CYANA target function.

The 20 energy-refined conformers were deposited in the Protein Data Bank with accession code 7OVC. The chemical shift assignments were deposited in the BioMagResBank (BMRB) database with accession code 34638.

Supplementary Materials: The following are available online at https://www.mdpi.com/article/10.3390/ijms22147390/s1, Figure S1: UFC1 ufmylation mediated by different UBA5 constructs, Figure S2: Interaction between UBA5 C-terminal constructs and UBA5 interacting proteins observed by ITC experiments, Figure S3: Interactions between UBA5 C-terminal constructs and UBA5-interacting proteins observed by NMR titration experiments, Figure S4: Structural features of the UFC1:UBA5 R3 complex, Figure S5: UFC1 binding to the full length UBA5 and UBA5~UFM1 complex, Table S1: NMR and refinement statistics for the UFC1:UBA5 $R3^{381-404W}$ complex.

Author Contributions: V.V.R. design and led the project. N.W., V.D. and V.V.R. wrote the paper. The ITC and NMR experiment were designed by V.V.R. and V.D. and were performed by N.W. and F.L. Constructs were designed by N.W. and V.V.R. and prepared by N.W. and N.R. Ufmylation assays were performed by N.W. with contribution of N.R. Structural calculations were performed by V.V.R. and N.W. All authors commented on final draft of the paper. All authors have read and agreed to the published version of the manuscript.

Funding: F.L., V.D. and V.V.R was supported by the DFG-funded Collaborative Research Centre on Selective Autophagy (SFB 1177 "Molecular and Functional Characterization of Selective Autophagy"), Germany. V.D. and V.V.R. also received funding from Structural Genomic Consortium. The Structural Genomics Consortium is a registered charity (no: 1097737) that receives funds from Bayer AG, Boehringer Ingelheim, Bristol Myers Squibb, Genentech, Genome Canada through Ontario Genomics Institute [OGI-196], EU/EFPIA/OICR/McGill/KTH/Diamond Innovative Medicines Initiative 2 Joint Undertaking [EUbOPEN grant 875510], Janssen, Merck KGaA (aka EMD in Canada and US), Pfizer and Takeda.

Institutional Review Board Statement: Not applicable.

Informed Consent Statement: Not applicable.

Data Availability Statement: The structure of UFC1:UBA5 R3 complex was deposited in the Protein Data Bank (https://www.rcsb.org/) (accessed on 7 July 2021) with accession code 7OVC. The chemical shift assignments were deposited in the BioMagResBank database (https://bmrb.io) (accessed on 7 July 2021) with accession code 34638.

Conflicts of Interest: The authors declare no conflict of interest.

Abbreviations

AD	UBA5 adenylation domain
ASC1	activating signal co-integrator 1
ATP	adenosine triphosphate
BEST	band-selective excitation short-transient
CSP	chemical shift perturbation
FL	full length
GABARAP	$GABA_A$-receptor-associated protein
HMQC	heteronuclear multiple quantum coherence
HSQC	heteronuclear single quantum coherence
ITC	isothermal titration calorimetry
LC3	microtubule-associated protein 1 light chain 3
LIR	LC3-interacting region
NEDD8	neural precursor cell expressed developmentally downregulated protein 8
NMR	nuclear magnetic resonance
NOESY	nuclear Overhauser and exchange spectroscopy
R1, R2, R3	UBA5 C-terminal regions R1, R2 and R3
SD	standard deviation
SOFAST	band-selective optimized-flip-angle short-transient
SUMO	small ubiquitin related modifier
TOCSY	total correlation spectroscopy
TROSY	transverse relaxation optimized spectroscopy
UBA5	UFM1-activating enzyme 5
UBL	ubiquitin-like
UFBP1	UFM1-binding protein 1

UFC1	UFM1-conjugating enzyme 1
UFIM	UFM1-interacting motive
UFL1	UFM1 ligase 1
UFM1	Ubiquitin fold modifier 1
UfSP1/2	UFM1-specific proteases 1 and 2

References

1. Komatsu, M.; Chiba, T.; Tatsumi, K.; Iemura, S.i.; Tanida, I.; Okazaki, N.; Ueno, T.; Kominami, E.; Natsume, T.; Tanaka, K. A novel protein-conjugating system for Ufm1, a ubiquitin-fold modifier. *EMBO J.* **2004**, *23*, 1977–1986. [CrossRef]
2. Sasakawa, H.; Sakata, E.; Yamaguchi, Y.; Komatsu, M.; Tatsumi, K.; Kominami, E.; Tanaka, K.; Kato, K. Solution structure and dynamics of Ufm1, a ubiquitin-fold modifier 1. *Biochem. Biophys. Res. Commun.* **2006**, *343*, 21–26. [CrossRef] [PubMed]
3. Banerjee, S.; Kumar, M.; Wiener, R. Decrypting UFMylation: How Proteins Are Modified with UFM1. *Biomolecules* **2020**, *10*, 1442. [CrossRef] [PubMed]
4. Daniel, J.; Liebau, E. The ufm1 cascade. *Cells* **2014**, *3*, 627–638. [CrossRef] [PubMed]
5. Ha, B.H.; Ahn, H.-C.; Kang, S.H.; Tanaka, K.; Chung, C.H.; Kim, E.E. Structural basis for Ufm1 processing by UfSP1. *J. Biol. Chem.* **2008**, *283*, 14893–14900. [CrossRef]
6. Ha, B.H.; Jeon, Y.J.; Shin, S.C.; Tatsumi, K.; Komatsu, M.; Tanaka, K.; Watson, C.M.; Wallis, G.; Chung, C.H.; Kim, E.E. Structure of ubiquitin-fold modifier 1-specific protease UfSP2. *J. Biol. Chem.* **2011**, *286*, 10248–10257. [CrossRef]
7. Kang, S.H.; Kim, G.R.; Seong, M.; Baek, S.H.; Seol, J.H.; Bang, O.S.; Ovaa, H.; Tatsumi, K.; Komatsu, M.; Tanaka, K. Two novel ubiquitin-fold modifier 1 (Ufm1)-specific proteases, UfSP1 and UfSP2. *J. Biol. Chem.* **2007**, *282*, 5256–5262. [CrossRef]
8. Bacik, J.-P.; Walker, J.R.; Ali, M.; Schimmer, A.D.; Dhe-Paganon, S. Crystal Structure of the Human Ubiquitin-activating Enzyme 5 (UBA5) Bound to ATP Mechanistic Insights into a Minimalistic E1 Enzyme. *J. Biol. Chem.* **2010**, *285*, 20273–20280. [CrossRef]
9. Gavin, J.M.; Hoar, K.; Xu, Q.; Ma, J.; Lin, Y.; Chen, J.; Chen, W.; Bruzzese, F.J.; Harrison, S.; Mallender, W.D. Mechanistic study of Uba5 enzyme and the Ufm1 conjugation pathway. *J. Biol. Chem.* **2014**, *289*, 22648–22658. [CrossRef] [PubMed]
10. Padala, P.; Oweis, W.; Mashahreh, B.; Soudah, N.; Cohen-Kfir, E.; Todd, E.A.; Berndsen, C.E.; Wiener, R. Novel insights into the interaction of UBA5 with UFM1 via a UFM1-interacting sequence. *Sci. Rep.* **2017**, *7*, 1–12. [CrossRef]
11. Mizushima, T.; Tatsumi, K.; Ozaki, Y.; Kawakami, T.; Suzuki, A.; Ogasahara, K.; Komatsu, M.; Kominami, E.; Tanaka, K.; Yamane, T. Crystal structure of Ufc1, the Ufm1-conjugating enzyme. *Biochem. Biophys. Res. Commun.* **2007**, *362*, 1079–1084. [CrossRef]
12. Tatsumi, K.; Sou, Y.-S.; Tada, N.; Nakamura, E.; Iemura, S.-I.; Natsume, T.; Kang, S.H.; Chung, C.H.; Kasahara, M.; Kominami, E. A novel type of E3 ligase for the Ufm1 conjugation system. *J. Biol. Chem.* **2010**, *285*, 5417–5427. [CrossRef]
13. Liang, J.R.; Lingeman, E.; Luong, T.; Ahmed, S.; Muhar, M.; Nguyen, T.; Olzmann, J.A.; Corn, J.E. A Genome-wide ER-phagy Screen Highlights Key Roles of Mitochondrial Metabolism and ER-Resident UFMylation. *Cell* **2020**, *180*, 1160–1177.e1120. [CrossRef]
14. Walczak, C.P.; Leto, D.E.; Zhang, L.; Riepe, C.; Muller, R.Y.; DaRosa, P.A.; Ingolia, N.T.; Elias, J.E.; Kopito, R.R. Ribosomal protein RPL26 is the principal target of UFMylation. *Proc. Natl. Acad. Sci. USA* **2019**, *116*, 1299–1308. [CrossRef]
15. Yang, R.; Wang, H.; Kang, B.; Chen, B.; Shi, Y.; Yang, S.; Sun, L.; Liu, Y.; Xiao, W.; Zhang, T. CDK5RAP3, a UFL1 substrate adaptor, is crucial for liver development. *Development* **2019**, *146*, dev169235. [CrossRef]
16. Yoo, H.M.; Kang, S.H.; Kim, J.Y.; Lee, J.E.; Seong, M.W.; Lee, S.W.; Ka, S.H.; Sou, Y.-S.; Komatsu, M.; Tanaka, K. Modification of ASC1 by UFM1 is crucial for ERα transactivation and breast cancer development. *Mol. Cell* **2014**, *56*, 261–274. [CrossRef] [PubMed]
17. Liu, J.; Guan, D.; Dong, M.; Yang, J.; Wei, H.; Liang, Q.; Song, L.; Xu, L.; Bai, J.; Liu, C. UFMylation maintains tumour suppressor p53 stability by antagonizing its ubiquitination. *Nat. Cell Biol.* **2020**, *22*, 1056–1063. [CrossRef] [PubMed]
18. Qin, B.; Yu, J.; Nowsheen, S.; Wang, M.; Tu, X.; Liu, T.; Li, H.; Wang, L.; Lou, Z. UFL1 promotes histone H4 ufmylation and ATM activation. *Nat. Commun.* **2019**, *10*, 1–13. [CrossRef] [PubMed]
19. Wang, Z.; Gong, Y.; Peng, B.; Shi, R.; Fan, D.; Zhao, H.; Zhu, M.; Zhang, H.; Lou, Z.; Zhou, J. MRE11 UFMylation promotes ATM activation. *Nucleic Acids Res.* **2019**, *47*, 4124–4135. [CrossRef] [PubMed]
20. Gak, I.A.; Vasiljevic, D.; Zerjatke, T.; Yu, L.; Brosch, M.; Roumeliotis, T.I.; Horenburg, C.; Klemm, N.; Bakos, G.; Herrmann, A. UFMylation regulates translational homeostasis and cell cycle progression. *bioRxiv* **2020**. [CrossRef]
21. Cai, Y.; Pi, W.; Sivaprakasam, S.; Zhu, X.; Zhang, M.; Chen, J.; Makala, L.; Lu, C.; Wu, J.; Teng, Y. UFBP1, a key component of the Ufm1 conjugation system, is essential for ufmylation-mediated regulation of erythroid development. *PLoS Genet.* **2015**, *11*, e1005643. [CrossRef] [PubMed]
22. Hu, X.; Zhang, H.; Song, Y.; Zhuang, L.; Yang, Q.; Pan, M.; Chen, F. Ubiquitin fold modifier 1 activates NF-κB pathway by down-regulating LZAP expression in the macrophage of diabetic mouse model. *Biosci. Rep.* **2020**, *40*, BSR20191672. [CrossRef] [PubMed]
23. Lemaire, K.; Moura, R.F.; Granvik, M.; Igoillo-Esteve, M.; Hohmeier, H.E.; Hendrickx, N.; Newgard, C.B.; Waelkens, E.; Cnop, M.; Schuit, F. Ubiquitin fold modifier 1 (UFM1) and its target UFBP1 protect pancreatic beta cells from ER stress-induced apoptosis. *PLoS ONE* **2011**, *6*, e18517. [CrossRef]

24. Roberts, A.M.; Miyamoto, D.K.; Huffman, T.R.; Bateman, L.A.; Ives, A.N.; Akopian, D.; Heslin, M.J.; Contreras, C.M.; Rape, M.; Skibola, C.F. Chemoproteomic screening of covalent ligands reveals UBA5 as a novel pancreatic cancer target. *ACS Chem. Biol.* **2017**, *12*, 899–904. [CrossRef] [PubMed]
25. Muona, M.; Ishimura, R.; Laari, A.; Ichimura, Y.; Linnankivi, T.; Keski-Filppula, R.; Herva, R.; Rantala, H.; Paetau, A.; Pöyhönen, M. Biallelic variants in UBA5 link dysfunctional UFM1 ubiquitin-like modifier pathway to severe infantile-onset encephalopathy. *Am. J. Hum. Genet.* **2016**, *99*, 683–694. [CrossRef]
26. Nahorski, M.S.; Maddirevula, S.; Ishimura, R.; Alsahli, S.; Brady, A.F.; Begemann, A.; Mizushima, T.; Guzmán-Vega, F.J.; Obata, M.; Ichimura, Y. Biallelic UFM1 and UFC1 mutations expand the essential role of ufmylation in brain development. *Brain* **2018**, *141*, 1934–1945. [CrossRef]
27. Tatsumi, K.; Yamamoto-Mukai, H.; Shimizu, R.; Waguri, S.; Sou, Y.-S.; Sakamoto, A.; Taya, C.; Shitara, H.; Hara, T.; Chung, C.H. The Ufm1-activating enzyme Uba5 is indispensable for erythroid differentiation in mice. *Nat. Commun.* **2011**, *2*, 1–7. [CrossRef]
28. Cappadocia, L.; Lima, C.D. Ubiquitin-like protein conjugation: Structures, chemistry, and mechanism. *Chem. Rev.* **2018**, *118*, 889–918. [CrossRef] [PubMed]
29. Oweis, W.; Padala, P.; Hassouna, F.; Cohen-Kfir, E.; Gibbs, D.R.; Todd, E.A.; Berndsen, C.E.; Wiener, R. Trans-binding mechanism of ubiquitin-like protein activation revealed by a UBA5-UFM1 complex. *Cell Rep.* **2016**, *16*, 3113–3120. [CrossRef]
30. Soudah, N.; Padala, P.; Hassouna, F.; Kumar, M.; Mashahreh, B.; Lebedev, A.A.; Isupov, M.N.; Cohen-Kfir, E.; Wiener, R. An N-terminal extension to Uba5 adenylation domain boosts Ufm1 activation: Isoform-specific differences in ubiquitin-like protein activation. *J. Mol. Biol.* **2019**, *431*, 463–478. [CrossRef] [PubMed]
31. Habisov, S.; Huber, J.; Ichimura, Y.; Akutsu, M.; Rogova, N.; Loehr, F.; McEwan, D.G.; Johansen, T.; Dikic, I.; Doetsch, V. Structural and functional analysis of a novel interaction motif within UFM1-activating enzyme 5 (UBA5) required for binding to ubiquitin-like proteins and ufmylation. *J. Biol. Chem.* **2016**, *291*, 9025–9041. [CrossRef]
32. Xie, S. Characterization, crystallization and preliminary X-ray crystallographic analysis of the human Uba5 C-terminus–Ufc1 complex. *Acta Crystallogr. Sect. F Struct. Biol. Commun.* **2014**, *70*, 1093–1097. [CrossRef]
33. Huber, J.; Obata, M.; Gruber, J.; Akutsu, M.; Löhr, F.; Rogova, N.; Güntert, P.; Dikic, I.; Kirkin, V.; Komatsu, M. An atypical LIR motif within UBA5 (ubiquitin like modifier activating enzyme 5) interacts with GABARAP proteins and mediates membrane localization of UBA5. *Autophagy* **2020**, *16*, 256–270. [CrossRef] [PubMed]
34. Colin, E.; Daniel, J.; Ziegler, A.; Wakim, J.; Scrivo, A.; Haack, T.B.; Khiati, S.; Denommé, A.-S.; Amati-Bonneau, P.; Charif, M. Biallelic variants in UBA5 reveal that disruption of the UFM1 cascade can result in early-onset encephalopathy. *Am. J. Hum. Genet.* **2016**, *99*, 695–703. [CrossRef]
35. Liu, G.; Forouhar, F.; Eletsky, A.; Atreya, H.S.; Aramini, J.M.; Xiao, R.; Huang, Y.J.; Abashidze, M.; Seetharaman, J.; Liu, J. NMR and X-ray structures of human E2-like ubiquitin-fold modifier conjugating enzyme 1 (UFC1) reveal structural and functional conservation in the metazoan UFM1-UBA5-UFC1 ubiquination pathway. *J. Struct. Funct. Genom.* **2009**, *10*, 127–136. [CrossRef] [PubMed]
36. Rogov, V.V.; Rozenknop, A.; Rogova, N.Y.; Löhr, F.; Tikole, S.; Jaravine, V.; Güntert, P.; Dikic, I.; Dötsch, V. A universal expression tag for structural and functional studies of proteins. *ChemBioChem* **2012**, *13*, 959–963. [CrossRef]
37. von Delbrück, M.; Kniss, A.; Rogov, V.V.; Pluska, L.; Bagola, K.; Löhr, F.; Güntert, P.; Sommer, T.; Dötsch, V. The CUE domain of Cue1 aligns growing ubiquitin chains with Ubc7 for rapid elongation. *Mol. Cell* **2016**, *62*, 918–928. [CrossRef] [PubMed]
38. Buetow, L.; Gabrielsen, M.; Anthony, N.G.; Dou, H.; Patel, A.; Aitkenhead, H.; Sibbet, G.J.; Smith, B.O.; Huang, D.T. Activation of a primed RING E3-E2–ubiquitin complex by non-covalent ubiquitin. *Mol. Cell* **2015**, *58*, 297–310. [CrossRef]
39. Dou, H.; Buetow, L.; Sibbet, G.J.; Cameron, K.; Huang, D.T. Essentiality of a non-RING element in priming donor ubiquitin for catalysis by a monomeric E3. *Nat. Struct. Mol. Biol.* **2013**, *20*, 982–986. [CrossRef]
40. Kumar, P.; Magala, P.; Geiger-Schuller, K.R.; Majumdar, A.; Tolman, J.R.; Wolberger, C. Role of a non-canonical surface of Rad6 in ubiquitin conjugating activity. *Nucleic Acids Res.* **2015**, *43*, 9039–9050. [CrossRef]
41. Eck, F.; Phuyal, S.; Smith, M.D.; Kaulich, M.; Wilkinson, S.; Farhan, H.; Behrends, C. ACSL3 is a novel GABARAPL2 interactor that links ufmylation and lipid droplet biogenesis. *J. Cell Sci.* **2020**, *133*, jcs243477. [CrossRef] [PubMed]
42. Farjon, J.; Boisbouvier, J.; Schanda, P.; Pardi, A.; Simorre, J.-P.; Brutscher, B. Longitudinal-relaxation-enhanced NMR experiments for the study of nucleic acids in solution. *J. Am. Chem. Soc.* **2009**, *131*, 8571–8577. [CrossRef] [PubMed]
43. Solyom, Z.; Schwarten, M.; Geist, L.; Konrat, R.; Willbold, D.; Brutscher, B. BEST-TROSY experiments for time-efficient sequential resonance assignment of large disordered proteins. *J. Biomol. NMR* **2013**, *55*, 311–321. [CrossRef]
44. Logan, T.M.; Olejniczak, E.T.; Xu, R.X.; Fesik, S.W. A general method for assigning NMR spectra of denatured proteins using 3D HC (CO) NH-TOCSY triple resonance experiments. *J. Biomol. NMR* **1993**, *3*, 225–231. [CrossRef] [PubMed]
45. Montelione, G.T.; Lyons, B.A.; Emerson, S.D.; Tashiro, M. An efficient triple resonance experiment using carbon-13 isotropic mixing for determining sequence-specific resonance assignments of isotopically-enriched proteins. *J. Am. Chem. Soc.* **1992**, *114*, 10974–10975. [CrossRef]
46. Löhr, F.; Hänsel, R.; Rogov, V.V.; Dötsch, V. Improved pulse sequences for sequence specific assignment of aromatic proton resonances in proteins. *J. Biomol. NMR* **2007**, *37*, 205–224. [CrossRef]
47. Brutscher, B.; Boisbouvier, J.; Pardi, A.; Marion, D.; Simorre, J.-P. Improved sensitivity and resolution in 1H– 13C NMR experiments of RNA. *J. Am. Chem. Soc.* **1998**, *120*, 11845–11851. [CrossRef]

48. Pervushin, K.; Riek, R.; Wider, G.; Wüthrich, K. Transverse relaxation-optimized spectroscopy (TROSY) for NMR studies of aromatic spin systems in 13C-labeled proteins. *J. Am. Chem. Soc.* **1998**, *120*, 6394–6400. [CrossRef]
49. Zwahlen, C.; Legault, P.; Vincent, S.J.; Greenblatt, J.; Konrat, R.; Kay, L.E. Methods for measurement of intermolecular NOEs by multinuclear NMR spectroscopy: Application to a bacteriophage λ N-peptide/boxB RNA complex. *J. Am. Chem. Soc.* **1997**, *119*, 6711–6721. [CrossRef]
50. Güntert, P. Automated structure determination from NMR spectra. *Eur. Biophys. J.* **2009**, *38*, 129–143. [CrossRef] [PubMed]
51. Berjanskii, M.V.; Neal, S.; Wishart, D.S. PREDITOR: A web server for predicting protein torsion angle restraints. *Nucleic Acids Res.* **2006**, *34*, W63–W69. [CrossRef] [PubMed]
52. Koradi, R.; Billeter, M.; Güntert, P. Point-centered domain decomposition for parallel molecular dynamics simulation. *Comput. Phys. Commun.* **2000**, *124*, 139–147. [CrossRef]

International Journal of *Molecular Sciences*

MDPI

Article

Pressure and Chemical Unfolding of an α-Helical Bundle Protein: The GH2 Domain of the Protein Adaptor GIPC1

Cécile Dubois [1], Vicente J. Planelles-Herrero [2], Camille Tillatte-Tripodi [1], Stéphane Delbecq [1], Léa Mammri [1], Elena M. Sirkia [2], Virginie Ropars [2], Christian Roumestand [1,*] and Philippe Barthe [1]

1 Centre de Biologie Structurale INSERM U1054, CNRS UMR 5048, Université de Montpellier, 34090 Montpellier, France; cecile.dubois@cbs.cnrs.fr (C.D.); camille.tillatte@gmail.com (C.T.-T.); stephane.delbecq@umontpellier.fr (S.D.); lea.mammri@gmail.com (L.M.); philippe.barthe@cbs.cnrs.fr (P.B.)

2 Structural Motility, Institut Curie, Paris Université Sciences et Lettres, Sorbonne Université, CNRS UMR144, 75248 Paris, France; vicente@mrc-lmb.cam.ac.uk (V.J.P.-H.); maria-elena.sirkia@curie.fr (E.M.S.); Virginie.ROPARS@cea.fr (V.R.)

* Correspondence: christian.roumestand@cbs.cnrs.fr

Abstract: When combined with NMR spectroscopy, high hydrostatic pressure is an alternative perturbation method used to destabilize globular proteins that has proven to be particularly well suited for exploring the unfolding energy landscape of small single-domain proteins. To date, investigations of the unfolding landscape of all-β or mixed-α/β protein scaffolds are well documented, whereas such data are lacking for all-α protein domains. Here we report the NMR study of the unfolding pathways of GIPC1-GH2, a small α-helical bundle domain made of four antiparallel α-helices. High-pressure perturbation was combined with NMR spectroscopy to unravel the unfolding landscape at three different temperatures. The results were compared to those obtained from classical chemical denaturation. Whatever the perturbation used, the loss of secondary and tertiary contacts within the protein scaffold is almost simultaneous. The unfolding transition appeared very cooperative when using high pressure at high temperature, as was the case for chemical denaturation, whereas it was found more progressive at low temperature, suggesting the existence of a complex folding pathway.

Keywords: protein folding; NMR; high hydrostatic pressure; thermodynamic stability; α-helical bundle

Citation: Dubois, C.; Planelles-Herrero, V.J.; Tillatte-Tripodi, C.; Delbecq, S.; Mammri, L.; Sirkia, E.M.; Ropars, V.; Roumestand, C.; Barthe, P. Pressure and Chemical Unfolding of an α-Helical Bundle Protein: The GH2 Domain of the Protein Adaptor GIPC1. *Int. J. Mol. Sci.* **2021**, *22*, 3597. https://doi.org/10.3390/ijms22073597

Academic Editor: Istvan Simon

Received: 11 March 2021
Accepted: 26 March 2021
Published: 30 March 2021

1. Introduction

Although small single-domain proteins are generally found to exhibit highly cooperative two-state unfolding transitions [1,2], the thousands of interactions that stabilize their 3D structure are unlikely to form simultaneously and folding intermediates should exist along their folding pathway. Nevertheless, especially in the case of small, fast-folding single-domain proteins, folding intermediates are generally low populated at equilibrium, and cannot be easily identified when using classical thermal or chemical perturbation in association with methods applied to a single probe in the 3D structure of the protein (for instance, intrinsic fluorescence of a tryptophan residue) or with methods giving global structural information (for instance, molar ellipticity in the case of circular dichroism (CD) study).

Multidimensional NMR spectroscopy is a particularly powerful tool to obtain high-resolution structural information about protein folding events because an abundance of site-specific probes can be studied simultaneously in a single spectrum. In the recent past, NMR combined with high-pressure perturbation has emerged as a powerful tool to explore in detail the folding landscape of small proteins, at a quasi-atomic resolution [3–6]. Indeed, contrary to chemical or thermal denaturation, which acts globally and depends on exposed surface area in the unfolded state, pressure denaturation depends on the elimination of the solvent-excluded internal voids, due to imperfect protein packing, by water penetration inside the core of the protein [7–9]. Thus, because the distribution of solvent-excluded voids depends on the protein structure, the pressure-induced unfolding originates from unique properties of the folded state.

High-pressure NMR unfolding studies have been applied to several single-domain proteins in order to characterize their folding landscape. Until now, these studies have concerned essentially all-β [10,11] or mixed-α/β [9,12,13] protein scaffolds, and similar studies are lacking for all-α structures although they represent a widespread assembly motif [14]. In the present manuscript, we report the NMR study of the folding of an α-helical bundle, the GH2 domain of the protein adaptor GIPC1 [15]. GIPC is an adaptor protein that binds and regulates vesicular trafficking of many transmembrane proteins [15]. The X-ray structure of GIPC1 has been solved [16] and shows that the full-length protein exists as a dimer in the crystal (Supplementary Materials, Figure S1). Each monomer contains three well-identified domains: a central PDZ domain flanked by an N-terminal GIPC-homology 1 (GH1) domain and a C-terminal GH2 domain [16]. The structure of the PDZ domain displays a typical PDZ fold with five β-strands and two α-helices. The GH1 domain adopts a ubiquitin-like fold composed of four β-strands and one α-helix, and the GH2 domain forms a four-helix globular fold. After solving the solution structure of the GIPC1-GH2 domain, we studied its folding/unfolding pathway at a residue-specific level using 2D NMR spectroscopy combined with high-pressure perturbation at three different temperatures, and with chemical perturbation. As a result, we found that whereas high-pressure NMR reveals the existence of a partial unfolding at low temperature, unfolding becomes highly cooperative at higher temperature or when using chemical perturbation.

2. Results

2.1. NMR Resonance Assignments and Solution Structure of GIPC1-GH2

Proton, nitrogen, and carbon NMR resonances of GIPC1-GH2, renumbered 1–79 for simplicity, have been assigned and its solution structure solved using essentially [^{1}H,^{15}N,^{13}C] triple-resonance and [^{1}H,^{15}N] double-resonance 3D NMR spectroscopy (see Section 4) with the classical sequential assignment strategy. ^{1}H and ^{15}N resonances have been assigned for all amide groups of nonproline (76) residues (Figure 1), and Cα, Cβ, C′ resonances for 96.2% residues. Resonance assignments have been deposited at the BMRB data bank (BMRB code 34609).

NOEs were measured on a 3D [^{1}H,^{15}N] NOESY-HSQC experiment and on a 2D [^{1}H,^{1}H] NOESY spectrum recorded in a deuterated buffer. Dihedral restraints (φ, ϕ, and χ_1) were obtained from TALOS-N [17] analysis of backbone atom chemical shifts. After conversion into distance restraints, these data sets were used with CYANA [18] to build the 3D structure of GIPC1-GH2. H-bond restraints were also used for the structure modeling. Usually, these restraints are deduced from hydrogen/deuteron (H/D) exchange experiments, yielding amide protons potentially involved in H-bonds as donor atoms. In the case of GIPC1-GH2, H/D exchange rates were unusually fast, even at low temperature (5 °C): all amide proton resonances disappeared during the few minutes needed for the setting of the experiments, preventing the use of this approach. Instead, we used CLEANEX-PM [19,20] experiments to determine which amide protons are solvent-exposed in the 3D structure. In these experiments, the water resonance was selectively excited, and water magnetization transferred to solvent-exposed amide protons with an appropriate spin-locking module. This sequence was incorporated in a conventional HSQC scheme to resolve amide peaks along the ^{15}N indirect dimension. Thus, amide corresponding cross-peaks which were not present in this experiment (Supplementary Materials, Figure S2) and which belonged to residues exhibiting φ, ψ values characteristic of secondary structure elements (α-helices, in the present case), as determined from TALOS analysis, were considered as involved in a regular H-bond, and the corresponding distance restraints were used for structure modeling.

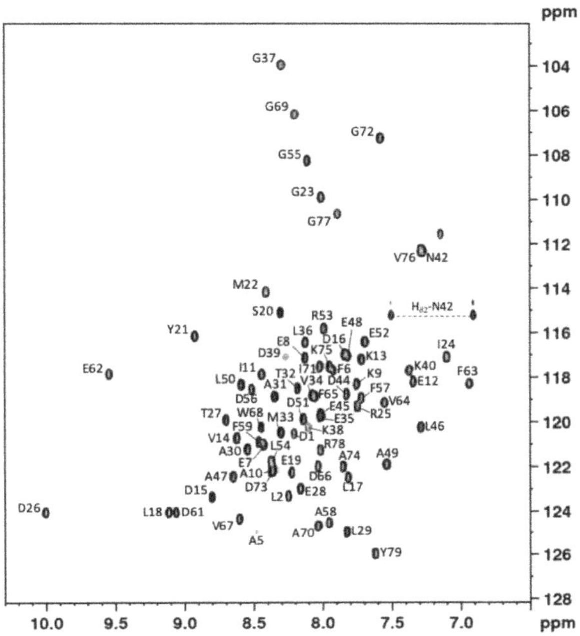

Figure 1. GIPC1-GH2 NMR fingerprint. [^{1}H-^{15}N] HSQC spectrum of GIPC1-GH2 at 800 MHz, 20 °C on a 0.5 mM, ^{15}N uniformly labeled sample dissolved in a 20 mM Tris-HCl pH 7.2, 150 mM NaCl buffer. Cross-peak assignments are indicated using the one-letter amino acid and number code.

A final set of 1219 restraints was used, and the pairwise rmsd calculated for backbone heavy atoms between the 20 best refined structures was 0.47 Å (residues 5–76) (Figure 2). Most of the residues (95.6%) fall in the most favored region of the Ramachandran plot, with no residue in the generously allowed or disallowed regions, highlighting the high quality of our model (see structural statistics, Supplementary Materials, Table S1). The solution structure of GIPC1-GH2 consists of four antiparallel amphipathic helices arranged in an α-helical bundle. It is virtually identical to the X-ray structure adopted by this domain in the full-length protein, as shown by their superimposition displayed in Figure 2. An rmsd value of 0.85 Å was measured between them for all backbone heavy atoms (residues 5–73). This value drops to 0.68 Å when considering only backbone heavy atoms involved in helices. The bundle can be divided in two subdomains, each one consisting of two antiparallel helices stapled against each other: the N- (helix I) and C-terminal (helix IV) helices form the first subdomain, while the α-hairpin made by helix II and III forms the second subdomain. These two subdomains make an angle of approximately 50° in the 3D structure of GIPC1-GH2. The structure coordinates of the NMR structure of GIPC1-GH2 have been deposited at the Protein Data Bank (PDB code 7NRN).

The intrinsic dynamics of GIPC1-GH2 were also investigated by measuring heteronuclear ^{15}N T_1, T_2 relaxation times and [^{1}H,^{15}N] heteronuclear nOes, and converting these parameters into $J(0)$, $J(\omega_N)$, and $<J((\omega_H)>$ spectral densities through Solomon equations [21] (see Section 4 and Supplementary Materials, Figure S3). Spectral densities were then fitted with model-free Lipari–Szabo equations [22] to extract the global (τ_c) and internal (τ_e) correlation times of the molecule, and generalized order parameters S^2 for each residue (Figure 3). A τ_c value of 6.09 ns was obtained from the fit of the relaxation parameters, in good agreement with the expected correlation time of this small protein at 20 °C.

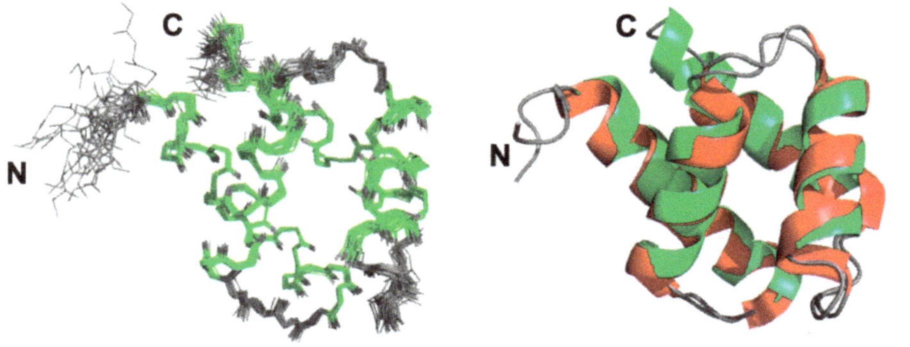

Figure 2. Solution structure of GIPC1-GH2. (**Left**) Overlay of the 20 best NMR structures (backbone atoms only) with lowest energy. The regular α-helices are colored in green. (**Right**) Superimposition of ribbon representations of the solution structure (with α-helices in green and the X-ray structure with α-helices in red) of the GH2 domain extracted from the structure of the full-length protein (PDB code: 5V6b).

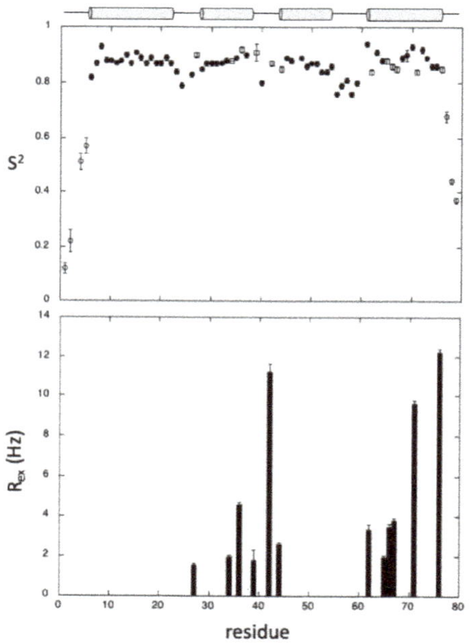

Figure 3. Intrinsic dynamics of GIPC1-GH2. (**Top**) Generalized order parameters S^2 obtained from Lipari–Szabo analysis plotted versus the protein sequence. The regular two-parameter spectral density function (Section 4, Equation (1)) has been used for residues plotted as filled circles, whereas the extended Lipari–Szabo formalism (Section 4, Equation (3)) has been used for residues plotted as open circles. Open squares correspond to residues for which $J(0)$ values have been corrected from exchange contributions (Section 4, Equation (2)). (**Bottom**) R_{ex} contributions obtained from this last equation are reported versus the sequence for residues exhibiting conformational exchange. The location of the four helices in the protein sequence is schematized with cylinders on top of the figure.

The regular Lipari–Szabo model was used for most of the residues involved in α-helices, yielding S^2 values close to 0.85, confirming that these secondary structure elements are well defined. The extended Lipari–Szabo model [23] was needed to fit the N- and

C-terminal residues, suggesting the existence of more complex motions in these flexible regions, as usually observed. Interestingly, adding exchange contribution (R_{ex}) to $J(0)$ was mandatory to fit some residues located in the loops connecting the four helices, but also in helix IV and, to a lesser extent, helix II, suggesting that these two helices are prone to conformational exchange.

2.2. GIPC1-GH2 Denaturation Studies

2.2.1. Pressure Denaturation

2D [^{1}H,^{15}N] HSQC spectra of ^{15}N uniformly labeled GIPC1-GH2 were recorded at variable pressures (1 to 2500 bar) and at 10, 20, and 30 °C (Figure 4). As usually found, the intensity of each native state peak decreases as a function of pressure, while the intensity of peaks corresponding to the unfolded state, centered around 8.5 ppm in the proton dimension, increases concomitantly. This supports a slow equilibrium on the NMR timescale for each residue between the native and unfolded state, and a two-state transition for each residue between their native/unfolded states during the unfolding process. Thus, even though the global protein unfolding does not likely conform to a two-state transition, locally this simple model can be used to interpret the loss of intensity for each native state cross-peak [6].

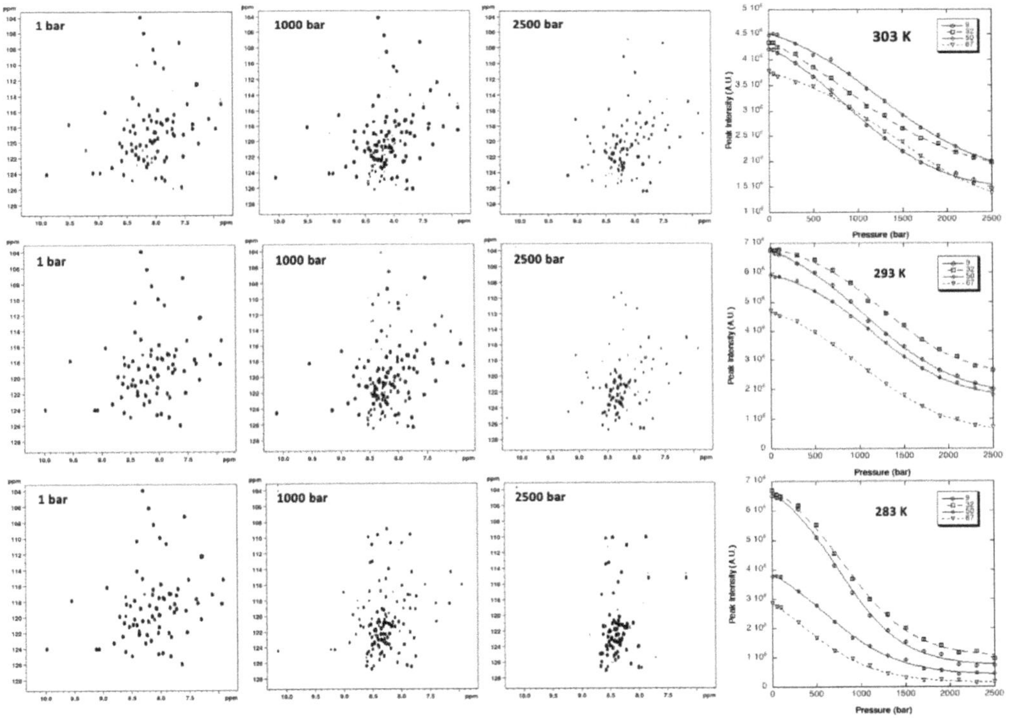

Figure 4. NMR detected high-pressure unfolding of GIPC1-GH2 at 30 °C, 20 °C, and 10 °C (from top to bottom). At each temperature, examples of [^{1}H,^{15}N] HSQC at 1, 1000, and 2500 bar are displayed from left to right. The rightmost panels report overlays of four (residues K9, T32, L50, and V67) residue-specific experimental denaturation curves obtained from the fits of the pressure-dependent sigmoidal decrease of the corresponding residue cross-peak intensities in the HSQC spectra with Equation (4).

A total of 44 residues (58% of the nonproline residues) gave overlapping cross-peaks neither in the folded state nor in between the folded and unfolded states at any of the temperatures used for the study. These residues displayed cross-peaks of sufficient intensity at atmospheric pressure to be accurately fitted to the two-state pressure-induced unfolding model described in the Section 4 (Equation (4)), Figure 4, giving a substantial number of local probes for the description of the GIPC1-GH2 folding pathway. At the residue level, the two-state model was adequate to fit all individual unfolding curves, but yielded significantly different values for apparent free energy ΔG_u^0 and apparent volume change ΔV_u^0 (Figure 5) of unfolding, suggesting a substantial deviation from a two-state behavior for the global unfolding of the protein, whatever the temperature of the study. The asymmetric distributions observed for apparent ΔG_u^0 and apparent ΔV_u^0 strongly support this assumption and suggest that partial unfolding of the molecule should appear when increasing the pressure (Supplementary Materials, Figure S4).

GIPC1-GH2 displayed a weak stability that appeared to be maximum at 20 °C, with an average value for the apparent free energy of unfolding $<\Delta G_u^0>$ of 1293 ± 62 cal/mol, and significantly decreased at higher ($<\Delta G_u^0>$ = 925 ± 57 cal/mol at 30 °C) or lower (($<\Delta G_u^0>$ = 903 ± 49 cal/mol at 10 °C) temperatures. Also, a linear decrease with temperature was observed for the average values (in magnitude mode) of apparent ΔV_u^0 ($<\Delta V_u^0>$ = −61 ± 4 mL/mol, −49 ± 7 mL/mol, and −38 ± 8 mL/mol at 10, 20, and 30 °C, respectively). The temperature-dependent decrease in ΔV_u^0 is a well-known effect due to the difference in thermal expansion between the folded and unfolded states [24].

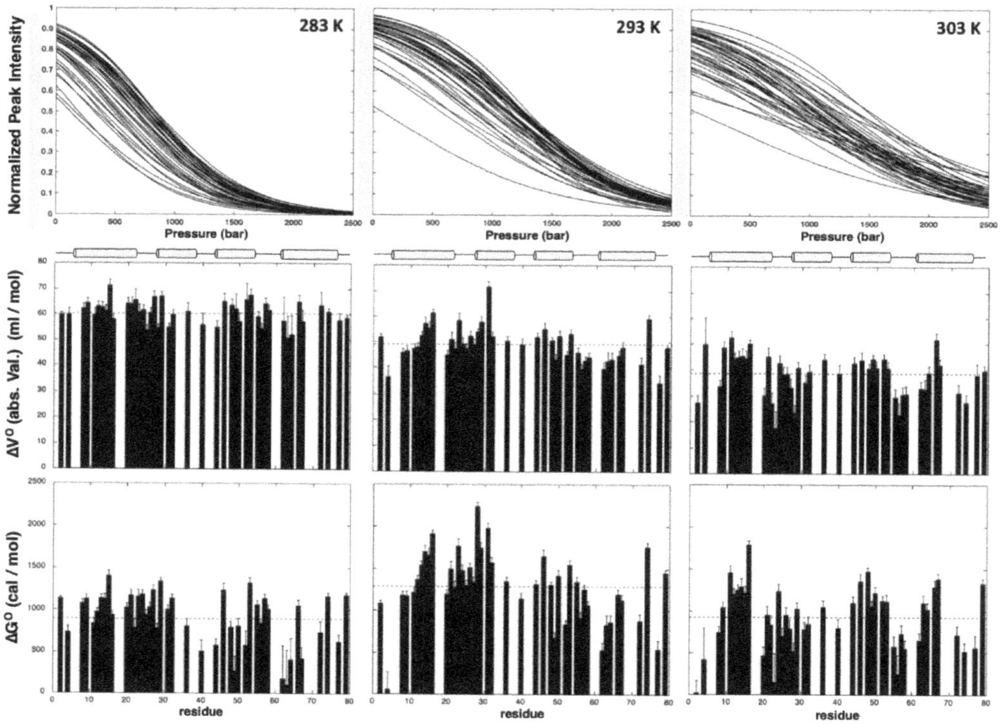

Figure 5. Steady-state thermodynamic parameters measured for GIPC1-GH2 at 10 °C, 20 °C, and 30 °C (from left to right) from residue-specific pressure denaturation curves. From top to bottom: overlay of the normalized residue-specific denaturation curves as obtained from the fit of the pressure-dependent sigmoidal decrease of the residue cross-peak intensities in the HSQC spectra with Equation (4); residue-specific values (absolute values) of the apparent volume change of unfolding ΔV_u^0 plotted versus the protein sequence; residue-specific values of the apparent free energy of unfolding ΔG_u^0 plotted versus the protein sequence. The dashed lines represent the mean values of the measured thermodynamic parameters. The location of the four helices in the protein sequence is schematized with cylinders on top of the graphics.

Average values of ΔG_u^0 calculated for each helix indicate differences in local stability (Supplementary Materials, Figure S5). Interestingly, this local stability depends on the temperature. At 10 °C and 20 °C, helix I and II appear slightly more stable than helix III and IV, helix II being the most stable at 20 °C whereas it has a similar stability as helix I at 10 °C. At 30 °C, the highest values of ΔG_u^0 are found in helix I and III. Importantly, the local thermal stability measured for each helix follows the thermal stability of the whole domain, with a maximum observed at 20 °C. Likewise, a similar decrease with temperature is observed for average ΔV_u^0 values calculated for each helix and for the average value calculated over the whole structure, but without any significant variations between the four helices.

Information brought by normalized residue-specific denaturation curves has been used to track and to characterize possible intermediates in the folding pathway of GIPC1-GH2 [6,8]. Thus, at a given pressure, the value of 1 measured for a given cross-peak ($I = I_f = 1$; Equation (4)) is associated with a probability P_i of 1 (100%) to find the corresponding residue "i" in the native state, while a residue "j" for which the corresponding cross-peak has disappeared ($I = I_U = 0$; Equation (4)) from the HSQC spectrum has a probability P_j equal to zero to be in a native state. Since these probabilities are related to the "native fraction" for a given residue, they are called fractional probabilities.

Given a pressure where these two residues i and j are in an intermediate situation ($0 < P_i$ and $P_j < 1$), and if these two residues are in contact in the native state (at atmospheric pressure), their fractional probability P_{ij} to be in contact at this pressure is given by the geometric mean of the two individual probabilities: $P_{ij} = \sqrt{P_i \times P_j}$ [12] (Figure 6).

At 20 °C, the temperature where GIPC1-GH2 exhibits the highest stability, the pressure dependence of the contact maps shows that helix III and IV are the first regions affected by an increase of pressure: tertiary contacts between these two helices are already significantly weakened at 500 bar ($P_{ij} \leq 50\%$), as well as secondary contacts characteristic of the helical structure (Figure 6). This partial unfolding concerns mainly these two helices up to 900 bar. Above this pressure, we observed a loss of contacts between helix IV and helix I, while helix I and II remain unaffected until 1100 bar. The unfolding of these two helices, as well as the loss of tertiary contacts between them, is observed at higher pressure (1300 bar, not shown). An identical scenario is observed at 10 °C, but with a shift to lower pressures, consistent with the lower stability of the protein at this temperature. At 700 bar, helix III and IV are unfolded ($P_{ij} \leq 50\%$), and local unfolding already concerns helix I and II, while all the contacts are lost at 900 bar. While a similar stability of the protein is observed at 30 °C, a rather different scenario is observed for unfolding. The structure remains stable until 900 bar, with little loss of tertiary contacts observed. At 1100 bar, a sharp unfolding transition is observed, that concerns both the tertiary contacts between the four helices and the secondary contacts in all the helices, simultaneously. Finally, we observed a global unfolding of the molecule at 1300 bar.

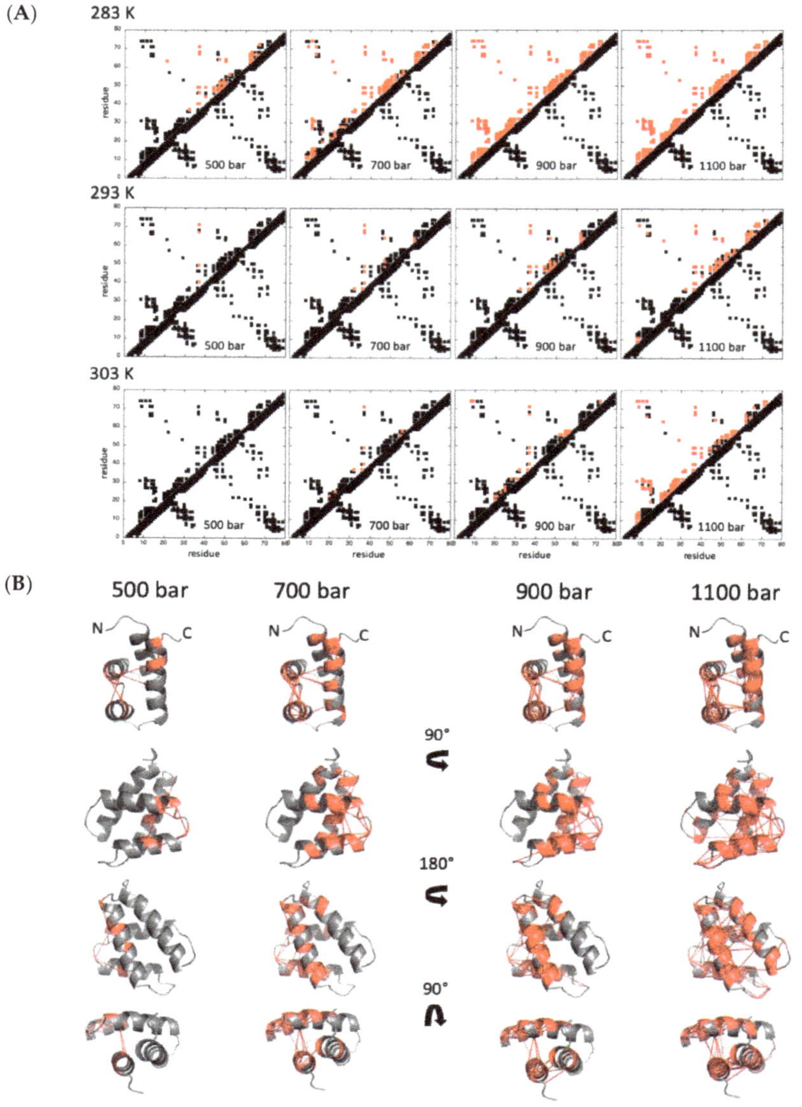

Figure 6. Pressure denaturation of GIPC1-GH2. (**A**) Contact maps built from the best solution structure obtained for GIPC1-GH2 at 500, 700, 900, and 1100 bar, and at 283, 293, and 303 K, as indicated. Contacts below the diagonal have been calculated with CMview (http://www.bioinformatics.org/cmview/; accessed on 25 March 2020): they correspond to residue where the distance to the corresponding Cα is lower than 9 Å. Above the diagonal, only the contacts for which fractional probability can be obtained have been reported. In addition, contacts have been colored in red when contact probabilities P_{ij} lower than 0.5 are observed. (**B**) Visualization of the probabilities of contact on ribbon representations of GIPC1-GH2 at 20 °C and at 500, 700, 900, and 1100 bar, as indicated. The red lines represent contacts that are significantly weakened ($P_{ij} \leq 0.5$) at the indicated pressure. Residues involved in these contacts are also colored in red. The arrows in the middle of the panel indicate the rotation between the different views.

2.2.2. Chemical Denaturation

2D [^{1}H,^{15}N] HSQC spectra of ^{15}N uniformly labeled GIPC1-GH2 were recorded at 20 °C, the temperature where the protein exhibits the highest stability, and at increasing urea concentrations.

A total of 49 residues (64% of the nonproline residues) gave overlapping cross-peaks neither in the folded state nor in between the folded and unfolded states, and these residues can be accurately fitted to the two-state pressure-induced unfolding model described in the Section 4 (Equation (5), Figure 7). As observed for pressure denaturation, the intensity of each native state peak decreases as a function of urea concentration in the NMR sample, while the intensity of peaks corresponding to the unfolded state increases concomitantly, supporting a slow equilibrium on the NMR timescale for each residue between the native and unfolded state during the unfolding process. As reported above for pressure denaturation, a two-state transition model has been used to interpret the loss of intensity for each native state cross-peak (see Section 4). The residue-specific values obtained for the apparent free energy ΔG_u^0 of unfolding and for the apparent m-values are displayed in Figure 8.

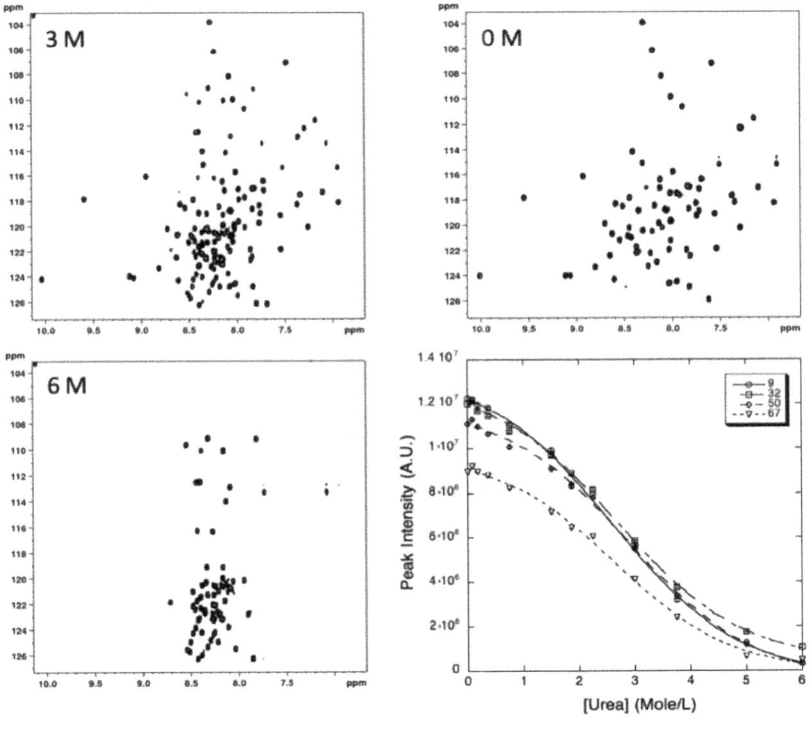

Figure 7. NMR detected chemical unfolding of GIPC1-GH2 at 20 °C. Examples of [^{1}H,^{15}N] HSQC at 0, 3, and 6 M urea are displayed. The last panel shows an overlay of four (residues K9, T32, L50, and V67) residue-specific denaturation curves obtained from the fits of the urea concentration-dependent sigmoidal decrease of the corresponding residue cross-peak intensities in the HSQC spectra with Equation (5).

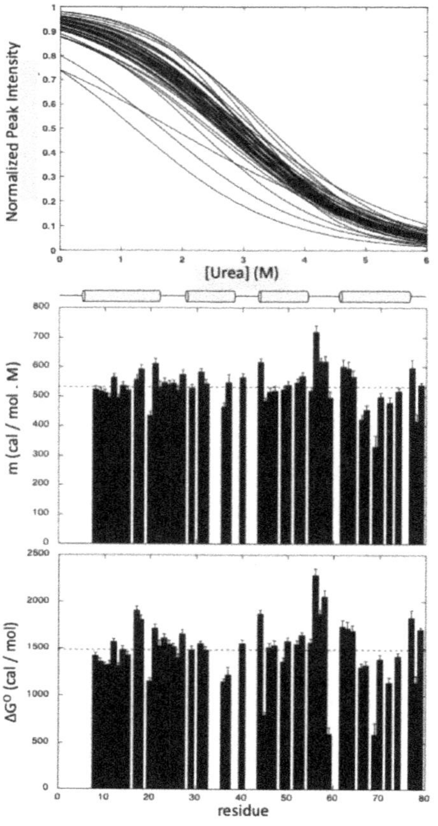

Figure 8. Steady-state thermodynamic parameters measured for GIPC1-GH2 at 20 °C from residue-specific urea denaturation curves. From top to bottom: overlay of the normalized residue-specific denaturation curves as obtained from the fit of the urea concentration-dependent sigmoidal decrease of the residue cross-peak intensities in the HSQC spectra with Equation (5); residue-specific values of the apparent m-values plotted versus the protein sequence; residue-specific values of the apparent free energy of unfolding ΔG_u^0 plotted versus the protein sequence. The dashed lines represent the mean values of the measured thermodynamic parameters. The location of the four helices in the protein sequence is schematized with cylinders on top of the graphics.

As previously observed, GIPC1-GH2 displays a weak stability at 20 °C, with an average free energy value for unfolding $<\Delta G_u^0>$ of 1484 ± 45 cal/mol, a value close to that measured from pressure denaturation curves at the same temperature. But contrary to what we observed with pressure denaturation, we did not observe significant variations in between the different helices when looking at the average values of ΔG_u^0 calculated for each helix (Supplementary Materials, Figure S6).

As in the case of pressure denaturation, we built fractional contact maps from probabilities of contact calculated from fractional probabilities of individual residues extracted from the normalized residue-specific chemical denaturation curves obtained at 20 °C (Figure 9).

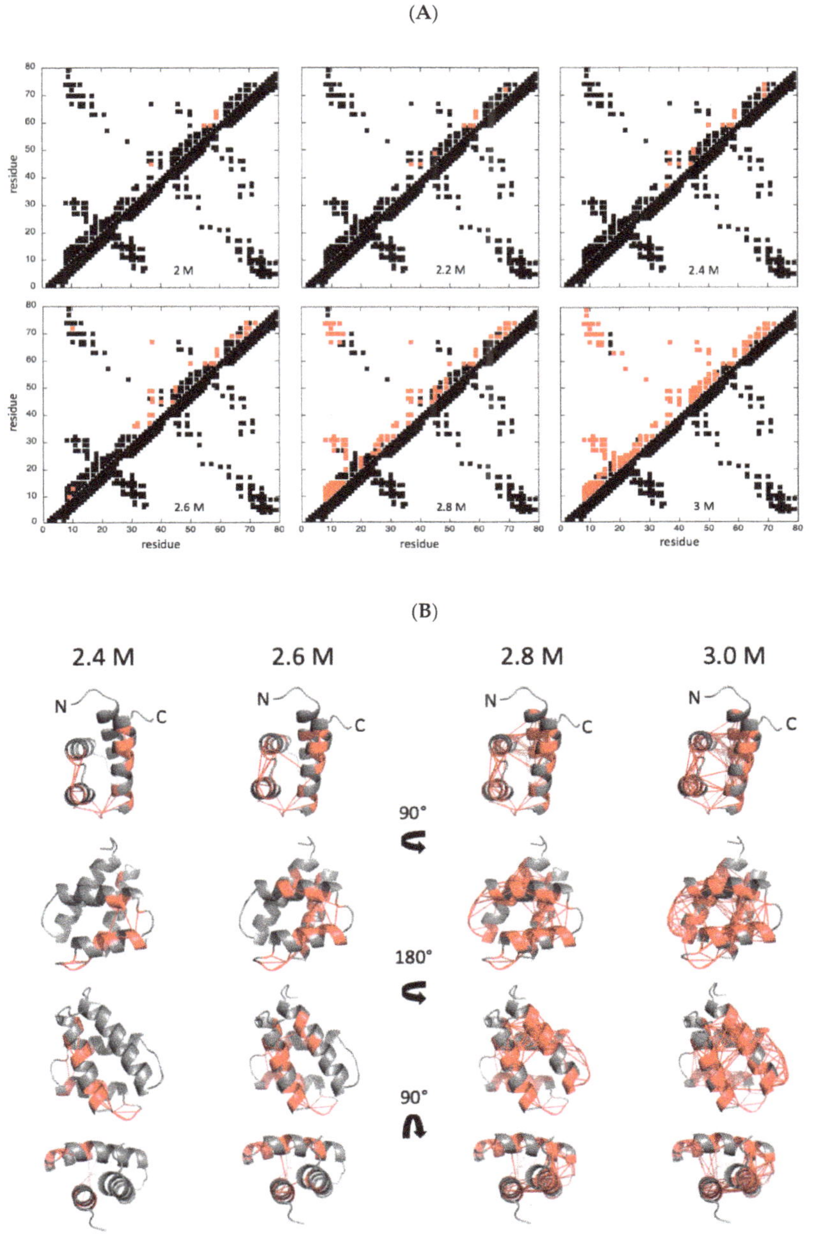

Figure 9. Chemical denaturation of GIPC1-GH2. (**A**) Contact maps built from the best solution structure obtained for GIPC1-GH2 at 20 °C, and at 2, 2.2, 2.4, 2.6, 2.8, and 3 M urea, as indicated. Contacts below and above the diagonal are displayed following the same rules as in Figure 6. (**B**) Visualization of the probabilities of contact on ribbon representations of GIPC1-GH2 at 20 °C and at 2.4, 2.6, 2.8, and 3 M urea, as indicated. The red lines represent contacts that are significantly weakened ($P_{ij} \leq 0.5$) at the indicated pressure. Residues involved in these contacts are also colored in red. The arrows in the middle of the panel indicate the rotation between the different views.

Interestingly, if some contacts are lost between the C-terminal end of helix IV and the N-terminal end of helix III at low urea concentrations ([urea] ≤ 2.4–2.6 M), a sharp unfolding transition takes place between 2.6 and 2.8 M urea that concerns both secondary and tertiary contacts in the four helices. In this view, the folding scenario looks like what was observed for pressure denaturation at 30 °C, where an increased unfolding cooperativity was observed when compared to 10 °C or 20 °C.

3. Discussion

The structure in solution of the GH2 domain of GIPC1 shows that this domain keeps its α-helical bundle fold outside the full-length protein context, even though it exhibits a rather low stability, probably due to the loss of intra- and intermolecular interactions that occur within the dimeric structure of the full-length protein. Indeed, this low stability is supported by the fast exchange rates exhibited by the amide protons that cannot be measured by regular H/D exchange NMR experiments. Nevertheless, CLEANEX-PM experiments showed that the amides involved in the H-bonds stabilizing the helical structures are not solvent-exposed, contrary to those located in the loops linking the different helices and in the flexible N- and C-terminal ends of the domain, supporting the idea that they are involved in the regular H-bonds expected in those regular elements of secondary structure. In addition, S^2 values close to 1, as obtained from the ^{15}N relaxation study, indicate that the helices are well structured, even though significant exchange contributions (R_2^{ex}) can be observed, especially in helix IV. This suggests that this helix is prone to conformational exchange. Notably, GIPC1-GH2 was found to be stable only in a limited range of temperature. The stability was found to be maximum at 20 °C, whereas it decreases significantly when increasing or decreasing the temperature. As evaluated from the ratio of the intensity of native/unfolded amide cross-peaks of representative residues measured on HSQC spectra, the fraction of native protein is about 70% at 40 °C and 60% at 0 °C (Supplementary Materials, Figure S7), indicating that the protein is sensitive both to thermal and to cold denaturation. GIPC1-GH2 is also very sensitive to high hydrostatic pressure, since it unfolds at 20 °C in the 1–2500 bar range without adding any sub-denaturant concentration of chaotropic reagents, as was usually observed for high-pressure denaturation of all-β or mixed-α/β structures in our previous study: sub-denaturant concentration ranging from 0.5 M [25] to about 2 M [8,10] of guanidinium chloride was used to tune the stability of these proteins into the pressure range allowed by the experimental set-up.

Looking closely to the unfolding pathways of GIPC1-GH2 under high hydrostatic pressure, we observed unexpected results. Indeed, since high-pressure denaturation is closely related to the presence of dehydrated internal solvent-excluded voids in the structure, we expected a two-step process starting with the loss of tertiary contacts within the 3D structure, followed by the loss of the secondary contacts yielding helices unfolding. This is because helices are well-packed secondary structure elements, without significant voids inside, while packing defaults are expected within the tertiary structure of the α-helical bundle. In fact, whatever the temperature of the study, local or global unfolding entails the quasi-simultaneous loss of the tertiary and secondary contacts in the concerned areas, meaning that helices are not stable outside the 3D scaffold context. Depending on the temperature used for the high-pressure denaturation study, we observed significant differences in the unfolding process. At 20 °C, the temperature where GIPC1-GH2 exhibits the highest stability, a partial unfolding of the molecule occurs first at helix III and helix IV, while helix I and helix II remain intact until approximately 1300 bar (note that for clarity of the discussion, we consider that a contact between two residues i and j is lost when $P_{ij} \leq 50\%$). The same scenario is observed at 10 °C, shifted at lower pressure: unfolding of helix III and IV starts below 500 bar instead of 900 bar at 20 °C, while helix I and II start to unfold at 700 bar. A different scenario is observed at 30 °C, where the four helices unfold almost simultaneously, with the onset of unfolding around 1000 bar. Interestingly, globally the protein appears to be more stable at high temperature (30 °C) than at low temperature: at 10 °C, the protein appears completely unfolded ($P_{ij} \leq 50\%$) at 900 bar while some residual structures are still present at 1100 bar and 30 °C.

This scenario described for high pressure denaturation at high temperature is very similar to what is observed for chemical denaturation of GIPC1-GH2 at 20 °C. We did not observe significant partial unfolding of any helix upon increase in urea concentration but rather a global unfolding of the molecule between 2.6 and 2.8 M urea. This is probably due to the different rules underlying high-pressure and chemical unfolding. As reported above, pressure unfolding is linked to the presence of solvent-excluded voids inside the 3D structure of the protein, and hence depends on the structure of the native state of the protein. On the contrary, the chemical denaturation process is driven by the increase of solvent-accessible area of the unfolded state with respect to the folded state, and is more dependent on the size of the protein [26–28]. This probably explains the difference that we observed in the folding pathway between high-pressure denaturation of GIPC1-GH2 at 20 °C or 10 °C, and its chemical denaturation at 20 °C. Note that thermal denaturation is also related to the solvent-accessible area of the unfolded state: this could explain the similar scenario observed for high-pressure denaturation at high temperature (30 °C) and for chemical denaturation. Indeed, at 30 °C, the stability of GIPC1-GH2 is decreased, and thermal denaturation probably competes with high-pressure denaturation, sweeping away the partial unfolding occurring at lower temperature.

4. Materials and Methods

4.1. Protein Expression and Purification

The construct GIPC1-GH2 domain (residues 255–333) was subcloned in pProEXHTB, allowing the expression of a 6xHis-TEV fusion protein, and was transformed into E. coli BL21-Gold (DE3) (Stratagene, Amsterdam, The Netherlands). Uniform ^{15}N or ^{15}N/^{13}C labeling was obtained by growing cells in minimal M9 medium containing ^{15}NH4Cl and/or ^{15}NH4Cl/^{13}C-u-labeled glucose as the sole nitrogen or carbon sources (Cortecnet). Protein was expressed overnight at 20 °C after induction with 0.2 mM IPTG. Cells were collected by centrifugation and suspended in lysis buffer comprising 20 mM Tris-HCl buffered at pH 7.5 and containing 150 mM NaCl, 2 mM imidazole, and a cOmplete™ EDTA-free tablet (Roche). Cells were lysed by sonication (1 s bursts for 4 min, at 30% amplitude with a large probe, Branson). Cell debris and insoluble materials were removed by centrifugation (Beckman Coulter Avanti J-20 XP centrifuge equipped with a 25.50 rotor, set at 20,000 rpm, at 6 °C). The supernatant was loaded through a benchtop peristaltic pump (Cytiva) onto a cOmplete™ His-Tag Purification Column (Roche, Basel, Switzerland) equilibrated with lysis buffer. After elution with lysis buffer supplemented with 200 mM imidazole, fractions containing the protein were dialyzed with homemade recombinant His tagged rTEV protease (mixed at 100:1 ratio) overnight at 4 °C in 20 mM Tris-HCl buffered at pH 7.5, 150 mM NaCl, and 1 mM imidazole. Cleavage was checked with SDS-PAGE and loaded again into a cOmplete™ His-Tag Purification Column equilibrated with the same buffer used for dialysis in order to remove the protease and the cleaved 6xHis tag. The GipC1-GH2 domain was finally injected through an AKTA system into a Superdex S75 16/60 (GE Healthcare) column, equilibrated with 20 mM Tris-HCl buffered at pH 7.2, 150 mM NaCl. The fractions containing the pure protein were pooled, concentrated to about 1 mM (protein concentration) (Vivaspin 15R, Sartorius). PMSF and EDTA were added (0.1 mM) to the samples that were then flash-frozen in liquid N2 and stored at −80 °C until NMR analysis.

4.2. NMR Assignments and Solution Structure

Protein samples were dissolved in 200 µL of aqueous buffer containing 20 mM Tris-HCl pH 7.2, 150 mM NaCl, and 0.1 mM PMSF and EDTA (5% D_2O for the lock) at a concentration of about 1 mM. Experiments were recorded at 20 °C on a Bruker AVANCE III 800 MHz (Bruker Biospin, Wissenbourg, France) equipped with a 5 mm Z-gradient TCI cryogenic probe head. ^{1}H chemical shifts were directly referenced to the methyl resonance of DSS, while ^{13}C and ^{15}N chemical shifts were referenced indirectly to the ^{13}C/^{1}H and ^{15}N/^{1}H absolute frequency ratios. All NMR experiments were processed with Gifa [29].

Backbone and Cβ resonance assignments were made using standard 3D triple-resonance HNCA, HNCACB, CBCA(CO)NH, HNCO, and HN(CA)CO experiments [30] and 3D [^{1}H,^{15}N] NOESY-HSQC (mixing time 150 ms) and TOCSY-HSQC (isotropic mixing: 60 ms) experiments performed on the ^{15}N,^{13}C-labeled GIPC1-GH2 sample. [^{1}H,^{15}N] NOESY-HSQC was used to extract the set of nOe's restraints used for structure modeling, completed by restraints obtained from a 2D homonuclear NOESY (mixing time 200 ms) recorded on a deuterated buffer. NOE cross-peaks were assigned through automated NMR structure calculations with CYANA 3 [18]. Backbone φ, ψ, and side-chain χ^1 torsion angle constraints were obtained from a database search procedure on the basis of backbone (^{15}N, HN, ^{13}C′, ^{13}Cα, Hα, ^{13}Cβ) chemical shifts using TALOS-N [17]. Hydrogen bond restraints were derived from the analysis of residue (φ,ψ) values and CLEANEX-PM experiments [19,20]. When identified, the hydrogen bond was enforced using the following restraints: ranges of 1.8–2.0 Å for d(N-H,O), and 2.7–3.0 Å for d(N,O).

The final list of restraints, from which values that were redundant with the covalent geometry were eliminated, was used for structure modeling. A total of 200 three-dimensional structures were generated using the torsion angle dynamics protocol of CYANA 3 from 1219 NOEs, 88 hydrogen bonds, and 164 angular restraints. The 20 best structures (based on the final target penalty function values) were minimized with CNS 1.2 according to the RECOORD procedure [31] and analyzed with PROCHECK [32]. The rmsds were calculated with MOLMOL [33]. Models are displayed with PyMOL [34]. All statistics are given in Supplementary Materials, Table S1.

4.3. Relaxation Studies

Relaxation rate constant measurements were performed on a 1 mM ^{15}N-labeled protein sample, at 14.1 T (600 MHz), using a Bruker AVANCE III spectrometer equipped with a 5 mm Z-gradient TXI probe head. The pulse sequences used to determine heteronuclear ^{15}N R_1, R_2 relaxation rates, and ^{15}N{^{1}H}NOE values were similar to those described [35–37], and experimental parameters and processing were previously reported in detail for other proteins studied in the laboratory [38–40]. The ^{15}N longitudinal relaxation rates R_1 were obtained from nine standard inversion-recovery experiments, with relaxation delays ranging from 18 ms to 1206 ms. The ^{15}N transverse relaxation rates R_2 were obtained from eight standard CPMG experiments, with relaxation delays ranging from 16 ms to 128 ms. Heteronuclear ^{15}N{^{1}H} NOE were determined from the ratio of two experiments, with and without saturation.

Relaxation data analysis: $J(0)$, $J(\omega_N)$, and $<J((\omega_H)>$ spectral densities were calculated from ^{15}N heteronuclear R_2, R_1, and ^{15}N{^{1}H} NOE using the so-called reduced spectral density mapping [35,36,41–44].

The model-free approach of Lipari and Szabo [22] was then used to further describe the mobility in terms of specific types of motion. This formalism makes the assumption that overall and internal motions contribute independently to the reorientation time correlation function of ^{15}N-^{1}H vectors and that internal motions occur on a much faster time scale than the global rotation of the molecule. For a protein with isotropic tumbling protein, one obtains:

$$J(\omega) = \frac{2}{5}\left\{S^2\frac{\tau_c}{1+(\omega\tau_c)^2} + \left(1-S^2\right)\frac{\tau}{1+(\omega\tau)^2}\right\} \quad (1)$$

where τ is the harmonics of the overall and the internal (fast) correlation time which pertains to each residue: $\tau^{-1} = \tau_c^{-1} + \tau_f^{-1}$. Fast internal motions are characterized by the square of a generalized order parameter S^2, which describes the relative amplitude of internal motions and ranges from 0 to 1, and an internal correlation time τ_f for the internal motions.

For some of the residues, the simple form of equation (1) turns out to be insufficient to fit the whole set of experimental data. This occurs for residues where observed $J(0)$ values are higher than expected, due to exchange contributions R_2^{ex}. In this case, the expression for the observed spectral density at 0 frequency is:

$$J(0) = \frac{2}{5}\left\{S^2\tau_c + \left(1 - S^2\right)\tau\right\} + \lambda R_2^{ex}\Big] \tag{2}$$

where λ is a scale factor. This occurs also when residues exhibit internal motions in a time window close to 1 ns. In this case, the expression for the spectral density function is extended to [23]:

$$J(\omega) = \frac{2}{5}\left\{S_f^2 S_s^2 \frac{\tau_c}{1 + (\omega\tau_c)^2} + S_f^2\left(1 - S_s^2\right)\frac{\tau}{1 + (\omega\tau)^2}\right\} \tag{3}$$

with $\tau^{-1} = \tau_c^{-1} + \tau_s^{-1}$, where S_f^2 and S_s^2 are the square of the partial order parameters for fast (picosecond time scale) and slow (τ_s, nanosecond time scale) internal motions, respectively. The square of the generalized order parameter S^2, defined as $S_f^2 S_s^2$, is a measure of the total amplitude of the internal motions. Note that S^2 equals S_f^2 in Equations (1) and (2). Equation (3) assumes that the contribution of the fastest motion to the spectral density function is negligible.

The values of the motional parameters of the individual residues can be derived from the fit of experimental $J(0)$, $J(60 \text{ MHz})$, and $\langle J(600 \text{ MHz})\rangle$ using Equations (1)–(3) implemented in the software DYNAMOF [45]. An iterative nonlinear least-squares algorithm [46] was employed to further minimize the error function. The "right" model was selected from χ^2 analysis.

4.4. Protein Unfolding

2D [^{1}H,^{15}N] HSQC were recorded on a Bruker AVANCE III 600 MHz spectrometer, at 3 temperatures (10, 20, and 30 °C) and 15 different hydrostatic pressures (1,50, 100, 300, 500, 700, 900, 1100, 1300, 1500, 1700, 1900, 2100, 2300, and 2500 bar) for pressure denaturation, and at 20 °C and 13 different urea concentrations (0, 0.1, 0.2, 0.375, 0.75, 1.125, 1.5, 1.875, 2.25, 3, 3.75, 5, and 6 M) for chemical denaturation. Samples with about 1 mM concentration of ^{15}N-labeled proteins were used on conventional 3 mm NMR tubes (200 μL of sample volume) for chemical denaturation, or in 5 mm o.d. ceramic tubes (330 μL of sample volume) from Daedelus Innovations (Aston, PA, USA) for pressure denaturation. Hydrostatic pressure was applied to the sample directly within the magnet using the Xtreme Syringe Pump also from Daedelus Innovations. Samples with different urea concentrations were prepared about 10 h before recording the NMR experiments used for chemical denaturation studies, although each pressure jump was separated by a 2-h relaxation time, to allow the protein to reach full equilibrium. In the case of pressure denaturation, relaxation times for the folding/unfolding reactions were estimated from a series of 1D NMR experiments recorded after 200 bar P-Jump, following the increase of the resonance band corresponding to the methyl groups in the unfolded state of the protein.

The cross-peak intensities for the folded species were measured at each pressure or each urea concentration, then fitted with a two-state model:

$$I = \frac{I_u + I_f e^{-(\Delta G_f^0 + p\Delta V_f^0)/RT}}{1 + e^{-(\Delta G_f^0 + p\Delta V_f^0)/RT}} \tag{4}$$

in the case of pressure denaturation, or:

$$I = \frac{I_u + I_f e^{-(\Delta G_f^0 + m[Urea])/RT}}{1 + e^{-(\Delta G_f^0 + m[Urea])/RT}} \tag{5}$$

in the case of chemical denaturation. In these equations, I is the cross-peak intensity measured at a given pressure or at a given urea concentration, and I_f and I_u correspond to the cross-peak intensities in the folded state (1 bar or 0 M urea) and in the unfolded state (2500 bar or 6 M urea), respectively. ΔG_f^0 stands for the residue-specific apparent free energy at atmospheric pressure or at 0 M urea. ΔV_f^0 corresponds to the residue-specific apparent volume of folding for pressure denaturation, while m is related to the steepness of the unfolding transition for chemical denaturation.

Native contact maps were obtained by using software CMView (http://www.bioinformatics.org/cmview/; accessed on 25 March 2020) with a threshold of 9 Å around the Cα of each residue, using the best structure obtained for GIPC1-GH2 among the 20 refined ones.

5. Conclusions

We demonstrate that combining NMR spectroscopy, which can bring information at an atomic resolution, with a mild and reversible method, such as high hydrostatic pressure, for protein unfolding can bring unprecedented details on the folding landscape of a protein. Here, we applied high-pressure NMR spectroscopy to the study of the folding/unfolding pathways of an α-helical bundle. Indeed, whereas similar studies have been widely applied to all-β or mixed-α/β 3D scaffolds, they are lacking for all-α helical structures. Unexpectedly, we found that the secondary and tertiary structures unfold simultaneously, although partial unfolding can occur. Importantly, this partial unfolding cannot be revealed with chemical denaturation, confirming the superiority of high pressure for exploring the folding landscape of a protein. Of course, whether these results obtained for GIPC1-GH2 can be generalized to other comparable α-helical bundles or whether more will have to be done remains an open question.

Supplementary Materials: The following are available online at https://www.mdpi.com/article/10.3390/ijms22073597/s1: Table S1: NMR restraints and refinement statistics for GIPC1-GH2 solution structures, Figure S1: Crystal structure of the full-length GIPC1 protein, Figure S2: Phase modulated CLEANEX-PM experiment recorded on 15N-labeled GIPC1-GH2, Figure S3: Relaxation measurements performed on 15N-labeled GIPC1-GH2., Figure S4: Distribution of the values of the thermodynamic parameters ΔG^0 and ΔV^0, Figure S5: Local pressure stability of GIPC1-GH2, Figure S6: Local chemical stability of GIPC1-GH2. Figure S7: Estimation of the folded/unfolded fraction of GIPC1-GH2.

Author Contributions: C.D., V.J.P.-H., C.T.-T., E.M.S., L.M., and S.D. prepared the unlabeled, 15N-labeled, and 13C/15N-labeled protein samples. C.D., C.T.-T. also carried out HP NMR experiments and analysis. P.B. made the NMR resonance assignment and the structural modeling of GIPC1-GH2. He also wrote analysis software for the analysis of the relaxation data and HP results. He finally participated in the coordination of the project with V.R., C.R. supervised the project, conceived experiments, participated in the interpretation, and wrote the article with P.B. All authors have read and agreed to the published version of the manuscript.

Funding: This research received no external funding.

Institutional Review Board Statement: Not applicable.

Informed Consent Statement: Informed consent was obtained from all subjects involved in the study.

Data Availability Statement: Data is contained within the article or Supplementary Materials.

Acknowledgments: This work was supported by French Infrastructure for Integrated Structural Biology (FRISBI) grant No. ANR-10-INSB-05. The authors are grateful to A. Nord (CBS, Montpellier) for careful proofreading and language correction of our manuscript.

Conflicts of Interest: The authors declare no conflict of interest.

Abbreviations and Nomenclature

GIPC1	GAIP Interacting Protein, C-terminus 1
GH2	C-terminal GIPC-homology 2
H(H)P-NMR	High (Hydrostatic) Pressure Nuclear Magnetic Resonance
nOe	nuclear Overhauser enhancement

References

1. Malhotra, P.; Udgaonkar, J.B. How cooperative are protein folding and unfolding transitions? *Protein Sci.* **2016**, *25*, 1924–1941. [CrossRef]
2. Sosnick, T.R.; Barrick, D. The folding of single domain proteins—Have we reached a consensus? *Curr. Opin. Struct. Biol.* **2011**, *21*, 12–24. [CrossRef]
3. Roche, J.; Royer, C.A.; Roumestand, C. Exploring the Protein Folding Pathway with High-Pressure NMR: Steady-State and Kinetics Studies. *High Press. Biosci.* **2015**, *72*, 261–278.
4. Roche, J.; Royer, C.A.; Roumestand, C. Monitoring protein folding through high pressure NMR spectroscopy. *Prog. Nucl. Magn. Reason. Spectrosc.* **2017**, *102*, 15–31. [CrossRef] [PubMed]
5. Roche, J.; Royer, C.A.; Roumestand, C. Exploring Protein Conformational Landscapes Using High-Pressure NMR. *Methods Enzymol.* **2019**, *614*, 293–320.
6. Dubois, C.; Herrada, I.; Barthe, P.; Roumestand, C. Combining High-Pressure Perturbation with NMR Spectroscopy for a Structural and Dynamical Characterization of Protein Folding Pathways. *Molecules* **2020**, *25*, 5551. [CrossRef] [PubMed]
7. Rouget, J.; Aksel, T.; Roche, J.; Saldana, J.; Garcia, A.E.; Barrick, D.; Royer, C.A. Size and sequence and the volume change of protein folding. *J. Am. Chem. Soc.* **2011**, *133*, 6020–6027. [CrossRef]
8. Roche, J.; Caro, J.A.; Norberto, D.R.; Barthe, P.; Roumestand, C.; Schlessman, J.L.; Garcia, A.E.; García-Moreno, B.E.; Royer, C.A. Cavities determine the pressure unfolding of proteins. *Proc. Natl. Acad. Sci. USA* **2012**, *109*, 6945–6950. [CrossRef]
9. Roche, J.; Dellarole, M.J.; Caro, A.; Guca, E.; Norberto, D.R.; Yang, Y.-S.; Garcia, A.E.; Roumestand, C.; García-Moreno, B.; Royer, C.A. Remodeling of the folding free-energy landscape of staphylococcal nuclease by cavity-creating mutations. *Biochemistry* **2012**, *51*, 9535–9546. [CrossRef] [PubMed]
10. Herrada, I.; Barthe, P.; Vanheusden, M.; DeGuillen, K.; Mammri, L.; Delbecq, S.; Rico, F.; Roumestand, C. Monitoring Unfolding of Titin I27 Single and Bi Domain with High-Pressure NMR Spectroscopy. *Biophys. J.* **2018**, *115*, 341–352. [CrossRef]
11. Saotome, T.; Doret, M.; Kulkarni, M.; Yang, Y.-S.; Barthe, P.; Kuroda, Y.; Roumestand, C. Folding of the Ig-Like Domain of the Dengue Virus Envelope Protein Analyzed by High-Hydrostatic-Pressure NMR at a Residue-Level Resolution. *Biomolecules* **2019**, *9*, 309. [CrossRef]
12. Fossat, M.J.; Dao, T.P.; Jenkins, K.; Dellarole, M.; Yang, Y.-S.; McCallum, S.A.; Garcia, A.E.; Barrick, D.; Roumestand, C.; Royer, C.A. High-Resolution Mapping of a Repeat Protein Folding Free Energy Landscape. *Biophys. J.* **2016**, *111*, 2368–2376. [CrossRef]
13. Zhang, S.; Zhang, Y.; Stenzoski, N.E.; Zou, J.; Peran, I.; McCallum, S.A.; Raleigh, D.P.; Royer, C.A. Pressure-Temperature Analysis of the Stability of the CTL9 Domain Reveals Hidden Intermediates. *Biophys. J.* **2019**, *116*, 445–453. [CrossRef]
14. Kohn, W.D.; Mant, C.T.; Hodges, R.S. α-Helical protein assembly motifs. *J. Biol. Chem.* **1997**, *272*, 2583–2586. [CrossRef] [PubMed]
15. Katoh, M. Functional proteomics, human genetics and cancer biology of GIPC family members. *Exp. Mol. Med.* **2013**, *45*, e26. [CrossRef]
16. Shang, G.; Brautigam, C.A.; Chen, R.; Lu, D.; Torres-Vázquez, J.; Zhang, X. Structure analyses reveal a regulated oligomerization mechanism of the PlexinD1/GIPC/myosin VI complex. *eLife* **2017**, *6*, e27322. [CrossRef]
17. Shen, Y.; Bax, A. Protein backbone ans side chain torsion angles predicted from NMR chemical shifts using artificial neural networks. *J. Biomol. NMR* **2013**, *56*, 227–241. [CrossRef]
18. Güntert, P. Automated NMR structure calculation with CYANA. *Methods Mol. Biol.* **2004**, *278*, 353–378. [PubMed]
19. Hwang, T.L.; Mori, S.; van Zijl, P.C. Application of phase-modulated CLEAN chemical EXchange spectroscopy(CLEANEX-PM) to detect water-protein proton exchange and intermolecular NOEs. *J. Am. Chem. Soc.* **1997**, *119*, 6203–6204. [CrossRef]
20. Hwang, T.L.; van Zijl, P.C.; Mori, S. Accurate quantitation of water-amide proton exchange rates using the phase-modulated CLEAN chemical EXchange (CLEANEX-PM) approach with a Fast-HSQC (FHSQC) detection scheme. *J. Biomol. NMR* **1998**, *11*, 221–226. [CrossRef]
21. Abragam, A. *Principles of Nuclear Magnetism*; Oxford, Science Publication, Clarendon Press: Oxford, UK, 1961.
22. Lipari, G.; Szabo, A. Model-free approach to the interpretation of nuclear magnetic resonance relaxation in macromolecules. *J. Am. Chem. Soc.* **1982**, *104*, 4546–4559. [CrossRef]
23. Clore, G.M.; Driscoll, P.C.; Wingfield, P.T.; Gronenborn, A.M. Analysis of the backbone dynamics of interleukin-1 beta using two-dimensional inverse detected heteronuclear 15N-1H NMR spectroscopy. *Biochemistry* **1990**, *29*, 7387–7401. [CrossRef] [PubMed]
24. Seemann, H.; Winter, R.; Royer, C.A. Volume, expansivity and isothermal compressibility changes associated with temperature and pressure unfolding of Staphylococcal nuclease. *J. Mol. Biol.* **2001**, *307*, 1091–1102. [CrossRef] [PubMed]
25. Kitahara, R.; Royer, C.A.; Yamada, H.; Boyer, M.; Saldana, J.L.; Akasaka, K.; Roumestand, C. Equilibrium and pressure-jump relaxation studies of the conformational transitions of P13MTCP1. *J. Mol. Biol.* **2002**, *12*, 609–628. [CrossRef]

26. Baase, W.A.; Liu, L.; Tronrud, D.E.; Matthews, B.W. Lessons from the lysozyme of phage T4. *Protein Sci.* **2010**, *19*, 631–641. [CrossRef] [PubMed]
27. Pace, C.N.; Fu, H.; Fryar, K.L.; Landua, J.; Trevino, S.R.; Shirley, B.A.; Hendricks, M.M.; Iimura, S.; Gajiwala, K.; Scholtz, J.M.; et al. Contribution of hydrophobic interactions to protein stability. *J. Mol. Biol.* **2011**, *408*, 514–528. [CrossRef]
28. Shortle, D. Staphylococcal nuclease: A showcase of m-value effects. *Adv. Protein Chem.* **1995**, *46*, 217–247.
29. Pons, J.L.; Malliavin, T.E.; Delsuc, M.A. Gifa V.4: A complete package for NMR data set processing. *J. Biomol. NMR* **1996**, *8*, 445–452. [CrossRef]
30. Sattler, M.; Schleucher, J.; Griesinger, C. Heteronuclear multi-dimensional NMR experiments for the structure determination of proteins in solution employing pulsed field gradients. *Prog. Nucl. Magn. Reson. Spectrosc.* **1999**, *34*, 93–158. [CrossRef]
31. Nederveen, A.J.; Doreleijers, J.F.; Vranken, W.; Miller, Z.; Spronk, C.A.; Nabuurs, S.B.; Güntert, P.; Livny, M.; Markley, J.L.; Nilges, M.; et al. RECOORD: A recalculated coordinate database of 500+ proteins from the PDB using restraints from the BioMagResBank. *Proteins* **2005**, *59*, 662–672. [CrossRef]
32. Laskowski, R.A.; Moss, D.S.; Thornton, J.M. Main-chain bond lengths and bond angles in protein structures. *J. Mol. Biol.* **1993**, *231*, 1049–1067. [CrossRef]
33. Koradi, R.; Billeter, M.; Wüthrich, K. MOLMOL: A program for display and analysis of macromolecular structures. *J. Mol. Graph.* **1996**, *14*, 51–55. [CrossRef]
34. Delano, W.L. PyMOL: An open-source molecular graphics tool. *CCP4 Newslett. Protein Crystallogr.* **2002**, *40*, 82–92.
35. Peng, J.W.; Wagner, G. Mapping of the spectral densities of N-H bond motion in eglin c using heteronuclear relaxation experiments. *Biochemistry* **1992**, *31*, 8571–8586. [CrossRef]
36. Peng, J.W.; Wagner, G. Mapping of the spectral density functions using heteronuclear NMR relaxation experiments. *J. Magn. Reson.* **1992**, *98*, 308–332.
37. Kay, L.E.; Nicholson, L.K.; Delaglio, F.; Bax, A.; Torchia, D.A. Pulse sequences for removal of the effects of cross correlation between dipolar and chemical-shift anisotropy relaxation mechanisms on the measurement of heteronuclear T1 and T2 values in proteins. *J. Magn. Reson.* **1992**, *97*, 359–375. [CrossRef]
38. Barthe, P.; Chiche, L.; Declerck, N.; Delsuc, M.A.; Lefèvre, J.F.; Malliavin, T.; Mispelter, J.; Stern, M.-H.; Lhoste, J.M.; Roumestand, C. Refined Solution Structure and Backbone Dynamics of 15N-Labeled C12A-p8MTCP1 Studied by NMR Relaxation. *J. Biomol. NMR* **1999**, *15*, 271–288. [CrossRef]
39. Guignard, L.; Padilla, A.; Mispelter, J.; Yang, Y.-S.; Stern, M.-H.; Lhoste, J.M.; Roumestand, C. Backbone dynamics and solution structure refinement of the 15N-labeled human oncogenic protein p13MTCP1: Comparison with X-ray data. *J. Biomol. NMR* **2000**, *17*, 215–230. [CrossRef] [PubMed]
40. Auguin, D.; Barthe, P.; Augé-Sénégas, M.T.; Stern, M.H.; Noguchi, M.; Roumestand, C. Solution structure and backbone dynamics of the pleckstrin homology domain of the human protein kinase B (PKB/Akt). Interaction with inositol phosphates. *J. Biomol. NMR* **2004**, *28*, 137–155. [CrossRef]
41. Farrow, N.A.; Zhang, O.; Szabo, A.; Torchia, D.A.; Kay, L.E. Spectral density function mapping using 15N relaxation data exclusively. *J. Biomol. NMR* **1995**, *6*, 153–162. [CrossRef] [PubMed]
42. Ishima, R.; Nagayama, K. Protein backbone dynamics revealed by quasi spectral density function analysis of amide N-15 nuclei. *Biochemistry* **1995**, *34*, 3162–3171. [CrossRef] [PubMed]
43. Ishima, R.; Nagayama, K. Quasi-spectral-density function analysis for nitrogen-15 nuclei in proteins. *J. Magn. Reson. B* **1995**, *108*, 73–76. [CrossRef]
44. Lefèvre, J.-F.; Dayie, K.T.; Peng, J.W.; Wagner, G. Internal mobility in the partially folded DNA binding and dimerization domains of GAL4: NMR analysis of the N-H spectral density functions. *Biochemistry* **1996**, *35*, 2674–2686. [CrossRef] [PubMed]
45. Barthe, P.; Ropars, V.; Roumestand, C. DYNAMOF: A program for the dynamics analysis of relaxation data obtained at multiple magnetic fields. *C. R. Chimie* **2006**, *9*, 503–513. [CrossRef]
46. Press, W.H.; Flannery, B.P.; Teukolsky, S.A.; Vetterling, W.T. *Numerical Recipes*; Cambridge University Press: Cambridge, UK, 1986.

International Journal of *Molecular Sciences*

MDPI

Article

Configurational Entropy of Folded Proteins and Its Importance for Intrinsically Disordered Proteins

Meili Liu [1,2,3], Akshaya K. Das [2,3], James Lincoff [2,4], Sukanya Sasmal [2,4], Sara Y. Cheng [2,3], Robert M. Vernon [5], Julie D. Forman-Kay [5,6] and Teresa Head-Gordon [2,3,4,7,*]

1 Department of Chemistry, Beijing Normal University, Beijing 100875, China; meililiu@berkeley.edu
2 Pitzer Center for Theoretical Chemistry, University of California, Berkeley, CA 94720, USA; akshaya.das@berkeley.edu (A.K.D.); jameslincoff@gmail.com (J.L.); sukanyasasmal@gmail.com (S.S.); sara.y.cheng.20@gmail.com (S.Y.C.)
3 Department of Chemistry, University of California, Berkeley, CA 94720, USA
4 Department of Chemical and Biomolecular Engineering, University of California, Berkeley, CA 94720, USA
5 Molecular Medicine Program, Hospital for Sick Children, Toronto, ON M5G 0A4, Canada; vernon.rm@gmail.com (R.M.V.); forman@sickkids.ca (J.D.F.-K.)
6 Department of Biochemistry, University of Toronto, Toronto, ON M5S 1A8, Canada
7 Department of Bioengineering, University of California, Berkeley, CA 94720, USA
* Correspondence: thg@berkeley.edu

Citation: Liu, M.; Das, A.K.; Lincoff, J.; Sasmal, S.; Cheng, S.Y.; Vernon, R.M.; Forman-Kay, J.D.; Head-Gordon, T. Configurational Entropy of Folded Proteins and Its Importance for Intrinsically Disordered Proteins. *Int. J. Mol. Sci.* **2021**, *22*, 3420. https://doi.org/10.3390/ijms22073420

Academic Editors: Istvan Simon and Csaba Magyar

Received: 7 March 2021
Accepted: 25 March 2021
Published: 26 March 2021

Abstract: Many pairwise additive force fields are in active use for intrinsically disordered proteins (IDPs) and regions (IDRs), some of which modify energetic terms to improve the description of IDPs/IDRs but are largely in disagreement with solution experiments for the disordered states. This work considers a new direction—the connection to configurational entropy—and how it might change the nature of our understanding of protein force field development to equally well encompass globular proteins, IDRs/IDPs, and disorder-to-order transitions. We have evaluated representative pairwise and many-body protein and water force fields against experimental data on representative IDPs and IDRs, a peptide that undergoes a disorder-to-order transition, for seven globular proteins ranging in size from 130 to 266 amino acids. We find that force fields with the largest statistical fluctuations consistent with the radius of gyration and universal Lindemann values for folded states simultaneously better describe IDPs and IDRs and disorder-to-order transitions. Hence, the crux of what a force field should exhibit to well describe IDRs/IDPs is not just the balance between protein and water energetics but the balance between energetic effects and configurational entropy of folded states of globular proteins.

Keywords: configurational entropy; force fields; intrinsically disordered proteins

1. Introduction

Intrinsically disordered peptides (IDPs) are a class of proteins that are defined as dynamic structural ensembles rather than a dominant equilibrium structure in solution [1]. Experimental methods such as nuclear magnetic resonance (NMR) spectroscopy [2], single-molecule fluorescence Förster resonance energy transfer (smFRET) [3], and small-angle X-ray scattering (SAXS) [4] can provide restraints on the structural ensemble of IDP systems but are unable to fully resolve important subpopulations of structure relevant for function [5]. Therefore, computational methods play a critical role by first generating putative structural ensembles [6] and secondly reconciling them with the highly averaged experimental information using Monte Carlo optimization [7,8] or, more recently, Bayesian formalisms [9–11]. In this work, we are concerned with the generation of IDP ensembles using physically motivated force fields and molecular dynamics simulations (MD) that model protein–protein, protein–water, and water–water interactions at the atomic level.

Nearly all MD simulations of IDP structural ensembles have been generated with pairwise additive force fields that have traditionally been parameterized to reproduce the folded states of proteins [12]. Nonetheless, atomistic force fields have struggled with issues ranging from biases in secondary structure conformations [13,14] or overly structured and collapsed ensembles that do not agree with experimental data on many IDP systems [15,16]. Additionally, IDPs are more solvent-exposed than folded globular proteins, thus the corresponding choice of water model used to simulate IDPs is critical for capturing the correct balance between protein–water and water–water interactions for folded and unfolded states and for disordered proteins [2,17,18]. The D.E. Shaw group was also the first to show that long standard MD simulations—on the order of hundreds of microseconds—are required to ascertain the ability of a force field to maintain the structural integrity of a globular protein [19,20]. We found that similar issues arise for IDPs that also require long simulations and/or accelerated sampling methods to better represent their structural properties [21].

To improve upon MD simulated predictions for IDPs, a few research groups have proposed energy parameter changes to standard force fields to bring them better in line with solution experiments. For the TIP4P-D water model [22], Piana et al. increased the C_6 dispersion coefficient of the Lennard–Jones parameter by ~50% to make London dispersion interactions more favorable, which, when combined with the Amberff99sb-ildn model [19] for the protein, resulted in more expanded IDPs with an improved agreement with experimental NMR and small-angle X-ray scattering (SAXS) data. Best and Mittal [23] introduced backbone parameter modifications of one of the Amber force fields combined with the TIP4P/2005 water model [24] to reproduce, for example, the temperature dependence of the helix–coil transition for the 15-residue peptide Ac-$(AAQAA)_3$-NH_2. The resulting A03WS/TIP4P/2005 is intended for use for IDPs but, when applied to poly-glutamine IDP in solution, was found to generate mostly featureless and highly extended conformations that do not correctly describe solution experiments [25]. Independently, Henriques et al. have shown that both Amberff99sb-ildn/TIP4P-D and A03WS/TIP4P/2005 reproduce better radius of gyration values for the disordered Histatin 5 (Hst 5) peptide, although both force fields exhibit more turn content for Hst 5 that creates more collapsed states [15]. Robustelli et al. performed extensive millisecond MD simulations on six different pairwise additive protein force fields on a range of fully disordered to folded globular protein systems [20]. These simulations revealed that none of these standard force fields agreed with experimental data for a number of IDP systems while also maintaining the ability to accurately model folded proteins [20].

Therefore, newer protein force fields and water model combinations have been proposed to capture the behavior of IDPs and folded proteins [12]. This is important for at least two reasons. First, they can be used when simulating interactions of IDPs with folded proteins [26], disorder-to-order transitions [27], and folded proteins with intrinsically disordered regions (IDRs) [28]; second, they satisfy the goal of any force field, which is transferability to new protein systems and other emerging problems such as liquid phase separation [29]. An example is the CHARMM36m protein model of Huang et al. that purports to better describe both IDPs and folded proteins using the same set of refined peptide backbone parameters and salt–bridge interactions and an increased Lennard–Jones (LJ) well depth to strengthen protein–water dispersion interactions [30]. These modifications led to a reduction in the percentage of predicted left-handed a-helices, as well as a better agreement with NMR scalar couplings and SAXS curves for folded proteins, although Huang et al. observed that no universal interaction strength parameter in the Lennard–Jones function could generate structural ensembles with good agreement with the experimental radius of gyration measurements for all IDP systems [30].

Hence, the logical next step is to consider more advanced potentials, albeit with a greater computational expense that can be made more accurate by including multipolar electrostatic interactions with many-body polarization that can respond to changes in the solvent conditions around biomolecules [31,32]. One purpose of this study is to ascertain how well the advanced many-body polarizable AMOEBA protein (AmPro13) [33] and water (AmW03) [34] force field performs against experiments across of range of folded proteins, IDRs and IDPs, when compared to a representative standard force field, AMBERff99sb/TIP3P(TIP4p-Ew), and recently modified fixed-charge force fields, CHARMM36(m)/TIP3P(m), where the parentheses refer to alternate protein and/or water model combinations.

The second important purpose of this work is to provide some easily ascertained measures of what constitutes a successful force field that can simultaneously describe both folded proteins and proteins with disorder. We hypothesized that a force field that provides the largest structural deviations and statistical fluctuations, which remains consistent with the experimental Rg of a folded globular protein, will better be able to capture the greater plasticity and match solution experiments for IDPs and IDRs. In fact, we consistently find that the polarizable model better reproduces the experimental Rg [35] for the disordered Hst 5 peptide exhibits a stronger temperature dependence in the disorder-to-order transition for the $(AAQAA)_3$ system due to an unusual $\alpha-$helical structure and maintains a folded core for the TSR4 domain while simultaneously exhibiting regions of disorder. By contrast, the fixed-charge force fields have Rg distributions that are in disagreement with SAXS intensity profiles and contain higher populations of turns for Hst 5 that contribute to a more collapsed state and show little change with temperature for $(AAQAA)_3$.

We emphasize that this work is not a quantitative benchmarking paper but to emphasize the qualitative importance of configurational entropy for folded states. By determining a range of metrics for its evaluation such as similarity/dissimilarity and Lindeman criteria, we note that better evidence of fluidity in the folded state will be predictive as to whether a force field will exhibit a better predictive capacity for IDPs/IDRs. This work better places theory as an equal partner to experiment in new areas of IDP studies such as liquid–liquid phase separation that are current and active areas of theory/experimental collaboration.

2. Results

The field of biomolecular modeling has historically relied on a simple representation of the potential energy surface of proteins and water based on the pairwise additive approximation of the nonbonded interactions [36].

$$U_{\text{nonbond}} = U_{\text{Pauli}} + U_{\text{Disp}} + U_{\text{Elec}} + U_{\text{Pol}} \tag{1}$$

The U_{Pauli} and U_{Disp} terms are combined within different force fields to formulate a Lennard–Jones 12-6 potential (as is done for Amber and Charmm force fields), whereas the AMOEBA model uses a buffered 14-7 functional form. The U_{Elec} interactions capture classical electrostatics in which Amber and Charmm use partial charges (monopoles), whereas AMOEBA uses a permanent multipole up through quadrupoles. Finally, only AMOEBA contains U_{Pol} for many-body polarization.

To compare these force fields for describing the behavior of both folded proteins and IDRs/IDPs, we first consider 7 globular proteins ranging in size from 130 to 266 residues, as shown in Figure 1. These proteins include: a serine protease (1arb), an n-acetyltransferase (1b6b), two hydrolases (beta-lactamase, 1bsg and xylanase, 4xq4), two isomerases (phosphoglycerate mutase, 1rii and cis-trans isomerase Cwc27, 4r3f), the sugar-binding protein DC-SIGN (2xr6), and finally the TSR4 domain (1VEX) as an intermediate class of protein with a small folded core dominated by IDRs.

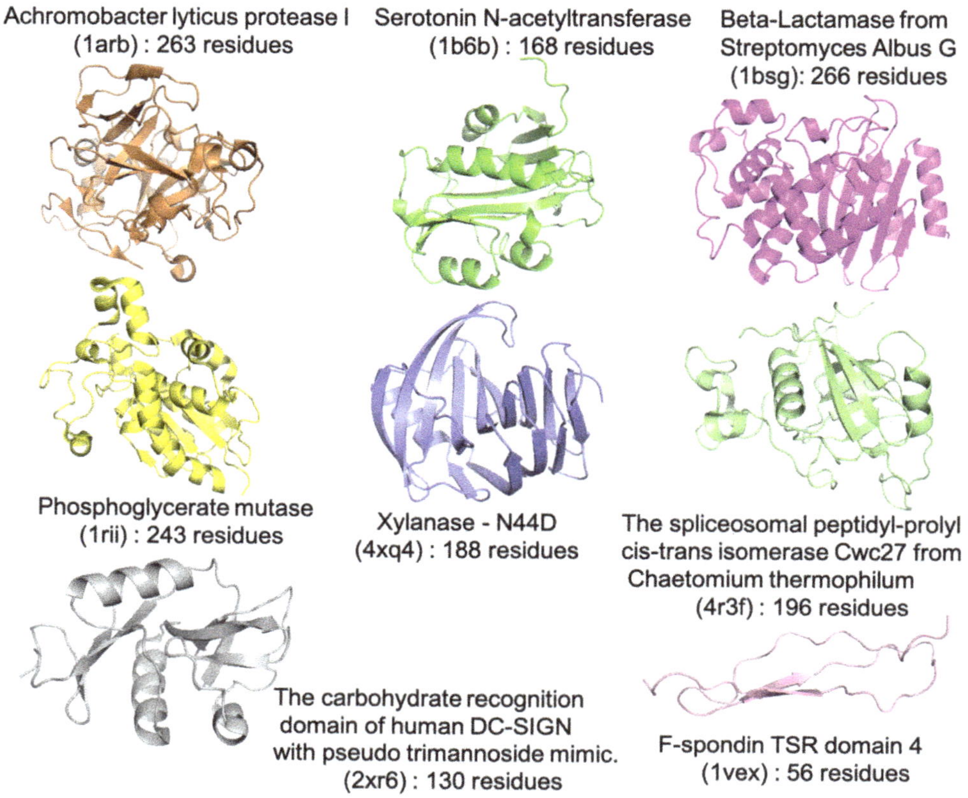

Figure 1. Seven folded proteins (PDB IDs: 1b6b [37], 1arb [38] 1bsg [39], 1rii [40], 2xr6 [41], 4r3f [42], and 4xq4 [43]) and one protein with intrinsically disordered regions (1vex [44]) simulated with polarizable and nonpolarizable force fields.

For any biomolecular force field comparison, it is typical to run molecular dynamics simulations of at least ~1 μs to measure protein stability by calculating global metrics such as the root mean square deviation (RMSD) and radius of gyration <Rg> [30]. Figure 2 and Supplementary Figure S1 report on the coordinate RMSD and <Rg> of the seven folded proteins over the 1 μs of MD simulation for each of the force field combinations. All seven folded globular proteins show no evidence of early unfolding events or significant degradation in a secondary structure with any force field, as shown in Supplementary Figure S2 for 1bsg and 1b6b. However, an important distinction is that the polarizable force field exhibits substantially larger root mean square deviations (RMSDs) than those of the nonpolarizable models, although all force fields maintain an average radius of gyration <Rg> in agreement with the experiment.

Although our 1 μs simulation timescales are typical of previous work on measuring protein stability [30], we consider additional metrics for acceptable deviations from the starting structures derived from the PDBs. Figure 2 reports a metric developed by Maiorov and Crippen that provides an empirical relationship to estimate structural similarity $D_{0,\,sim}$ and dissimilarity $D_{0,\,dis}$ for globular proteins (see Supplementary Tables S1 and S2) [45]. Values below or at the similarity measure defines a valid ensemble of structures for which loop regions may reconfigure while not significantly shifting the <Rg> and core fold, while values at or above the $D_{0,\,dis}$ metric distinguish the dissimilarity between a reference structure and its mirror image and thus any large shifts in <Rg> and conformation. In this work, we measure Rg from both the PDB structure for each protein and from polymer scaling law estimates parameterized by PDB structures (see Supplementary Table S2) under poor solvent conditions and structural variations of globular proteins of the same size [46,47]. The larger Rg values from the polymer scaling laws relative to the PDB structure are well within the expectations from solution experiments [48], and consistent with crystal structures differing somewhat from NMR [49] and SAXS [50] ensembles for folded states (Supplementary Tables S1 and S2).

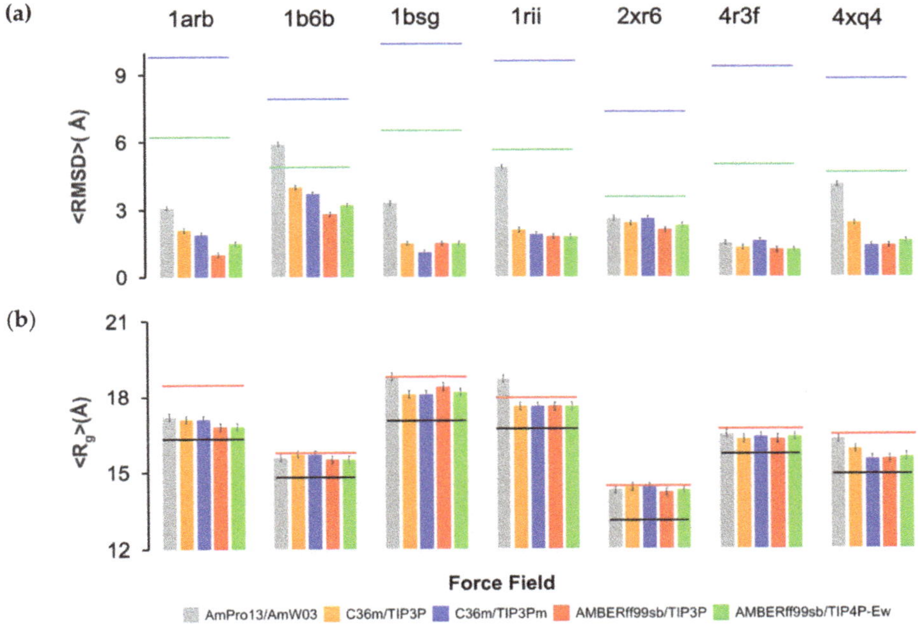

Figure 2. Measures of protein stability when simulated with polarizable and nonpolarizable force fields. (**a**) Root mean square deviation (RMSD) for 1 μs MD simulations for AmPro13/AmW03, C36m/TIP3P, C36m/TIP3Pm, ff99SB/TIP3P, and ff99SB/TIP4P-Ew. The black line is the value of the $D_{0,\,sim}$ metric and the red line the metric and the red line the $D_{0,\,dis}$ metric. (**b**) <Rg> for all force fields and comparison to the Rg of the PDB structure (black) or polymer scaling laws (Supplementary Table S2) as a measure of solution (red). Proteins characterized are 1arb [38] 1b6b [37], 1bsg [39], 1rii [40] 4xq4 [43], 4r3f [42] and 2xr6 [41]

As seen in Figure 2, all force fields yield RMSDs within the range of the $D_{0,\,sim}$ metric for the seven folded proteins. With the exception of 1b6b, for which the <RMSD> using AmPro13/AmW03 is within the $D_{0,\,sim}$ by ~0.5 Å, all models have not fully reached allowed values of the $D_{0,\,sim}$ metric, and no force field exhibits unfolding or instability as measured by $D_{0,\,dis}$ (see Supplementary Figure S2). However, just as importantly, it is also evident that the fixed-charge force fields generally yield folded states with much smaller <RMSD> values, whereas the polarizable force field model is closest to the upper bound of the similarity metric for the globular proteins. In addition, the <Rg> for the pairwise additive models are more often closer to the PDB structure, while the <Rg> values for the polarizable model are more in line with biopolymer scaling law estimates (Supplementary Table S1).

Because values of RMSD correlate directly with root mean square fluctuations (RMSF) [51], Figure 3 shows that the <RMSF> by residue for the seven folded proteins is largest on average for the polarizable model relative to the fixed-charge force fields, although large regions of structural stability are evident throughout the structure. The question one might ask is whether the larger <RMSF> by residues of the polarizable model is physically sound and correct, and are the fixed-charge models thus overly rigid?

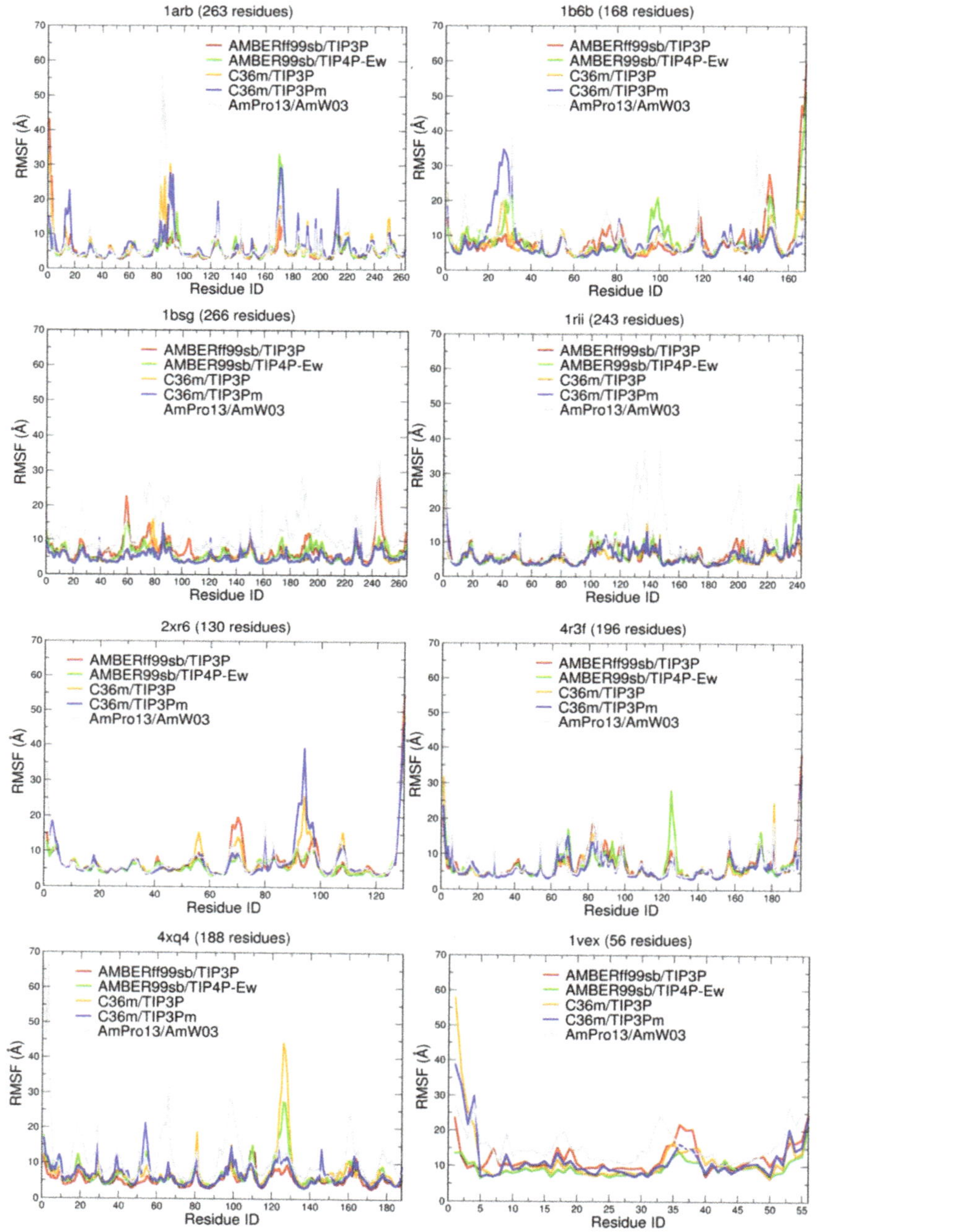

Figure 3. Average root mean square fluctuation for each residue in the simulated trajectories averaged over the last 100 ns. For 1arb [38] 1b6b [37], 1bsg [39], 1rii [40] 4xq4 [43], 4r3f [42] and 2xr6 [41].

In order to answer that question, we consider using the Lindemann criterion developed originally for the melting of a solid crystal [52]. The Lindemann value $\Delta_L = RMSF/a$ has been adapted to the case of proteins by replacing the crystal lattice constant, a, with an average nonbonded distance [53,54]. Katava et al. provided experimental estimates of the Δ_L from inelastic neutron scattering for hen egg white lysozyme (HEWL) and, assuming a = 4.75 Å, found a Lindemann value at the protein melting temperature (T_m) of $\Delta_L^{exp}(T_m)$~0.17–0.18, driven by the mixing in of a greater proportion of unfolded state fluctuations [54]. Below (T_m), the contributions from unfolded state fluctuations diminish as temperature, of course, decreases, but Zhou et al. showed that the folded-state fluctuations comprise an interior protein core that is suppressed and solid-like (Δ_L^{core} ~0.05–0.1) whereas the protein surface is quite fluid (Δ_L^{core} ~0.15–0.2) [53,55], which, in part, explains the overall experimental value for the HEWL protein near 300 K of $\Delta_L^{exp}(300\ K)$~0.15–16 in water solvent [54]. Because Katava and coworkers found similar results for myoglobin, crambin, hemoglobin, and BSA, they expect these results to be universal values for any folded state of a globular protein of average size, and hence we rely on comparisons to $\Delta_L^{exp}(300\ K)$ in our simulations of the seven folded proteins analyzed here.

By contrast, the polarizable Table 1 reports the corresponding $\Delta_L^{sim}(300\ K)$ values for each protein, assuming a value a = 4.375 Å, which is an average taken among all previous work [53–55], but with the RMSF calculated from the fixed-charge and many-body force field simulations (Figure 3 and Supplementary Table S3). Averaged over all of the folded proteins, the nonpolarizable force fields yield Lindemann values $\Delta_L^{sim}(300\ K)$ of ~0.12; to put this value into perspective for the fixed-charge force fields, this value is close to $\sim \Delta_L^{exp}(230\ K)$ for HEWL. By contrast, the polarizable force field predicts <RMSF> values that are ~30% larger than those of the fixed charge models, with values of $\Delta_L^{sim}(300\ K)$~0.16 that are in good agreement with the experimental value at room temperature. Supplementary Table S4 shows that all force fields have a very solid structural core, $\Delta_L^{core}(300\ K) \sim 0.09$ for the fixed charge force fields and ~0.12 for the polarizable model and that their total simulated averages are thus dominated by their surface fluctuations, $\Delta_L^{surf}(300\ K)$, which are largest for the many-body potential (0.155 vs. 0.21). The lower $\Delta_L^{sim}(300\ K)$ values from the fixed-charge force fields are thus indicators that they will generally overestimate the melting temperature and/or the amount of native structure in the unfolded state, an undesirable feature of standard force fields noted previously [54,56–58]. From the perspective of the Lindemann criteria, this is because they do not fully activate their allowed thermal vibrations permitted by $D_{0,\ sim}$ in the fully populated folded state, requiring much higher temperatures to exceed the RMSF threshold to realize the larger collective modes for unfolding.

Table 1. Lindemann values for 7 folded proteins at 300 K. A value of α = 4.375Å and <RMSF> averaged over all residues (Figure 3, Supplementary Table S2) were used to calculate $\Delta_L^{sim}(300\ K)$.

Force Field/Proteine	$\Delta_L^{sim}(300\ K)$							Ave.
	1arb	**1b6b**	**1bsg**	**1rii**	**2xr6**	**4r3f**	**4xq4**	
ff99sb/TIP3P	0.10	0.14	0.14	0.13	0.11	0.11	0.12	0.12
ff99sb/TIP4P-Ew	0.10	0.13	0.12	0.12	0.11	0.12	0.10	0.11
C36m/TIP3P	0.11	0.14	0.11	0.12	0.11	0.12	0.14	0.12
C36m/TIP3Pm	0.12	0.18	0.11	0.13	0.14	0.12	0.12	0.13
AmPro13/AmW03	0.13	0.16	0.18	0.22	0.13	0.16	0.17	0.16

We therefore anticipate that T_m values using the polarizable force field will be in better agreement with the experiment because large surface fluctuations are evident by their $D_{0,\ sim}$ values that approach the estimated upper bound [45] while remaining consistent with the folded R_g. We thus conclude from the folded protein class that force fields should exhibit, in addition to a balance between protein–protein and protein–water energetics, a good balance between energy and configurational entropy in order to realize $\Delta_L^{sim} \sim \Delta_L^{exp}$.

We carry this idea further to predict that the force fields with $\Delta_L^{sim} \sim \Delta_L^{exp}$ for folded proteins will be better suited to representing the structural ensembles of IDRs and IDPs as well; by corollary, force fields with $\Delta_L^{sim} < \Delta_L^{exp}$ for folded states will not be able to describe the greater plasticity of intrinsically disordered states. To test the extrapolation from folded proteins, we now consider the TSR4 domain (1vex), which comprises a small β-sheet core stabilized by a network of pi-contacts, with large loops that have been classified as intrinsically disordered regions [59]. For TSR4 (1vex), the <RMSD> values for all force fields (Table 2) are well outside the $D_{0,\,sim}$ metric (1.34 Å) and in better agreement with the $D_{0,\,dis}$ value (4.49 Å) given the presence of significant segments of disorder. Figure 3 shows that <RMSF> per residue for TSR4 (1vex) is larger on average relative to the folded protein case for all force fields. For the TSR4 domain, all force fields have a less solid structural core than for the folded proteins, Δ_L^{core}~0.16–0.18, and are dominated by large surface fluctuations, Δ_L^{surf}~0.18–0.29, that exceed those of the folded proteins. There are no direct-solution experimental data to validate against, but these results support the expectation that the Lindemann criteria value for globular proteins is not universal and cannot be extended to IDRs and IDPs. Even so, we find that the Amber force fields yield the most suppressed $\Delta_L^{sim}(300\ K)$ values, while the C36 and C36m force fields fluctuate more, and the polarizable model yields the largest $\Delta_L^{sim}(300\ K)$ value for the TSR4 domain.

Table 2. Fluctuation properties of the TSR4 domain at 300 K. <RMSD> is the average root mean square distance to the starting structure of TSR4. A value of a = 4.375Å and <RMSF> averaged over all residues of TSR4 were used to calculate the total Lindemann value, Δ_L^{sim}. Δ_L^{core} was evaluated from the β-sheet core residues; Δ_L^{surf} was calculated from all protein residues not characterized as core residues.

Force Field	$\langle RMSD \rangle$	Δ_L^{core}	Δ_L^{surf}	Δ_L^{sim}
ff99sb/TIP3P	3.8	0.16	0.18	0.17
ff99sb/TIP4P-Ew	3.5	0.16	0.20	0.18
C36m/TIP3P	3.1	0.17	0.23	0.20
C36m/TIP3Pm	3.0	0.18	0.24	0.21
AmPro13/AmW03	5.5	0.20	0.29	0.25

These significant $\Delta_L^{sim}(300\ K)$ differences for the TSR4 domain would lead to substantial differences among the force fields with complete disorder. We therefore next consider Histatin 5, a cationic IDP, for which it has been challenging using fixed-charge force fields to achieve agreement with the reported experimental data. These include SAXS form factors that measure a $\langle R_g \rangle = 13.8 \pm 2.2$ Å [35] and solution CD and NMR [60,61] measurements, showing that Hst 5 lacks significant secondary structure in aqueous solution, although Hst 5 prefers α-helical conformations in nonaqueous solvents. From Figure 4, we see that the pairwise additive force fields ff99SB/TIP3P, C36m/TIP3Pm, and C36m/TIP3P predict a more narrow R_g distribution around compact structures with $\langle R_g \rangle$~10.0–11.0 Å, with higher populations of turns that likely account in part for these collapsed states. The ff99SB/TIP4P-Ew model predicts a bimodal distribution of collapsed and expanded states, but this is in disagreement with the SAXS form factor. The AmPro13/AmW03 potential, with no force field modifications, predicts a more expanded $\langle R_g \rangle$~14.0–14.5 Å in good agreement with the SAXS observable and NMR and CD experiments.

Finally, we consider the very challenging temperature dependence of the $(AAQAA)_3$ peptide, in which NMR experiments have previously ascertained a (partial) disorder-to-order transition as the temperature is lowered. There are several issues that are not sufficiently discussed in the literature regarding this peptide and previous simulation attempts to reproduce its behavior. The first is that the NMR experiment was designed to determine the ^{13}C-carbonyl shift at each residue, providing an experimental measure of the helicity at each residue for comparison to a helix–coil model that predicts the helicity at each residue [62]. Hence, an overall percentage averaged across all 15 residues is not the correct measure as the NMR shifts are residue-specific values, yielding estimates of 0% to 25%, depending on position, with the N-terminus being more helical. This is in contrast to the highly symmetric prediction of the helix–coil model [62]. Previous studies found that alanine peptides are unusually enriched [63,64] with the π-helix in particular, while the ^{13}C-carbonyl chemical shifts are not generally able to differentiate among all three helix categories, especially for fluctuating states. Note that there are statistically different shifts for the stable α−helix and 3_{10} helix [65], suggesting that comparison of structural ensembles to the standard NMR experiment should combine the propensities of the different helix types.

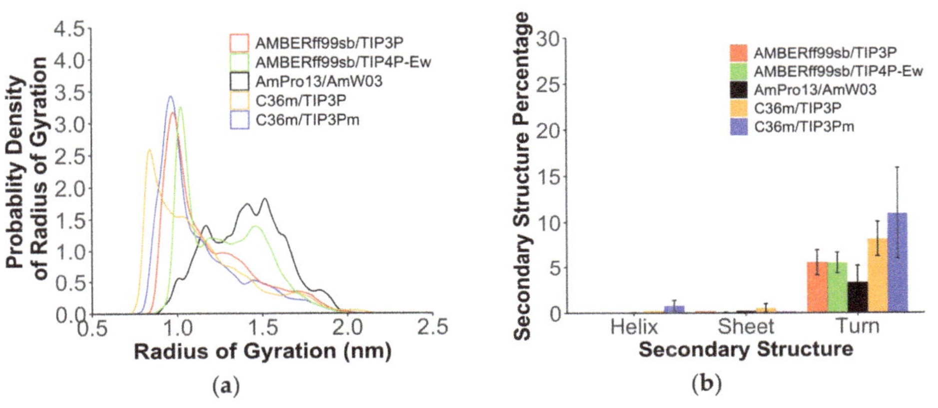

Figure 4. Structural properties for Hst 5 using polarizable and nonpolarizable force fields. (**a**) Probability density estimates of the radius of gyration and (**b**) average percentages of different secondary structures features for the disordered Hst 5 peptide.

We first investigate the definition of an α−helix percentage used by previous research groups, defined as three consecutive residues residing in a broad α−helix basin of the Ramachandran plot (labeled sequential in Figure 5). Unlike most recent studies, we provide individual residue percentages for the (AAQAA)$_3$ peptide (Figure 5a,b and Supplementary Figure S3) [66]. As determined by Boostra and coworkers [67], the C36m results depend critically on the "right" water model, i.e., the standard TIP3P water model must be used, to predict the higher helical content at low temperatures, with little helical content observed using TIP3Pm at any temperature. We support that result using TCW sampling in which C36m/TIP3Pm yields ~5% α−helix at 300 K (Figure 5a), as do the other fixed-charge force fields (Supplementary Figure S3), and they all exhibit a flat temperature dependence (Supplementary Table S5) in very good agreement with Robustelli et al. using 20 µs MD simulations [20]. The AmPro13/AmW03 polarizable model gives α−helical percentages that are similar to the Amber and CHARMM force fields for (AAQAA)$_3$ peptide, i.e., <~5% with no disorder-to-order transition (Figure 5b and Supplementary Table S5).

Instead, we consider an alternative definition of helical percentages in which the (AAQAA)$_3$ peptide might adopt not only α−helix, but π−helix and 3_{10} helix configurations [63] as well based on values of ψ(i) and φ(i+1) values (which we label pairwise in Figure 5). Figure 5a,b and Supplementary Figure S3 show that, when using this definition, the fraction of helical percentages for each residue increases for all force fields and temperatures, ~15–20%, but with important differences between the polarizable and nonpolarizable models. It is seen that the fixed-charge models (Figure 5c and Supplementary Figure S3) have no temperature dependence, with nearly the same helical percentages at 300 and 360 K. By contrast, the AmPro13/AmW03 model shows some temperature dependence, with a loss of helical structure at 360 K relative to 300 K as seen in Figure 5d. This supports our hypothesis that fixed-charge force fields that are overly stabilized for folded proteins will manifest as too inflexible for disordered states, in this case due to the inability to melt the N-terminal helix of $(AAQAA)_3$ at high temperatures, unlike the polarizable model, which exhibits a better temperature dependence for the configurational entropy. This result has also addressed a long-standing problem with the characterization of the $(AAQAA)_3$ peptide with temperature using simulation that must emphasize not only standard the $\alpha-$helix the but $\pi-$helix and 3_{10} helix categories as well and characterize not average helix percentages over the whole peptide but the residue-by-residue average helical percentage values instead.

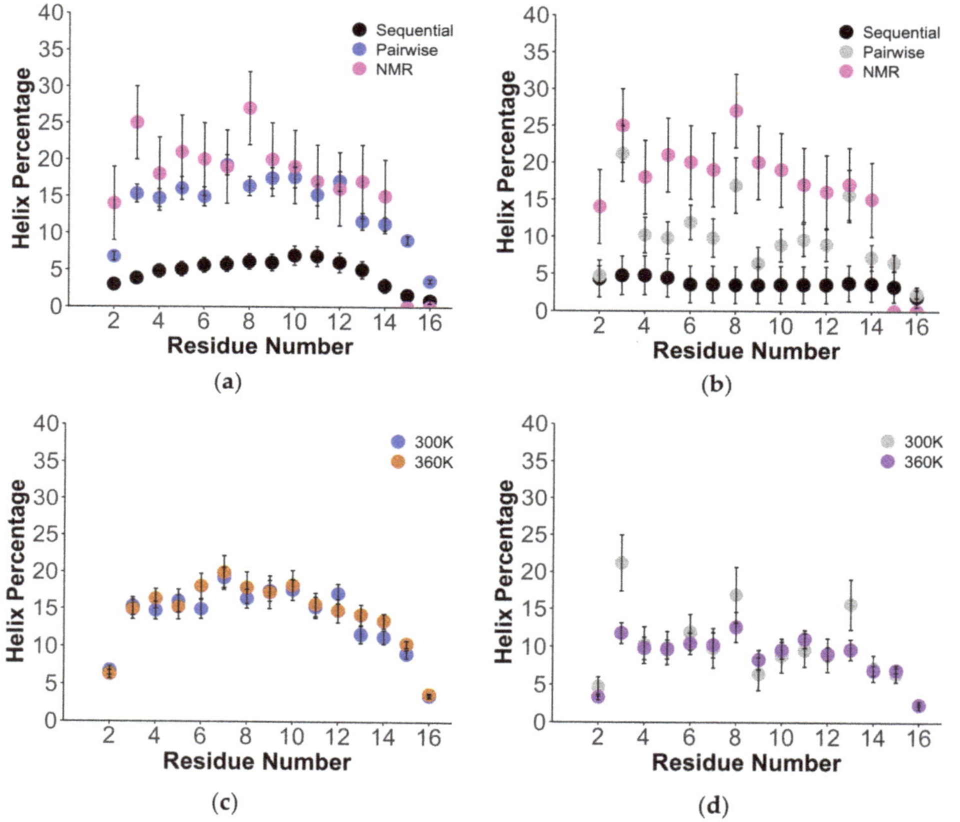

Figure 5. Structural properties for $(AAQAA)_3$ using polarizable and nonpolarizable force fields. Comparison of estimated helical propensities from NMR (pink), average $\alpha-$helix from the simulation assuming 3 sequential residues (black), and pairwise average over any presence of $\alpha-$helix, $\pi-$helix, and 3_{10} helix for (**a**) C36m/TIP3Pm (blue) and (**b**) AmPro13/AmW03 (gray) at 300 K. Comparison of changes in helix propensity with temperature at 300 and 360 K for (**c**) C36m/TIP3Pm and (**d**) AmPro13/AmW03.

3. Discussion

We have presented a comparison of a range of pairwise additive force fields and the many-body force field AMOEBA to test their ability to simultaneously describe the stable folded states of seven globular proteins, proteins with regions of disorder illustrated with the TSR4 domain, the Hst 5 IDP, and the partial disorder-to-order transition as the temperature is lowered for the $(AAQAA)_3$ peptide. We find that the fixed-charge force fields yield small RMSD differences from the PDB structures of the folded globular proteins, whereas the polarizable model has larger RMSD values that are within the expectations from solution experiments [48–50] on folded states. However, we have also shown that force fields that generate the largest RMSDs that are still consistent with the experimental R_g, thus exhibiting larger statistical fluctuations on average, are better able to simultaneously describe the plasticity of proteins with regions of complete structural disorder, as shown for the TSR4 domain, Hst 5, and the $(AAQAA)_3$ peptide.

In particular, the polarizable AMOEBA force field presents a significant advantage over a fixed-charge force field for IDP simulations, even those that have been specifically modified to better reproduce IDP behavior, as it does not require any problem-specific parameterization for IDPs and can be used as a general force field for different types of IDPs and their complexes. Our analysis indicates that fixed-charge force fields uniformly describe overly collapsed and rigid structural ensembles of the folded proteins, whereas the polarizable model is inherently more fluid with greater configurational entropy that captures both the folded structure and structural ensembles of IDPs. Finally, we note that other force fields tested previously on $(AAQAA)_3$ should be reevaluated to consider both π-helices and 3_{10} helices in addition to the α-helix, with a metric that evaluates the helical content on a residue-by-residue basis as the C-terminal end remains unstructured at any temperature [62] We also note that more current state-of-the-art estimates of helical structure based on NMR shifts could be used to obtain a better experimental reference for this peptide [68,69].

We believe that the analysis we have presented here offers several new ideas on force field validation criteria. The first is to measure the ability of a force field to more systematically approach the full value permitted by the structural similarity $D_{0,\ dis}$ metric for globular proteins [45] , as well as a Lindemann criteria values Δ_L^{sim} that are close to that determined from inelastic neutron scattering experiments and that are touted to be universal criteria for any folded protein in water [54]; a related metric is the ability to reproduce the melting temperature of folded proteins. These measures are best at assessing the balance between energetic effects and configurational entropy and what a force field should exhibit to equally well describe IDRs/IDPs and folded states of globular proteins. While this study has concluded that the polarizable AMOEBA force field is better by these structural and dynamical metrics, it is still an open question as to whether some fixed-charge force fields are capable to the same extent or can be made more capable in this regard. While we found that the pairwise additive force field combinations examined here are not fully sufficient, further evaluation and fitting to reproduce the dynamical criteria introduced can provide good guidance to improving force fields in general.

4. Materials and Methods

The Hst 5, TSR4 domain, and the 7 folded protein systems were modeled with the following force field combinations: Amberff99sb (ff99SB) [70] with TIP3P [71] and TIP4P-Ew [72], CHARMM36m (C36m) [30] with TIP3P [71] and Charmm-modified TIP3P (TIP3Pm), and AmPro13 [33] with Amoeba Water03 (AmW03) [34]. We used 1 µs standard MD simulations for the folded proteins, the TSR4 domain, and the Hst 5 system with the OpenMM [73] package for the fixed-charge force fields and the Tinker-OpenMM platform [74] for AMOEBA. We also developed a modified version of the OpenMM [73] and Tinker-OpenMM platforms [74] to perform calculations on graphics processing units (GPUs) with Temperature Cool Walking (TCW) [21,75,76] to further improve the sampling of the $(AAQAA)_3$ systems. For $(AAQAA)_3$, we considered the force field combinations of ff99sb/TIP4P-Ew, ff99sb-ildn/TIP4P-D, C36m/TIP3Pm, C36/TIP3Pm, and AmPro13/AmW03 models.

4.1. System and Simulation Preparation

Initial disordered-state structures for Hst 5 and Ace-$(AAQAA)_3$-Nme were generated using the tleap function in the AMBER MD engine [77]. The initial coordinates of the TSR4 and seven folded proteins were taken from their PDB structures. Solvation of these systems were performed using tleap for simulations using the ff99sb force fields, VMD or the online CHARMM-GUI for simulations using the C36m force field [78], and TINKER 8 for simulations using the AmPro13 force field [79]. All simulations were performed on systems with the addition of Na^+ or Cl^- counter-ions to maintain net zero charge.

The Hst 5 system was equilibrated according to the following procedure. First, the fully extended peptide was solvated using a 10 Å buffer, and the system was simulated at 500 K for 1 nanosecond (ns) in the NVT ensemble to collapse the peptide. Second, the peptide was resolvated using a smaller cubic box with side lengths of 59.1 Å, with a total of 6166 water molecules. The resolvated peptide was equilibrated with NVT conditions at 500 K for 1 ns, followed by 1 ns of NVT at 300 K. Finally, the peptide was run in the NPT ensemble at 300 K to equilibrate the size of the simulation box. The initial structure for production NVT MD simulations was chosen based on the maximum probable density.

For the $(AAQAA)_3$ system, the peptide we also started from an α-helix and solvated using a 10 Å buffer for the fixed-charge force fields, and the heavy atoms of the protein backbones were harmonically restrained with a spring constant of 10 kcal/mol/Å^2 during a 1 ns simulation in the NPT ensemble over a temperature range that captures the transition (300, 320, 340, 360, or 380 K). Second, 100 ps of NPT simulations were run where the position restraints of the protein backbone were relaxed from 10.0 to 0.0 kcal/mol/Å^2, reducing the spring constant by 1.0 kcal/mol/Å^2 every 10 ps. Finally, 20 ps of NPT simulations were run with no restraints on the protein backbone.

Finally, the larger protein systems were energy minimized to a potential energy tolerance of 0.5 kJ/mol with a nonbonded cutoff of 9.4 Å. The heavy atoms in the protein backbones were harmonically restrained with a spring constant of 10 kcal/mol/Å^2, and the system was heated in the NVT ensemble from 10 to 300 K at a rate of 1 K/ps using a Langevin integrator with a 1 fs timestep. Once the systems reached 300 K, a 1 ns simulation was run in the NPT ensemble with an rRESPA multi-timestep integrator with a 4 fs timestep for fixed-charge force fields and 2 fs timestep for polarizable force fields, using an Andersen Thermostat at 300 K with a collision frequency of 50 ps^{-1}. A Monte Carlo Barostat was used with a target pressure of 1.01325 bar and an exchange attempt frequency made every 50 fs.

4.2. Production Simulation Details and Analysis

For the solvated TSR4 and folded proteins, we performed 1 μs molecular dynamics simulations in the NVT ensemble at 300 K with the Bussi thermostat using the RESPA integrator and heavy-hydrogen mass repartitioning with a 3 fs time step. Ewald cutoffs of 7 Å and van der Waals cutoff of 12 Å were used. A pairwise neighbor list for partial-charge and polarizable multipole electrostatics and for van der Waals interactions was used. A grid size of 64 × 64 × 64 Å was used for PME summation and a 10^{-4} Debye convergence criterion for self-consistent induced dipoles. Frames were saved every 10 ps and used to perform further analysis. For $(AAQAA)_3$, the TCW simulations were performed in the NVT ensemble with the Andersen Thermostat and velocity verlet integrator with a 2 fs timestep to propagate the target temperature (300, 320, 340, 360, or 380 K) and high-temperature (456 K) walkers. Frames from the low-temperature replica were saved every 1 ps and used to perform further analysis.

Supplementary Figure S1 shows the raw RMSD and RMSF over the 1 μs trajectory for the folded proteins. Analyses of the trajectories were performed using Amber Tools and in-house analysis scripts to analyze the secondary-structure propensity for Hst 5, radius of gyration for Hst 5 and the folded proteins, and/or RMSDs and RMSFs of the protein–water systems using block averaging over ~50 ns blocks over the last 800 ns of the trajectory. For the $(AAQAA)_3$ system, a residue was classified as being in a helical conformation using two different definitions when compared with NMR chemical shift data from experiments [62]. The first definition is defined as a series of three consecutive residues where the φ dihedral angle was between −160° and −30° and the ψ angle was between −67° to −7° [64]. The second definition more directly targeted different types of helices; when the first and last residue pairs are excluded, the ψ dihedral angle of one residue and the φ dihedral angle of the next residue sum to $-125° \pm 10°$ for the π-helix, $-75° \pm 10°$ for the 3_{10} helix, whereas that for the α-helix is $-105° \pm 10°$.

Supplementary Materials: The following are available online at https://www.mdpi.com/article/10.3390/ijms22073420/s1.

Author Contributions: Conceptualization, T.H.-G.; software, M.L., A.K.D., S.Y.C.; formal analysis, J.L., S.S., S.Y.C., T.H.-G.; writing—original draft preparation, J.L., S.S., S.Y.C., T.H.-G.; writing—review and editing, J.D.F.-K., R.M.V.; funding acquisition, T.H.-G., J.D.F.-K. All authors have read and agreed to the published version of the manuscript.

Funding: The Berkeley and Toronto investigators thank the National Institutes of Health for support under Grant 5R01GM127627-03. M. Liu thanks the China Scholarship Council for a visiting scholar fellowship. J.D.F.-K. also acknowledges support from the Natural Sciences and Engineering Research Council of Canada (NSERC) grant RGPIN-2016-06718 and the Canada Research Chairs program. This research used the computational resources of the National Energy Research Scientific Computing Center, a DOE Office of Science User Facility supported by the Office of Science of the U.S. Department of Energy under Contract No. DE-AC02-05CH11231.

Data Availability Statement: Data available upon request.

Conflicts of Interest: The authors declare no conflict of interest. The funders had no role in the design of the study; in the collection, analyses, or interpretation of data; in the writing of the manuscript, or in the decision to publish the results.

References

1. Wright, P.E.; Dyson, H.J. Intrinsically disordered proteins in cellular signalling and regulation. *Nat. Rev. Mol. Cell Biol.* **2015**, *16*, 18–29. [CrossRef] [PubMed]
2. Fawzi, N.L.; Phillips, A.H.; Ruscio, J.Z.; Doucleff, M.; Wemmer, D.E.; Head-Gordon, T. Structure and dynamics of the Aβ(21–30) peptide from the interplay of NMR experiments and molecular simulations. *J. Am. Chem. Soc.* **2008**, *130*, 6145–6158. [CrossRef]
3. Gomes, G.-N.W.; Krzeminski, M.; Namini, A.; Martin, E.W.; Mittag, T.; Head-Gordon, T.; Forman-Kay, J.D.; Gradinaru, C.C. Conformational Ensembles of an Intrinsically Disordered Protein Consistent with NMR, SAXS, and Single-Molecule FRET. *J. Am. Chem. Soc.* **2020**, *142*, 15697–15710. [CrossRef]
4. Svergun, D.; Barberato, C.; Koch, M.H.J. CRYSOL—A Program to Evaluate X-ray Solution Scattering of Biological Macromolecules from Atomic Coordinates. *J. Appl. Crystallogr.* **1995**, *28*, 768–773. [CrossRef]
5. Bhowmick, A.; Brookes, D.H.; Yost, S.R.; Dyson, H.J.; Forman-Kay, J.D.; Gunter, D.; Head-Gordon, M.; Hura, G.L.; Pande, V.S.; Wemmer, D.E.; et al. Finding Our Way in the Dark Proteome. *J. Am. Chem. Soc.* **2016**, *138*, 9730–9742. [CrossRef] [PubMed]
6. Ozenne, V.; Bauer, F.; Salmon, L.; Huang, J.-R.; Jensen, M.R.; Segard, S.; Bernadó, P.; Charavay, C.; Blackledge, M. Flexible-meccano: A tool for the generation of explicit ensemble descriptions of intrinsically disordered proteins and their associated experimental observables. *Bioinformatics* **2012**, *28*, 1463–1470. [CrossRef]
7. Krzeminski, M.; Marsh, J.A.; Neale, C.; Choy, W.Y.; Forman-Kay, J.D. Characterization of disordered proteins with ENSEMBLE. *Bioinformatics* **2013**, *29*, 398–399. [CrossRef] [PubMed]
8. Ball, K.A.; Wemmer, D.E.; Head-Gordon, T. Comparison of structure determination methods for intrinsically disordered amyloid-beta peptides. *J. Phys. Chem. B* **2014**, *118*, 6405–6416. [CrossRef]
9. Brookes, D.H.; Head-Gordon, T. Experimental Inferential Structure Determination of Ensembles for Intrinsically Disordered Proteins. *J. Am. Chem. Soc.* **2016**, *138*, 4530–4538. [CrossRef] [PubMed]
10. Köfinger, J.; Stelzl, L.S.; Reuter, K.; Allande, C.; Reichel, K.; Hummer, G. Efficient Ensemble Refinement by Reweighting. *J. Chem. Theory Comput.* **2019**, *15*, 3390–3401. [CrossRef]
11. Lincoff, J.; Haghighatlari, M.; Krzeminski, M.; Teixeira, J.M.C.; Gomes, G.-N.W.; Gradinaru, C.C.; Forman-Kay, J.D.; Head-Gordon, T. Extended experimental inferential structure determination method in determining the structural ensembles of disordered protein states. *Commun. Chem.* **2020**, *3*, 74. [CrossRef] [PubMed]
12. Nerenberg, P.S.; Head-Gordon, T. New developments in force fields for biomolecular simulations. *Curr. Opin. Struct. Biol.* **2018**, *49*, 129–138. [CrossRef]
13. Nerenberg, P.S.; Head-Gordon, T. Optimizing protein-solvent force fields to reproduce intrinsic conformational preferences of model peptides. *J. Chem. Theory Comput.* **2011**, *7*, 1220–1230. [CrossRef] [PubMed]
14. Chong, S.-H.; Chatterjee, P.; Ham, S. Computer Simulations of Intrinsically Disordered Proteins. *Annu. Rev. Phys. Chem.* **2017**, *68*, 117–134. [CrossRef]
15. Henriques, J.; Cragnell, C.; Skepö, M. Molecular dynamics simulations of intrinsically disordered proteins: Force field evaluation and comparison with experiment. *J. Chem. Theory Comput.* **2015**, *11*, 3420–3431. [CrossRef]

16. Siwy, C.M.; Lockhart, C.; Klimov, D.K. Is the Conformational Ensemble of Alzheimer's Aβ10-40 Peptide Force Field Dependent? *PLoS Comput. Biol.* **2017**, *13*, e1005314. [CrossRef]
17. Wickstrom, L.; Okur, A.; Simmerling, C. Evaluating the performance of the ff99SB force field based on NMR scalar coupling data. *Biophys. J.* **2009**, *97*, 853–856. [CrossRef] [PubMed]
18. Zaslavsky, B.Y.; Uversky, V.N. In Aqua Veritas: The Indispensable yet Mostly Ignored Role of Water in Phase Separation and Membrane-Less Organelles. *Biochemistry* **2018**, *57*, 2437–2451. [CrossRef]
19. Lindorff-Larsen, K.; Piana, S.; Palmo, K.; Maragakis, P.; Klepeis, J.L.; Dror, R.O.; Shaw, D.E. Improved side-chain torsion potentials for the Amber ff99SB protein force field. *Proteins Struct. Funct. Bioinform.* **2010**, *78*, 1950–1958. [CrossRef]
20. Robustelli, P.; Piana, S.; Shaw, D.E. Developing a molecular dynamics force field for both folded and disordered protein states. *Proc. Natl. Acad. Sci. USA* **2018**, *115*, E4758–E4766. [CrossRef]
21. Lincoff, J.; Sasmal, S.; Head-Gordon, T. The combined force field-sampling problem in simulations of disordered amyloid-beta peptides. *J. Chem. Phys.* **2019**, *150*, 104108. [CrossRef]
22. Piana, S.; Donchev, A.G.; Robustelli, P.; Shaw, D.E. Water dispersion interactions strongly influence simulated structural properties of disordered protein states. *J. Phys. Chem. B* **2015**, *119*, 5113–5123. [CrossRef] [PubMed]
23. Best, R.B.; Zheng, W.; Mittal, J. Balanced Protein-Water Interactions Improve Properties of Disordered Proteins and Non-Specific Protein Association. *J. Chem. Theory Comput.* **2014**, *10*, 5113–5124. [CrossRef] [PubMed]
24. Abascal, J.L.F.; Vega, C. A general purpose model for the condensed phases of water: TIP4P/2005. *J. Chem. Phys.* **2005**, *123*, 234505. [CrossRef] [PubMed]
25. Fluitt, A.M.; de Pablo, J.J. An Analysis of Biomolecular Force Fields for Simulations of Polyglutamine in Solution. *Biophys. J.* **2015**, *109*, 1009–1018. [CrossRef]
26. Wang, Y.; Chu, X.; Longhi, S.; Roche, P.; Han, W.; Wang, E.; Wang, J. Multiscaled exploration of coupled folding and binding of an intrinsically disordered molecular recognition element in measles virus nucleoprotein. *Proc. Natl. Acad. Sci. USA* **2013**, *110*, E3743–E3752. [CrossRef]
27. Moritsugu, K.; Terada, T.; Kidera, A. Disorder-to-order transition of an intrinsically disordered region of sortase revealed by multiscale enhanced sampling. *J. Am. Chem. Soc.* **2012**, *134*, 7094–7101. [CrossRef]
28. Wells, M.; Tidow, H.; Rutherford, T.J.; Markwick, P.; Jensen, M.R.; Mylonas, E.; Svergun, D.I.; Blackledge, M.; Fersht, A.R. Structure of tumor suppressor p53 and its intrinsically disordered N-terminal transactivation domain. *Proc. Natl. Acad. Sci. USA* **2008**, *105*, 5762–5767. [CrossRef]
29. Rauscher, S.; Pomès, R. The liquid structure of elastin. *eLife* **2017**, *6*, e26526. [CrossRef]
30. Huang, J.; Rauscher, S.; Nawrocki, G.; Ran, T.; Feig, M.; de Groot, B.L.; Grubmuller, H.; MacKerell, A.D., Jr. CHARMM36m: An improved force field for folded and intrinsically disordered proteins. *Nat. Methods* **2017**, *14*, 71–73. [CrossRef]
31. Ponder, J.W.; Wu, C.; Ren, P.; Pande, V.S.; Chodera, J.D.; Schnieders, M.J.; Haque, I.; Mobley, D.L.; Lambrecht, D.S.; DiStasio, R.A., Jr. Current status of the AMOEBA polarizable force field. *J. Phys. Chem. B* **2010**, *114*, 2549. [CrossRef]
32. Demerdash, O.; Wang, L.-P.; Head-Gordon, T. Advanced models for water simulations. *Wiley Interdiscip. Rev. Comput. Mol. Sci.* **2018**, *8*, e1355. [CrossRef]
33. Shi, Y.; Xia, Z.; Zhang, J.; Best, R.; Wu, C.; Ponder, J.W.; Ren, P. The Polarizable Atomic Multipole-based AMOEBA Force Field for Proteins. *J. Chem. Theory Comput.* **2013**, *9*, 4046–4063. [CrossRef] [PubMed]
34. Ren, P.; Ponder, J.W. Polarizable atomic multipole water model for molecular mechanics simulation. *J. Phys. Chem. B* **2003**, *107*, 5933–5947. [CrossRef]
35. Cragnell, C.; Durand, D.; Cabane, B.; Skepö, M. Coarse-grained modelling of the intrinsically disordered protein Histatin 5 in solution. Monte Carlo simulations in combination with SAXS. *Proteins Struct. Func. Bioinform.* **2016**, *84*, 777–791. [CrossRef]
36. Lifson, S.; Warshel, A. Consistent Force Field for Calculations of Conformations, Vibrational Spectra, and Enthalpies of Cycloalkane and N-Alkane Molecules. *J. Chem. Phys.* **1968**, *49*, 5116–5129. [CrossRef]
37. Hickman, A.B.; Klein, D.C.; Dyda, F. Melatonin Biosynthesis: The Structure of Serotonin N-Acetyltransferase at 2.5Å Resolution Suggests a Catalytic Mechanism. *Mol. Cell* **1999**, *3*, 23–32. [CrossRef]
38. Tsunasawa, S.; Masaki, T.; Hirose, M.; Soejima, M.; Sakiyama, F. The primary structure and structural characteristics of Achromobacter lyticus protease I, a lysine-specific serine protease. *J. Biol. Chem.* **1989**, *264*, 3832–3839. [CrossRef]
39. Dideberg, O.; Charlier, P.; Wéry, J.P.; Dehottay, P.; Dusart, J.; Erpicum, T.; Frère, J.M.; Ghuysen, J.M. The crystal structure of the 3-lactamase of Streptomyces albus G at 0.3 nm resolution. *Biochem. J.* **1987**, *245*, 911–913. [CrossRef]
40. Muller, P.; Sawaya, M.R.; Pashkov, I.; Chan, S.; Nguyen, C.; Wu, Y.; Perry, L.J.; Eisenberg, D. The 1.70 angstroms X-ray crystal structure of Mycobacterium tuberculosis phosphoglycerate mutase. *Acta Crystallogr. D Biol. Crystallogr.* **2005**, *61 Pt 3*, 309–315. [CrossRef] [PubMed]
41. Sutkeviciute, I.; Thepaut, M.; Sattin, S.; Berzi, A.; McGeagh, J.; Grudinin, S.; Weiser, J.; Le Roy, A.; Reina, J.J.; Rojo, J.; et al. Unique DC-SIGN clustering activity of a small glycomimetic: A lesson for ligand design. *ACS Chem. Biol.* **2014**, *9*, 1377–1385. [CrossRef] [PubMed]
42. Ulrich, A.; Wahl, M.C. Structure and evolution of the spliceosomal peptidyl-prolyl cis-trans isomerase Cwc27. *Acta Crystallogr. D Biol. Crystallogr.* **2014**, *70 Pt 12*, 3110–3123. [CrossRef]

43. Wan, Q.; Parks, J.M.; Hanson, B.L.; Fisher, S.Z.; Ostermann, A.; Schrader, T.E.; Graham, D.E.; Coates, L.; Langan, P.; Kovalevsky, A. Direct determination of protonation states and visualization of hydrogen bonding in a glycoside hydrolase with neutron crystallography. *Proc. Natl. Acad. Sci. USA* **2015**, *112*, 12384–12389. [CrossRef] [PubMed]
44. Paakkonen, K.; Tossavainen, H.; Permi, P.; Rakkolainen, H.; Rauvala, H.; Raulo, E.; Kilpelainen, I.; Guntert, P. Solution structures of the first and fourth TSR domains of F-spondin. *Proteins* **2006**, *64*, 665–672. [CrossRef]
45. Maiorov, V.N.; Crippen, G.M. Significance of Root-Mean-Square Deviation in Comparing Three-Dimensional Structures of Globular Proteins. *J. Mol. Biol.* **1994**, *235*, 625–634. [CrossRef]
46. Kolinski, A.; Godzik, A.; Skolnick, J. A general method for the prediction of the three dimensional structure and folding pathway of globular proteins: Application to designed helical proteins. *J. Chem. Phys.* **1993**, *98*, 7420–7433. [CrossRef]
47. Dima, R.I.; Thirumalai, D. Asymmetry in the Shapes of Folded and Denatured States of Proteins. *J. Phys. Chem. B* **2004**, *108*, 6564–6570. [CrossRef]
48. Yang, L.-W.; Eyal, E.; Chennubhotla, C.; Jee, J.; Gronenborn, A.M.; Bahar, I. Insights into equilibrium dynamics of proteins from comparison of NMR and X-ray data with computational predictions. *Structure* **2007**, *15*, 741–749. [CrossRef]
49. Andrec, M.; Snyder, D.A.; Zhou, Z.; Young, J.; Montelione, G.T.; Levy, R.M. A large data set comparison of protein structures determined by crystallography and NMR: Statistical test for structural differences and the effect of crystal packing. *Proteins Struct. Funct. Bioinform.* **2007**, *69*, 449–465. [CrossRef]
50. Hura, G.L.; Hodge, C.D.; Rosenberg, D.; Guzenko, D.; Duarte, J.M.; Monastyrskyy, B.; Grudinin, S.; Kryshtafovych, A.; Tainer, J.A.; Fidelis, K.; et al. Small angle X-ray scattering-assisted protein structure prediction in CASP13 and emergence of solution structure differences. *Proteins Struct. Funct. Bioinform.* **2019**, *87*, 1298–1314. [CrossRef] [PubMed]
51. Pitera, J.W. Expected Distributions of Root-Mean-Square Positional Deviations in Proteins. *J. Phys. Chem. B* **2014**, *118*, 6526–6530. [CrossRef]
52. Lindemann, F. The calculation of molecular vibration frequencies. *Z. Phys.* **1910**, *11*, 609–612.
53. Zhou, Y.; Vitkup, D.; Karplus, M. Native proteins are surface-molten solids: Application of the lindemann criterion for the solid versus liquid state. *J. Mol. Biol.* **1999**, *285*, 1371–1375. [CrossRef]
54. Katava, M.; Stirnemann, G.; Zanatta, M.; Capaccioli, S.; Pachetti, M.; Ngai, K.L.; Sterpone, F.; Paciaroni, A. Critical structural fluctuations of proteins upon thermal unfolding challenge the Lindemann criterion. *Proc. Natl. Acad. Sci. USA* **2017**, *114*, 9361. [CrossRef] [PubMed]
55. Zhou, Y.; Karplus, M. Folding thermodynamics of a model three-helix-bundle protein. *Proc. Natl. Acad. Sci. USA* **1997**, *94*, 14429–14432. [CrossRef] [PubMed]
56. Freddolino, P.L.; Harrison, C.B.; Liu, Y.; Schulten, K. Challenges in protein-folding simulations. *Nat. Phys.* **2010**, *6*, 751–758. [CrossRef]
57. Sosnick, T.R.; Hinshaw, J.R. How Proteins Fold. *Science* **2011**, *334*, 464. [CrossRef] [PubMed]
58. Piana, S.; Lindorff-Larsen, K.; Shaw, D.E. How Robust Are Protein Folding Simulations with Respect to Force Field Parameterization? *Biophys. J.* **2011**, *100*, L47–L49. [CrossRef] [PubMed]
59. Alowolodu, O.; Johnson, G.; Alashwal, L.; Addou, I.; Zhdanova, I.V.; Uversky, V.N. Intrinsic disorder in spondins and some of their interacting partners. *Intrinsically Disord. Proteins* **2016**, *4*, e1255295. [CrossRef] [PubMed]
60. Brewer, D.; Hunter, H.; Lajoie, G. NMR studies of the antimicrobial salivary peptides histatin 3 and histatin 5 in aqueous and nonaqueous solutions. *Biochem. Cell Biol.* **1998**, *76*, 247–256. [CrossRef] [PubMed]
61. Raj, P.A.; Marcus, E.; Sukumaran, D.K. Structure of human salivary histatin 5 in aqueous and nonaqueous solutions. *Biopolymers* **1998**, *45*, 51–67. [CrossRef]
62. Shalongo, W.; Dugad, L.; Stellwagen, E. Distribution of Helicity within the Model Peptide Acetyl(AAQAA)3amide. *J. Am. Chem. Soc.* **1994**, *116*, 8288–8293. [CrossRef]
63. Shirley, W.A.; Brooks, C.L., III. Curious structure in "canonical" alanine-based peptides. *Proteins Struct. Funct. Bioinform.* **1997**, *28*, 59–71. [CrossRef]
64. Huang, J.; MacKerell, A.D., Jr. Induction of peptide bond dipoles drives cooperative helix formation in the (AAQAA)3 peptide. *Biophys. J.* **2014**, *107*, 991–997. [CrossRef] [PubMed]
65. Mayo, A.; Yap, K. *Empirical Analysis of Backbone Chemical Shifts in Proteins*; Ikura Laboratory, Department of Medical Biophysics, University of Toronto, Division of Molecular and Structural Biology, Ontario Cancer Institute: Toronto, ON, Canada, 2001.
66. Best, R.B.; Hummer, G. Optimized molecular dynamics force fields applied to the helix-coil transition of polypeptides. *J. Phys. Chem. B* **2009**, *113*, 9004–9015. [CrossRef]
67. Boonstra, S.; Onck, P.R.; van der Giessen, E. CHARMM TIP3P Water Model Suppresses Peptide Folding by Solvating the Unfolded State. *J. Phys. Chem. B* **2016**, *120*, 3692–3698. [CrossRef] [PubMed]
68. Marsh, J.A.; Singh, V.K.; Jia, Z.; Forman-Kay, J.D. Sensitivity of secondary structure propensities to sequence differences between α- and γ-synuclein: Implications for fibrillation. *Protein Sci.* **2006**, *15*, 2795–2804. [CrossRef] [PubMed]
69. Camilloni, C.; De Simone, A.; Vranken, W.F.; Vendruscolo, M. Determination of Secondary Structure Populations in Disordered States of Proteins Using Nuclear Magnetic Resonance Chemical Shifts. *Biochemistry* **2012**, *51*, 2224–2231. [CrossRef]
70. Hornak, V.; Abel, R.; Okur, A.; Strockbine, B.; Roitberg, A.; Simmerling, C. Comparison of multiple Amber force fields and development of improved protein backbone parameters. *Proteins* **2006**, *65*, 712–725. [CrossRef]

71. Jorgensen, W.L.; Chandrasekhar, J.; Madura, J.D.; Impey, R.W.; Klein, M.L. Comparison of simple potential functions for simulating liquid water. *J. Chem. Phys.* **1983**, *79*, 926–935. [CrossRef]
72. Horn, H.W.; Swope, W.C.; Pitera, J.W.; Madura, J.D.; Dick, T.J.; Hura, G.L.; Head-Gordon, T. Development of an improved four-site water model for biomolecular simulations: TIP4P-Ew. *J. Chem. Phys.* **2004**, *120*, 9665–9678. [CrossRef]
73. Eastman, P.; Friedrichs, M.S.; Chodera, J.D.; Radmer, R.J.; Bruns, C.M.; Ku, J.P.; Beauchamp, K.A.; Lane, T.J.; Wang, L.-P.; Shukla, D. OpenMM 4: A reusable, extensible, hardware independent library for high performance molecular simulation. *J. Chem. Theory Comput.* **2013**, *9*, 461. [CrossRef]
74. Harger, M.; Li, D.; Wang, Z.; Dalby, K.; Lagardere, L.; Piquemal, J.P.; Ponder, J.; Ren, P. Tinker-OpenMM: Absolute and relative alchemical free energies using AMOEBA on GPUs. *J. Comput. Chem.* **2017**, *38*, 2047–2055. [CrossRef] [PubMed]
75. Brown, S.; Head-Gordon, T. Cool walking: A new Markov chain Monte Carlo sampling method. *J. Comput. Chem.* **2003**, *24*, 68–76. [CrossRef]
76. Lincoff, J.; Sasmal, S.; Head-Gordon, T. Comparing generalized ensemble methods for sampling of systems with many degrees of freedom. *J. Chem. Phys.* **2016**, *145*, 174107. [CrossRef] [PubMed]
77. Roe, D.R.; Cheatham, T.E., 3rd. PTRAJ and CPPTRAJ: Software for Processing and Analysis of Molecular Dynamics Trajectory Data. *J. Chem. Theory Comput.* **2013**, *9*, 3084–3095. [CrossRef] [PubMed]
78. Jo, S.; Lim, J.B.; Klauda, J.B.; Im, W. CHARMM-GUI Membrane Builder for mixed bilayers and its application to yeast membranes. *Biophys. J.* **2009**, *97*, 50–58. [CrossRef]
79. Rackers, J.A.; Wang, Z.; Lu, C.; Laury, M.L.; Lagardere, L.; Schnieders, M.J.; Piquemal, J.P.; Ren, P.; Ponder, J.W. Tinker 8: Software Tools for Molecular Design. *J. Chem. Theory Comput.* **2018**, *14*, 5273–5289. [CrossRef]

MDPI
St. Alban-Anlage 66
4052 Basel
Switzerland
Tel. +41 61 683 77 34
Fax +41 61 302 89 18
www.mdpi.com

International Journal of Molecular Sciences Editorial Office
E-mail: ijms@mdpi.com
www.mdpi.com/journal/ijms

www.ingramcontent.com/pod-product-compliance
Lightning Source LLC
Chambersburg PA
CBHW041942260926
51850CB00018B/423
9783036567815